INTRODUCTION TO FIRE PROTECTION

LAW

INTRODUCTION TO FIRE PROTECTION

LAW

By

Donna L. Rosenbauer, M.Ed., J.D.

Chapter VIII By

Vincent M. Brannigan, J.D.

NATIONAL FIRE PROTECTION ASSOCIATION

470 Atlantic Ave., Boston, MA 02210

The outline and plan for this text was field tested by Vincent M. Brannigan
at the University of Maryland

NFPA No. TXT-5
ISBN: 0-87765-121-3
Library of Congress No.: 78-50930
Printed in U.S.A.

DEDICATION

For

MARY CONNELLY ROSENBAUER

CONTENTS

ACKNOWLEDGMENTS

No work is ever completed alone. To the following people, and also to the many unnamed people who have formed my life and shaped my thoughts, I wish to express my gratitude:

To Chester I. Babcock of the National Fire Protection Association for his vision and his guidance.

To Keith Tower of the National Fire Protection Association whose unlimited talent has touched every aspect of this text.

To the staff at the National Fire Protection Association — especially Amy E. Dean, for her enthusiasm and dedication to this project.

To Frank Lucas, artist and friend, for preparing the illustrations and cover design for this text.

To Peter J. Connelly, Jr., member of the New York Bar Association, whose love of the law has been such a beacon in my life, for reviewing this text. And to Dorothy Hart Connelly, for adding so many "grace notes," as usual.

To Robert A. Barton, Professor of Law, New England School of Law, for his advice and guidance when he reviewed the chapters on Criminal Law and Criminal Procedure.

To Timothy J. Cronin, Jr., Professor of Law, New England School of Law, for his careful review of the chapters on Administrative Law and Administrative Procedure.

To Michael Wheeler, Professor of Law, New England School of Law, for his advice and encouragement.

To Vincent M. Brannigan, Jr., J.D., faculty member of the Department of Fire Protection Engineering and the Department of Textiles and Consumer Economics at the University of Maryland, for field testing the outline and plan for this text at the University of Maryland, and also for preparing the chapter on Administrative Procedure.

To Marcia C. Sullivan, whose patience, monumental efforts, and flawless precision has contributed so greatly to this text.

To Joseph Paige, whose encouragement and ability to put things in perspective has been a constant source of strength.

To Mary T. Danehy, whose help always comes just at the right time.

To Nancy Keegan, for her much-appreciated help.

To my family, especially to Kathleen, whose constant encouragement, loving patience and understanding, and many sacrifices, both large and small, have become such an important part of this text.

To all those publishers and representatives of the fire service and legal profession who gave permission to use their works and excerpts from their publications for reproduction in this text, and to the many people and organizations who contributed helpful suggestions, ideas, and materials.

And finally, to the editors, proofreaders, and production personnel for their cooperation and efforts.

Donna L. Rosenbauer, September, 1978

TO THE STUDENT

The basic goal of *Introduction to Fire Protection Law* is to help members of the fire service, members of fire service related occupations, and those who aspire to a career in the fire service become aware of the legal dimensions that can and do affect their professional lives. Because the fire service is a public service concerned with the public welfare, additional legal considerations are involved that are not generally present in employment in the private sector.

Introduction to Fire Protection Law has been organized and written not to produce lawyers, but to give a new dimension to fire service careers. As such, it serves two purposes: (1) to help make students more aware of some of the legal considerations involved in a career in the fire service (thus the ''law book look''), and (2) to help increase student awareness of the general areas of the law they may have to deal with (thus the ''global and general presentation'').

It is, of course, impossible within the limits of one textbook to give a detailed, in-depth presentation of every nuance and aspect of the general branches of law presented in this text. Therefore, while each chapter of the text could properly be the subject of an entire course in itself, each attempts to provide an overview of a particular interest to readers involved in the fire service and related fields.

Although this text was written to give you an overview of major branches of American law, it is not to be considered as a ''law bible.'' The study of law is the study of the rights and duties of people in a society. As such, interpretations of law change as the society changes and develops. For example, just before this text went to press, a new ruling on the need for search warrants in some arson investigations was handed down by the Supreme Court of the United States. This decision has been incorporated into the text for your information and consideration. A year from now, an additional ruling may be handed down based on new sets of circumstances or facts. It will be up to you to become aware of these new rulings — either through the newspaper, television, or organizations such as the National Fire Protection Association. From what you learn in this course, you will be better able to interpret what these rulings mean to you as a member of the fire service. Thus, another major goal of this text is to give you a background and a springboard for determining what professional actions you will take when there are legal questions involved.

CONTENT

This text presents the basics of three major areas of law: (1) civil law, (2) criminal law, and (3) administrative law. In addition to these, other applicable legal subject areas are also included. For example, because constitutional law affects virtually every branch of American law, consideration of aspects of constitutional law are interwoven throughout the text.

Emphasis is placed primarily on these three major areas because they affect fire service personnel performing their official duties and because there is increasing need for awareness of these areas of the law when fire service decisions are made. Each of the major areas is presented in general and then related to the special areas of attention pertinent to the fire service. This relationship is made by presenting judicial decisions that involve fire service-related incidents, thus helping the student to become more aware of the legal and judicial context of these incidents.

PREVIEWING THE TEXT

The following suggestions are intended to help you gain maximum benefit from this text.

Once you have carefully read the "Introduction," thumb through this book and notice its format:

• Because it is a general overview of the law, many quotations from legal authorities are included.

• Look at the "Table of Contents" and the chapters themselves. Consider how these subject areas are important to you.

• Because it is a law text, many judicial decisions are included to emphasize and further clarify the areas of law you will be studying.

• Notice how some cases are preceded by questions: these questions will apply the knowledge gained from reading the case to specific fire service activities.

CASE STUDIES

Judicial decisions have been included in this text to serve several major functions, chief of which is to further elaborate some aspect of the law presented in the text. Decisions were chosen primarily to demonstrate how the law affects the fire service and how both the law and the fire service may be treated by the courts. Virtually all of the judicial decisions presented are decisions handed down as a result of an appeal of a lower court decision. This means that the case was originally heard by a lower court — usually a trial court — at which evidence was presented, witnesses heard, and a first decision reached by a judge or jury. When a case is granted a hearing at the appeal level, the attorney for the party who lost at the lower-court level is given the opportunity to argue that the lower-court judge erred on a point or points of law at the trial. If *certiorari* is granted by the higher court, all the records of the lower-court proceedings are called up by the higher court for review. However, *the case is not retried*. Only those points of law in dispute are argued. It should be remembered that, during a jury trial, the judge rules on matters of law only — it is up to the jury to make decisions regarding the facts. Thus, in decisions based on appeals of lower-court decisions, the facts in a case are not an issue — only points of law. When you read a decision based on an appeal, it is important to keep this fact in mind. Many decisions handed down are based on disputes over several points of law; some parts of such decisions that are not relevant to what you are studying have been omitted. These omissions are noted in the text.

READING A DECISION

The following suggestions may be helpful regarding the decisions in this text:
- Before you read a case, read the brief summary or headnote at the beginning of the decision. Some headnotes are included in the body of the text. Headnotes for each case appear in Appendix E.
- Note the section in the chapter in which the case is placed. This will give you an indication of the area of law that the decision is being used to illustrate or elaborate.
- Note the date the decision was handed down by the court.
- Note whether the decision was unanimous or was handed down by a divided court in those cases where more than one judge was involved in the decision. If a decision is handed down by a divided court, the force of the decision may be weakened. This is also true if the judges involved in the decision agree, but for substantially different reasons. It has been stated that strong reasoning in a dissenting opinion weakens a case. However, a strong dissenting opinion as well as a strong majority opinion also indicates that the case was thoroughly and carefully considered, and thus can strengthen a case as an authority.
- Several dissenting opinions are included in this text; they should be read and considered as carefully as the majority opinions themselves. One important reason for this recommendation is that dissenting opinions read in conjunction with majority opinions give a full and complete picture of the points of law being considered and the reasons why they were interpreted as they were. Very often, points you might raise about a majority opinion are considered and discussed in the dissenting opinion.
- Read each decision once, twice, or as many times as you need to in order to feel comfortable with the facts, the issues, and the decision.
- Look up legal terms you do not fully understand in the "Glossary of Terms."

BRIEFING THE CASE

Once you are familiar with the case, it is suggested that you "brief the case" as law students do. Essentially, this means that you will isolate the core of the case and the pure principle of law for which it stands by going beyond the style and language used by the judge who wrote the decision. Briefing a case allows you to quickly review the points of a case and also helps you check to see if you have isolated the major points in the decision. The technique used in briefing a case is also an important skill that can be applied to areas of your studies not directly connected with law — a technique that can help you get to the heart of many matters.

The major elements of briefing a case are determining facts, issue, decision, reasoning, and dissent (if any).

Facts. Write a brief paragraph describing *only* those facts or actions between the parties that affected the legal decision.

Issue. Write one (no more than two) sentences in question form that will ask the legal question the court has been called upon to decide. For example, the court might be asked to decide the following questions: "Was the evidence obtained through an illegal search?" or "Did the actions of the plaintiff contribute to the injury caused?"

Decision. This step answers the question posed in the preceding step. The answer can be a single word or a phrase.

Reasoning. Write a few sentences giving the judge's reasons for the decision handed down. This is an important step; it gives the boundaries of the decision based on the facts of a particular case. It is important to remember that a change in the facts or circumstances of a case might elicit a very different decision.

Dissent. If a dissent is given, briefly state the reasons for the dissenting opinion.

ADDITIONAL REFERENCES

Additional aids are provided for you in the footnotes and end-of-chapter "Bibliographies." Further references appear at the end of this text and consist of the following:

The Constitution of the United States (Appendix A) is a major reference. If a section of the Constitution is referred to in the text, read the appropriate part of the Constitution to see how the reference applies.

Federal Laws Pertaining to Equal Employment (Appendix B) presents the sections of the Civil Rights Act of 1964, as amended by the Equal Employment Act of 1972, that are of particular interest to members of the fire service.

Administrative Procedure Act (Appendix C) is a reference for Chapters VII and VIII, and will also serve as a future reference when you have finished this course. It is suggested that you read this Act to familiarize yourself with the rules and guidelines included in order to gain maximum benefit from Chapters VII and VIII.

Federal Agencies Involved in Fire Protection (Appendix D) lists the various federal agencies involved in firesafety. All of the agencies have headquarters in the Washington, D.C., area; most have eight to ten regional offices in major cities and field offices throughout the country.

Headnotes for Cases Included in This Text (Appendix E) is a summary of the decisions of cases used in this text. Suggestions for its use have already been made earlier in this "Introduction" in the section titled "Reading a Decision."

The "Glossary of Terms" is placed after the "Appendices" and before the "Subject Index." Use of the "Glossary" has already been suggested in this "Introduction." Some of the terms do not appear in the text; they are included for your information and future reference.

One further note: the presence of the pronoun "he" in the quotations and judicial decisions reprinted in this text does not reflect the policies of the NFPA or this writer, but the need to reproduce quotations exactly as written.

D. L. Rosenbauer

I

INTRODUCTION
TO AMERICAN LAW

OVERVIEW

American law can clearly be seen to have its basis in the common laws of Great Britain. The early colonists brought with them to America not only their language, customs, religion, and traditions, but also a legal background that extended back into the days of feudalism. Thus, the laws of any country, including fire protection laws and regulations, can clearly be seen to reflect the general origins and personality of the people who developed them.

THE BASIS OF FIRE PROTECTION REGULATIONS

The first laws that established fire departments or that granted some associations the right to organize for the purpose of fire fighting have developed into today's overwhelming mass of legislation pertaining to the fire service. Because of this, fire service members are involved on a daily basis in situations that require an understanding of law and its processes. Municipal and county ordinances as well as federal laws and regulations are involved in the operation of the fire services, and are used as a basis for interpreting legal cases such as those concerning the authority and responsibility of both the municipality and the fire department, the hiring and regulating of fire service personnel, and the enactment and enforcement of fire regulations.[1]

---

[1]H. Newcomb Morse, *Legal Insight*, 2nd Ed., NFPA, Boston, 1975, p. v.

1

Fire protection laws are only a small part of the laws that govern the fire service. Federal, municipal, and county ordinances affect the daily operation of all fire service organizations and their members. An understanding of laws related to the fire services cannot be complete without an understanding of American law, its evolution, its concepts, and its components.

THE FOUNDATIONS OF AMERICAN LAW

The United States of America is the oldest federal government in the world today. The heritage and sources of the laws that govern this country probably account for the longevity of this particular system of government. The longevity of this system of government is often looked upon as a tribute to the many individuals who culled the essence of law and government from ages past to form a new, yet old, government.

In order to understand the nature of American law, how it affects citizens, and how citizens affect it, it is important to understand both the heritage and major sources from which this legal structure has been drawn, and the basic components of the American legal system. A study of the evolution of American law also serves as a reminder that American law is a "living" discipline that is still evolving to fit the needs of the people it serves.

THE CONCEPT OF LAW

The concept of "law" can be defined simply in much the same way that the concept of "fire service" can be defined simply. "Fire service" can be defined as an organization committed to fire protection and fire prevention. Similarly, "law" can be defined as the rules by which people live so that the group involved — whether it be a family, an organization, a city, or a nation — can function in a somewhat uniform manner that results in the same general goals for all the members of the group. However, as members of the fire service well know, beneath the simple concept of the words "fire service" are interwoven many procedures, rules, techniques, and goals. These various procedures and disciplines combine to result in a unified concept of "fire service." Although the essence remains the same, these procedures and disciplines are constantly amended and updated so that the fire service can keep pace with and meet the needs of modern times. Thus, although the major goals of the fire service are much the same as they were a hundred years ago, the "personality" of the fire service has definitely changed.

The law of this nation has undergone similar "personality" changes because, like the fire service, it is based on a simple concept "tailored to fit" the needs of the people and society it serves.

Standards of Law: In primitive and ancient times, society was considered to be a collection of families. Rules and laws were made that were based on the assumption that the family was the basic unit of society. Today, the basic unit of society is considered to be the individual — a fact that is reflected in current laws and judicial decisions. Perhaps the most obvious example of the importance of the individual's rights in current American law can be found in the constitutional rights mandated to a person accused of a crime. In current American law, an individual can truly be "a majority of one."

Even though the basic unit of American society is considered to be the individual, the overriding goal of the laws and judicial decisions in this nation is still directed to the general welfare of society as a whole. The following comments by Justice Oliver Wendell Holmes address this point:[2]

> The standards of the law are standards of general application. The law takes no account of the infinite varieties of temperament, intellect, and education which make the internal character of a given act so different in different men. It does not attempt to see men as God sees them for more than one sufficient reason. In the first place, the impossibility of nicely measuring a man's powers and limitations is far clearer than that of ascertaining his knowledge of law, which has been thought to account for what is called the presumption that every man knows the law. But a more satisfactory explanation is, that, when men live in society, a certain average of conduct, a sacrifice of individual peculiarities going beyond a certain point, is necessary to the general welfare. If, for instance, a man is born hasty and awkward, is always having accidents and hurting himself or his neighbors, no doubt his congenital defects will be allowed for in the courts of Heaven, but his slips are no less troublesome to his neighbor than if they sprang from guilty neglect. His neighbors accordingly require him, at his proper peril, to come up to their standard, and the courts which they establish decline to take his personal equation into account.
>
> The rule that the law does, in general, determine liability by blameworthiness, is subject to the limitation that minute differences of character are not allowed for. The law considers, in other words, what would be blameworthy in the average man, the man of ordinary intelligence and prudence, and determines liability by that. If we fall below the level in those gifts, it is our misfortune; so much as that we must act at our peril, for the reasons just given. But he who is intelligent and prudent does not act at his peril, in theory of law. On the contrary, it is only when he fails to exercise the foresight of which he is capable, or exercise it with evil intent, that he is answerable for the consequences.

Historical Background: A review of the law from primitive and ancient times to the time when laws of other nations overwhelmingly affected American law is contained in Richard W. Nice's book titled *Treasury of the Rule of Law.* Following is a summary from Nice's review:[3]

> Since the beginning of time men have sought to establish rules and regulations in order to control their conduct within their particular societies. Some of the earliest attempts to regulate human behavior sprang from powerful, primitive religious influences that governed matters of ethics and morals. Examples of these

[2]Oliver Wendell Holmes, Jr., *The Common Law,* Harvard Ed., Little, Brown, and Company, Boston, 1963, pp. 86-87.

[3]Richard W. Nice, ed., *Treasury of the Rule of Law,* Littlefield, Adams, & Co., Totowa, NJ, 1965.

types of ancient legal systems may be found in the Code of Hammurabi and the Mosaic Code.

The first "pure" law can be traced back through classical history to Solon and to Lycurgus, as well as to the Law of the Twelve Tables, which was the forerunner of Roman law. Under the Emperor Justinian, Roman law became an elaborate, highly developed system of government that had an immeasurable effect upon the evolution of modern Western law.

The advances and influences of Roman law subsequently contributed heavily to the development of the complicated canon law of the Catholic Church (although Mosaic law remained at the heart of Catholic dogma) and even more heavily to later feudal law, which placed most of its emphasis on the individual's rights and privileges rather than on a concern for the state.

As European civilization progressed into the Renaissance and international commerce became an important facet of governmental life, legal reform was concentrated on advancements in maritime law and resulted in such documents as the *Consulado del Mar,* The Law of Oleron, and the English Black Book of the Admiralty.

Following the Norman conquest, feudal law was eventually replaced by the authority of royal courts; *i.e.,* the King's Bench. From this developed common law, which adhered primarily to the concepts of precedent and equity.

THE "PERSONALITY" OF AMERICAN LAW

Individuals have personalities of their own. Families and other groups drawn together for a common purpose can also be said to have personalities that distinguish them as groups and make them special in their own right. However, individuals, families, and groups do not "invent" their own personalities. These personalities result from generations of building, changing, and adapting to situations and circumstances at hand. Those changes and adaptations that are only "for the moment" and that suit immediate needs are dropped as generations succeed generations. Thus, certain customs and traits are passed on and, in the passing, are adapted to new circumstances and situations so that the old becomes "new." It must be remembered, however, that the "new" almost always builds on the "old." "New personalities" are the result of centuries of building and adapting; the "newness" is a result of new combinations or new adaptations of old customs, theories, habits, and laws.

In the same manner, the legal structure or legal "personality" of a nation is the result of a combination of many of the rules and laws that were built to reflect the customs and habits of a nation. Various sources of law gradually evolved into legal structures that can be said to define and describe the nations that had built them. Often, if a law does not fit the habits, customs, and temperament of the people it governs, the law is amended or rescinded. It is important, therefore, to understand how the "personality" of American law first took shape, and how it reflects the "personality" of the United States.

Roman Law: One of the major contributors to American law was Roman law. Virtually from the time Rome was founded in 753 B.C., Roman law was known for its formality, both in the statements of the law and in the procedures by which it was practiced. A suit could be lost if the procedural forms of charge and denial were not strictly followed. The strict and formal procedures, controlled by a body of patrician priests called the College of Pontiffs, were finally committed to writing in 250 B.C. so that everyone, patrician and plebian (common person) alike, could be aware of these strict rituals. It was at this point that the plebians could become involved in Roman law other than as defendants.

About 100 B.C., an important aspect of Roman law evolved — the law of the *praetors,* or chief magistrates. On assuming office, the chief magistrate would announce the principles that would govern his decisions. For determining the issues between the parties involved, the chief magistrates used a formula that eliminated ritualistic speeches and addressed itself to the specific issue at hand. When the exact nature of the dispute was determined, the matter was brought before a private *judex,* or "judge," who considered the evidence and rendered judgment.

After the foundation of the Roman Empire, the development of Roman law passed in large measure into the hands of the emperor. Laws called constitutions were promulgated in abundance by the emperor or through the senate. At this point, the legal situation became so complex that a specially trained group of scholars was enlisted to guide judges in making decisions. These specially trained scholars, or jurists, were instrumental in bringing procedure and some measure of uniformity to Roman law.

By the 4th century most of the branches of Roman law were sufficiently developed to respond to and meet legal needs as well as local customs. However, one major need still had to be met. The gradual evolution of Roman law had resulted in such contradictions, confusion, and repetition that a code of Roman law was required. This task culminated in the *Corpus Juris Civilis,* completed in 535 A.D., during the reign of the emperor Justinian; it is also known as the Justinian Code. Today, American law is summarized in much the same manner in the *Corpus Juris* and *Corpus Juris Secundum.*

The *Corpus Juris Civilis* forms the basis for modern civil law in a large part of the world today. In common usage today, the term civil law is used to refer to those laws that govern private legal matters and, when used in this connotation, is meant to differentiate civil law from public law and criminal law.

English Common Law: Another major contributor to American law is the system of English common law. It is, of course, to be expected that countries colonized by the English would bring English laws to the colonies. English common law is, in many ways, a contrast to Roman law. The term common law is derived from the medieval theory that laws promulgated by the king's courts represented the common custom of the land. In his book titled *Comparative Legal Cultures,* Henry W. Ehrmann explains:[4]

In England the King's itinerant judges amalgamated on the circuits the widely

---

[4]Henry W. Ehrmann, *Comparative Legal Cultures,* Prentice-Hall, Inc., Englewood Cliffs, NJ, 1976, p. 22.

differing local customs into a common custom. After a time, the common law of the realm was used synonymously with customary law and an increasingly influential bench and thus far did their part not only in creating a national legal culture but in welding together a nation. Although it was stretched out over time and was continuously seeking its justification in the past, the creation of a common law was in its effects as revolutionary as comprehensive codification elsewhere. What was repeated again and again, either by the members of the society or by the elite sitting in judgment of the acts of common men, was deemed good and beneficial. In this way a judge-made law was given the appearance of being intimately tied to the people's habits, their wisdom, and their consent. The function of the judiciary was to make customs concrete and explicit by applying "legal reason" to the cases brought before them. Custom and court decisions were and have remained singly or intertwined a primary, though not the only, source of law in common-law systems.

The legitimation which the judges gave to their activity by building their decisions on a seemingly uninterrupted sequence of cases could not have been maintained if each situation coming before the courts had been looked at afresh, on its own terms. Hence the binding force of *precedent,* an obligation to subsume new decisions under the principles of previous ones, was, at least in theory, a keystone to the viability of the common law.

The common law thus became an oral, or, as it is often called, an "unwritten law" because it was evolved from individual judicial decisions based on a judge's interpretation of the law. These individual interpretations had the effect of law by precedent on later decisions. Judges, in effect, thus became the lawmakers of the land.

It was inevitable that confusion would arise under this type of judicial system. Ten judges could rule in ten extremely varying ways — all on the same law. In his book titled *History of Law,* Gleason L. Archer, former Dean of Suffolk Law School, states:[5]

The custom arose of printing some of the important decisions of the great judges of the land. The common law, as we know it today, then began to take definite shape, for the printing of judicial decisions preserved to us the hitherto unwritten common law. The decisions thus reported were of course a most haphazard collection. Each case decided some controversy before the court according to ancient wisdom of the common law, and this printed decision became a precedent of great importance. In time, nearly all important features of the common law had thus been recorded.

The common law, as we Americans understand it today, is the law embodied in acts of Parliament and the reported decisions of the English courts prior to the American Revolution. When the colonies became independent, they wisely decided that the English common law was as much their birthright as the language itself. It was the common property of Englishmen everywhere.

---

[5]Gleason L. Archer, *History of the Law,* Suffolk Law School Press, Boston, 1928, p. 239.

BASIC COMPONENTS OF AMERICAN LAW

Just as the basis of American law evolved from the heritage and "personalities" of the early settlers from England, so did the basis of our early fire laws. One of the first tasks of the early settlers in the Boston area was to build shelter against the harsh New England winters. Using local materials, they constructed wooden houses with thatched roofs similar to the ones they were accustomed to in their homeland. The chimneys of these houses were made from wooden frames covered with mud or clay. However, such structures invited catastrophe through fire when exposure to the elements dried out the thatch and blew away the mud or clay in the chimneys. Because of this, the town fathers of the Bay Colony outlawed thatched roofs and wooden chimneys and levied a ten shilling fine on anyone who had a chimney fire. In effect, the first fire law was thus established and enforced.

This first fire law, and other fire laws that followed it, can be seen as the basis for many of our present fire laws. In the United States today, all of our laws are based on three major components: (1) the Constitution, (2) legislative acts, or statutory law, and (3) the common law and the law of equity.

THE CONSTITUTION OF THE UNITED STATES

The Constitution is the supreme law of the land. All laws in this country can be said to flow from the Constitution, since all public authority must be exercised and administered as mandated in the Constitution, and no law may violate the rights of any citizens. (See Appendix A.) The Constitution is, as Justice Marshall stated, the "fundamental and paramount law of the nation."

The Constitution defines the basic framework of the government of the United States and prescribes the extent of public authority, as this excerpt from *Van Horne's Lessee* v. *Dorrance* explains:[6]

> The constitution fixes limits to the exercise of legislative authority, and prescribes the orbit within which it must move. In short . . . the constitution is the sum of the political system, around which all legislative, executive and judicial bodies must revolve. Whatever may be the case in other countries, yet in this, there can be no doubt, that every act of the legislature, repugnant to the constitution, is absolutely void.

The second major goal of the Constitution is to protect the rights of the citizens within the legal framework it has established. As excerpted from *Hanson* v. *Vernon:*[7]

> It grants no rights to the people, but is the creature of their power — the instrument of their convenience. Designed for their protection in enjoyment of the

---

[6]*Van Horne's Lessee* v. *Dorrance*, 2 Dall. 304, 310.

[7]*Hanson* v. *Vernon*, 27 Stiles 28, 74.

rights and powers which they possessed before the Constitution was made, it is but the framework of the political government, and necessarily based upon pre-existing condition of laws, rights, habits, and modes of thought.

The Constitution has been used by some fire department members as a means to contest departmental orders that they feel are in violation of their rights as citizens. For example, a volunteer fire fighter was suspended from membership in his fire department because his district's Board of Fire Commissioners charged that the length of his hair and sideburns violated departmental rules. The fire fighter filed a proceeding in the Supreme Court of New York against the Board of Fire Commissioners, and the Court held that the department's rules were in violation of the fire fighter's constitutional rights. He was therefore reinstated as a member of the department.[8]

STATUTORY LAW

Legislative acts, also called statutory law, are formal, written enactments mandated by the authorized powers of a government through its legislature. These acts are generally promulgated by federal or state legislatures. Statutory law is often contrasted with common law. Common law consists of "unwritten" legal precedents derived from judicial decisions, policy, and custom, while statutory laws are formal, written documents often embodied in a code. Although common law retains great importance as a source of law in the United States, increased government regulation has resulted in an immense growth of the body of statutory law in this country. Charles W. Bahme, in his book titled *Fire Service and the Law*, states:[9]

> Laws that have been adopted by Congress as federal statutes, and laws that have been passed by the state legislatures are generally termed *statutory laws*. When used in the broadest sense, statutory law means *written law* (as distinguished from unwritten law) and includes local ordinances.
>
> Many states incorporate their statutes in various codes. A *code* is a systematic compilation of both statutory law and the law handed down by the judges in their decisions. Typical examples of the state codes in use today are the probate, civil, penal, labor, political, administrative, educational, military and veteran's, health and safety, civil procedure, business and professional, and government. Local governments also frequently adopt codes relating to firesafety, such as the electrical, building, plumbing, heating and ventilating, refrigeration, and fire codes.

One example of a statutory law affecting the fire service is exemplified in the state of Wisconsin where the role and interaction of the fire service during disaster operations has been legally defined by state statutes. Within the state's Division of Emergency Government, there is the position of Director of Fire and Rescue

---

[8]Morse, *Legal Insight*, pp. 22-23.

[9]Charles W. Bahme, *Fire Service and the Law*, NFPA, Boston, 1976, p. 2.

Services. The responsibilities of this position are to develop an operational fire and rescue capability statewide to minimize fire loss from an enemy attack or other emergencies.[10]

THE COMMON LAW

Common law, or the law of judicial precedent and custom, still holds a very important position in American law. The doctrine of *stare decisis* (from the Latin meaning "let what is decided stand") is the source of the power of judicial precedent. *Stare decisis* means that the decisions of the highest court in the jurisdiction are binding on all other courts in that jurisdiction.

Court decisions have been made that affect paid and volunteer fire officers and fire fighters, and cover such cases as those involving residency requirements for fire fighters, promotions, equipment and apparatus safety, and vehicular accidents involving fire apparatus, to name a few. Under common law, decisions reached in these incidents serve as a background for similar cases and, even if the decisions are subsequently overturned, the original decision still has weight as a part of American common law.

Common law has not kept pace with the vast array of legal problems and situations in modern life. As a result, statutes play an increasingly important role in this country's legal developments. Administrative and criminal law are prime examples of areas in which statutory law has superseded common law. However, statutory decisions often have their basis in common-law decisions.

In the United States, all states except Louisiana follow the common law. Most states have enacted statutes that allow the common law as a source of law and authority. In Canada, all provinces but Quebec accept the common law.

THE LAW OF EQUITY

The law of equity has been so closely connected with common law — and is still such an integral part of American law — that brief mention of it seems appropriate at this point. Like the common law, it is English in origin. When claims did not fall within the prescribed forms of English law, but were still considered worthy, the claimant could appeal directly to the English Chancellor, who was considered "the keeper of the king's conscience." The basic premise of this procedure was that no wrong should be without an adequate remedy.

Where judgments at common law resulted in money damages, damages at equity led to injunctions and the remedy of contempt of court citations. Courts of equity cut through legal technicalities and decided what was fair and just without reference to *stare decisis*. Eventually the court of equity (also known as the court of chancery) became a rival of the court of common law; it could decide any claim of jurisdiction that the courts of common law disputed. Courts drew on Roman law

---

[10]Ron Reuter and Anthony Testolin, "Disaster Operations," *Wisconsin Fire Journal*, Vol. 3, No. 3, Fall 1977, p. 8.

and canon law for precedents and thus amassed their own body of precedents; these laws gave rise to the laws of mortgage and trust.

The law of equity is particularly significant to the fire service because it carries the force of law when there is no other law. For example, when a fire hazard such as an abandoned building threatens the safety of nearby dwellings and the property owner refuses to remove it, a fire department can obtain an injunction through a court of equity, thus forcing the owner to take action. The law of equity, therefore, assists a fire department when there is no other force of law, and serves as a fire prevention method in removing hazards from a community.

Courts of equity can issue injunctions to prevent injury, to enforce the performance of contracts, and to correct or cancel written agreements. Equity trials are usually conducted without a jury. Norman Fetter gives one reason why:[11]

> Whenever a court of law is competent to take cognizance of a right, and has power to proceed to a judgment which affords a plain, adequate, and complete remedy, the plaintiff must proceed at law, because the defendant has a constitutional right to a trial by jury.

In earlier American legal history, the courts of common law were considered courts of legal jurisprudence; the courts of equity were considered courts of equitable jurisprudence. In the United States today, only a few states have separate equity courts. Judges generally perform the double function of presiding over courts that have jurisdiction over law as well as equity.

THE BASIS OF FIRE SERVICE-RELATED COURT DECISIONS

Legal data that affects fire service members have their basis in the judicial system of the United States government; judgments made on cases brought before the federal and state courts serve to define the scope of fire department authority and provide legal measures for given situations such as liability, firesafety, fire fighting equipment safety, fire department relationship to municipal government, and personnel decisions in such areas as suspension, compensation, pensions, and appointments. Although this data is of direct and vital significance to the fire service, of equal significance is an understanding of the federal and state judicial systems and the roles of federal and state courts in formulating this data.

THE AMERICAN JUDICIAL SYSTEM

An integral part of this country's legal system is the judicial system. There are two judicial systems in the United States — the federal judicial system and the state judicial system. The multiplicity of courts is a salient feature in the American

---

[11]Norman Fetter, *Handbook of Equity Jurisprudence*, West Publishing Co., St. Paul, MN, 1895, p. 10.

legal system. For example, at the beginning of the 20th century, the state of Nebraska had more judges than all of England.

The government of the United States is a dual government, with both federal and state legislatures. As has been previously stated, the Constitution limits the powers of the federal government and allows all other powers to reside with the states. This is reflected in the dual judicial systems in the United States.

Dual Judicial Systems: The federal judicial system is limited in its powers — it can exercise only those powers granted to it by the Constitution. The state judicial system has virtually unlimited powers to decide cases. *The United States Courts*, a publication of the U.S. Government Printing Office, states:[12]

> Throughout the United States there are two sets of judicial systems. One set is that of the State and local courts established in each State under the authority of the State government. The other is that of the United States courts set up under the authority of the Constitution by the Congress of the United States.

> The State courts have general, unlimited power to decide almost every type of case, subject only to the limitations of State law. They are located in every town and county and are the tribunals with which citizens most often have contact. The great bulk of legal business concerning divorce and the probate of estates and all other matters except those assigned to the United States courts is handled by these State courts.

> The United States courts, on the other hand, have power to decide only those cases in which the Constitution gives them authority. They are located principally in the larger cities. The controversies in only a few carefully selected types of cases set forth in the Constitution can be heard in the United States courts.

The Judiciary Act of 1789: One of the initial laws enacted by the first Congress was the Judiciary Act of 1789. This act established a federal court system and duplicated the court system of the states with specific restrictions placed on the federal court system. As stated in *The United States Courts:*[13]

> The controversies which can be decided in the United States courts are set forth in section 2 of Article III of the United States Constitution. These are first of all "Controversies to which the United States shall be a party," that is, cases in which the United States Government itself or one of its officers is either suing someone else or is being sued by another party. Obviously it would be inappropriate that the United States Government depend upon the State governments for the courts in which to decide controversies to which it is a party.

> Secondly, the United States courts have power to decide cases where State courts are inappropriate or might be suspected of partiality. Thus, Federal judicial power extends "to Controversies between two or more States; between a State and

---

[12]*The United States Courts,* U.S. Government Printing Office, Washington, DC, 1975.
[13]*Ibid.*

Citizens of another State; between Citizens of different States; between Citizens of the same State claiming Lands under Grants of different States, ..." If the State of Missouri sues the State of Illinois for pollution of the Mississippi River, the courts of either Missouri or Illinois would be inappropriate and perhaps not impartial forums. These suits may be decided in the United States courts. At various times State feeling in our country has run high, and it has seemed better to avoid any suspicion of favoritism by vesting power to decide these controversies in the United States courts.

State courts are also inappropriate in "Cases affecting Ambassadors, other public Ministers and Consuls" and in cases "between a State, or the Citizens thereof, and foreign States, Citizens, or Subjects." The United States Government has responsibility for our relations with other nations, and cases involving their representatives or their citizens may affect our foreign relations so that such cases should be decided in the United States courts.

And, thirdly the Constitution provides that the judicial power extends "to all Cases, in Law and Equity, arising under this Constitution, the Laws of the United States, and Treaties made, or which shall be made, under their Authority" and "to all Cases of admiralty and maritime jurisdiction." Under these provisions the United States courts decide cases involving the Constitution, laws enacted by Congress, treaties, or laws relating to navigable waters.

Some cases may be tried in both state and federal courts. In this instance, if a state court rules on federal law, its decision may be reviewed by the United States Supreme Court. Congress has also provided that, in those cases that may be tried in either state or federal court, only those cases where the amount involved exceeds $10,000 may be subject to being tried in federal court.

THE FEDERAL COURT SYSTEM

The Constitution provides that "The Judicial Power of the United States, shall be vested in one supreme Court, and in such inferior Courts as the Congress may from time to time ordain and establish." Thus, with the exception of the Supreme Court, all other federal courts have been established by Act of Congress for practical reasons. The number of cases requiring federal jurisdiction became increasingly large, and the physical difficulty for Supreme Justices to travel to hear cases in federal jurisdiction became virtually impossible. By 1974, in addition to the Supreme Court, the federal system included ninety four district courts and eleven courts of appeal. All federal judges are appointed by the President, subject to congressional approval. In addition, there are five federal courts with specialized jurisdictions: the Court of Claims, which has jurisdiction over certain claims against the government; the Tax Court; the Customs Court, which reviews certain administrative decisions by customs officials; the Court of Customs and Patent Appeals; and the Court of Military Appeals. (See Fig. 1.1.) *The United States Courts* explains:[14]

---

[14]*Ibid.*

THE UNITED STATES COURT SYSTEM

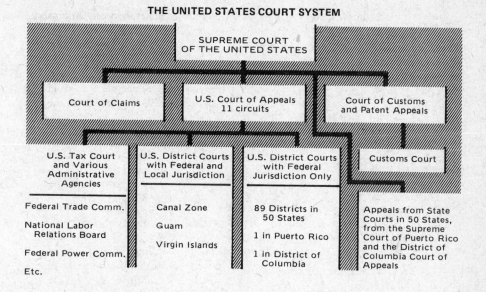

Fig. 1.1.   Representation of the present United States Court system.

At the present time the United States court system may be likened to a pyramid. At the apex of the pyramid stands the Supreme Court of the United States, the highest court in the land. On the next level stand the United States courts of appeals, 11 in all. On the next level stand the United States district courts, 94 in all, including the United States District Courts for the District of Columbia and Puerto Rico and the district courts in the Canal Zone, Guam, and the Virgin Islands. The United States Tax Court and, in a sense, certain administrative agencies may be included here because the review of their decisions may be directly in the courts of appeals. Some agency reviews, however, are handled by the district courts.

A person involved in a suit in a United States court may thus proceed through three levels of decision. His case will be heard and decided by one of the courts or agencies on the lower level. If either party is dissatisfied with the decision rendered, he may usually have review of right in one of the courts of appeals. Then, if he is still dissatisfied, but usually only if his case involves a matter of great national importance, he may obtain review in the Supreme Court of the United States.

This pyramidal organization of the courts serves two purposes. First, the Supreme Court and the courts of appeals can correct errors which have been made in the decisions in the trial courts. Secondly, these higher courts can assure uniformity of decision by reviewing cases where two or more lower courts have reached different results.

The Supreme Court: Originally there were six Supreme Court justices; the number has gone as high as ten. Presently, the Supreme Court consists of nine justices who have been appointed for life by the President with congressional approval. The figure has been kept at nine since 1869.

Unlike most other courts, the Supreme Court can decide which cases it will hear. The Court receives about 5,000 cases per year, most of which are determined to be not proper subject matter or are of insufficient importance to warrant full Court review. These cases are therefore not granted *certiorari,* which is a writ or order from a higher court to call up the records of a lower court so that the case may be formally reviewed. *Certiorari* is the normal method of obtaining review by the Supreme Court. Each year approximately 250 cases are decided; of these, about half are published in full decisions.

The District Courts: The trial court in the federal judicial system is the district court. There are ninety four district courts — eighty nine in the fifty states, and one each in the District of Columbia, the Canal Zone, Guam, Puerto Rico, and the Virgin Islands. Each state has at least one court; Texas, California, and New York have four. For convenience and efficiency, a district may be divided so that a case may be heard in one of several places in a state. As of 1975, district courts were receiving about 103,530 civil cases, 38,000 criminal cases, and 189,000 bankruptcy cases every year. As stated in *The United States Courts:*[15]

> The work of the United States district courts is partly reflected in the statistics on the types of cases which are filed every year. In the fiscal year ended June 30, 1974, exactly 103,530 civil cases were filed in these courts. Of these, 27,585 involved the United States either as a party plaintiff or party defendant. The United States as plaintiff commenced about 462 cases under the food and drug laws, 1,624 cases to collect money due on promissory notes, 515 civil commitment proceedings under the Narcotic Addict Rehabilitation Act, 1,505 cases to enforce the fair labor laws, 2,205 cases to collect money owed on mortgage loans, and about 5,350 other cases. The United States was a defendant in about 1,700 habeas corpus cases and 1,800 cases involving motions to vacate criminal sentences, 1,850 Tort Claims Act cases, 1,400 tax cases, 3,585 suits to review social security benefit awards and many others. These figures show the important role the district courts play in enforcing Federal statutes and in achieving justice even when the United States Government is involved in a law suit.
>
> Private parties brought into the district courts during the 1974 fiscal year 75,945 cases of which 46,797 involved questions arising under Federal laws. The largest category of these cases (about 13,400) was that involving petitions for writs of habeas corpus by persons held in custody who have alleged violations of Federal constitutional rights. About 27,000 cases came into the district courts in suits between private persons from different States. Of this number about 6,600 were concerned with personal injuries arising out of motor vehicle accidents. An additional 2,185 civil cases of a local nature were filed in the territorial district courts.
>
> Of the 37,667 criminal cases brought into the district courts during the year ended June 30, 1974, 4,685 involved embezzlement or fraud, 4,360 forgery and counterfeiting, 1,790 the interstate transportation of a stolen automobile, 641

[15]*Ibid.*

failure to pay the tax on alcoholic beverages and 1,008 the violation of the Selective Service laws. The remainder involved the violation of a host of other laws. In these criminal cases the United States courts share with the officers of the law and the attorneys the enforcement of criminal laws for the protection of all citizens.

An important work performed by the United States district courts is the supervision by probation officers of persons convicted of crime, but placed on probation rather than sent to prison. During the fiscal year 1974 over 19,450 persons were placed on probation by the district courts.

The Court of Appeals: The intermediate appellate courts in the federal judicial system are the Court of Appeals. Ten Court of Appeals are apportioned to include three or more states, and the eleventh Court of Appeals is designated for the District of Columbia. Although the principal function of the Court of Appeals is to review cases decided by the district courts, they also review orders of federal administrative agencies for errors of law.

THE STATE JUDICIAL SYSTEM

Every state has its own three- (and sometimes two-) level system of courts created by its constitution and statutes. The majority of the nation's legal business is conducted in the state courts.

Although court names may differ from state to state, the state legal hierarchy is much the same as the federal hierarchy. For example, in New York the main trial court is called the Supreme Court. Above the Supreme Court is the Appellate Division, and the highest court in the state is called the Court of Appeals. Probably the best way to trace the hierarchy of the state judicial structure is from "bottom to top." (See Fig. 1.2.)

Inferior Courts: Inferior courts deal with "petty" cases in which small monetary amounts or minor criminal penalties are involved. Inferior courts are generally not "courts of record" in that no detailed record of the proceedings is involved. Procedure at this level may be relatively informal.

Inferior courts include township justice courts, police courts, traffic courts, small claims courts, and city courts. In *The American System of Government,* Ferguson and McHenry state:[16]

> . . . Most common are municipal courts set up for handling problems peculiar to congested areas. These are usually headed by several elective judges, each of whom presides over one or more divisions specializing in criminal, civil, traffic, domestic relations, or juvenile cases. In addition to handling disputes arising from municipal law, these nearly always have concurrent jurisdiction with the usual district or county courts. This means that for violating a city ordinance one would be prosecuted in the municipal court, but for a state offense committed within the city's limits one might be tried in either the municipal court, a justice's court,

---

[16] John H. Ferguson and Dean E. McHenry, *The American System of Government,* McGraw-Hill Book Company, Inc., New York, 1950, pp. 859-860.

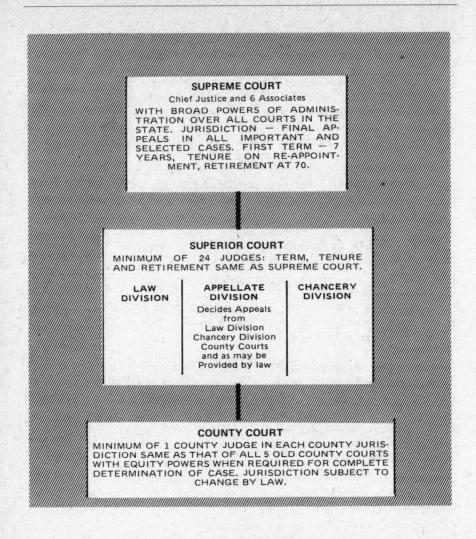

SUPREME COURT

Chief Justice and 6 Associates

WITH BROAD POWERS OF ADMINIS-
TRATION OVER ALL COURTS IN THE
STATE. JURISDICTION — FINAL AP-
PEALS IN ALL IMPORTANT AND
SELECTED CASES. FIRST TERM — 7
YEARS, TENURE ON RE-APPOINT-
MENT, RETIREMENT AT 70.

SUPERIOR COURT

MINIMUM OF 24 JUDGES: TERM, TENURE
AND RETIREMENT SAME AS SUPREME COURT.

LAW DIVISION	APPELLATE DIVISION	CHANCERY DIVISION
	Decides Appeals from Law Division Chancery Division County Courts and as may be Provided by law	

COUNTY COURT

MINIMUM OF 1 COUNTY JUDGE IN EACH COUNTY JURIS-
DICTION SAME AS THAT OF ALL 5 OLD COUNTY COURTS
WITH EQUITY POWERS WHEN REQUIRED FOR COMPLETE
DETERMINATION OF CASE. JURISDICTION SUBJECT TO
CHANGE BY LAW.

INFERIOR COURTS

NOT ABOLISHED BY REVISION BUT MAY BE
ESTABLISHED, ALTERED OR ABOLISHED BY LAW
APPEALS THEREFROM AS PROVIDED BY LAW.

County Traffic Courts	District Courts	Police, Recorder & Family Courts	Juvenile & Domestic Relations Courts	Surrogate Courts	Criminal Judicial District Courts	Small Cause Courts	Justice of Peace Courts

**ALL JUDGES APPOINTED BY GOVERNOR WITH APPROVAL
OF SENATE, EXCEPT MUNICIPAL JUDGES AND SURROGATES**

Fig. 1.2.   Representation of the state of New Jersey's judicial system. (Adapted from *Journal of the American Judicature Society*, Vol. 31, No. 5, Feb. 1948, p. 143).

or a county (or district) court. Needless to say, this is often quite confusing, especially where cities have spread out to include an entire county or more. Other special courts frequently found include those for handling small monetary claims, domestic relations, and juvenile cases.

Trial Courts: The intermediate court in the state hierarchy is the court in which the suit or trial originated, commonly known as the trial court of general original jurisdiction. This court is identified by different titles in different states — for example, superior court, district or circuit court, supreme court — and hears both civil and criminal cases. Specialized state trial courts handle probate, divorce, and juvenile cases. State trial courts are courts of record, that is, a record is made of the entire judicial proceeding. Court procedure at this level is formal.

The trial courts handle appeals from inferior courts although these cases are generally heard as completely new trials (trials *de novo*). In addition, state trial courts may be authorized to review decisions of state administrative agencies.

Appellate Courts: The highest court in the state is the appellate court. Its function is almost exclusively to hear cases appealed from the lower courts. Appellate courts neither ''try'' cases, nor hear the introduction of new evidence; rather they review the record and the law and sustain, reverse, or modify decisions made by the lower courts.

Appellate court decisions are written and published. As the highest state courts, appellate courts are the ultimate and final interpreters of state constitution and laws unless federal law or some aspect of federal law is involved.

CIVIL PROCEDURE

Proceedings in a court are called lawsuits, which are either civil or criminal. Criminal procedure and administrative procedure will be addressed later in this text. The following steps are generally followed in a civil suit:[17]

The principal steps in a civil suit are:
1. A complaint is filed with the court by the plaintiff.
2. A summons is issued by the court which gives the defendant a certain number of days to answer the complaint.
3. The defendant must file a pleading within the prescribed time: this may be an ''answer'' or a *demurrer*. Where a demurrer is interposed the court rules on the question of law as to the sufficiency of the pleading.
4. A trial is conducted before a judge and a jury of twelve people, though the constitution of some states permits less than twelve if the parties agree.
    (a) Plaintiff's attorney opens the case by calling witnesses and presenting evidence to prove plaintiff's contentions.
    (b) Defendant's attorney cross-examines plaintiff's witnesses in an attempt to bring out facts damaging to plaintiff's case.
    (c) Defendant's attorney then puts defense witnesses on the stand, and they in turn are cross-examined by the plaintiff's counsel.

---

[17]Bahme, *Fire Service and the Law*, p. 20.

(d) Attorneys give their closing arguments if they so desire. (Attorney for the plaintiff has the right to give the first and last of such arguments.)

5. The judge instructs the jury in matters of law; the jury decides all questions of fact in the case, and applies the law as instructed. If the trial has been conducted without a jury, the judge takes the case under advisement.

6. The jury brings in a *verdict*; at least three-fourths of the jurors must be in agreement in a civil matter. When there is no jury the judge renders a judgment.

7. There may be a request for a new trial because of some serious error committed during the process, or an apppeal may be made to a higher court because of some errors in law, or error in judging the evidence.

8. If the plaintiff is successful there will be an attempt to collect the amount set forth in the judgment or verdict. This may require the levying of an *execution* against the defendant's property, whereby the property may be seized by the sheriff and sold to satisfy the amount awarded.

SUMMARY

The primary source of early fire protection laws and American law is English law, which was brought to this country by the English colonists and adapted to fit the needs of the new colonies as they were established. English law and the Roman civil law both came under the mandates of the Constitution of the United States, once the republic was formed. American law gradually grew and evolved into today's legal structure, and is still undergoing a process of evolution in order to meet the needs and complexities of modern-day society.

Because American society is becoming more and more complex and changes are more fast-paced, statutory or legislative law is becoming one of the largest components of the American legal system. Statutory law is more able to keep pace with the changing American scene than the more gradual pace of the American court system.

Based on civil law, common law, and constitutional law, many major branches of American law have evolved: administrative law, admiralty law, commercial law, contract law, corporate law, criminal law, family law, international law, labor law, patent and copyright law, property law, and tax law, to name but a few. All of these branches of law have evolved from the increasingly complex needs of the people and the mandate of the Constitution "to promote the general welfare."

It is important to understand the dual nature of the judicial system of the United States for several reasons. The judicial system provides the framework for articulating and interpreting the laws of the United States, and the legal data generated from the judicial system has a direct bearing on the fire service and in defining fire service responsibilities to the citizens it protects. For the fire science student, it is important to be aware of the "forum" in which a case is presented — state or federal; trial or appellate. The judicial forums of a nation belong to each citizen, but they are useless and empty unless each citizen is aware of the judicial safeguards and tribunals that are available.

ACTIVITIES

1. (a) Discuss with a group of your classmates: When are laws necessary? Why are laws necessary?
   (b) As a result of your discussion, give three reasons why you might need laws if you formed a group, and three laws that you would mandate for this group.
2. How would the basic unit of society as considered in primitive and ancient times, and as considered in modern American times, affect the nature of the laws in these two conceptions of society?
3. Review Oliver Wendell Holmes, Jr.'s, comments regarding the standards of the law in terms of those who break laws. Based on his comments, write a brief summary regarding the standards of law that might be applied to a compulsive arsonist.
4. (a) Compare Roman law and English law. If you were to found a country, which kind of law would you choose and why?
   (b) What are the major contributions of Roman law and English law to American law? What terms are still used in American law?
5. (a) What are the three basic components of American law?
   (b) Why is each component important to the fire service?
6. At the turn of the century, the state of Nebraska had more judges than all of England. This situation was fairly typical of all of the United States. Why do you think this was so? Discuss with your classmates why this might have been necessary.
7. (a) Describe the federal judicial system and tell how it is similar to and different from the state judicial system.
   (b) Describe your state's judicial structure.
8. (a) Why do you think the case load of the United States Supreme Court is so heavy?
   (b) Give examples from your own knowledge or experience of how a case heard in a state court could also qualify to be reviewed by the Supreme Court of the United States.
9. Why are only about half of the Supreme Court decisions published as full decisions?
10. A member of your fire department has been sued in a civil action. What procedures would the court follow in this civil suit?

BIBLIOGRAPHY

Archer, Gleason L., *History of the Law,* Suffolk Law School Press, Boston, 1928.

de Funiak, William Q., *Handbook of Modern Equity,* 2nd Ed., Little, Brown, and Company, Boston, 1956.

Ehrmann, Henry W., *Comparative Legal Cultures,* Prentice-Hall, Inc., Englewood Cliffs, NJ, 1976.

Ferguson, John H. and McHenry, Dean E., *The American System of Government,* McGraw-Hill Book Company, Inc., New York, 1950.

Field, Richard H. and Kaplan, Benjamin, *Materials for a Basic Course in Civil Procedure,* The Foundation Press, Inc., Mineola, NY, 1973.

Fleming, James, Jr., *Civil Procedure,* Little, Brown, and Company, Boston, 1965.

Gray, John Chipman, *The Nature and Sources of the Law,* Beacon Press, Boston, 1963.

Holmes, Oliver Wendell, Jr., *The Common Law,* Harvard Ed., Little, Brown, and Company, Boston, 1963.

Mermin, Samuel, *Law and the Legal System,* Little, Brown, and Company, Boston, 1973.

Nice, Richard W., ed., *Treasury of the Rule of Law,* Littlefield, Adams & Co., Totowa, NJ, 1965.

II

MUNICIPAL CORPORATIONS

OVERVIEW

Fire service personnel, whether serving in a volunteer or paid capacity, will probably be working for or be connected in some manner with a corporation. It is not unusual for a fire department itself to be a corporation or a quasi corporation. As such, members of the fire service should know, for example, whether they are legal members of the corporation or under private contract to the corporation. Both situations are possible, and both affect the legal status and liability of those involved in fire protection and fire prevention. Even if a fire department is not directly connected with a corporation, it can be involved with the corporate nature of the fire service in a variety of ways — water supply, mutual aid, etc. Specific rights and liabilities for specific acts will be covered in detail in later chapters of this text. This chapter will concentrate on the corporation as the source of fire protection and fire prevention services, and the role the fire service — its members and its departments — plays in such a corporate structure.

CORPORATIONS IN GENERAL

A corporation is an "artificial person" created by law. Today, a corporation is created by statute, charter, or any other act of a legislature to accomplish a specified goal, whether it be idealistic, monetary, or governmental.

The powers granted to a corporation by the state are usually granted in the form of a charter. This charter becomes the basic law of the corporation: it establishes the structure of the corporation, the conditions under which the corporation may operate, and the powers that may be exercised by the corporation. The charter also specifies the general means by which the corporation may exercise the powers the state has granted to it. Once a charter is granted to a corporation by the state, it becomes, in effect, a contract between the corporation and the state.

EARLY HISTORY

Corporations can be traced back as far as early Greek and Roman times, when groups were allowed to combine and associate for a common purpose. The concept of corporations was transferred to Britain after the Roman conquest, and ultimately to colonial America where, in the early days of the United States, corporations, both public and private, were created and allowed to exist. Again, organizations, including early fire protection organizations, were incorporated for both public and private purposes.

NATURE OF A CORPORATION

A corporation, once established and authorized, has a legal personality of its own and, as such, changes in membership cannot disband or alter the corporation in much the same way that changes in personnel within the fire services do not alter the basic purpose and goals of the fire services.

The following are generally considered to be some of the major attributes that describe the ordinary powers of any corporation:

1. Perpetual succession, *i.e.*, change in membership does not alter or disband a corporation.

2. The ability to sue and be sued (as a "legal person").

3. As a "legal person," to receive and grant powers and to purchase land and make contracts.

4. To have a corporate or common seal.

5. To make by-laws.

These attributes originate from the one essential attribute that alone can give a corporation existence: a corporation must be able to exist and function — within the limitations of the powers it is granted — as a separate "legal person" or entity, independent of the individuals who comprise its membership. This is the sole attribute from which all corporations spring. Without this separate legal personality, there can be no corporation.

CORPORATE LIABILITY

Because a corporation is a separate "person," its members are not liable for torts (violations of the personal or private rights of others) or breaches of duty committed in the name of the corporation unless, of course, individual members of the corporation have been involved in such corporate breaches. Conversely, torts committed against a corporation are not considered as torts committed against individual members of a corporation. Moreover, when an individual is not acting as a duly authorized agent of the corporation, the corporation is not responsible for such acts of the individual. This is an important distinction: when is a member of the fire service acting as an agent of a corporation, and when is a member of the fire service acting as a private individual?

A corporation is liable in its attribute as a "legal person" for the civil wrongs committed by its authorized representatives or agents when these representatives or agents are acting within the authority conferred on them as employees or authorized representatives. The scope of this corporate liability is very broad. A corporation can, in some instances, be considered liable for the acts of its agents, even if such acts are committed by disregarding instructions, or if the acts appear to be wanton, malicious, or even criminal in nature. However, if a representative or agent of a corporation acts beyond the scope of delegated authority, the corporation is not liable — the individual is.

PUBLIC AND PRIVATE CORPORATIONS

A corporation can be considered either public or private, and the distinction between the two is not always easy to distinguish. Most corporations can be considered a combination of both public and private corporations, depending on their various goals and purposes. Basically, public corporations are considered to be those that are created or formed by the people or a government for governmental or public purposes. Such corporations as cities, towns, fire districts, and hospitals are considered to be essentially public corporations. However, each of these can also have some of the characteristics of a private corporation.

Private corporations are basically created for private purposes, as opposed to public purposes, and for the private benefit of the members of the corporation. Ordinary business organizations, airlines and the like, are considered private corporations, although some of the characteristics of these corporations could qualify them to be described also as public corporations. A further distinction: public corporations can be considered as instruments or extensions of the state, while private corporations in their public dealings can be considered as private, separate entities that have entered into a contract with the state.

A private corporation is formed by the voluntary request of every member of the group wishing to form the corporation. Generally, each member of the group knows in advance of this request both the exact nature of the corporation to be founded and the legal rights and responsibilities of such a corporation. This is not often true when a public corporation is created.

A public corporation may be created with or without the consent of the members involved. Even when the voluntary consent of all the members of a public corporation is involved, all the members are usually not fully or equally aware of the specific rights, duties, and responsibilities of the public corporation that is being created. This is probably because the ultimate goals of both public and private corporations are different: a private corporation is generally motivated by profit, while a public corporation is generally founded in response to a public demand or need.

The state exercises different kinds of control over public and private corporations. While the creation of a private corporation in effect forms a contract between the state and the corporation which cannot be altered and only rarely can be rescinded, the creation of a public corporation does not result in a similar contractual relationship in that the agreement between the state and a public corporation can be amended, modified, altered, or even rescinded.

Public and private corporations differ in the scope and range of activities that are allowed by the state. A private corporation can branch its activities into any area it desires, so long as it does not violate the law. A public corporation can only perform those activities that are authorized by the state. If a public corporation wishes to branch out its activities, it cannot do so unless the state specifically authorizes it to do so.

MUNICIPAL CORPORATIONS

Municipalities that have been incorporated, whether they are called cities, villages, boroughs, or towns, are legal entities by virtue of their incorporation. Because of their status as legal entities, municipal corporations possess certain legal powers that are not possessed by other local government units.

EARLY HISTORY

It can be said that cities are the oldest "public corporations" in history. The following is a description by Stuart MacCorkle of the origin of cities:[1]

> History reveals that defense was doubtless the earliest motive for the establishment of local groupings of population. Ancient peoples held together by family ties, living in caves or rudely constructed stockades, developed a spirit of cooperation. In this humble manner, civilization and the city began life together, and they have since remained inseparable. As time passed, those villages which were the most capable of defense accumulated larger populations and grew in wealth and power. Athens, for example, clustered close about the steep slopes of the Acropolis; and Rome, entrenched in the marshes of the Tiber, soon outdistanced other villages in population, wealth, and military power.

From very early times, cities, as municipal corporations, have contributed a great deal to civilization. One of the major reasons for the vitality that engendered such major contributions was the fact that these municipalities had the right to govern themselves locally. The Latin word *municipium* identified those towns and cities that had special rights and privileges. The contributions of the cities of ancient Rome — perhaps the greatest of all municipal corporations — are certainly

[1]Stuart A. MacCorkle, *American Municipal Government and Administration*, D. C. Heath and Company, Boston, 1948, p. 20.

not unknown today. In fact, the early Roman municipal structure became the prototype for many of today's municipal corporations.

The characteristics of the ancient Roman cities contributed in large measure to the successful development of English cities. By the fifteenth century, the English borough was an incorporated municipality. This concept of the incorporated English borough was brought to America by the early colonists. It was adapted to serve the purposes of the colonists, and became the foundation upon which the concept of American municipalities was built. Even today, American municipalities still reflect in many ways the incorporated English boroughs of the 15th century.

CHARACTERISTICS OF MUNICIPAL CORPORATIONS

As has been previously stated, a corporation is a group of individuals formed into a "legal person" for one purpose or another. A municipal corporation is a public corporation that is usually created by the state at the request of, or with the consent of, its inhabitants. Generally, a municipal corporation consists of a group of inhabitants in a geographic area — usually a city or a town. The main purpose of a modern municipal corporation is the same as it was in ancient Roman times — for the purpose of local self-government.

Following are the elements necessary for a municipal corporation, as described by Eugene McQuillan in *The Law of Municipal Corporations*:[2]

> 1. Incorporation as such pursuant to the Constitution of the state or to a statute.
> 2. A charter.
> 3. A population and prescribed area within which the local civil government and corporate functions are exercised. . . .
> 4. Consent of the inhabitants of the territory to the creation of the corporation, with certain exceptions (actually, the legislature in most states may act without consulting the local residents).
> 5. A corporate name.
> 6. The right of local self-government, although in most states this is held to be not an inherent right. Unless otherwise provided by statute, a test as to whether an organization is a municipal corporation, using the term in its strict sense, is whether it has the power of local government as distinguished from merely possessing powers which are merely executive and administrative in their character. The characteristic feature of a municipal corporation beyond all others is the power and right of local self-government.

In *Commentaries on the Law of Municipal Corporations,* Judge John F. Dillon, who carved and defined many of the legal guidelines for municipal corporations still followed today, summarizes as follows:[3]

---

[2]Eugene McQuillan, *The Law of Municipal Corporations,* 3rd Ed., Vol. 1, Sec. 2.07, Callaghan and Company, Chicago, 1971.

[3]John F. Dillon, *Commentaries on the Law of Municipal Corporations,* 5th Ed., Little Brown & Company, Boston, 1911.

Municipal corporations owe their origin to, and derive their powers from, the legislature. It breathes into them the breath of life, without which they cannot exist. As it creates, so it may destroy. If it may destroy, it may abridge the control. Unless there is some constitutional limitation on the right, the legislature might, by a single act, if we can suppose it capable of so great a folly and so great a wrong, sweep from existence all of the municipal corporations of the state, and the corporations could not prevent it. We know of no limitation on this right so far as the corporations themselves are concerned. They are, so to phrase it, the mere tenants at will of the legislature. (*City of Clinton* v. *Cedar Rapids and Missouri Railroad Company*, 24 Iowa 455 [1868]).

MUNICIPAL CORPORATIONS
AND
QUASI-MUNICIPAL CORPORATIONS

Municipal corporations, or municipalities, are commonly thought of as urban cities — usually large urban cities. Under American law, a municipal corporation is any subordinate public authority that has been created by the state and vested with the legal rights of a corporation. Thus, a municipal corporation can be a city, a village, a town, a county, a special district, or even a fire department. Such public corporations are designated as either municipal corporations or quasi-municipal corporations.

As has been previously stated, municipal corporations are usually created at the request of, or certainly with the knowledge and consent of, the inhabitants to be included in the municipality. The main purpose for such a request is primarily to ensure a more ready response to local needs. When the state creates such a municipality, the municipal corporation becomes an agent of the state and aids the state in its governmental role. Quasi-municipal corporations are not created at the request of, or with the consent of, the inhabitants. They are created by the state to aid it in its administrative functions. Although in some states the distinction between municipal corporations and quasi-municipal corporations is not clearly defined, the general body of law makes the distinction quite clear. The distinction revolves primarily around the nature of the powers of municipal corporations and quasi-municipal corporations and, as such, has bearing on the rights and liabilities involved. In general, municipal corporations are more liable for torts and contract liabilities, have broader law-making powers, and are less bound to state control and regulation than are quasi-municipal corporations.

Because they are separate "legal persons," true municipal corporations, in addition to being creations of the state, are liable (especially) for acts of negligence from which quasi-municipal corporations may be immune because of their particular relationship with the state. This is a vital distinction. Although this distinction does not give those affiliated with quasi-municipal corporations liberty and license to neglect their duty as members of such corporations, it is important to be aware of the "legal setting" in which one functions. An example of the importance of this distinction is presented in the following excerpts from a decision handed down by Judge Haymond of the Supreme Court of Appeals of West Virginia. (Read the following two paragraphs of background material prior to reading the excerpts. Then, as you read the excerpts from Judge Haymond's decision, determine why the distinction was an important one and on what basis the distinction was made.)

A West Virginia statute provided that every municipal corporation publish annually a sworn statement regarding its financial condition. Failure to do so would result in a misdemeanor charge against every member of the city council and, upon conviction, would result in a fine being levied against each and every member convicted. Thus, the statute — because of the misdemeanor charges involved — is a penal statute. One of the powers conferred on the board of park commissioners was the power to levy taxes on all taxable property within its jurisdiction. Thus, the involvement of the state tax commissioner.

The circuit court ruled in favor of the board, and the state appealed the decision to the Supreme Court of Appeals. Judge Haymond agreed with the circuit court ruling because the board of park commissioners was not a true municipal corporation.

STATE ex rel. KOONTZ, State Tax Com'r, v. BOARD OF PARK COM'RS OF CITY OF HUNTINGTON et al.

Supreme Court of Appeals of West Virginia.
April 27, 1948.
47 S.E.2d 689

There have been many attempts to define a municipal corporation. In *Brown* v. *Gates,* 15 W.Va. 131, this Court quoting with approval Dillon on Municipal Corporations, says: "Municipal corporations are bodies politic and corporate . . . established by law to share in the civil government of the country, but chiefly to regulate and administer the local or internal affairs of the city, town or district which is incorporated," and that "A municipal corporation is also defined to be 'an investing the people of a place with the local government thereof' ." In *State ex rel. Thompson* v. *McAllister,* 38 W.Va. 485, 18 S.E. 770, 774, 24 L.R.A. 343, it is said that a municipal corporation is "the legislative grant of local self-government to the inhabitants within a certain designated territory, which is known as the 'city,' 'town,' or 'village,' and corporate powers granted are exercised by its inhabitants in its corporate name." In Dillon on Municipal Corporations, 5th Ed., Vol. 1, Section 31, the learned author says: "A *municipal corporation* in its strict and proper sense, is the body politic and corporate constituted by the incorporation of the inhabitants of a city or town for the purposes of local government thereof." One recognized legal treatise uses this language with respect to municipal corporations: "In this country, municipal corporations are usually classified as cities, towns and villages. . . . While the term 'municipal corporation' is sometimes used, in its broader meaning, to include such public bodies as the state and each of the governmental subdivisions of the state — such as counties, parishes, townships, hundreds, etc. — it ordinarily applies only to cities, villages and towns which are organized as full-fledged public corporations." 37 Am.Jur., Municipal Corporations, Sections 5 and 6. Another standard work says: "A municipal corporation is commonly called a 'municipality'. . . . The term includes cities of all classes as well as towns. . . . Public corporations are classified as municipal, quasi-municipal, and public-quasi-corporations. Public corporations include not only municipal corporations, but all other incorporated agencies of government of whatever size and form or degree. . . . While all municipal corporations are public corporations, all public corporations are not municipal corporations." 43 C.J., pages 66, 67, 72, and 73. Cities, towns and villages are regarded as true municipalities whereas counties, highway districts, and school districts are considered not to be municipalities but "quasi-municipalities," or public corporations. *[Citations omitted.]* Though such public bodies as school districts, boards of education, boards of water commissioners and boards of park commissioners have been recognized by some courts as municipal corporations *[Citations omitted]*, they are not municipal corporations or municipalities in the proper sense. . . .

. . . Careful consideration of the provisions of the act of 1925 leads to the conclusion that the board of park commissioners of the City of Huntington, which by its terms is designated as a body corporate, is not a municipal corporation or a municipality in the proper or ordinary sense of any of those terms. It differs in many ways from the incorporated city, town or village which constitutes the true municipal corporation under the laws of this State. A number of these differences could be pointed out but mention of only two of them will suffice. Under the general statute, both before and since the adoption of the Home Rule Amendment to the Constitution, Article VI, Section 39(a), and under the legislation, as amended, which makes that amendment effective, the creation of a municipal corporation depends upon the voluntary acts and the approval of the voters of the municipality, whereas the existence of the board of park commissioners of the City of Huntington resulted from the independent action of the Legislature without regard to any official expression of the inhabitants of the district concerning its establishment. A municipal corporation, created under the statutes of this State, is vested with the power and the authority to exercise and perform the functions of local self-government generally, whereas the board of park commissioners of the City of Huntington was created not for the purpose of general government but for a limited and special purpose. It is a different kind of corporation from a city, a town or a village and its specific powers are limited to the operation, the maintenance, the management and the control of a system of parks. The only ordinances it is authorized to enact are those which are proper to regulate and police the park system, and its power to levy taxes is limited to the purposes of the system of parks for which the act makes provision. The power of local government, which the statute creating the board of park commissioners does not confer upon it, has been said to be the distinctive purpose and the distinguishing feature of a true municipal corporation. 37 Am.Jur., Municipal Corporations, Section 6.

CLASSIFICATIONS OF MUNICIPAL AND QUASI-MUNICIPAL CORPORATIONS

The following have generally been considered to be the basic classifications of municipal and quasi-municipal corporations, although variations occur in many states:

Municipal Corporations	*Quasi-municipal Corporations*
Cities	Counties
Villages	Townships
Boroughs	Towns
Towns	Special districts

Judge Dillon, in his treatise on municipal corporations, explains why these distinctions were made:[4]

> It has been customary in the past to classify public corporations functioning at the local level as either municipal corporations or public quasi-corporations. Incorporated cities, towns, villages, and boroughs constituted the first class; everything else fell within the second. At the root of this distinction were four factors: (1) the notion that counties and other "quasi-corporations" were almost exclusively state agencies administering matters of state concern as distinguished from the large sphere of local business with which municipalities dealt; (2) the assumption that the agencies to be labeled "quasi-corporations" were less complete forms of corporate organisms; (3) the notion that municipalities, unlike a county, for example, possessed, at least in a limited sense, a private as well as public character; and (4) the theory that while municipalities were established only at the request or with the consent of the inhabitants, counties and other quasi-corporations were created by the state without the consent of the inhabitants.

Cities: Municipalities, whether they are called cities, boroughs, villages, or towns, are the most important general-purpose units of local government. Municipalities exist in all fifty states. Although the state of Illinois has over 1,200 municipalities, most states contain far fewer than this number. Municipalities are not as inclusive in territory as counties; they do, however, exercise more powers and provide a greater variety of services than those of county governments.

The actual chartering of cities began in the early 1650s in Boston with the peti-

---

[4]*Ibid.*, §§35-39.

tion of the citizenry to become a corporation, and in 1653 the first municipal charter in America was granted to New Amsterdam. In his book titled *Urban Politics,* Murray S. Stedman, Jr. explains the evolution of the need for municipal corporations:[5]

> Since every city is in a county, the question may arise as to why the county — as a subdivision of a state — cannot itself exercise the functions normally associated with a city. Why, in short, is it necessary to have municipal corporations? Or are they, in fact, superfluous? Why can't county governments do what city governments do?
>
> The answers to these questions are deeply rooted in American history. To begin with, it must be noted that the county was originally conceived as a unit of rural government. As such, its powers were those which befitted the conditions of the time and of the area. This is perfectly understandable when one recalls that only about three percent of the American people lived in cities at the time the Constitution was written in 1787. Residents of urban areas were obviously the exception rather than the rule, in 1787.
>
> But persons who did live in urban areas had special needs and requirements which rural counties were not equipped to handle. For example, in the colonial era cities required, as they do now, special arrangements for public safety, economic regulation, and certain minimal public services. Since such services could be provided for neither by the existing county nor through an unincorporated (*i.e.,* legally powerless) settled area — it was necessary to seek a special legal status for sizeable settlements. The device which was used was the municipal charter, given by the colonial legislature, which created a municipal corporation and allocated the necessary governmental authority to meet the felt needs of the area. There was nothing at all new about this practice, for English cities had been created for hundreds of years — at first through royal charters, and later through acts of Parliament.
>
> In short, special conditions required a special governmental form — the municipal corporation — on both sides of the Atlantic. In the United States, the function of incorporation is no longer carried out by a special act of a state legislature. The common practice is to provide by statute that urban areas which meet certain specifications regarding population will automatically be incorporated.

Villages: A village is a small municipality that has been incorporated and whose government parallels the governments of large cities. Villages can be found in every part of the United States. Although they are usually defined by size of population, the criterion for determining what constitutes a village varies among the states. For example, in Ohio municipalities that have populations below 5,000 are called villages, while in Minnesota municipalities that have populations that

---

[5]Murray S. Stedman, Jr., *Urban Politics,* Winthrop Publishers, Inc., Cambridge, MA, 1975, pp. 3-4.

range from 100 to 10,000 may be called villages. Although the governmental organization in villages is a fairly simple one, it differs from governmental organization in cities in degree, not in function, as explained by George S. Blair in his book titled *American Local Government*:[6]

> It is difficult to evaluate village government in general terms since the villages range in population from only a few persons to 2,500. They can be a part of the satellite system of a larger municipality, or be a small independent community in the center of a rural area. Similarly, the number and level of services rendered by village governments range from only limited police protection to a number of efficiently operated urban-type services. Generally, however, the services provided are few and inadequate when measured against any standards for determining satisfactory levels of service. Since the governmental structure is simple in organization and part-time in nature, it is difficult to praise or condemn its operation. The needs of citizens in villages are not great in terms of services, and village government seems reasonably responsive to the demands ordinarily made on it.

Boroughs: Boroughs, although sometimes used synonymously with the term "towns," are considered as distinct legal entities. However, much of what will be stated with regard to incorporated towns or townships also may apply to boroughs. Like villages, boroughs are small communities that have outgrown their rural status to the extent that specific public services are required and "personal" local government is necessary to meet the requirements of the population. Incorporation allows the powers necessary for boroughs to meet this need.

Towns: When a town is created by the state as a result of the voluntary assent of its inhabitants, it is, of course, a municipal corporation. A town is defined as "any large collection of houses and buildings, public and private." This definition highlights the "size" aspect that distinguishes an incorporated town. As in the case of villages and boroughs, towns are incorporated to meet public service needs in communities that have outgrown their rural status. Towns other than those in the New England states are usually incorporated.

Counties: Counties, as creations of the state on an "involuntary" basis, function as quasi-municipal corporations. Counties, as units of local government, include more territory than any other unit of local government and exist in every state but Alaska, Connecticut, and Rhode Island. In Louisiana, counties are called parishes. Because of their quasi-municipal status, counties usually do not exercise the powers, provide the variety of services, or have the legal liability of true municipal corporations.

The county as a unit of American local government was first established in Virginia in 1634. The principal role of the county involves judicial administration in both civil and criminal matters. The court approves town by-laws, directs tax

---

[6]George S. Blair, *American Local Government,* Harper & Row, Publishers, New York, 1964, pp. 235-236.

apportionments among the towns in a county, registers land titles, records deeds, and erects prisons.

The county is the most inclusive unit of local government. As previously stated, with the exception of Alaska, Connecticut, and Rhode Island, there are over 3,000 counties in the United States.

Bernard F. Hillenbrand, Executive Director of the National Association of Counties, describes the importance and impact of counties on local government considerations, including the assumption in urban areas of county responsibility for fire protection:[7]

> Many people have mistaken ideas about county government. They have a mental picture of fat politicians sitting around a pot-bellied stove, spraying tobacco juice into a copper spittoon, and plotting how to grease the political machine. Actually, have you been in a courthouse lately? One is more likely to find that it is a modern, air-conditioned building with automatic data-processing machines in the basement; an ultra efficient, two-way sheriff's radio on the roof; and everything in between just as modern — symbols of 20th century progress.

> County government will be the dominant unit of local government in the United States in the next decade. . . . Counties are growing like adolescents. A portion of the increase in the importance of county government is reflected in the expansion of traditional county government services, due both to the population increase and the traditional American demand for improvement and expansion of existing services. These demands have brought spectacular county improvements in election administration (automatic ballot-counting); penal administration (honor farms); administration of justice (streamlined court procedure and use of special service personnel, psychiatrists, etc.); roads and highways (use of modern earth-moving equipment); record keeping (up-to-the-minute machines and techniques); education (student aptitude testing, special counseling, etc.); health and welfare (out-patient clinics for the mentally ill, and spotless hospitals).

> The really tremendous growth of county government, however, has come in urban areas where the existing units of government have demonstrated that they are not capable of solving area-wide problems. Here one finds counties assuming responsibility for police and fire protection, planning and zoning, water supply, sewage disposal, civil defense, industrial development, air pollution control, airports, traffic control, parks and recreation, urban renewal and development, and finance administration.

> As a solution to local problems, county governments offer many innate advantages that theoreticians sometimes overlook. In the first place, counties have a long and honorable history of service, dating from the earliest times in America and before that in Great Britain. . . .

> Second, counties provide the territorial limits for the organization of many

[7]Bernard F. Hillenbrand, "County Government Is Reborn," *Public Administration Survey,* May 1960, pp. 1-8 (from Robert L. Morlan, *Capitol, Courthouse, and City Hall,* 5th Ed., Houghton Mifflin Company, Boston, 1977, pp. 259, 261).

nongovernmental as well as governmental activities. Medical societies are nearly always countywide, as are bar associations. Nearly all of the nation's agricultural and rural service programs are based upon the county as the primary unit. A large part of our educational systems is county-oriented. The national census uses the county as the basic accounting unit. Virtually all of the country's systems of courts and administration of justice are county-oriented. Conservation and soil conservation districts are usually coterminous with a single county.

A large number of city-dwellers fail to see how county governments serve them, and think that such governments only benefit small town residents and people who live in rural areas. Hillenbrand goes on to address this issue:

> Perhaps the greatest advantage of a county is that everyone in the state is served by a county government. Whether a voter lives in a city or in the rural portion of the county, he is represented on the county governing body. The notion that county functions are beyond the control of the city resident is, of course, false since the city resident is required to contribute to the financial support of the county [and] since he participates in the election of representatives to the county governing body just as the rural resident does. It is true, however, that very often a rural resident has a stronger voice in county affairs because he has only one unit of government to keep an eye upon and therefore is more vigilant in county affairs. The city person has both his city and county governments to watch; and because his attention is divided, he may be less knowledgeable about his county government. This problem, however, can be remedied. In a democracy every citizen has the positive obligation to participate fully and intelligently in the affairs of his governments no matter how many there are. We certainly concede that actual participation in local affairs is increasingly difficult — particularly for the poor citizen who is served by a city, school district, multiple-service-district or authority, and by his state and Federal governments as well.
>
> Finally, the county serves as the political base upon which our two-party system is built. The county is the fundamental organizational unit of both major parties and is their basic strength both state-wide and nationally. Because the parties are based on the county, they are controllable by the electorate. This political arena is the one place where all of the interests of the community are represented. Many decisions about local affairs are and should be made at this level because all interests are represented. The decision as to whether limited community funds are to be used to build a school or a bridge is, in this sense, political; and typically it is debated (or mutually endorsed) by the two parties in two-party areas and by opposing factions of the same party in one-party areas. . . .

In his survey, Hillenbrand further describes the increasing importance of the functions that counties are called upon to assume:

> . . . To say that counties have a bright present and an even brighter future is not to say that they do not have problems. Chief among these is the absence of home rule (the right of local people to decide local affairs for themselves). Originally (and presently, for that matter) counties were established as local administrative districts of the state. Their responsibilities were quite simple in the beginning,

enabling the state to establish a uniform system of county organization and to spell out in precise detail, in statute or constitution, exactly how the counties were to discharge these responsibilities. Most counties, however, are still forced to operate under these same rules in spite of changed circumstances which have brought on new responsibilities. As a result, counties now find themselves in a veritable straitjacket of state control.

The problems created by this rigid control are numerous. Most county officials are severely restricted in establishing local salary scales for county employees. In Massachusetts, for example, the state legislature has complete control over local county budgets, personnel and all. In order for a county official to purchase a typewriter, the item and the specific cost must be included in the county's budget and approved by the state legislature.

Increasingly the functions that counties are called upon to assume require endless special state statutes, and yet all but a handful of state legislatures meet only once every two years to consider substantive legislation. Many of the restrictions that are most disruptive of orderly, sensible local determination, moreover, are spelled out in the state constitution — an extremely difficult document to amend.

Most areas in the United States, whether municipal or rural, also come under the jurisdiction of a county government. Although a large percentage of counties are classified as rural, the number of urban counties is increasing. Both urban and rural counties have the same general governmental structure. Virtually all counties have a general governing board, county officers, and boards and commissions assigned to various tasks.

The powers of a county are limited in comparison to the powers conferred by the state on municipalities. However, the inclusive nature of counties in the United States, coupled with the rapid expansion and mobility of American life, is resulting in trends towards an increasing revitalization in county government. Some of these trends are described as follows by George S. Blair:[8]

1. The county continues to increase in importance. One reason for this is the use of the county as a base for a number of federal grant-in-aid programs in such areas as agriculture, health, and welfare. A second reason is the expansion of population in suburban counties surrounding our major cities. As suburban communities increase in both size and number, the role of the county in coordinating some services and in providing others is enhanced.

2. The services rendered by counties are increasing in number and importance. The traditional functions are improved through programs of federal and state grants, and new functions are encouraged by such programs. Citizen demands for new and improved services are also falling on more receptive ears at the county level.

3. County reorganization, while advancing slowly to be sure, continues to move ahead. County executives continue to grow in number, and county boards

[8]Blair, *American Local Government*, pp. 194-95.

seem increasingly aware of their opportunities and responsibilities.

4. The problem of state-county relations seems to be emerging as a cooperative relationship rather than one primarily of state control and supervision. State legislatures appear more willing to permit flexibility among counties and to enact permissive rather than mandatory legislation in many fields.

5. The role of the county governing body is becoming increasingly important. As the county becomes a more important unit, the new strength is reflected in the powers and functions of the county board. With this expanded role of the board, it will become more obvious that many of the presently elective positions should be appointive, and this power will be gradually transferred to the county board.

6. The cost of county government is increasing to meet the costs of the expanded and added services rendered. As the property taxes increase and additional taxes are levied, citizens become more aware of their county government and may keep a closer watch on its operations.

7. The county is less a stereotype than it was formerly as it becomes increasingly recognized that there are several types of counties. The rural county (which once was the form of all counties) has been joined by its more urban counterparts serving densely populated counties. Cities were never expected to fit into a common mold, but counties for much too long were so treated by state control and supervision. The greater flexibility among counties will enable some to become outstanding units of local government and will erase the popular stereotype of the county as an inefficient unit of government.

Townships: Townships and towns that are created by the state for convenience in state administration are also quasi-municipal corporations. These units were created primarily to serve rural areas and are found primarily in the northeastern, central, and north central parts of the United States.

The governmental structure of unincorporated towns and townships is generally very simple, primarily because as units, unincorporated towns and townships perform relatively few governmental functions. Towns in the New England states are quasi-municipal corporations.

Importance of Fire Protection to Special Districts: Special district governments exist in all 50 states and are the most numerous of all types of local government. In 1962, the Bureau of the Census reported a total of 53,001 special and school districts. Of these, 34,678 were independent school districts, while the other 18,323 were special districts that served a variety of single or multiple purposes ranging from air pollution control to water supply, storage, conservation, or maintenance. (See Table 2.1.) The three major types of special districts are those for fire protection, soil conservation, and drainage. By far, special districts dealing with fire protection outnumber any other type of special district. (See Fig. 2.1.) This is undoubtedly due to the importance of fire protection to communities in safeguarding lives and property from the threat of fire.

Table 2.1 Units of Government, 1952-1962*

Governments	1952	1962	Percent change
United States	1	1	
States	48	50	4.2
Counties	3,052	3,043	— 0.3
Municipalities	16,807	17,997	7.1
Townships and towns	17,202	17,144	— 0.3
School districts	67,355	34,678	—48.5
Special districts	12,340	18,323	48.5

*From U.S. Bureau of the Census.

Figure 2.1 also shows the variety of functions of special districts, such as sewerage, flood control, housing, and parks and recreation. However, these special districts are unevenly distributed among the states. One of the major reasons for their unequal distribution is that a representative number of special districts are multifunctional districts. For example, because fire protection is in a great part dependent upon water supply, many urban water supply districts combine their functions with those of fire protection.

Special districts are formed to serve four basic areas, as described by George S. Blair in his book titled *American Local Government:*[9]

> First, metropolitan or regional districts are created to solve or ameliorate areawide problems encompassing a number of separate governmental units. An example is the Metropolitan Water District of Southern California which supplies water to municipalities in six counties, or the Metropolitan Sanitary District of Greater Chicago. Second, there are coterminous special districts which have boundaries identical to those of an existing general purpose local government. Such districts are often created for such functions as housing or parks. Third, special districts are often established in urban fringe areas bordering on cities to provide municipal-type services for the residents of these unincorporated areas. Such districts are created for such varied purposes as water, sewerage, sanitation, street lighting, and fire protection. Fourth, in rural areas special districts are established to meet agricultural needs and provide such functions as soil conservation, drainage, and irrigation.

According to Victor Jones in his book titled *Metropolitan Government,* the growth of special districts reflects the immediacy of their advantages.[10] Some of the advantages of special districts over other forms of metropolitan organization are: (1) their creation does not do away with the corporate identity of any of the existing units of local government; (2) they can cross county and state boundary lines; (3) they may circumvent constitutional and statutory limitations on debt and tax capacities of local units, and enable the financing of needed capital im-

---

[9]*Ibid.,* pp. 252-54.

[10]Victor Jones, *Metropolitan Government,* University of Chicago Press, Chicago, 1942, p. 129.

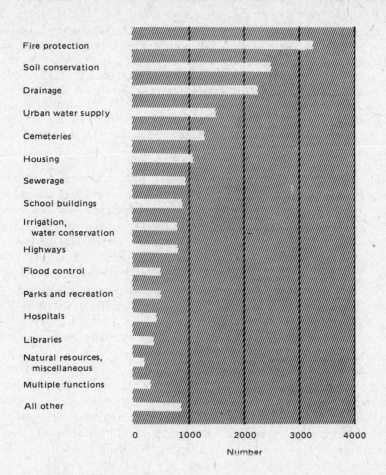

Fire protection
Soil conservation
Drainage
Urban water supply
Cemeteries
Housing
Sewerage
School buildings
Irrigation,
    water conservation
Highways
Flood control
Parks and recreation
Hospitals
Libraries
Natural resources,
    miscellaneous
Multiple functions
All other

0   1000   2000   3000   4000

Number

Fig. 2.1. Number of special districts, by function: 1962. (Adapted from the U.S. Bureau of the Census, *The Census of Governments, 1962 — Governmental Organization,* Washington, DC)

provements; and (4) they are easily established in most states. The growing acceptance of special districts is explained by Charles R. Adrian and Charles Press in *Governing Urban America:*[11]

There are several reasons for the trend toward special districts. Citizens learn that their state or other states authorize districts to perform certain services. The approach usually has strong appeal. Citizens do not resist it as they do attempts at governmental consolidation, since they expect that it will be less expensive and that it will preserve the independence of their local government. Interest groups that may want services performed by government (amateur pilots wanting an air-

[11]Charles R. Adrian and Charles Press, *Governing Urban America*, 3rd Ed., McGraw-Hill Book Company, New York, 1968, pp. 268-269.

port, physicians wanting a hospital), together with the professional administrators of particular functions, characteristically want their special problems handled in a special way by a special organization. School administrators have long ago convinced the American public that they should be independent of the rest of local government, and the school district is both the most common and the best-known special district. Others have followed the lead of the educators and the interest groups supporting them. The special district has also had great appeal to the wistful who would take their pet governmental function "out of politics."

Role of Fire Protection in Metropolitan Areas: Although standard metropolitan areas are neither governmental units nor legal entities, these large urban areas are described here because of their special importance to members of the fire service.

The Bureau of the Census defines a standard metropolitan area as an area that contains a county or a group of contiguous counties and at least one city with a population of 50,000 or more. Adjacent counties are included in the metropolitan area if they are densely populated, contain a large number of nonagricultural workers, and are socially and economically integrated with the center city. The Bureau of the Census identified 212 standard metropolitan areas in the United States in the year 1960.

An old and congested city is usually at the core of a metropolitan area. Surrounding this core city are a number of politically independent suburban governments. Social and economic ties bind all of these units into one metropolitan area. Because metropolitan areas are neither governmental units nor legal entities, a resultant lack of overall government can cause confusion. This lack of overall formal government is usually compensated for through informal or contractual agreements between local governmental units in a metropolitan area.

It is becoming increasingly common for such intergovernmental arrangements to be made for the disposal of sewage, garbage, and rubbish; to share police radio networks; or to have the core-city police radio supply the suburban departments. Unusual police and fire problems often necessitate formal agreements or informal understandings concerning emergency standby assistance. It is quite common for fire departments of one community to attend large fires and conflagrations in other communities.

A large number of core cities sell water to the suburbs on a contractual basis, either directly to the suburbanite, or to the suburb itself at a master meter. Many cities have understandings of an informal type concerning traffic-flow patterns — always of importance to fire departments — including such necessities as the establishment of one-way streets through two or more communities. There are thousands of other cooperative arrangements that have been worked out in various parts of the nation.

Mutual Aid Agreements for Fire Protection in Metropolitan Areas: Mutual aid agreements with regard to fire protection are increasingly common occurrences between local governments. Figure 2.2 shows a mutual aid plan between local governmental units in the Dayton, Ohio area.

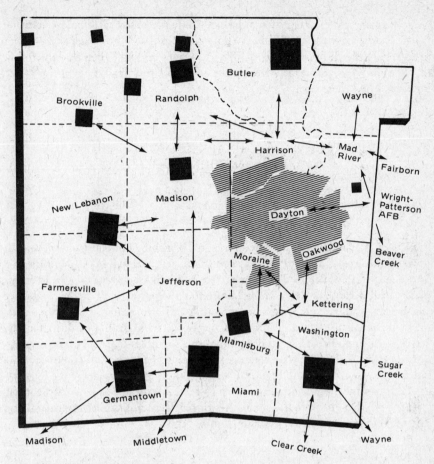

Fig. 2.2. Mutual-aid contracts for fire protection, Montgomery County local governments, 1958. (Adapted from John C. Bollens and others, *Metropolitan Challenge,* Metropolitan Community Studies, Inc., Dayton, OH)

The following are recommendations from the National Fire Protection Association with regard to mutual aid agreements:[12]

True "mutual aid" is a relationship in which each member is prepared to render assistance to the parties of the agreement. In many places, there are programs of "outside aid" whereby communities or individual properties known to be deficient in fire fighting resources contract in advance for certain fire fighting assistance in case of need. In some instances, the contract covers basic first alarm response, and in other cases additional assistance for fighting major fires. Both in the case of "mutual aid" and "outside aid," definite agreements should be made in advance in line with the legal requirements governing fire department operations outside its normal jurisdiction.

---

[12]*Fire Protection Handbook,* 14th Ed., NFPA, Boston, 1976, p. 9-29.

Refer again to the excerpts from the case study presented earlier in this chapter, titled *State* v. *Board of Park Commissioners*. Consider: If a fire department were in a special district, how would this affect its legal standing? Then read the following case study: as you read, compare the results of your reasoning with the findings contained therein.

CULLOR et ux. v. JACKSON TOWNSHIP, PUTNAM COUNTY et al.

Supreme Court of Missouri, Division No. 2.
June 9, 1952.
249 S.W.2d 393

Richard A. Cullor and wife sued Jackson Township, Putnam County, and the township trustee and board members for damages by flooding of plaintiffs' land as the result of an increased flow of surface water diverted onto such land from a public road because of defendants' failure to provide sufficient ditches and levees and raising of the height of road in repairing and reconstructing it. From a judgment of the Circuit Court of Putnam County, V. C. Rose, Jr., J., dismissing the action with prejudice on defendants' motion, plaintiffs appealed. The Supreme Court, Leedy, P. J., held that defendant township, being a quasi-corporation and political subdivision of the State, exercising purely governmental functions, was clothed with the State's immunity from tort liability, in the absence of a contrary statute and that individual defendant's duties were not purely ministerial, but discretionary, so as to absolve them from liability.

Judgment affirmed.

LEEDY, Presiding Judge.

Plaintiffs appeal from a judgment of dismissal with prejudice rendered upon the sustention of defendants' motion (raising the objection of failure of the petition to state a claim upon which relief can be granted) and the declination of plaintiffs to amend or further plead. Of necessity, then, the facts about to be stated, and by which the sufficiency of the petition is to be tested are those gleaned from the mooted pleading itself. The action was brought in two counts (one for damages, and the other praying a mandatory injunction), both being bottomed on the same facts; but the questions briefed are limited to those arising under the first (or damage) count, so the injunction feature passes out of the case, and will not be further noticed.

Plaintiffs, husband and wife, are owners of a tract of land in Jackson Township, Putnam County. Defendants are the township (which is alleged to be "a body corporate and a general road district") and Vernon Parrish, John McVey and Orville

Dickerson, who are alleged to be, respectively, the trustee and board members of the township, as well as road overseers therein. Plaintiffs' land is on the west side of a public road running north and south, which road, it is alleged, "is under the jurisdiction, care and maintenance of the said defendant township, acting through and by its duly elected and qualified officials, to-wit: Vernon Parrish, Trustee, John McVey and Orville Dickerson, Board Members." Plaintiffs deeded to the county a right of way for road purposes, and thereafter, in October, 1949, defendant township started reconstruction of the road above mentioned under the so-called King Road Law, sections 231.440-231.500 RSMo 1949, V.A.M.S., and "under special plans, specifications and minimum requirements set up and devised under the 'King Road Law' statute;" which "special plans, specifications and minimum requirements set out in detail the type of ditches and drainage systems that must be constructed for roads built under the provisions of this specific road law." In this connection it was alleged that "the above-named individual defendants in their official capacity had the duty to oversee the construction and maintenance of this said public road in accord with the special plans, specifications and minimum requirements set forth by law, and further that the said defendants assumed and undertook to carry out this duty."

The petition next alleged the adequacy of the ditches and levees and absence of surface water flowing over or accumulating upon plaintiffs' land prior to the reconstruction work. Then, after alleging certain conclusions as to the duty cast by law upon the township and its officers and agents in the premises with respect to surface water, and the flow and accumulation thereof, the petition proceeds thus: "but that defendant Jackson Township and the above defendant officials and agents . . . through their negligence and carelessness while repairing and reconstructing the above mentioned public road adjacent to property owned by these plaintiffs and through the negligent and careless disregard for the special plans, specifications and minimum requirements covering such construction, caused to be accumulated and gather alongside said road a greatly increased quantity and volume and flow of water and negligently and carelessly failed and refused to provide sufficient ditches and levees to carry off the increased flow and volume of surface water caused by said repairs and reconstruction and carelessly built up the said public road to such height that the flow of surface water and the volume thereof was increased thereby and diverted into and onto and caused to flood plaintiffs' above described real estate and to accumulate and stand thereon" to the damage of plaintiffs in the sum of $1,500. The prayer was that plaintiffs "recover from the defendant Jackson Township and/or the defendants Vernon Parrish, John McVey and Orville Dickerson, the sum of $1,500.00 as damages and the costs of this action."

Jurisdiction of the appeal is in this court because Jackson Township, as an organized township in a county under township organization, is a political subdivision of the State, under Art. V, § 3, Const. of Mo. 1945, V.A.M.S. *[Citations omitted.]*

Plaintiffs' contention as to defendant township is that it is liable in tort under the facts of this case on the same principle that such liability is imposed on municipal corporations for damages to property resulting from water being caused to flow over and accumulate thereon due to the negligence of the municipality in the construction or maintenance of a street. In the consideration of the question thus posed it is important to bear in mind the distinction between municipal corporations (in the strict and proper sense), such as cities, towns and villages, and quasi-corporations, such as counties and townships. Municipal corporations exercise both governmental and proprietary (sometimes called corporate) functions. Their liability or nonliability in tort depends on the character of the function involved as being governmental on the one hand, or proprietary on the other.

Plaintiffs invoke this court's language concerning the exercise *by a municipal corporation* of its discretionary powers of a public or legislative character: "When the exercise of these powers ceases to be discretionary, when it no longer depends upon the will of the municipal legislature, but upon the paramount will of the charter-making body, then only does the neglect of the municipality subject it to an action for damages." *Cassidy* v. *City of St. Joseph,* 247 Mo. 197, 206, 152 S.W. 306, 309. It is argued that under this pronouncement the township is liable because it had no discretion in the building of ditches on a road project under the King Road Law, but that an absolute duty rested upon it to improve the road under the minimum standards and specifications directed to be set up by the act just mentioned, and hence "in accordance with the mandate of the paramount will of the charter-making body." But the quoted language was not used in reference to quasi-corporations whose nonliability the court had previously recognized and made clear in the very paragraph from which the above excerpts were taken, viz.: "Neither the State nor those quasi-corporations consisting of political subdivisions which, like counties and townships, are formed for the sole purpose of exercising purely governmental powers, are, in the absence of some express statute to that effect, liable in an action for damages either for the nonexercise of such powers, or for their improper exercise, by those charged with their execution. This applies alike to the acts of all persons exercising these governmental functions, whether they be public officers whose duties are directly imposed by statute, or employees whose duties are imposed by officers and agents having general authority to do so." In that case the court reaffirmed the doctrine of nonliability of a city for negligence in the exercise of its discretionary powers. Plaintiffs' other cases likewise deal with municipal, and not quasi, corporations.

In the early case of *State ex rel. Jordan* v. *Haynes,* 72 Mo. 377, 379, this analogy between counties and organized townships was drawn: "The directors are the general officers of the township, occupying toward it similar relations to those borne by the justices of the county court toward the county, and possessing also similar powers." The present statute authorizing township organization denominates it as "an alternative form of county government." Section 65.010 RSMo 1949 V.A.M.S.

"According to the prevailing rule, *counties are under no liability in respect of torts,* except as imposed (expressly or by necessary implication) by statute. They are political subdivisions of the State created for convenience, and are usually regarded not to be impliedly liable for damages suffered in consequence of neglect to repair a county road or bridge; such a liability, unless declared by statute, is generally, but not quite universally, denied to exist. On the same grounds, such organizations as townships, school-districts, road-districts, and the like, though possessing corporate capacity and power to levy taxes and raise money, for their respective public purposes, have been very generally considered *not to be liable in case, or other form of civil action, for neglect of public duty, unless such liability be created by statute." [Citations omitted.]*

No difference in principle is apprehended between the case at bar and *Hausgen* v. *Elsberry Drainage Dist.,* . . . That was a tort action against a quasi-corporation and the individuals composing its Board of Supervisors for negligence in constructing a levee and otherwise negligently carrying out the district's plan of drainage as adopted pursuant to the requirements of the statute, and approved by the circuit court. Demurrer to the petition was sustained, judgment entered accordingly, which, on appeal to the St. Louis Court of Appeals, was affirmed; held, no liability, in the absence of a statute, a drainage district being a governmental agency. This court granted *certiorari,* in quashing which, on final hearing, the court *en banc* said, 298 Mo. loc.cit. 458 and 250 S.W. loc. cit. 907: "The functions exercised by drainage districts being purely governmental in character, the question is whether or not they are liable for negligence of their agents in the prosecution of the reclamation plan, where damages are sought by one whose lands are within the district. These districts have no private or proprietary functions to perform, and in this their powers are not as broad as cities, towns, and villages. Their functions are governmental and public. The very foundation stone of their structure is public necessity, or public convenience, or public welfare."

In *State ex rel. McWilliams* v. *Little River Drainage Dist.,* 269 Mo. 444, 457, 190 S.W. 897, 900, it was held that in improving, repairing and dealing with the public highways, the several counties and the county courts thereof are but agencies, or agents of the state, acting by delegated authority as legal subdivisions of the State. Under the authorities cited, Jackson Township is not a municipal corporation in the sense that that term is used herein, but, on the contrary, is a quasi-corporation and political subdivision of the state exercising purely governmental functions, and, as such, is clothed with the state's immunity from tort liability, in the absence of a statute to the contrary.

The other and remaining question is as to the personal liability of the individual defendants. Plaintiffs' brief says that "under ordinary circumstances the duties of the individual defendants would be considered discretionary, but in this particular case the duties were ministerial. Under the law they had to obey instructions and there was no room for any special discretion, judgment or skill." In other words, does the circumstance that those "special plans, specifications and minimum requirements" contemplated by the 'King Road Law' as applicable to an incident of the work in question, have the effect of converting into mere ministerial duties those functions of the township officers which would otherwise be regarded as governmental in nature? The township was empowered under section 231.480 RSMo 1949, V.A.M.S., to "perform the work provided for in the specifications" only if no bids were received, or, if received, they exceeded the estimated costs, so that there was undoubtedly some measure of discretion devolving upon its officers, to be exercised by them as necessities occasioned by the doing of the work arose; and to say to the contrary — that such duties became purely ministerial under the circumstances — is not tenable.

There has been a gradual tendency toward weakening the doctrine of sovereign immunity which has produced some interesting, and (in this writer's personal view) highly desirable changes which probably presage even more sweeping reforms. But the changes thus far wrought have been not so much through judicial decision as by legislative act. Notable in this field are the Federal Tort Claims Act, 28 U.S.C.A. §§ 1346, 2671 et seq., and the New York Court of Claims Act. The former has particular significance because none of the federal government's functions are regarded as being proprietary in nature. By judicial decision the state's waiver of immunity under the New York act was held to put an end to the "legal irresponsibility" (as an extension of the exemption possessed by the state) theretofore enjoyed by such governmental units as counties, cities, towns and villages. Bernardine v. City of New York, 294 N.Y. 361, 62 N.E.2d 604, 161 A.L.R. 364. But there has been no such change in the public policy of this state in that regard, either through statutory enactment or constitutional revision. The principle has become so deeply rooted in the jurisprudence of this state, as the cases herein referred to readily disclose, that any partial abandonment of it as to quasi-corporations, as here sought, should come through the legislative process rather than by judicial decision.

The judgment is affirmed.

All concur.

MUNICIPAL LIABILITY

The area of municipal liability has always been difficult to define. In addition to the fact that various and varying kinds of municipal liability are described and defined by legislative statute, the historical position held by municipalities as units of local government has affected judicial decisions with regard to municipal liability. The doctrine of sovereign immunity, the dual nature and functions of municipal corporations, and the identity of municipal corporations as true municipal corporations or quasi-municipal corporations all affect municipal liability.

A brief mention of the types of municipal liability is in order at this point. Questions of municipal liability usually occur in the areas of contract and tort. Municipalities can be held responsible for violations in these two areas.

Municipalities enter into countless contracts with individuals and companies and are responsible for true violations of a contract. Major contract problems generally occur in three areas: problems arising from interpretation of the contract; problems arising from determining when the contract has been fulfilled; and problems arising from compensation when conditions make the contract impossible to fulfill. As Adrian and Press point out:[13]

> The principal problem that faces the courts in the interpretation of contracts to which the city is a party is that of determining the conditions under which the contract is valid. In general, a valid contract is one that the city is authorized to make, that is made by the proper officer according to law, and that has been adopted according to proper procedures. (For example, the law may require that certain contracts be let only after advertising for bids, and then only to the lowest responsible bidder.)

A tort is a violation of a personal right that is protected by law. The law of torts and its relevance to fire protection will be presented in detail in the next chapter. Generally, problems in tort arise when injury to persons or property is involved. It is often in the area of torts that municipal liability becomes involved and enmeshed in those factors (sovereign immunity doctrine, dual nature and function of municipal corporations, municipal and quasi-municipal corporations) that affect the entire area of municipal liability.

Although the fundamental rules regarding municipal liability are found in the common law, these rules are subject to statutory modification and/or change by each state. Because of this, the law regarding municipal liability is not the same in any of the states. Thus, all of the following factors regarding municipal liability must be considered in terms of their pertinence to the various state laws and of their effect on municipal liability in a given area.

THE DOCTRINE OF SOVEREIGN IMMUNITY

The doctrine of sovereign immunity, which states that a government cannot be sued without its consent, was based on the English theory that "the King can do no wrong." This doctrine or rule was incorporated into American law with little apparent thought to the effect it would have in a democracy. Municipal corporations (while functioning in a strictly governmental capacity) as well as governmental agencies traditionally fell under the protection of this immunity.

Although the various states and the federal government are not truly "sovereign" — meaning that they are responsible to no one — the rule evolved in

---

[13]Adrian and Press, *Governing Urban America*, p. 178.

American legal thought that the state could not be sued for wrongs or violations without its consent. The reason for this doctrine was stated by Justice Holmes in *Kawananokoa* v. *Polyblank*:[14]

> A sovereign is exempt from suit, not because of any formal conception of obsolete theory, but on the logical and practical ground that there can be no legal right as against the authority that makes the law on which the right depends.

This premise that there is no legal right against the state that makes the laws was restated more graphically by Justice Holmes, also in *Kawananokoa* v. *Polyblank*, as follows:[15]

> It seems to me like shaking one's fist at the sky, when the sky furnishes the energy that enables one to raise the fist.

The Federal Tort Claims Act of 1946 consented to suits filed against the federal government with regard to torts. Many of the states soon began to enact similar statutes.

The Federal Tort Claims Act of 1946 greatly narrowed the federal government's immunity from liability in tort. States soon began to enact similar statutes. Following this trend, by 1974 about one half of the states had abolished immunity from tort liability at the state and local government levels. In 1976 the Administrative Procedure Act was amended to completely eliminate the doctrine of sovereign immunity with regard to the federal government.

In recent years, courts have increasingly refused to recognize this doctrine or aspects of it so that in more and more states various kinds of immunity are based on statutes and not on the old common-law doctrine.

Because immunity as provided by statute varies from state to state, it is important to be aware of the statutory provisions of various or particular states as they apply to particular interests. It is also important to be aware that, because the doctrine of sovereign immunity has been so firmly enmeshed in the fabric of American law, there is still some confusion as to the application or nonapplication of this doctrine.

The following represents some of the aspects of the difficulty that arises from applying and adapting this doctrine.

(It is always a good practice to read the dissenting opinion in a judicial opinion. The dissenting opinion, drawn from the same set of facts, represents a kind of balance that can be used to weigh and assess the facts and that reflects both the importance of the issue under consideration and the vitality of American legal thought. The 3-2 decision in the following case regarding sovereign immunity had a direct effect on the law as related to fire protection. As you read the case, consider: What does this decision reflect about the state of judicial thought regarding sovereign immunity in 1945?)

---

[14]*Kawananokoa* v. *Polyblank*, 205 U.S. 349, 353 (1907).
[15]*Ibid.*

STEITZ et al. v. CITY OF BEACON

Court of Appeals of New York
Dec. 7, 1945
64 N.E.2d 704

Appeal from Supreme Court, Appelate Division, Second Department.

Action by Charles Steitz and others against the City of Beacon for damages suffered as the result of a fire which destroyed plaintiffs' property. From a judgment of the Appellate Division, 268 App. Div. 1008, 52 N.Y.S.2d 788, affirming a judgment of the Supreme Court, 52 N.Y.S.2d 812, on an order at Special Term, Coyne, J., granting defendant's motion to dismiss the complaint under Rules of Civil Practice, rule 106, as failing on its face to state facts sufficient to constitute a cause of action, plaintiffs appeal.

Affirmed.

THACHER, Judge.

The complaint in this action was dismissed at Special Term for failure to state facts sufficient to constitute a cause of action. The Appellate Division affirmed unanimously and we granted leave to appeal.

The action is brought to recover damages suffered as a result of a fire which occurred in the city of Beacon on December 26, 1942. Under section 24 of that city's charter (L.1913, ch. 539, as amended by L.1920, ch. 171, § 6) the city "may construct and operate a system of waterworks," and the same section provides that "it shall maintain fire, police, school and poor departments." Pursuant to these powers the city caused a system of waterworks and mains to be constructed to supply water for private and public use, including fire protection. As part of this system a pressure and flow regulating valve was constructed on a water line and water main located near the plaintiffs' property upon which there were certain buildings used in conducting the business of raising chickens.

It is alleged that the fire broke out on these premises and that plaintiffs' property was destroyed because of the carelessness and negligence of the city in failing to create and maintain a fire department, including fire equipment and protection for the benefit of plaintiffs' property and the properties of others located nearby. It is also alleged that the city negligently failed to keep in repair the pressure and flow regulating valve located near the plaintiffs' property and that it negligently operated a certain manually operated valve, and that by reason of such negligence an insufficient quantity of water was provided to combat effectively the fire in question. Plaintiffs' freedom from contributory negligence and damages in the sum of $27,900 are also alleged and the plaintiffs demand judgment for that amount.

The waiver of sovereign immunity by section 8 (formerly § 12-a) of the Court of Claims Act has rendered the defendant municipality liable, equally with individuals and private corporations, for the wrongs of its officers and employees. In each case, however, liability must be "determined in accordance with the same rules of law as applied to actions in the supreme court against individuals or corporations." Accordingly the city is governed and controlled by the rules of legal liability applicable to an individual sued for fire damage under the circumstances alleged in the complaint. The question is whether the facts alleged would be sufficient to constitute a cause of action against an individual under the same duties as those imposed upon the city solely because of failure to protect property from destruction by fire which was started by another. There is no such liability known to the law unless a duty to the plaintiff to quench the fire or indemnify the loss has been assumed by agreement or imposed by statute. There was no agreement in this case to put out the fire or make good the loss, and so liability is predicated solely upon the above-quoted provisions of the city's charter defining its powers of government. Quite obviously these provisions were not in terms designed to protect the personal interest of any individual and clearly were designed to secure the benefits of well-ordered municipal government enjoyed by all as members of the community. There was indeed a public duty to maintain a fire department, but that was all, and there was no suggestion that for any omission in keeping hydrants, valves or pipes in repair, the people of the city could recover fire damages to their property.

An intention to impose upon the city the crushing burden of such an obligation should not be imputed to the Legislature in the absence of language clearly designed to have that effect. Language similar to that found in the Charter of the City of Beacon may be found in many municipal charters. *[Citations omitted.]* Furthermore, many of these charters antedate the enactment of section 12-a, Laws 1929, c. 467, the predecessor of section 8, of the Court of Claims Act, as does the charter in this case. As was said in *Moch Co.* v. *Rensselaer Water Co.*, 247 N.Y. 160, 166, 159 N.E. 896, 898, 62 A.L.R. 1199; "If the plaintiff is to prevail, one who negligently omits to supply sufficient pressure to extinguish a fire started by another assumes an obligation to pay the ensuing damage, though the whole city is laid low. A promisor will not be deemed to have had in mind the assumption of a risk so overwhelming for any trivial reward." *A fortiori* the Legislature should not be deemed to have imposed such a risk when its language connotes nothing more than the creation of departments of municipal government, the grant of essential powers of government and directions as to their exercise.

Such enactments do not import intention to protect the interests of any individual except as they secure to all members of the community the enjoyment of rights and privileges to which they are entitled only as members of the public.

Neglect in the performance of such requirements creates no civil liability to individuals. *[Citations omitted.]*

Our decision in *Foley* v. *State of New York,* 294 N.Y. 275, 62 N.E.2d 69, was not governed by this rule because the duty neglected in that case was imposed for the sole purpose of protecting the interests of the plaintiffs and others similarly situated against the particular hazard from which the plaintiffs suffered. In that case the Vehicle and Traffic Law, Consol.Laws, c. 71, imposed upon the State the duty to maintain traffic control lights upon State highways and a majority of the court held that we could not say as matter of law that the plaintiffs' injuries were not proximately caused by the negligence of agents of the State in failing to replace a burned out traffic light bulb. The sole purpose of requiring the State to maintain such lights was to protect individuals using the highways from collision damage. The violation of such a duty, resulting in damage, gives rise to an action in tort, if, but only if, the intent of the statutory enactment is to protect an individual against an invasion of a property or personal interest. *[Citations omitted.]*

The case at bar is governed by our decision in the Moch case, *supra.* There the defendant water company was under a far more specific statutory duty than any to be found in the Charter of the City of Beacon. It was a public service corporation subject to the provisions of the Transportation Corporations Law (L.1909, ch. 219, § 81, Consol.Laws, c. 63), which imposed upon it the duty to furnish water at reasonable rates upon demand by the city through hydrants or in public buildings ''for the extinguishment of fires and for sanitary and other public purposes.'' In accordance with this statute the water company contracted with the city to deliver water at the fire hydrants. It was alleged that while the defendant was under this duty a building caught fire, the flames from which spread to the plaintiff's warehouse, destroying this warehouse and its contents. The defendant according to the complaint, was promptly notified of the fire but neglected after such notice to furnish sufficient water under adequate pressure to extinguish the fire before it reached plaintiff's warehouse, although equipped to do so and thus to prevent the spread of the fire and the destruction of plaintiff's property. Judgment was demanded for the resulting damage. The action for damages was claimed to be maintainable as an action for breach of contract within *Lawrence* v. *Fox,* 20 N.Y. 268, as a cause of action for a common-law tort within *MacPherson* v. *Buick Motor Co.,* 217 N.Y. 382, 111 N.E. 1050, L.R.A. 1916F, 696, Ann.Cas.1916C, 440, and as a cause of action for the breach of a statutory duty, predicated upon the provisions of the Transportation Corporations Law. It was held that the action could not be maintained for a tort at common law or for a breach of statutory duty because the duty was owing to the city and not to its inhabitants and because the failure to furnish an adequate supply of water was at most the denial of a benefit and was not the commission of a wrong.

The Moch case is controlling here because it has judicially determined that a corporation under a positive statutory duty to furnish water for the extinguishment of fires is not rendered liable for damages caused by a fire started by another because of a breach of this statutory duty.

Accordingly, the judgments of the courts below were correct both upon principle and authority and there is no statute justifying the imposition of liability upon the city in this case.

The judgment should be affirmed, with costs.

DESMOND, Judge (dissenting).

The courts below dismissed this complaint on the ground that it fails to state facts sufficient to constitute a cause of action. According to the complaint, a fire broke out in plaintiff's building in the city of Beacon, New York, on the night of December 26, 1942. In the course of the fire the building burned down. It is alleged by plaintiff that the destruction of his property was due to the neglect of defendant City of Beacon in that, according to the complaint, the city failed to keep in repair a pressure and flow regulating valve in the city's water lines and negligently allowed that valve to be and remain in a state of disrepair. It is further alleged that the city was at fault in its operation of another, or hand valve, in the water line, also. By reason of such carelessness of the city, alleges plaintiff, there was no water, or at least not sufficient water, available for the effective fighting of the fire. Particularly to be noted is the allegation in the complaint that the city, under its charter (see L.1920, ch. 171, § 6 amending § 24) was required by law to create and maintain a fire department. That same charter authorizes the city to construct and maintain a water system. Special Term, in ordering the complaint dismissed, relied on *Hughes* v. *State of New York,* 252 App.Div. 263, 299 N.Y.S. 387. In the Hughes case the court, among other things, held that, since the extinguishment of fires is a ''governmental'' function of a city, no liability can be predicated on such a corporation's failure to furnish adequate fire protection. There are innumerable cases throughout the United States, some of them in this State, so holding, but they all rest in whole or in part on the theory that, in furnishing protection against fire and in extinguishing fires, a city acts in its ''governmental'' and not in its ''proprietary'' capacity. That traditional theory of immunity from liability for faulty performance of ''governmental'' activities, including fire protection is, however, only an aspect of the larger rule of the common law that a sovereign could not be held to respond in damages. *[Citations omitted.]* The State of New York, however has, by section 8 of the Court of Claims Act, waived that immunity from liability and action and has assumed such liability, and consented to have the same determined, ''in accordance with the same rules of law as applied to actions in the supreme court against individuals or corporations.'' Pointing out that no city of this State ''has any independent sovereignty'' and that the immunity formerly enjoyed by cities ''was nothing more than an extension of the exemption from liability which the State possessed,'' we have held that the City of New York would be liable for negligence in connection with the conduct of its Police Department, even without General Municipal Law, section 50-b, Consol.Laws, c. 24. *Bernardine* v. *City of New York,* 294 N.Y. 361, 365, 62 N.E.2d 604, 605. The common-law theory of nonliability of municipal corporations in connection with their police departments was precisely the same as with respect to their fire departments. We conclude that the cities of this State have, as a result of the State's waiver of its immunity, become liable at least for negligent maintenance of such facilities and appliances for fire protection as they possess. We limit ourselves herein to saying that an actionable wrong is stated in this complaint insofar as it alleges that the city, having installed certain hydrants, valves, etc., negligently allowed them to fall into disrepair and disuse. We need not now decide what, if any, liability there would be in a case where the alleged fault of the city was not the negligent upkeep of its firefighting plant, but a failure to provide adequate fire-fighting facilities or to use them efficiently. *[Citations omitted.]* Our upholding this complaint would go only so far as to say that the statements therein that the city failed to keep its fire equipment in good order, with resulting damage to the property of the citizen, allege an actionable wrong. Such damage need not be by physical impingement, but may be such as results from a failure to keep protective devices working. *[Citation omitted.]*

This complaint discloses another ground on which liability might be predicated. As above pointed out, an applicable State statute commands defendant City of Beacon to maintain a fire department. Even before the State of New York relinquished its sovereign immunity, the cities of this State were held liable for their defaults in connection with such State-mandated services. That obligation follows from the contract with the State, implied from the acceptance of the city's charter, that the city will discharge the duties imposed by the charter. *[Citations omitted.]* That idea of an implied contract is well stated at 43 Corpus Juris, page 927, as follows: "In many jurisdictions the rule is laid down that a municipal corporation, when charged in its corporate character with the performance of a municipal function, the duty being absolute or imperative and not merely such as under a grant of authority is entrusted to the judgment and discretion of the municipal authorities, is civilly liable for injuries resulting from misfeasance or nonfeasance with respect to such duty; and this rule has been especially applied in cases relating to streets, and to sewers and drains." As the above quotation from Corpus Juris points out, most of the cases enforcing this rule involve streets and sewers and follow *Conrad* v. *Trustees of the Village of Utica,* 16 N.Y. 158 (and *Weet* v. *Trustees of Village of Brockport* which in effect is incorporated into this court's opinion in the Conrad case — see reporter's note at page 161 of 16 N.Y.). While we find no reported cases containing like holdings as to a city's negligence in connection with maintaining its fire-extinguishing paraphernalia, we see no reason against such a holding. It follows that this complaint should not have been dismissed for insufficiency.

It is not presently important that the failure was of a valve or hydrant in a waterline, rather than of the fire-fighting machinery itself. The charter's mandate to the city to have and operate a fire department necessarily includes the furnishing of water therefore.

Although liability for failure to supply fire protection was denied in both *Springfield Fire & Marine Insurance Co.* v. *Village of Keeseville [Citations omitted]*, neither case is a compelling authority for affirmance here. Both those cases were, of course, decided before the State gave up its sovereign immunity. Furthermore, the Keeseville and Moch opinions carefully pointed out that neither the Village of Keeseville nor the City of Rensselaer, was under any such legal duty as has been imposed on this defendant by its legislative charter to supply its inhabitants with protection against fire.

The judgment should be reversed and a new trial ordered, with costs to abide the event.

LEWIS, CONWAY, and DYE, JJ., concur with THACHER, J.
DESMOND, J., dissents in opinion in which LOUGHRAN, C. J., concurs.

LEHMAN, C. J., deceased.

Judgment affirmed.

DUAL NATURE OF MUNICIPAL CORPORATIONS

Municipal corporations are created by the state. As such, they function as agents of the state and have only those powers granted to them by the state. When acting as agents of the state, municipal corporations are considered to be acting in their *governmental* or public capacity. As such, they are performing the duties delegated to them by the state. Municipal corporations are required to perform these duties as agents of the state that created them. Under the heading of promoting the general welfare, these duties usually include the maintenance of law and order and provisions for health, fire protection, education, sanitation, etc. When functioning in their governmental capacity, municipal corporations have usually fallen under the doctrine of the state's sovereign immunity.

Because a municipal corporation also has a corporate nature, it can perform private or *proprietary* functions as a "legal person."

In its proprietary capacity, a municipal corporation performs functions that address the local needs and requirements of the people living within the municipal boundaries. It performs services for the comfort and convenience of the immediate community, as explained by Adrian and Press in the following excerpt from their book titled *Governing Urban America:*[16]

> In theory, the city acts as an agent of the state whenever it performs a function
> in which the state as a whole has a certain interest; for example, when it enforces

---

[16] Adrian and Press, *Governing Urban America*, pp. 162-3.

the law, or maintains public health standards, or collects taxes. On the other hand, the city may perform some tasks purely for the comfort and convenience of the local inhabitants, in theory at least. This classification includes such things as the operation of a water supply or public transportation system. . . .

. . . Property owned for carrying on governmental functions is in theory held for the state, which may dispose of it without consulting the city and without payment to it. On the other hand, property owned by the city for carrying on its proprietary functions cannot ordinarily be taken by the state without compensation, on the theory that constitutional guarantees of property rights extend to property held by the city in its proprietary capacity.

The courts are not clear in drawing the distinction between what is a governmental (or public) function and what is a propietary (or corporate) function. This is often compounded by the confusion of the inhabitants of a municipal corporation. Generally, inhabitants can see no reason why a municipal corporation is considered responsible for its acts in one situation and not responsible for its acts in another, as illustrated in the following extract from *American Municipal Government and Administration* by Stuart MacCorkle:[17]

It is difficult for John Smith who is hit by a police car to understand why he is unable to recover from the city, when William Jones who is struck by a street maintenance truck is able to do so. If Smith has a suit of clothes ruined through the negligent operation of a street sprinkler, he must be more puzzled when he is informed that while recovery is impossible for him, his friend who has had a similar accident because of a street flusher operator has a good opportunity to do so. Certainly he will not understand why he is unable to recover for an accident sustained by a fall on the defective steps of the city hall which operates no utilities, while his friend will probably be successful in obtaining relief for a similar accident occurring when he went to pay his light bill in the city hall of his city.

Although the doctrine of sovereign immunity has been gradually eroded, state legislatures are reluctant to pass statutes making cities liable for all acts of negligence. The feeling that governmental agencies have unlimited funds may lead juries to award monetary judgments that could, in time, escalate the tax rate or even exceed a town's annual budget.

When a municipal corporation (acting in its public capacity) provides benefits to its inhabitants, but through private means, it is considered as a public or municipal corporation transacting private business and, as such, is considered to have a private (or quasi-private), not municipal, character. This is an important distinction for those connected with the fire service. For example, is a particular fire protection organization in a given area or situation a private agency hired by the municipal corporation, or is the fire protection organization part of the public municipal structure itself? The two-fold character of a municipality is vital to an

---

[17]MacCorkle, *American Municipal Government and Administration*, p. 84.

understanding of the rights and liabilities of members of the fire service in this regard. When a municipal corporation acts in a private or quasi-private capacity, it is subject to laws and rules governing private individuals or corporations.

DISCRETIONARY AND MINISTERIAL ACTS

Generally, municipal liability falls into two categories: (1) breach of contract, and (2) torts. The aspect of tort law with which municipalities are most often concerned is called "acts of negligence." Acts of negligence are of particular concern to fire protection interests.

In addition to determining whether or not a municipal corporation was acting in a governmental or proprietary capacity, it is important to draw another major distinction. This distinction revolves around the question, "Was the act complained of a discretionary act or a *ministerial* act?" A discretionary act, while not specifically mandated by the state, is an action that a municipal corporation decides properly fulfills a particular municipal responsibility. A ministerial act addresses the manner in which an act or policy was carried out or performed. This distinction is emphasized in the following explanation from *Governing Urban America*, by Adrian and Press:[18]

> Discretionary acts involve the making of public policy decisions, and they are *not* subject to tort action. Ministerial acts involve the carrying out of established policy, and they are normally the only type of acts that will allow for a legal action. For example, if the city council decides not to extend a water main or not to build a bridge across a stream, or if a police officer (acting in good faith and upon probable cause) decides to arrest a person who subsequently proves to be innocent of any crime, there could be no maintainable action, since these all involve decision making. On the other hand, once a bridge is built, the public has a right to expect the ministerial function of maintaining it to be carried out, and once the city decides to operate a bus line, the public has a right to expect that individuals will not be injured while riding on the buses.

In addition to the distinctions already mentioned concerning a municipal corporation's liability, two further considerations have important implications with regard to a municipality's liability for the actions of its officers and employees: (1) the doctrine of *respondeat superior*, and (2) actions that are *ultra vires*.

THE DOCTRINE OF RESPONDEAT SUPERIOR

It is a general rule in American law that an employer is held responsible for those acts of an employee that are performed in the ordinary course of employment. This rule, called the doctrine of *respondeat superior* ("let the master answer"), is based on the common-law doctrine that a master is responsible for

---

[18]Adrian and Press, *Governing Urban America*, p. 180.

the actions of a servant, providing, of course, that the servant's actions can be reasonably said to follow the master's orders. The doctrine of *respondeat superior* and its implications for employees is discussed in greater detail in the later chapters of this textbook: its mention in this chapter is as one aspect of the considerations involved in municipal liability. It should be kept in mind that the city can be considered liable for the actions of its officers and employees only insofar as the officers and employees perform actions within the scope and requirements of their particular jobs.

ULTRA VIRES ACTIONS

If an officer or employee of a municipality performs actions that cannot be said to fall within the scope of requirements of a given job, then these actions are said to be *ultra vires* and the municipality incurs no responsibility of liability for them, as explained in the following excerpt from *Governing Urban America*:[19]

> . . . Since an individual is responsible for his own actions where there is no master-servant relationship to protect him, this means that the citizen who is wronged may sometimes be able to sue the *individual* city official or employee when he cannot sue the city itself. This may sometimes protect the injured person when he is wronged in the course of the city's performing a governmental function. But the individual, it must be remembered, is not responsible when the city cannot be sued — unless his actions are *ultra vires*. If a city policeman arrests an innocent man against whom he holds a grudge, or if a fire-truck driver operates his vehicle in a grossly negligent manner, the individual who is harmed may be able to collect from the officer or employee as an individual. The chances are excellent, however, that even if he gets a court judgment, the city employee will not earn enough from his job to pay for such damages.

The following brief summary of some rulings regarding municipal liability highlights the aspects of municipal liability when governmental functions are involved — in this case, the governmental requirement to provide fire protection.[20]
Why do you think the courts ruled as they did in the following situations in which the city itself was sued? How do you think the doctrine of *respondeat superior* or *ultra vires* actions affected the decisions?

> • If the fire fighters in a certain station refuse to heed a fire call on a cold night in the winter when they are cosily enjoying themselves at a poker game, the owner of the house that is destroyed by fire has no recourse against the city, although the owner pays taxes to support the fire department. The owner's only hope of compensation is a suit against the fire fighters themselves.
> • If the fire fighters, in fighting a bad fire, start a backfire and destroy property

---

[19]*Ibid.*

[20]Harold Zink, *Government of Cities in the United States,* The Macmillan Company, New York, 1948, p. 136.

some distance from the scene of the original conflagration, the owners of that property cannot hold the city liable.

• If a fire truck runs into a parked car, the owner of the car must not expect the city to pay for the damages. Even if a fire truck runs down people who are standing, as they imagine safe from danger, on a sidewalk, the city has usually been held not liable. However, a few states make the distinction between acts of fire trucks on the way to fires and returning from fires, and in the latter case classify such a function as proprietary and consequently involving municipal liability.

In the following decision by the Supreme Court of Idaho, the Court was required to address the concept of the dual nature of municipal corporations, the doctrine of *respondeat superior,* and also whether the superintendent in charge of construction was responsible for his own actions *(ultra vires)* in order to decide who was liable for negligence. Be prepared to briefly give the reasons for the Court's ruling regarding all three concepts.

STRICKFADEN et al. v. GREEN CREEK HIGHWAY DIST. et al. (No. 4506.)

(Supreme Court of Idaho. July 10, 1926.)
248 P. 456

Appeal from District Court, Idaho County; Miles S. Johnson, Judge.

Action for damages for personal injuries by Charles H. Strickfaden and another against the Green Creek Highway District, Ed. Dasenbrock, and B. A. Baerlocker and others, Highway District Commissioners. Judgment of nonsuit in favor of the commissioners, and judgment for plaintiffs against first-named defendant and first-named defendant appeals, and plaintiffs cross-appeal. Affirmed. . . .

. . . GIVENS, J. Appellant, a regularly organized highway district under the laws of the state of Idaho, through its employees, had left crosswise of the highway an open ditch or excavation in which was to be placed a culvert. Loose rock and dirt from the excavation was piled to the height of about 2½ feet along the banks of the trench. Respondent and cross-appellant, and family approaching this excavation or ditch down a slight grade at night, struck a plank placed across the road on the ground as a warning, which so interfered with the management of the automobile as to prevent its being stopped before the car had run against, and partly into, the rocks and dirt piled along the side of the excavation, which resulted in respondent, his wife, and children being severely injured. Respondent and cross-appellant brought suit for damages against the highway district, the three commissioners of the district, and one Dasenbrock, appellant's foreman or superintendent in charge of the construction work. At the conclusion of plaintiffs' case, the court granted a nonsuit in favor of the three commissioners. The jury returned a verdict in favor of respondent and cross-appellant and against the highway district, but not against Dasenbrock. Appellants appealed from the judgment based on such verdict, and respondents appealed from the judgment of nonsuit in favor of the commissioners, and also because the verdict was in favor of Dasenbrock.

Appellants contend that a highway district is a quasi-public corporation, having the particular characteristics of a county, and that it is therefore not liable for the consequences herein detailed. Respondents, on the other hand, contend that a highway district is a quasi-municipal corporation such as a city, town, or village, and is therefore liable.

Highway districts as created by the Highway District Law of Idaho (C. S. § 1490 *et seq.*) are a comparatively new organization, and the law with regard to their liability for torts is in this state an open question.

Counties may be said to be true public corporations. They are local organizations, which for the purposes of civil administration are invested with a few functions characteristic of a corporate existence. They are legal political subdivisions of the state, created or superimposed by the sovereign power of the state of its own sovereign will, without any particular solicitation or consent of the people within the territory affected. *[Citations omitted.]*

Cities, towns, and villages may be classified as true municipal corporations, voluntarily organized under the general law at the request and with the concurrent consent of their members, and, in addition to the exercise of the functions of self-government, transact matters of a quasi-private or business character not governmental in their nature, but rather proprietary, or for the acquisition of private gain for the municipality and its citizens. *[Citations omitted.]*

Highway districts, as intended by the Highway District Law, *supra,* cannot be said to correspond identically with either public corporations, or counties, or municipal corporations, or cities, towns, and villages. They are quasi-municipal corporations, not political municipalities, and are not created for purposes of government, but, instead, for a special purpose; namely, that of improving the highways within the district.

"A highway district as intended by this act is not a political municipality. It is not created for the purpose of government. It is an entirely different kind of municipality from that of a city, town or village. Its powers are specially limited to the construction of highways upon lines of benefits to the inhabitants and the property within the territory embraced within the district. It is made a taxing district and consists of such territory as may be determined by the county commissioners in creating the same. It is contemplated by the provisions of the statute that the property and the people of the entire district are interested in the construction and improvement of the public highways of the district, and it is created for a special purpose, to wit, the assessment of property within the district for the sole and only purpose of improving the highways within the district." *Shoshone Highway Dist.* v. *Anderson,* 22 Idaho, 109, 125 P. 219. . . .

. . . It is well settled that, in the absence of an express statute to that effect, the state is not liable for damages either for nonperformance of its powers or for their improper exercise by those charged with their execution. Counties are generally likewise relieved from liability, for the same reason. They are involuntary subdivisions or arms of the state through which the state operates for convenience in the performance of its functions. In other words, the county is merely an agent of the state, and, since the state cannot be sued without its consent, neither may the agent be sued. *[Citations omitted.]*

It is well recognized that there are two kinds of duties imposed or conferred upon municipal corporations; those termed public governmental functions, where the municipality performs certain duties as an agent or arm of the state, and those other municipal activities which are sometimes termed administrative, ministerial, corporate, private, or proprietary functions, performed for the municipality's own benefit, or for the benefit of its citizens, and, while acting in the performance of its governmental functions or in a public capacity as an arm or agency of the state, the municipality is not liable for its failure to exercise these powers or for their negligent exercise, unless such liability has been imposed by statute. *[Citations omitted.]* In its capacity as a private corporation a municipality stands on the same footing as would an individual or body of persons upon whom a like special franchise had been conferred. Hence it is liable in the same manner as such individual or private corporation would be under like circumstances. *[Citations omitted.]* Thus municipalities are held liable for their torts in the performance of ministerial, private, corporate, or proprietary functions. *[Citations omitted.]*

In those jurisdictions where redress is allowed against municipalities, the general rule is that a municipality is bound to exercise ordinary care to keep its streets in a condition of reasonable safety for the use of the public *[Citations omitted]*, but from an analysis of the decisions supporting this doctrine it is apparent that the reasoning for such rule falls within two divergent classes; the first class being those cases which have held that the care and maintenance of the streets is a ministerial, corporate, or private duty the city owes to its inhabitants, and not a public or governmental duty.

"The duty of the city in connection with the maintenance of its streets is ministerial and corporate, and for its negligence in that connection it is liable. *[Citations omitted.]*

"In *Sutton* v. *Snohomish,* 11 Wash. 24, 39 P. 273, 48 Am. St. Rep. 847, we decided that the duty of a city to keep streets in repair was not a governmental but a ministerial duty, and for a breach thereof an action will lie in favor of a person injured as a result of such negligence. In the course of the opinion it was said

'There is undoubtedly a want of harmony among the decisions of the courts upon this question; but we believe the decided weight of authority, as well as sound reason, is in favor of the view above expressed' — citing many leading cases to sustain the announcement. . . .

". . . This class of cases is in entire harmony with those decisions announcing the rule that municipalities are liable for their proprietary, corporate, or ministerial functions, but not liable for public or governmental functions.

"The second class of decisions are those which class the maintenance and care of the streets by a municipality as governmental or public functions, and at the same time hold the city liable for misfeasance or nonfeasance in the exercise of such governmental duty. This class of decisions practically without exception recognize that such conclusion is illogical, and not based upon sound reason, when the nonliability for governmental functions is considered, and state that such liability is an exception to the general rule. The reason most generally assigned for making a distinction between such cases and those governed by the general rule is that such municipalities, having been given the exclusive control over their streets with ample power to provide funds to care for and maintain them, are chargeable with the duty to keep them safe for travel, and that it follows by implication therefrom that they are liable for failure to perform such duty. [Citations omitted.] . . .

". . . As suggested by Judge Keller in *Cousins* v. *Butler County*, 73 Pa. Super. Ct. 86, 94, the explanation of the highway cases may more probably be found in historic reasons than in general rules, or by the process of legal or logical deduction therefrom. This is quite possibly correct, for, when the latter methods are pursued, it will be seen that, if a breach of duty imposed be accepted as the sole or controlling criterion by which to judge questions of liability in cases of negligence toward others on the part of municipal employees, then, since so far as legislative mandates are concerned, the duty to organize and maintain police and fire service is just as mandatory on cities (at least on those of the first class) as to construct and maintain highways, such a breach could as well give rise to the right to sue for damages growing out of the negligence of the police or fire fighters as of employees in charge of, or working on, the construction or repair of highways; yet it has been repeatedly held that, except where a right to recover is expressly given by act of assembly, no suit lies against municipalities for negligence of their police or fire departments." *Scibilia* v. *City of Philadelphia*, 279 Pa. 549, 124 A. 273, 32 A. L. R. 981.

In either event, whether the duty of maintaining and constructing streets is said to be a governmental duty or a private or ministerial duty, the conclusion reached in both lines of decisions has been the same; namely, that municipalities are liable for negligent construction or maintenance of their streets. Idaho has followed the former, and what appears to be the better rule, based upon logic and good reason, holding that a municipality in the care and maintenance of its streets is performing a ministerial or private duty which it owes to the individuals it impliedly invites to travel over its streets, and for negligence in performing such functions the municipality is liable. . . .

. . . That there is a distinction between certain acts of a municipality to the effect that certain classes of acts are deemed governmental in character while others are deemed merely corporate or private in character is not questioned by any of the authorities, and the authorities are not so very inharmonious in describing the qualities which determine whether an act is governmental or corporate, but they quite frequently disagree when it comes to applying the principles of law to a concrete case. It has been said that the distinction between governmental and quasi-private acts of a municipality may be determined by this, namely, if the power conferred on the municipality be granted for public purposes exclusively, it is governmental, but, if for private advantage and emoluments, though the public derive a common benefit therefrom, the corporation *quo ad hoc* is to be regarded as a private corporation. *[Citations omitted.]* A great majority of cases determine whether the act is governmental or proprietary in character from the inherent nature of the act; that is, whether truly governmental in character or an act which may be performed by an individual. "Where duties are cast on a municipality by mandatory provisions of law, or by permissive provisions adopted by the municipality, and are not in their nature governmental, but could be carried on by individuals, and relate to the convenience, pleasure, or welfare of individual citizens, the municipality is acting as a legal individual, in a ministerial and administrative capacity, and under the obligation to perform such duties with reasonable care and vigilance." *Van Dyke* v. *City of Utica, supra.* . . .

. . . From a study of the cases, it is apparent that, in determining whether an act is governmental or corporate, all rest upon the inherent nature of the act and to whom the benefits of such acts inure. . . .

. . . Since it has been held in this state that the building of roads is not a governmental duty, but a ministerial one *(Carson* v. *Genesee, supra),* and that a highway district does not perform governmental duties *(Shoshone Highway District* v. *Anderson, supra),* it does not appear necessary to lay down any exact rule for the determination of what is a governmental or proprietary function. . . .

. . . Highway officers as well as other public officers are in general liable for injury in consequence of their malfeasance or nonfeasance in the performance of ministerial duties.

"'Highway officers charged with the performance of a ministerial duty,' says a text-writer, 'are, in general, liable for negligently performing it to one to whom the duty is owing and upon whom they inflict a special injury.' Elliot, Roads & Streets (2d Ed.) § 674. Any person to whom such obligation is due and who sustains a special injury in consequence of the malfeasance or nonfeasance of public officers in the performance of purely ministerial duties, may obtain redress against such officer or officers by a private action adapted to the nature of the case." *[Citations omitted.]*

The rule of *respondeat superior* does not apply where public officers are sought to be bound by the negligence of subordinate officers or employes, unless there has been a failure to exercise due care in the selection of such subordinates, or the officers have knowledge of the negligent acts of the inferior officers. In other words, their liability or nonliability rests upon their knowledge or lack of knowledge of the negligence of their employês or inferior officers who are actually performing the work, and whether or not they have used due care in selecting competent employés for the work.

In *Doeg* v. *Cook, supra,* it appeared that the trustees had actual knowledge of the dangerous condition of the highway, and because of such knowledge they were held liable; the court saying: "It is proper to join as defendants in an action for damages for injury resulting from a defective highway the town marshal, his bondsmen, and the board of town trustees, when the injury resulted from the separate, but concurrent, negligence of the marshal and trustees. . . ."

. . . Whether or not the act in question was discretionary or whether the district had funds to perform the work does not arise in this case, for the reason that the work had already been commenced, and was being performed at the time the accident happened. It is not alleged that the commissioners of the district had failed to exercise due care in securing a competent man as Director of Highways, and there is no evidence that they had actual knowledge of the negligent manner in which the work was being performed or that they participated in such work. The court, therefore, did not err in granting a nonsuit in their favor.

It is urged that Dasenbrock was the party responsible for the negligence, if any, and that he was an independent officer. The rule thus contended for appears in *Altvater* v. *Mayor and City Council of Baltimore,* 31 Md. 462:

"It is but just that responsibility for the proper discharge of duty should result from the power to perform the duty, and if a party is to be held responsible for the conduct of another party charged with the performance of duty, over whom he can exercise no control, some law creating such responsibility ought to be shown." *Dupuis* v. *City of Fall River,* 223 Mass. 73, 111 N. E. 706. . . .

. . . In *Sievers* v. *City and County of San Francisco,* 115 Cal. 648, 47 P. 687, 56 Am. St. Rep. 153, it is emphasized that a duty is imposed upon the officer by express law, and not by order of the municipality, concluding, therefore that the municipality was relieved from liability. The converse of this would be that, if the order had been given by the municipality, which of course can act only through the board of commissioners, the officer would have been acting for the municipality, and not as an independent officer. Bearing this proposition in mind, turning to the answer we find it alleged in the second and separate defense under paragraph 8:

"That, before the happening of any of the events mentioned in plaintiff's complaint, and on or about the month of June, 1922, that certain highway mentioned and described in plaintiff's complaint was in a condition deemed unsafe for travel by the said board of highway commissioners, in that the plank covering of that certain culvert mentioned in plaintiff's complaint had become decayed and broken, and that certain other repairs were deemed necessary to said highway by said board of commissioners, acting officially in the capacity aforesaid, and not otherwise, authorized and directed the defendant, Dasenbrock, to remove said wooden culvert, and to install in its place a culvert of rock and concrete. . . .

. . . "That on or about the 27th day of June, 1922, the defendant, Ed. Dasenbrock, acting as an employē, foreman, and superintendent of the defendant highway district, and by authority of the said commissioners of said highway district so regularly and officially given to him, and authorizing the work and improvements, removed the said defective wooden culvert, and began excavation across the said highway for the purpose of installing and constructing the said concrete culvert.

"Defendants further allege that the installation of said culvert was proper and necessary repair upon said highway; that the same was done lawfully and by authority of the said highway district and its officers. . . ."

. . . The defendants thus specifically alleged that Dasenbrock was an officer of the highway district, and that the highway district was to do the work, and it would seem, therefore, that the authorities referred to are beside the point. . . .

. . . Appellants do not assign the instruction telling the jury they might find against one defendant and in favor of the other as error, and simply urge that the verdict and judgment in favor of Dasenbrock should stand, but do urge that the verdict against the district should be set aside as inconsistent therewith. On the other hand, respondents and cross-appellants ask that the verdict against Dasenbrock be set aside because contrary to the evidence.

In *Texas & P. Ry. Co.* v. *Huber* (Tex. Civ. App.) 95 S. W. 568, the court stated very sane reasons why a verdict in favor of the employē should not vitiate a verdict against the master, and said:

"We also overrule the tenth assignment, which contends that a motion in arrest of judgment ought to have been sustained because the verdict in favor of Oliphant was a finding that he was free from negligence, and, he being the agency through which the railway company committed the negligence, if any was committed, the verdict against this appellant was unfounded. The question was considered in *Railway* v. *James,* 73 Tex. 12, 10 S. W. 744 [15 Am. St. Ry. 743] where it was held, in a case of similar character, that, although such a verdict has the appearance of being based on inconsistent and contradictory findings of the jury, this is not of itself enough to require the reversal of a judgment against the passive defendant; the reason being that the finding in favor of the defendant whose act constitutes the negligence complained of, and the finding against the other in the same case by the same jury, can be attributed to improper conduct of the jury in arbitrarily exonerating the former and not necessarily to a finding that there was no negligence on his part. It has often been held that, where the servant or agent who was the real and active wrongdoer has been sued and a judgment has been rendered in his favor, it will be a bar to a judgment against the employer or principal. This rule is not deemed to be applicable where the verdict in favor of the agent or servant nevertheless bears intrinsic evidence that the jury found that the wrongful act complained of had been committed by the agent or servant. The finding against appellant indicates this.''

In the case of *Whitesell* v. *Joplin & P. Ry. Co. et al.,* 115 Kan. 53, 222 P. 133, the Supreme Court of Kansas said:

''It must be conceded that the collision was the proximate cause of the injury, and that the crowded condition of the car contributed thereto by compelling the plaintiff to stand in the vestibule of the car, where he was struck by the controller box of the car on which he was riding. Although there was no verdict against the motorman, it is not seen how it can be successfully contended that he was not guilty of negligence in running his car into the car on which the plaintiff was riding. If the company had been sued alone, the jury would have been warranted in finding the company guilty of negligence, and in returning a verdict in favor of the plaintiff on that finding. The fact that the jury failed to return a verdict as to the defendant motorman is no reason why a verdict against the company, based on a finding of the jury that the motorman was guilty of negligence, should not stand. The failure of the jury to return a verdict against the motorman cannot be used as a reason for setting aside the judgment rendered against the railroad company. . . .''

. . . The verdict of a jury can only be set aside on appeal for want of substantial evidence to support that particular verdict, and not because the verdict may seem inconsistent with another verdict, or because another verdict is wrong, or some other party has been discharged or exonerated. It would be a subversion of the ends of justice to allow a defendant to complain of a just judgment against it — one supported by the overwhelming weight of evidence — because its joint tortfeasor was not also held. Again, it if be contended by appellant that the verdicts are inconsistent, and one of them is therefore capricious, or the result of prejudice, what more reason is there for believing that the caprice or prejudice was directed against the appellant than to suppose that it was directed in favor of Dasenbrock and against respondent?

The evidence is amply sufficient to sustain the verdict, and the judgment is therefore affirmed, and it is so ordered. Costs awarded to respondents on the appeal of the highway district, and in favor of the highway district on the cross-appeal by respondents and cross-appellants.

WILLIAM A. LEE, C. J., and BUDGE, J., concur.

WM. E. LEE and TAYLOR, JJ., dissent.

The following summary from E. M. Borchard's "Governmental Liability in Tort"[21] summarizes some of the arguments and judicial theory that have played an important part in judicial considerations regarding municipal liability. As you read the summary, consider how it applies in general to what you have learned about municipal liability from this chapter, and how the arguments in it apply specifically to the fire service.

> In the effort to distinguish governmental from corporate functions of municipal corporations, the courts have drawn in aid various criteria or justifications that seemed to them controlling or persuasive. Thus, aside from the argument derived from the sovereign immunity of the city as agent of the state, the immunity has been placed on the ground that the city derives no pecuniary benefit from the exercise of public functions; that in the performance of public governmental duties the officers are agents of the state and not of the city, and that therefore the doctrine of *respondeat superior* does not apply; that cities cannot properly perform their functions if they are made liable for the torts of their employees; that the city should not be liable for negligence in the performance of duties imposed upon it by the legislature, but only in the case of those voluntarily assumed under general powers; that in determining whether or not to undertake an act the function is governmental, but the execution of the decision in practice is corporate or ministerial, that powers exercised for the benefit of the public at large are governmental, but those conferred for its own benefit, and by reason of its nature as a municipal corporation are corporate.

MUNICIPAL OFFICERS, AGENTS, AND EMPLOYEES

Officers in a municipal corporation are usually either elected or appointed by the corporation. Municipal employees are usually not considered municipal officers. The reverse is also true. This distinction is ordinarily drawn by the duties assigned to be performed.

Generally, a municipal officer holds an office that is created by legislation,

---

[21]E. M. Borchard, "Governmental Liability in Tort," 34 Yale L.J., p. 127, pp. 132-133 (1924).

while a municipal employee's relation to a municipality is based solely on some form of contract. A municipal officer often takes an oath of office and has fixed or predetermined public duties and functions that are governmental in nature. A municipal employee, on the other hand, does not take an oath and does not have fixed public duties prescribed by law.

An agent or an employee of a municipal corporation has a legal status different from that of an officer of a municipal corporation. If an agent or employee is connected with a municipality on a contractual basis, then the relationship can be said to be that of an individual dealing with the private corporation aspect of a municipality. A municipal employee is considered different from a municipal agent. A municipal agent can represent a municipal corporation in dealings with third parties. For example, a municipal agent can enter into private contracts in behalf of the municipality. A municipal employee's function is considered ministerial. This is the basic distinction between one who serves the municipal corporation and one who represents it as an agent. The scope of an employee's job is determined by the functions and duties performed by the employee with the employer's knowledge and consent. An employee may vary or alternate, within reason, the means and methods of performing a task and still function within the scope of the employee's employment. The relationship of an employee to a municipal corporation can also be fixed by statute and/or defined by either the state or the municipality. Very often this relationship is founded on civil service regulations.

Every unit of state government and virtually every unit of local government is involved in employment: Government employment is usually referred to as civil service employment, and the term "civil servant" means a civilian who is employed by the government. Civil service or similar recruitment and employment systems attempt to ensure that competent employees are chosen for appointment in government. Such systems also attempt to eliminate politics from hiring and firing practices. Because of the possible political influence involved in municipal employment, many cities either control or prohibit the political involvement of their employees.

In some cities, limitations on employees' political activities are far-reaching. When the manager of Skokie, a suburb of Chicago which adopted the council-manager plan in 1957, undertook to prepare the city's 250 employees (many of whom were active and frank political partisans) for service in a nonpartisan system, he issued a directive that specified in considerable detail what would and what would not be permitted. The content of the directive, reproduced here as Table 2.2, is generally considered as representative of the efforts that many municipalities have taken to ensure that municipal employment will not be affected by political overtones.

The following case study, in which the complainants are five fire fighters, illustrates the problems that courts have been asked to address with regard to political involvement on the part of municipal employees. The case study defines the role of fire fighters as civil servants and what personal actions of theirs may or may not qualify as political campaigns.

<hr />

**Dan R. HUDSON, as Chairman of the
Personnel Board of Jefferson County,
Alabama, et al.**

v.

Billy GRAY et al.

Supreme Court of Alabama.
March 26, 1970.
Rehearing Denied May 15, 1970.
234 So. 2d 564

Declaratory judgment action in equity by members of fire department of city against chairman of personnel board of county and others to determine right to secure signatures to petition addressed to city regarding adoption of ordinance reducing duty hours of firemen. The Circuit Court, Jefferson County, William C. Barber, J., entered judgment in favor of firemen and respondents appealed. The Supreme Court, McCall, J., held that members of city fire department in securing signatures to the petition were not taking part in a "political campaign," within civil service law prohibiting civil service employees from taking part in political campaigns.

Affirmed.

W. Gerald Stone, Bessemer, for appellants.

Wingo, Bibb, Foster & Conwell, Birmingham, for appellees.

McCall, Justice.

The complainants are five firemen of the City of Birmingham who as such are subject to its civil service law, Vol. 14, Appx., §§ 645-672, Code of Alabama, 1940, Recompiled 1958, as amended. The relevant part of this statute, § 669, as amended, prohibits any employee in the classified service of the city from taking part in any *political campaign,* except privately to express his political opinion and to vote. "Provided that nothing in this section shall be construed so as to deny the right of a public servant to petition his city, county, state or national government."

The complainants who are the appellees, propose to petition the city pursuant to Tit. 62, § 636, Code of Alabama, 1940, to adopt an ordinance to reduce the on duty hours of firemen, except the chief of the department, to 40 hours each calendar week, with the exceptions applicable to emergencies, or, to call an election on the question of whether such ordinance should be adopted. The ordinance would beome effective immediately on its enactment, but would not become operative for twelve months to allow time to organize and prepare for its operation. Section

Table 2.2 Regulations of Political
Activities by Village of Skokie Employees*

| | Whether permitted or not | | | |
Type of activity	Partisan elections	Nonpartisan elections	On village time	On employee's time
Membership in a political party, club, or organization	Yes	—	No	Yes
Officer or committee chairperson for a political party or organization	No	—	No	No
Attendance at political rallies or meetings, as spectator only**	Yes	Yes	No	Yes
Speak at political meetings, make endorsements or appear on behalf of any candidate or proposal	No	Yes	No	Yes
Circulate petitions, distribute printed matter or badges, or sell tickets for any candidate or party	No	No	No	No
Sign a petition	Yes	Yes	Yes	Yes
Solicit or accept money from any person for any political purpose	No	No	No	No
Serve as precinct captain or party worker for any political organization	No	No	No	No
Assist in getting voters to polls on election day	No	No	No	No
Act as poll watcher for a political party	No	—	No	No
Make contributions to political party or organization	No	No	No	No
Be a candidate for public office***	Yes	—	No	Yes
Use or threaten to use influence of position to coerce or persuade vote	No	No	No	No
Participate in nonpartisan voter registration campaigns	—	Yes	No	Yes
Participate in partisan voter registration campaigns	Yes	—	No	Yes
Be a delegate to a political convention	No	No	No	No
Cast a vote	Yes	Yes	Yes	Yes

*From *City Politics,* by Edward C. Banfield and James Q. Wilson, Joint Center for Urban Studies Publications Series, Harvard University Press, Boston, © 1963, p. 211.

**Not permitted in uniform.

***Must take leave of absence during campaign and term of office.

636 requires that any proposed ordinance may be submitted to the city council by a petition signed by at least 5,000 qualified electors of the city with their places of residence given. The petition shall contain the proposed ordinance in full and

have printed thereon the names and addresses of at least five electors who shall be officially regarded as filing the petition and shall constitute a committee of the petitioners for the purpose of circulating the petition to secure the signatures in compliance with § 636. The director of the Civil Service Board promulgated a letter stating that the appellees' activity was political and warning them and other firemen that anyone who canvassed for signatures on the petition or circulated it, or caused it to be circulated, would forfeit his position. This action was affirmed by the Board.

Under the bill of complaint, filed by the appellees in the circuit court for a declaratory judgment, Tit. 7, § 156 *et seq.,* Code of Alabama, 1940, as amended, the trial court held in favor of the complainants that their activity did not constitute taking part in a political campaign under the civil service law, § 669. The appellants, the members and the director of the Civil Service Board, have appealed from this final decree. . . .

. . . The bill presents justiciable issues between the parties under the Declaratory Judgment Act, Tit. 7, § 157, Code of Alabama, 1940, as amended, by seeking a declaration of the rights, status, or other legal relations between the parties under both of the statutes, § 669, the civil service law, and § 636, the initiative and referendum statute, and may be brought in equity. There was no error therefore in overruling the appellants' demurrer to the bill of complaint.

This appeal is submitted on a stipulation of facts contained in the record. The question presented is: Does serving as a committee under § 636, petitioning the city council to pass and adopt the proposed ordinance, or, if not, to call an election on the question constitute taking part in a political campaign?

The case of *Hawkins* v. *City of Birmingham,* 248 Ala. 692, 29 So.2d 281, holds that a similar municipal ordinance, fixing the hours of firemen, was properly the subject of the initiative and referendum statute applicable to Birmingham, Tit. 62, § 636, Code of Alabama, 1940. Substantial compliance with the provisions of this initiative statute is a condition precedent to submitting the petition to the city council for official consideration. *[Citations omitted.]*

In *Heidtman* v. *City of Shaker Heights,* 163 Ohio St. 109, 126 N.E.2d 138, the question was whether city firemen were taking part in "politics" when they circulated parts of an initiative petition seeking enactment of an ordinance to establish the three-platoon system in the fire department. The Ohio court held that the word "politics" as used in their civil service law meant dealing with political affairs in a party sense, and that circulating parts of the initiative petition did not constitute taking part in politics as that term is used in the civil service law. The court arrived at its conclusion on the premise that since the statute refers to the solicitation of funds for political parties or candidates as well as political organizations, the expression "take part in politics" was intended to cover only politics embraced in the party sense. The other reason was that prevalent politics had controlled the police and fire departments, and it was to prevent abuses and resulting evils in this field that the civil service law was passed, thus showing an intention to give the term "politics" the narrower meaning, which is consistent with the objective to be achieved by the civil service law.

In *Hawkins* v. *City of Birmingham*, 248 Ala. 692, 29 So.2d 281, this court held that it was appropriate to construe § 636 in *pari materia* with the civil service law applicable to Birmingham. This holding should be applied here and in so doing, the rules set forth in *Birmingham Paper Co.* v. *Curry*, 238 Ala. 138, 140. 190 So. 86, 88, will be useful also. There the court said:

> The Court in determining the Legislature's intention in construing statutes, in the establishment of uniformity in the law, and to give rational construction to doubtful meanings of words and phrases employed, may look to other provisions of the same act, consider its relation to other statutory and constitutional provisions, view its history and the purposes sought to be accomplished and look to the previous state of law and to the defects intended to be remedied. *[Citations omitted.]*

Keeping in mind that the initiative statute § 636 was enacted to afford the people of Birmingham the power, where appropriate, to compel the city council to enact desired ordinances, or to have them submitted to a vote of the people, *Geller* v. *Dallas Ry. Co.* (Tex.Civ.App.), 245 S.W. 254, 257, and that the civil service law, § 669, as amended, was enacted to promote efficiency and economy in government and security in employment *[Citations omitted]*, the fullest application possible should be given both statutes, while avoiding any conflict. *[Citations omitted.]* In doing this, we think it compatible with the operation of both statutes to conclude that circulating and filing the petition as a condition precedent to engaging the power of initiative action is not to be construed as political activity or taking part in a political campaign under the civil service law, § 669, as amended, and that this statute was not intended to take away from those in the classified service, the right to petition effectively under the initiative statute in order to bring a matter on for consideration. We hold that the appellees are not thereby taking part in a political campaign.

Further § 669 was amended so that nothing therein shall be construed to deny the right of a public servant to petition his city. If to petition requires signatures, then the amendment gives the appellees the right to solicit the necessary number of signers. We interpret the amendment to mean that a public servant has the right to petition his city in a manner to comport with legal requirements so as to be heard.

Act No. 229, Acts of Ala., Regular Session 1967, p. 598, approved August 16, 1967, states that no person shall hold employment as a fireman who is a member of an organization of employees that asserts the right to strike against any municipality in the state. The constitution of the Birmingham Fire Fighters Association contains a "no strike" clause, as does the constitution of the Uniformed Fire Fighters Association of Alabama. The constitution and bylaws of the International Association of Fire Fighters contained a clause that no member or subordinate union of the association shall strike, but this clause was repealed in August 1968. We hold that repealing the "no strike" clause does not amount to an assertion of a right to strike against the municipality. The absence of a "no strike" clause, is not in itself an assertion of a right to strike.

The decision of the trial court is affirmed.

Affirmed.

LIVINGSTON, C. J., and SIMPSON, COLEMAN and BLOODWORTH, JJ., concur.

MUNICIPAL
FIRE PROTECTION

The fire department of a municipality is regulated by statutes in response to the public duty of a municipality to provide adequate fire protection to the individuals within the municipality. As such, both fire prevention codes and rules and regulations governing the operation and conduct of both the fire department and its members are considered law, whether or not these laws are expressly stated or can be implied. While additional rules and regulations cannot modify a statute, they will not be considered null and void if they are reasonable and if their intent is to provide more efficient and effective fire protection services. The following are some rules that have not been considered unreasonable by some courts:
   • Prohibition against members of the fire service engaging in outside private business or employment.
   • Prohibition against disrespect and insubordination toward a superior officer.
   • Prohibition against members of the fire service giving out information relative to the conduct of business of the department.
   • Prohibition against liquor on the premises of a fire department.
   The preceding rules are but some examples; the individual circumstances surrounding the cases in which the validity of these rules was tested have not been described here.
   Providing adequate fire protection is considered a governmental or public function. Although the main goal of municipal corporations is local self-government, the creation and operation of fire departments is generally considered to be a statewide concern also.

FIRE DEPARTMENTS

The sources of the power to create and maintain a fire department are varied. Among them are the following:
- Legislated by the state.
- Mandated by the state and legislated by the municipality.
- Mandated and legislated by the municipality alone.

Some statutes have created the fire department as a legal entity separate from the municipality, while other statutes provide that the fire department is independent of the legislative department of a municipality. It is not uncommon for a fire department to be considered a quasi-public or municipal corporation by itself.

The fire department of a municipal corporation has the powers that are prescribed to it by law. The powers conferred on a fire department may be such that it becomes a quasi-corporation and, as such, can sue or be sued. This is particularly important in suits regarding negligence. In one instance, a fire department was held liable for the negligence of its employees for injuries to persons who were lawfully in the street.

In its function of guarding the public welfare, a municipality may maintain an auxiliary fire service in addition to its fire department.

VOLUNTEER FIRE COMPANIES

A municipal corporation may also provide fire protection through the use of volunteer fire departments. Such volunteer fire departments are sometimes regarded as part of the municipal government, although its members are not considered municipal employees.

Compare the decisions in the following two case studies regarding volunteer status in some fire department situations.

STEFFY v. CITY OF READING et al.

Supreme Court of Pennsylvania.
March 25, 1946.
46 A.2d 182

Appeal No. 131, January term, 1945, from order of Court of Common Pleas, Berks County, No. 39, October term, 1943; Shanaman, Judge.

Action by Henry L. Steffy against the City of Reading and others for a writ of mandamus to compel plaintiff's reinstatement as a paid driver in a volunteer fire company. The trial court gave binding instructions to the jury in favor of defendants and refused plaintiff's motion for judgment *non obstante verdicto,* and the plaintiff appeals.

Affirmed.

HORACE STERN, Justice

The first question here presented is one of law: If a paid driver in a volunteer company which constitutes part of the fire department of a third class city is not a city employee is he nevertheless entitled to the civil service protection provided by the Act of May 31, 1933, P.L. 1108, 53 P.S. § 8479 *et seq.*? The second is a question of fact: Was plaintiff an employee of the City of Reading or only of the volunteer fire company whose apparatus he drove?

Plaintiff served as a paid fireman from May 11, 1923 to July 6, 1943, attached to the "Oakbrook Fire Company" in Reading. He was originally appointed by the trustees of that company, and, on the latter date, was discharged by them without any charges being preferred against him and without any trial or hearing. He thereupon instituted proceedings for a writ of mandamus against the Mayor, the City Council and the Civil Service Board of the city, to compel his reinstatement. He relies upon the Act of May 31, 1933, P.L. 1108, which provides for the appointment, promotion, reduction, removal and reinstatement of paid officers, firemen and employees of fire departments in cities of the second and third class;[22] from the date of its enactment no person was to be either appointed or discharged as a paid member of any fire department in such cities in any manner other than as prescribed in the act. In *Lehman* v. *Hazleton,* 135 Pa.Super. 410, 5 A.2d 646, it was held that this act related to volunteer as well as to paid fire departments, and that therefore paid fire drivers in volunteer companies are entitled to its benefits; in that case, however, the driver was admittedly a municipal employee. The present question is whether a paid driver of a volunteer company is within the civil service protection of the statute *if he is not on the payroll of the municipality.* We do not find in the act any indication of an intention on the part of the legislature to provide civil service benefits for any firemen except those in the employ of the city. If it were held that a nonmunicipal employee is protected by the act against an arbitrary discharge it would necessarily follow that he comes under the statute also in regard to the method of his appointment and it is scarcely conceivable that the act contemplates the appointment by city officials of the employes of volunteer companies. Section 16 provides that "All paid firemen . . . *in the employ of any city* upon the effective date of this act, shall continue to hold their positions subject to the provisions of this act." This provision — and it is the one upon which plaintiff necessarily relies — is an indication that the statute has only municipal employees in contemplation.

As to the second question, it is clear that plaintiff was never in fact an employee of the City of Reading. He was not appointed by the City Council; his position was not established by any councilmanic ordinance; he was not paid by the city nor discharged by it. Only the City Council has the power to appoint and dismiss city employees and prescribe their number, duties and compensation: Third Class City Law of 1931, P.L. 932, §§ 901, 902, 53 P.S. §§ 12198-901, 12198-902. It is

---

[22]By the Act of June 27, 1939, P.L. 1207, sec. 9, 53 P.S. § 9395, the Act of May 31, 1933, P.L. 1108, was repealed in so far as it applies to cities of the second class.

true that there are fourteen volunteer companies, of which Oakbrook Fire Company is one, which are recognized by the City of Reading as constituting its fire department; it is also true that the city owns and keeps in repair all the fire apparatus, that it owns or leases all the fire stations, and that it makes annual appropriations to the volunteer companies by way of gratuities to cover their expenses of operation including the payment of their salaried employees. But none of these facts establishes a relationship of employer and employees between the city and the paid drivers of the volunteer companies. The members and employees of such companies are not subject to the control of any city official except that when an actual fire occurs they must, of course, obey the directions of the Fire Chief; even if they fail in that respect, however, they can be discharged only by the trustees of the company to which they belong or by which they are employed. The companies are supervised and managed, not by the city, but by their own trustees, and although the practice is to have applicants for positions examined by a board of which some of the members are city officials this is purely a voluntary system; the examining board is selected by the representatives of the companies and the trustees make the actual appointments without being bound in any way by the results of such examinations. While, in the Act of June 21, 1939, P.L. 566, 77 P.S. § 1201 *et seq.* which amends the Workmen's Compensation Act, all members of volunteer fire companies of the various cities, boroughs, incorporated towns, and townships are declared to be "employees" of such cities, boroughs, incorporated towns and townships for all the purposes of *that act,* and entitled to receive compensation for injuries received while actually engaged as firemen or while going to or returning from a fire, that "declaration" has no bearing upon the Act of 1933; on the contrary, the inference is clear that without such an express provision the members of volunteer companies would not, merely by reason of such membership, be regarded as municipal employees entitled to receive workmen's compensation. The same observation applies to the Acts of June 22, 1931, P.L. 751, and June 29, 1937, P.L. 2329, amending section 619 of the Vehicle Code, 75 P.S. § 212, which make a municipality liable for damage caused by the negligence of any member of a volunteer fire company of such municipality while operating fire department equipment in going to, attending, or returning from a fire.

The Court properly gave binding instructions to the jury in favor of defendants and refused plaintiff's motion for judgment *n.o.v.*

The order appealed from is affirmed.

As you read the following case study, which also concerns volunteer fire department status, compare its decision with the decision in the preceding case study. Then, consider the following questions:

- Why are volunteer fire departments sometimes regarded as part of the municipal government?

- Why are volunteer fire departments sometimes not regarded as part of the municipal government?

SHINDLEDECKER v. BOROUGH OF NEW BETHLEHEM et al. and three other cases.

Superior Court of Pennsylvania.
June 30, 1941.
20 A.2d 867

Appeals Nos. 11-14, April term, 1941, from judgments of Court of Common Pleas at Nos. C-104, C-105, C-106, and C-107, August Term, 1939, Clarion County; Harry M. Rimer, President, Judge.

Proceedings under the Workmen's Compensation Act by W. Kenneth Shindledecker, Florence Marie Slaugenhoup, widow of William Paul Slaugenhoup, deceased, Robert F. Mateer, Jr., and George A. Cowan, respectively, to recover compensation for personal injuries and decedent's death, opposed by the Borough of New Bethlehem, employer, and the United States Fidelity & Guaranty Company, insurance carrier. From judgments on awards of compensation by the Workmen's Compensation Board, the borough, and by the insurance carrier appeal.

Reversed, and entry of judgments for defendants directed.

HIRT, Judge.

The question in these appeals is whether the members of a volunteer fire company were protected by the amending Act of May 14, 1925, P.L. 714, the legislation then in force, making a borough liable for compensation for accidental injury to members of its volunteer fire company within the definite limitations set forth in the act.

The facts are not in dispute. Decedent, the husband of Florence M. Slaugenhoup, and the other claimants were members of a volunteer fire company of the Borough of New Bethlehem. This company was associated with ten other like volunteer companies of neighboring towns and boroughs in an organization known as the Mutual Firemen's Association, the purpose of which was to further the interests and promote the efficiency of its constituent member companies. An annual convention was of importance in the association's program as a means of accomplishing its purpose and the borough of New Bethlehem was designated as the place for the convention of 1936. By the rules of the association, the sponsoring borough assumed the entire financial responsibility and in return was entitled to the net financial profits of the convention. To stimulate advance interest in the convention and to insure a large attendance and a corresponding financial profit, the New Bethlehem Company had directed its convention committee to visit each of the companies in the association. William Paul Slaugenhoup and other members of that committee, in the early morning of March 9, 1936, were returning from a meeting with the volunteer fire company at Emlenton, when the automobile in which they were riding crashed in the fog into another car. In the accident William Paul Slaugenhoup was killed and the other claimants, Robert F. Mateer, Jr. and George A. Cowan were injured. The convention held later in the Borough of New Bethlehem made a profit of $1,185.45 for the benefit of the volunteer company of the borough.

The referee denied compensation but the board reversed, on its substituted finding that these members were "then actually engaged as firemen." The lower court took the view that because they were engaged solely in company affairs at its direction at the time of the accident, these members in effect were statutory employees of the borough entitled to compensation under the act and accordingly entered judgments on the awards. We are unable to agree with the members of the board that the application of the act can be so extended, nor with the court as to the test to be applied.

Many kinds of gratuitous services are accepted by a borough from its citizens, among them the services of its volunteer firemen. Parades, exhibitions, conventions and the like, all, directly or indirectly, contribute to the efficiency of a volunteer fire company and to the benefit of the borough. But they who thus give of their service do not, thereby, become employees of the borough. Liability for compensation without negligence on the part of the borough can arise only by virtue of statutory enactment. There is no public policy of the Commonwealth as to compensation except as expressed in the compensation statutes. *DeFelice* v. *Jones & Laughlin Steel Corp.*, 137 Pa.Super. 191, 8 A.2d 465. Courts can sustain only such compensation awards as are within the scope of the liability so imposed by the legislature. *Versellesi* v. *Board of Elizabeth Tp. et al.*, 136 Pa.Super. 362, 7 A.2d 381. By the amending act of 1925 members of volunteer fire companies of boroughs were included within the definition of "employees" under the Compensation Law, not generally, but only "while actually engaged as firemen or while going to or returning from any fire which the fire companies of which they are members shall have attended." Injury, to be compensable, must have occurred within the restricted circumstances which alone give rise to the right under the act. Casual injuries incident to other activities of the volunteer fire company were not contemplated by the act, except to be excluded by necessary implication as risks which must continue to be assumed by the members.

There is evidence in subsequent amendments that the legislature itself so construed the act. The amendment of June 4, 1937, P.L. 1552, § 104(b), 77 P.S. 22, extended its protection to volunteer firemen, "while performing any other duties of such companies." But the legislature in 1939, 77 P.S. § 22a, repealed this provision and by re-enacting the 1925 act again imposed liability for injury within the same narrow limits applicable to these appeals.

The act has been liberally construed, as it must be, to accomplish the purpose of the legislation. *[Citations omitted.]* But even a most liberal construction cannot bring claimants within the benefits of the act. Its clear language necessarily excludes liability for compensation, as applied to these appeals, except to members "while actually engaged as firemen." One promoting a convention is not so acting and though these members undoubtedly were furthering the interests of their company and of the borough, there is no foundation in law for the awards in these cases.

Judgments reversed and directed to be entered for the defendants.

Because the relationship between a municipality and a volunteer company is purely voluntary, the municipality cannot compel the company to perform services and can decline to accept the services of such a company at any time. Members of a volunteer fire department cannot be called upon to perform duties unrelated to the extinguishment of fire or the preservation of property from damage by fire. However, because membership in such a fire department is purely voluntary and can be terminated by the members at any time, there are instances in which the department itself is not considered a volunteer department, and the resignation of all of the members of a department does not disband the department itself. Various statutory or charter provisions can modify the essential legal nature of a voluntary fire department and the legal position of its members.

TYPES OF FIRE PROTECTION ORGANIZATIONS

The following types of fire protection organizations, as described in the NFPA's *Fire Protection Handbook,* represent various legal frameworks in which members of the fire service might function:[23]

Types of Organizations

The types of public fire protection organizations in existence are widely varied. One of the most familiar types common in most large municipalities is the public *fire department* — a department of municipal government — with the head of the department directly responsible to the chief administrative officer of the municipality.

Less common is a *fire bureau* in a department of public safety. In this organization, the time of the public safety department head must be divided between several important functions, frequently including police service, and fire protection matters are thus likely to receive less adequate attention.

An increasingly important type of fire force is the *county fire department,* which has considerable acceptance in metropolitan areas. Under this organization numerous small suburban municipalities can enjoy the benefits of a large, profes-

----

[23]*Fire Protection Handbook,* 14th Ed., NFPA, Boston, 1976, p. 9-6.

sionally administered, public fire department, with its staff and service facilities, which, ordinarily, few small communities could individually afford. Frequently, this begins with a county fire prevention office and a fire communications system.

Another type of public fire service organization is the *fire district,* which is organized under special provisions of state or provincial law. It is, in effect, a separate unit of government, having its own governing body in the form of commissioners or trustees and commonly supported by a district tax levy. Usually, it is organized following a favorable vote of the property owners in the proposed district. The fire district may include portions of one or more townships or other governmental subdivisions. Its fire force is frequently termed a "fire department" although in many cases it is the only department operated by that unit of government.

A fifth common type of fire-protection authority is the *fire protection district,* which in some states is a legally established, tax-supported unit for the purpose of buying fire protection through contract from a nearby fire department or even from a voluntary fire association. The fire protection district is a means of giving fire protection service to rural and suburban areas by providing a source of extra income and special rural fire apparatus to some small municipal fire department which contracts to protect the area. This type of organization provides the equivalent of municipal fire protection in a rural district where it might be difficult to maintain an experienced and effective fire fighting force. On the other hand, the municipality generally retains priority in the use of personnel and equipment. Fire prevention functions may be neglected in fire protection district organizations except when persons in the district participate in a municipal fire prevention campaign, or when fire prevention is provided by a separate "fire marshals" organization operated by the state, county, or township.

An early type of public fire service organization is the volunteer fire company or association that raises its own funds by public activities and subscriptions, frequently with contributions of funds or equipment from interested units of government. There are many voluntary fire associations maintaining excellent equipment and stations, which also frequently serve as centers for various community activities. In many instances the membership prefers to retain its independence from government, especially when purchasing equipment, etc. Many volunteer organizations have a long and successful history that may antedate the municipal governments in their areas. Frequently, the activities of independent fire organizations are coordinated through special associations and governmental advisory boards. It is not uncommon for several volunteer companies to join in providing fire protection for a sizable municipality.

The volunteer fire department is the most common type of organization for small towns and rural areas. There are four main implementations of volunteer fire protection plans for small towns and rural areas:[24]

(1) Part paid and/or volunteer departments.

(2) Cooperative fire districts.

(3) Township or county fire departments.

(4) Agreement with the nearest governmental unit for fire protection services.

---

[24]Blair, *American Local Government,* p. 416.

FIRE DEPARTMENT ORGANIZATION

In most municipalities, fire departments are organized in much the same manner as police departments. Table 2.3 shows the typical chain of command in a fire department, and Figures 2.3, 2.4, and 2.5 show the typical organization of small, medium, and large fire departments. In the following excerpt from *American Local Government*, George S. Blair explains how the police and fire departments are combined in some areas:[25]

> In some areas, police and fire departments are combined into a bureau or department of public safety.
>
> While police and fire services still have some essential differences, the recent strong emphasis on prevention has drawn the two departments closer together. In a number of cities a recognition of this development has resulted in the establishment of a single department of public safety. One form of integration is administrative only, and within a single department of public safety headed by a lay commissioner will be a bureau of police and a bureau of fire headed by professional careermen. A second type of integration recognizes that some duties are so similar that particular duties are integrated. A policeman walking the beat will also check on fire hazards and men on patrol function will be on guard against the outbreak of both an act of violence or a fire. In communities with this system, the patrol car will often be a patrol wagon equipped with a fire extinguisher and other basic fire fighting equipment. A third type of integration results in positions known as public safety officers — men who are trained to perform both police and fire fighting functions.

Board of Fire Commissioners: In larger municipalities, a board of fire commissioners is empowered by constitution, statute, or charter to control the fire departments. This board may be considered in some instances as a branch of the municipal government or as a separate legal entity. In some instances, board members are not even required to be members of the fire department. The board of fire commissioners receives its power by law and often must act in the manner of a legal tribunal, *i.e.,* act as a body and speak by its records. Such boards may be given legislative power and broad discretion in matters pertaining to fire protection and prevention. A fire commissioner may be empowered with the duties of a building inspector and make and enforce fire protection regulations.

Chief or Superintendent: The chief or superintendent of a municipal fire department may be appointed by the mayor, the city council, the trustees of a village, or a board of fire commissioners. In some areas, the chief of the fire department may be appointed without reference to civil service rules and classifications and, if the appointee is in a civil service position, will not be considered as such during the period the appointee functions as chief. Even if the position of chief or superintendent of the fire department is placed in the civil service, open competitive examinations may be held in place of promotional examinations.

---

[25]*Ibid.,* p. 419.

Table 2.3. Typical Chain of Command in a Fire Department

Top Administrative Officer:	Commissioner, Director, Chief of Department
Executive Officer:	Chief of Department, Assistant Fire Chief, Deputy Fire Chief, Chief Fire Marshal
Administrative Division Heads:	Assistant Fire Chiefs, Deputy Fire Chiefs, Superintendents
Fire Fighting Divisions:	Deputy Fire Chiefs, Assistant Fire Chiefs, Division Fire Marshals
Fire Fighting Districts:	District Fire Chiefs, Battalion Fire Chiefs, Deputy Fire Chiefs, Battalion Fire Marshals, Assistant Fire Chiefs
Fire Companies:	Fire Captain
Company Platoons or Work Groups:	Captain, Junior Fire Captain, Fire Lieutenant, Fire Sergeant

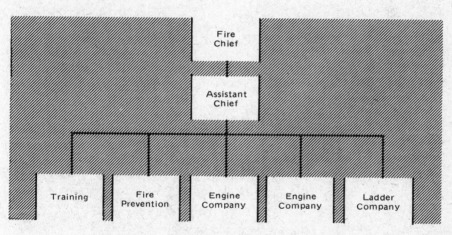

Fig. 2.3. Typical organizational structure of a small-sized fire department.

The powers and duties of a fire chief are prescribed by law but are not limited to the law as expressed. Many duties may be implied from the law resulting in broad discretionary power which the chief may use in the performance of duties related to fire protection and prevention. The typical chain of command from the fire chief on down is explained in the following excerpt from *Introduction to Municipal Government and Administration*, by Arthur W. Bromage:[26]

> The basic unit within the fire department is the company. At the top is the chief serving either as department head or as a subordinate of a commissioner or board. In a large city, the chain of command passes downward from the fire chief to an assistant chief to district chiefs, who control five or six stations. Within each station, there will be at least an engine company or a pumper-ladder company.

---

[26]Arthur W. Bromage, *Introduction to Municipal Government and Administration*, Appleton-Century-Crofts, Inc., New York, 1950, pp. 540-541.

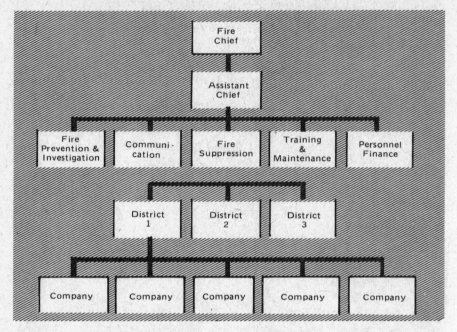

Fig. 2.4.   *Typical organizational structure of a medium-sized fire department.*

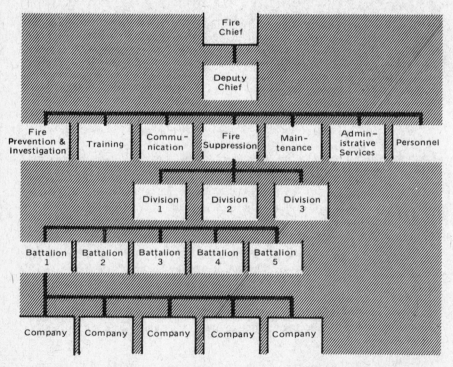

Fig. 2.5.   *Typical organizational structure of a large-sized fire department.*

Some of the stations will house both engine and ladder companies. The commanding officer of each station will report to the district chief. Smaller cities are not divided into districts with district chiefs. Where there are only five or six station houses, the chief and assistant chiefs can span the entire fire fighting force without an intermediate administrative level of district chiefs.

In addition to the assistant chief who supervises the fire fighting companies in the larger cities, other assistant chiefs head the bureau of fire prevention; the training school; and the fire alarm system. Provision must also be made for record keeping and maintenance of equipment. No single formula for administrative organization can be fixed, since fire departments range from volunteer companies with a few paid firemen to large city departments with thousands of firemen and no volunteers.

As in any administrative organization, the fire chief must command and coordinate the organization. He must hold conferences with his subordinates (assistant chiefs and district chiefs) to explain policies and programs. Any reorganization of the department should be taken only after adequate consultation and exploration of the advantages and disadvantages. Through his immediate subordinates, the chief must maintain a trained and alert organization. The chief who reports directly to mayor or manager has the responsibility for budget preparation, and for the presentation of the department's needs to the municipal executive. He has a major duty in interdepartmental cooperation with other city departments under policies set by the municipal executive.

A deputy fire chief is subordinate to the chief and, in the chief's absence, is vested with the chief's authority and power.

The assistant chief's jurisdiction is limited to district authority rather than city-wide authority. In some circumstances an assistant chief may perform the duties of a chief, but only with the powers of an acting chief rather than with the full powers that would accrue to a deputy chief in the same situation. The assistant chief generally discharges the duties of a captain at a station.

Fire Fighter: A fire fighter is generally defined as a person engaged or employed to fight fires. Whether a fire fighter is an officer or employee is often determined by the legislature. As previously stated, an officer is considered to be part of the governmental function and power of a municipality, while an employee does not have powers conferred by law as an officer does, but by contract.

Basically, the duties of fire fighters consist of following the instructions of superior officers in the department when these instructions relate to fire protection and prevention. A fire fighter is required to perform all duties in a reasonably careful manner, and can be held personally responsible for personal negligence. In *Principles of Fire Protection*, Percy Bugbee describes the responsibilities of fire fighters as follows:[27]

A fire fighter's main responsibility is to protect the public from the dangers of fire, to prevent loss of life from fire, and to minimize the property damage that

[27]Percy Bugbee, *Principles of Fire Protection*, NFPA, Boston, 1978, pp. 309-310.

fire can cause. This is accomplished through highly organized teamwork under the supervision of a commanding officer. Thus, the career of the fire fighter involves a group work activity: some fire fighters are assigned to forcible entry tasks, others to water apparatus, and others to the operation and placement of ladders. After the extinguishment of a fire, fire fighters remain on the scene to ensure that no further danger exists from rekindling. Therefore, while fire fighters may have specialized tasks on the fire scene, it is through the work of each individual that the entire unit accomplishes its task.

Between fires, fire fighters may be assigned to inspect public buildings for compliance with fire codes and safety regulations. During these inspections, fire fighters may apply their knowledge of firesafe buildings to ensure that exits, passageways, and hallways are not blocked, that combustibles are stored safely, and that any hazards or inoperable firesafety devices are corrected.

Fire Marshal: A fire marshal is not a fire fighter in the normally accepted sense of the word and, in some municipalities, is not considered as a member of the fire department. If the fire marshal's office is created by statute, and the fire marshal's duties and salary are fixed by law, then this office is considered a public office.

The duties of a fire marshal do not include the fighting and extinguishment of fires but duties of an administrative or investigative nature, such as investigating the sources and circumstances of fire.

A supervisor of a fire department can be employed by the board of fire commissioners to advise both the board and the fire chief on more efficient methods which can be employed in a fire department. In the following excerpt from *Introduction to Municipal Government and Administration*, Arthur Bromage emphasizes the importance of the fire marshal's functions as follows:[28]

In the field of fire prevention and investigation, an important organization at the state level is the fire marshal. The state fire marshal's activities may be classified under such categories as education of the public; issuance of regulations governing fire hazards; and investigation of fires of suspicious origin. Many states having state fire marshal laws authorize municipal fire department officials to conduct investigations either by making them assistant fire marshals, or by authorizing the state fire marshal to deputize them to act. One of the most important functions of the fire marshal is the establishment of uniform state regulations governing such hazards as explosives, dry cleaning plants, handling and storage of flammable liquids.

The following excerpt from Percy Bugbee's *Principles of Fire Protection* further explains the functions of state and provincial fire marshal offices as follows:[29]

Many states in the United States and all the provinces in Canada have a state or provincial fire marshal. The fire marshal's responsibilities vary in different states and provinces; however, the primary function of most fire marshal offices is to

---

[28]Bromage, *Introduction to Municipal Government and Administration*, p. 553.
[29]Bugbee, *Principles of Fire Protection*, p. 308.

sponsor and promote fire prevention programs and to investigate fires, particularly for arson.

Most state fire marshals receive their authority from the state legislature, and thus work under the auspices of the governor, a high state officer, or a commission created to oversee the operation of the office of the state fire marshal. Few state fire marshal offices operate as agencies separate from other departments of the state; most are a division of the state insurance department, state police, state building department, state commerce division, or other state agency.

SUMMARY

Business corporations and municipal corporations are similar in many respects. One of the major differences between the two is the assessment of success. For example, where a business corporation is measured by the yardstick of profits made and dividends paid, the municipal corporation's measure of success is judged by the quality of the services rendered.

Municipal corporations are subject to and subordinate to the state that created them. Thus, they can be considered as agents of the state, although the chief purpose and function of a municipal corporation is local government — which gives it its corporate nature.

A municipal corporation possesses only such power as the state grants it. The power of a municipal corporation is not separate from the state, but flows from it. Therefore, although municipal corporations have power of local self-government as granted by the state, they are also part of the governmental power of the state and, as such, are subject to any restrictions the state may impose.

In recent years, the rapid growth of new and varying lifestyles and their resultant needs as well as the tremendous expansion of the "business of government" has led to a gradual erosion of the distinctions between municipal corporations and quasi-municipal corporations. However, these distinctions still have bearing in court decisions as well as local-government unit decisions and powers.

Most questions of municipal liability are concerned with the area of contracts or torts, particularly actions of negligence in the field of torts. Many factors are involved in the question of municipal liability, *i.e.*: Does the question of liability involve the municipal corporation in a governmental or private and corporate capacity? Were the actions performed ones in which the municipality was responsible for the actions of its employees?

Although the doctrine of sovereign immunity is no longer recognized by the federal government and by many state governments, the notion of sovereign immunity in governmental matters is slow to disappear because it has been so firmly entrenched in the law for so many centuries. Variations on the doctrine of sovereign immunity are now mandated by legislative statute and vary from state to state. Therefore, it is important for members of the fire service to be fully aware of the statutory requirements of the states they are affiliated with.

In many areas of the United States, fire protection comes under the form of local government known as special districts. The importance of fire protection to all types of communities — cities, towns, and rural areas — may be summized from the fact that, according to the U.S. Bureau of the Census, there were over 3,000 special fire protection districts in 1962.

Fire departments must adhere to the fire prevention codes and rules that municipalities set up in order to provide adequate fire protection. Such rules and regulations cover fire department members' lives inside and, to some extent, outside the fire station. However, the rules governing actions outside the fire station are only those that courts have not deemed unreasonable.

ACTIVITIES

1. In a written description, compare a private corporation with a municipal corporation by explaining the difference between the two. Be sure to include the five powers of a corporation in your comparison.
2. Explain how a municipal corporation differs from a quasi-municipal corporation.
3. Describe your community's fire service organization with relation to its corporate structure or affiliation.
4. Explain the dual nature of municipal corporations.
5. Describe the factors that can affect municipal liability. Include examples in your description.
6. Describe the status of municipal officers, agents, and employees, and tell how this status affects your community's fire service organization.
7. Describe the source of the governmental power structure in your community's fire service organization.
8. Discuss your community's type of fire protection organization from a legal point of view.
9. Compare the organization of your community's fire department with that of a municipal corporation as described in this chapter.
10. (a) Briefly describe the distinction between a city, a village, a borough, a town, and a county.
    (b) How does each of these affect a fire department located within its boundaries?

BIBLIOGRAPHY

Anderson, William and Weidner, Edward W., *American City Government*, Rev. Ed., Henry Holt and Company, New York, 1950.

Blair, George S., *American Local Government*, Harper & Row, Publishers, New York, 1964.

Bollens, John C., *Special District Governments in the United States*, University of

California Press, Berkeley and Los Angeles, CA, 1957.

Colman, William G., *Cities, Suburbs, and States,* The Free Press, A Division of Macmillan Publishing Co., Inc., New York, 1975.

Kneier, Charles M., *City Government in the United States,* Rev. Ed., Harper and Brothers, New York, 1947.

MacCorkle, Stuart A., *American Municipal Government and Administration,* D. C. Heath and Company, Boston, 1948.

Macdonald, Austin F., *American City Government and Administration,* Thomas Y. Crowell Company, New York, 1951.

Robertson, James C., *Introduction to Fire Prevention,* Glencoe Fire Science Series, Glencoe Press, Beverly Hills, CA, 1975.

Zink, Harold, *Government of Cities in the United States,* The Macmillan Company, New York, 1948.

For detailed in-depth study of the topics presented in this chapter, the following is recommended:

Dillon, J. F., *Commentaries on the Law of Municipal Corporations,* 5th Ed., Little Brown & Company, Boston, 1911.

III

THE LAW OF TORTS

OVERVIEW

A tort is the violation of the personal or private rights of another. It is the legal duty of all to respect and honor those personal and private rights that are protected by law. A breach of this legal duty to respect the rights of others is called a tort. This chapter presents those aspects of tort law that are of greatest concern to members of the fire service in the performance of their duties. Municipal corporations that strictly adhere to the doctrine of sovereign immunity can still be sued in their private capacities as corporations. Because many states have eliminated the doctrine of sovereign immunity — either entirely, or to a limited extent — more and more complaints and suits are being initiated against state and local governments and their agents and employees. The vast majority of these suits are civil actions with regard to contract or tort. Even the federal government acknowledges its possible liability in tort, and consents to be sued under the provisions of the Federal Torts Claims Act.

THE LAW OF TORTS

The law of torts is a relatively "young" branch of law. As practiced today, formal tort law has existed for only about one hundred years. In this brief period of time, tort law has become a major branch of the law.

In addition, the global aspects of human relationships and rights with which the law of torts is concerned has touched and affected the lives of virtually every citizen in America. Basically, the law of torts is concerned with the personal and private rights of individuals. In *Handbook of the Law of Torts,* noted authority on tort law William L. Prosser describes some of the personal and private concerns of individuals to which the law of torts might be addressed:[1]

> . . . Men wish to be secure in their persons against harm and interferences, not only as to their physical integrity, but as to their freedom to move about and their peace of mind. They want food and clothing, homes and land and goods, money, automobiles and entertainment, and they want to be secure and free from disturbance in the right to have these things, or to acquire them if they can. They want freedom to work and deal with others, and protection against interference with their private lives, their family relations, and their honor and reputation. They are concerned with freedom of thought and action, with opportunities for economic gain, and with pleasant and advantageous relations with their fellow men.

WHAT IS A TORT?

The word "tort" comes originally from the Latin word "tortus," meaning "twisted," from the French word meaning "injury" or "wrong." A tort is harm or injury inflicted in violation of a duty imposed by law with regard to the person or property of another.

Everyone has a right to personal safety and personal liberty. Also, all people have a right to use and enjoy their personal property without interference by others. The law protects rights such as these, and imposes a legal duty on all to respect these rights. If any of these rights are violated, one party can sue another party in tort.

There is a further aspect to the law of torts: in addition to protecting the legal rights of others, it attempts to provide relief if a personal or private right has been violated. Therefore, in addition to the violation of a person's personal rights, some damage, injury, or harm must result from this violation. Recovery or relief cannot be provided if some form of harm or injury has not resulted.

BACKGROUND OF TORT LAW

The law of torts is of relatively recent origin because the infinite variety of the possible kinds of torts and the considerations of society in general (and moral overtones in particular) tended to make the field of tort law too cumbersome to harness as a branch of the law. Courts historically did not wish to be forced to determine a person's state of mind with regard to intent and malice believing that, in the words of Chief Justice Bryan, "The thought of man shall not be tried, for the devil himself knoweth not the thought of man."

The courts were (and are) reluctant and unwilling to pass "moral judgments" on parties to a suit. However, in the interests of keeping the peace between individuals by providing an alternative to private vengeance, courts became increas-

---

[1]William L. Prosser, *Handbook of the Law of Torts*, 3rd Ed., West Publishing Co., St. Paul, MN, 1964, p. 15.

ingly involved in tort litigation and, as a result, in setting general social goals. Prosser's *Handbook of the Law of Torts* states:[2]

> On the other hand, there are still many immoral acts which do not amount to torts, and the law has not yet enacted the golden rule. It is impossible to afford a lawsuit for every deed of unkindness or betrayal, and there is much evil in the world which must necessarily be left to other agencies of social control. The basest ingratitude is not a tort, nor is a cruel refusal of kindness or courtesy, or a denial of aid. The rich man is under no compulsion to feed his starving neighbor, and it is still the law that the owner of a boat who sees another drowning before his eyes may rest on his oars and let him drown — although perhaps in so extreme a case it is a reproach to the law that it is so. Petty insults, threats, abuse, and lacerated feelings must be endured in a society not many centuries removed from the law of the club. To what extent the moral ideas of a future day may yet create new torts to deal with such misconduct, it is now impossible to say.

Despite its slow beginnings, the branch of law concerned with torts is now firmly established as a major branch of the law.

TORT LAW AND CRIMINAL LAW

A tort is a breach of civil law as opposed to a breach of criminal law, in that a tort is considered a private wrong. As such, the injured party can sue for redress and be compensated for the injury or injuries suffered at the expense of the wrongdoer. A crime, on the other hand, is a public wrong. Violation of a criminal law is not considered a violation against an individual, but a violation against both the community and public authority.

Criminal law is considered a breach of law against the public welfare and, as such, is redressed by the state in criminal proceedings. Although some states and local governments are considering providing statutory compensation to victims of crime, the victim of a crime cannot collect damages from the wrongdoer in a criminal action. However, because the same act may be both a criminal and a tortious act, both a civil tort action and a criminal prosecution may result, as further explained by Prosser:[3]

> A tort is not the same thing as a crime, although the two sometimes have features in common. The distinction between them lies in the interests affected and the remedy afforded by the law. A crime is an offense against the public at large, for which the state, as the representative of the public, will bring proceedings in the form of a criminal prosecution. The purpose of such a proceeding is to protect and vindicate the interests of the public as a whole, by punishing the offender or eliminating him from society, either permanently or for a limited time, by reforming him or teaching him not to repeat the offense, and by deterring others from imitating him. A criminal prosecution is not concerned in any way with compensation of the injured individual against whom the crime is committed, and his only part in it is that of an accuser and a witness for the state. So

---

[2]*Ibid.*, p. 18.
[3]*Ibid.*, p. 7.

far as the criminal law is concerned, he will leave the courtroom empty-handed.

The civil action for a tort, on the other hand, is commenced and maintained by the injured person himself, and its purpose is to compensate him for the damage he suffered, at the expense of the wrongdoer. If he is successful, he receives a judgment for a sum of money, which he may enforce by collecting it from the defendant. The state never can sue in tort in its political or governmental capacity, although as the owner of property it may resort to the same tort actions as any individual proprietor to recover for injuries to the property, or to recover the property itself. It has been held, . . . that the state as a government has no cause of action against an escaped convict for the expenses incurred in recapturing him.

TORT LAW AND CONTRACT LAW

Where the concerns of tort law versus criminal law are those of personal harm or injury, the distinctions between tort law and contract law are centered on the concept of duty. Generally, the concept of duty in tort is created by law. The concept of duty in contract is created by the terms of the contract itself.

Breaches of contract are usually not considered torts. There has been, however, some confusion on the part of the courts in maintaining this distinction at civil law. Generally, it is considered that an action in tort cannot arise out of the terms of the contract itself (a breach of contract), but can arise out of a violation of the *relationship* established by the contract, as explained in the following excerpt from the case of *Montgomery Ward & Co.* v. *Scharrenbeck:*[4]

A party to contract owes a common-law duty to perform with care, skill, and reasonable expedience and faithfulness the thing agreed to be done, and a negligent failure to observe such duty is a "tort," as well as a breach of contract.

ELEMENTS OF A TORT

As has been previously stated, there are many personal injustices that do not amount to torts. Certain elements must be present to comprise an actionable tort, that is, a breach of legal duty for which one party can sue another. Each of these elements must be able to be proved and although each element can overlap with the other, all elements must be present. The following definition includes the elements necessary for an action in tort:

- A tort is the *violation* of a *legal duty* by one party to another with a *resultant injury, damage,* or *harm* that can be *remedied at law.*

LEGAL DUTY

A major element of a tort is the fact that a legal duty of one party to another must be present. This duty can be prescribed by statute, state and/or federal constitutions, or by common law.

---

[4]*Montgomery Ward & Co.* v. *Scharrenbeck,* 204 S.W.2d 508, 510, 146 Tex. 153.

A violation of legal duty in tort is founded in a disregard of the legal rights of others. In order for a violation to be shown, it must be proven that a legal duty exists. It must be noted that a party generally has no legal duty to *benefit* another party, but does have a legal duty not to interfere with the rights of others; *i.e.*, there can be no "harmful interference with the interests of others." For members of the fire service, responding to emergency situations does not involve suspending the legal duty of respecting the rights of others, even other fire service personnel responding to the same emergency, as the decision in the following case points out.

HORSHAM FIRE COMPANY NO. 1
v.
FORT WASHINGTON FIRE COMPANY
NO. 1, Appellant.

Supreme Court of Pennsylvania.
Jan. 3, 1956.
119 A.2d 71

Action for damages resulting from intersectional collision between fire trucks of two different volunteer fire companies which were responding to same fire alarm when driver of defendant company's truck drove through stop sign at admittedly dangerous blind intersection at speed of about sixty miles per hour. Following verdict for plaintiff company, defendant company moved for judgment notwithstanding verdict and new trial. The Court of Common Pleas of Montgomery County, as of June Term, 1952, No. 433, E. Arnold Forrest, J., refused judgment notwithstanding verdict and granted new trial, and defendant company appealed. The Supreme Court, No. 170, January Term, 1955, Musmanno, J., held that whether conduct of driver of defendant company's truck constituted reckless disregard of rights of others on the highway was question for jury, even though defendant company, as well as plaintiff company, had certain privileges under Motor Vehicle Code.

Affirmed.

Before STEARNE, JONES, BELL, MUSMANNO and ARNOLD, JJ.

MUSMANNO, Justice.

This case is unique in that it has to do with two fire companies which responded to an alarm to fight the common enemy and ended up by fighting each other. On October 9, 1951, the fire truck of the Horsham Fire Company collided with the fire truck of the Fort Washington Fire Company at the intersection of Welsh Road and Butler Pike in Montgomery County. The Horsham Fire Company sued the Fort Washington Fire Company in trespass and recovered a verdict of $10,331. . . .

. . . It is to be observed at the outset that the owners of emergency vehicles such as ambulances, police wagons, and fire department vehicles, despite privileges allowed them under the Motor Vehicle Code, 75 P.S. § 1, *et seq.* are responsible in damages for reckless disregard of the rights of others. *[Citations omitted.]*

In the case at bar the driver of the defendant Fort Washington fire truck drove through a stop sign on Butler Pike at a speed of from 55 to 60 miles per hour into an intersection which he knew to be a dangerous one, aware that drivers approaching from his right on Welsh Road could not see him because of a field of standing corn, as indeed it also shut off his vision of traffic coming from the east on that thoroughfare. It became a question for the jury to determine whether this conduct did not constitute a reckless disregard of the rights of others on the highway, even though the defendant truck was responding to a fire alarm and even though it had certain privileges under the Motor Vehicle Code.

The driver of the plaintiff Horsham fire truck, although also entitled to the same privileges enjoyed by the Ft. Washington fire truck, was circumspect in his approach to the perilous intersection. While not ignoring the demands for reasonable dispatch in the fulfillment of his appointed mission to reach the conflagration to which his fire company had been summoned, he adjusted his movement to the circumstances which confronted him. Taking heed of a bus which had stopped on Welsh Road he decelerated his speed to 20 to 25 miles per hour and swung around the bus as he approached Butler Pike. Aware of the Stop sign on Butler Pike, which should halt all traffic before crossing Welsh Road, and knowing that he was approaching the intersection from the right, he had reason to assume that the crossing would be clear for him to pass. He could not see the approaching defendant truck because of the field of unharvested corn which screened the lateral view of Butler Pike from his vision, nor could he hear the siren and bell of the defendant's truck because of the din being made by his own warning, noise-making devices as he prudently hurried on his errand. Suddenly a "big red flash and a blur" loomed before him, and the crash followed. The speed of the defendant truck carried it 81 feet beyond the point of the impact, the plaintiff truck drifted 31 feet before it turned over.

If the vehicles involved in this accident had been ordinary pleasure cars or business trucks, it is perhaps unlikely that the defendant would be seeking judgment *n.o.v.* since the issue of negligence, which was not a complicated one, was properly submitted to the jury which could find, as it undoubtedly did, that the defendant vehicle proceeded into the intersection at an excessive rate of speed after ignoring stop signs. It is urged, however, on behalf of the defendant, that this encounter was not the usual traffic accident since the defendant vehicle was an apparatus enjoying certain immunities under the Motor Vehicle Code.

It is true that fire department cars are not bound by certain prohibitions in the code. Sec. 501(f) of the Motor Vehicle Code, 75 P.S. § 501, provides that "The speed limitation set forth in this section [50 miles per hour] shall not apply to vehicles, when operated with due regard for safety . . . to fire department . . . vehicles when traveling in response to a fire alarm.''

Section 1016(d) of the code, 75 P.S. § 591(d), declares that this section [requiring stoppage at '' 'Thru Traffic Stop' '' signs] ''shall not apply to vehicles, when operated with due regard for safety . . . to fire department or fire patrol vehicles responding to a fire alarm.''

Section 1013 of the Motor Vehicle Code, 75 P.S. § 572, directs that when two vehicles approach an intersection at approximately the same time, ''the driver of the vehicle . . . on the left, shall yield the right of way to the vehicle . . . on the right.'' Section 1014(b), 75 P.S. § 573, however, provides an exception to this rule, namely,

''The driver of a vehicle upon a highway shall yield the right of way to . . . fire department vehicles.''

In view of these legalized immunities in behalf of fire department vehicles from requirements laid down for the ordinary passenger and commercial vehicles by the Code, counsel for the appellant contends that his client cannot be held liable in damages for doing what the law permits. But it is to be noted in this connection that while the law — in view of the vital and urgent missions of fire department vehicles — wisely allows them certain privileges over other vehicles, it does not assign to them absolute dominion of the road. The appellant's fire truck was engaged in a praiseworthy enterprise: it was on its way to extinguish a conflagration, it had the right to proceed through red lights, to surpass speed limits, and to take the right of way over ordinary vehicles. No words can be too laudatory in extolling the merit and sacrifice of members of volunteer fire companies who expend time, money, energies and often health and impairment of body without any recompense except the satisfaction of serving one's community and one's fellow man. The defendant here is a volunteer fire company, but it is to be kept in mind that the plaintiff is *also* a volunteer fire company. Its fire truck was also on its way to a fire, in fact, the same fire as that which cried for the services of the defendant company. The plaintiff company thus also had the right to discard speed limitations and to pass up stop signs.

It is, of course, regrettable that the vehicles of these estimable organizations should have met in collision. It is unfortunate that the very attribute which sounds the highest praise for fire companies, namely, speed, should have been the very demon which brought about their undoing. The need for celerity of movement on the part of fire engines and fire trucks is not only traditional but objectively demonstrable. The sooner the fire-extinguishing apparatus arrives at the conflagration, the less chance there is that human lives and valuable property will be lost in the all-consuming blaze. But speed which is uncontrolled, is capable of wreaking as much havoc and causing as much sorrow as fire itself.

What happens when two vehicles with equally assumed privileges insist on the right of way? The answer was dramatically provided in this case: collision, destruction, and disaster. It was providential that, considering the massive weight and size of the two trucks, no one was killed or injured. The most poignant type of regret is that which is based on a misfortune which could have been avoided, and it requires no sapience to conclude that when the paths of two fire engines cross, one of them must stop. This is not only a matter of common sense, but a proposition of imperious self-preservation.

The authors of the Motor Vehicle Code, in providing exemptions for fire vehicles, police cars, and ambulances, obviously intended those exemptions to operate as against ordinary nonprivileged vehicles, and not as against each other. Certainly they did not plan to set up at street crossings an arena of combat between police motorcycles and ambulances, hook-and-ladder vehicles and police wagons, between fire trucks and fire engines. Where parity is involved, elemental judgment dictates that normal rules must apply. Two police cars as against each other enjoy no privileges under the code. Their common purpose wipes out priority. If two police cars chasing a bandit arrive at an intersection precisely at the same moment, and communication between the two cannot decide sequence of passage, the drivers should accept the normal rule of the road, and allow the one approaching from the right to take priority — if the bandit is not to escape because of self-entanglement of the pursuers. If two fire trucks arrive simultaneously at intersecting streets, one being confronted with a stop sign and the other with a through highway beacon, the fire truck facing the stop sign should stop to permit the other to proceed. It would be absurd for both vehicles to insist on precedence because they happen to enjoy similar privileges under the code. As a matter of fact, when one studies the whole blueprint of the code, it will be seen that the privileges are not actually similar. Circumstance can always befog a picture so that movement must depend on current adjustment to what is best for all travelers concerned. If insistence is to ignore facts and impetuosity is to plunge forward blindly, the resulting conflict will be injurious not only to vehicles and drivers, but to the public welfare in addition.

Appellant's counsel complains in his brief that to sustain the lower court would mean that the driver of every fire truck "must be prepared to slow down at every intersection to yield the right of way to some other hypothetical fire department vehicle that might just possibly be approaching from his right." But the Horsham fire truck was not a hypothetical fire department vehicle. There is nothing on the road that can be less hypothetical than a fire truck. Its massive volume and elongated dimensions, its numerous chariot wheels, ponderous equipment, fiery color, screaming siren, clanging bells and flashing warning lights make it as conspicuous as a herd of trumpeting elephants on a rampage.

Maxon, the defendant truck's driver, knew that Welsh Road and Butler Pike formed a dangerous crossway. He testified that he saw one car proceeding east on

Welsh Road before he arrived there. Despite defendant's counsel's argument that it would be impracticable for the driver of a fire truck to be prepared to slow down at an intersection, Maxon was actually slowing down to stop at Welsh Road:

"Q. You say you were slowing down? A. I was slowing down, to stop.

"Q. To what? A. To stop.

"Q. To come to a stop? A. That is correct.

"Q. For what reason were you going to bring your truck to a stop? A. I have always figured no fire is worth getting killed over."

Maxon knew this was a dangerous intersection:

"Q. Well, just tell us what was your intention or your purpose in stopping your truck at the Welsh Road intersection, or intending to stop it? A. Having lived in that section for many years, I know it is a very dangerous intersection. I wouldn't normally go through it without stopping.

"Q. Well, of course 'normally' is a word of some significance. Was this a normal or an abnormal situation? A. To me, it was a normal situation. I would have stopped.

"Q. Normal for a fire truck? A. Regardless."

What caused Maxon finally not to stop was the waving of a passenger on the bus which Maxon interpreted as an invitation for him to proceed. But this ambiguous hand waving was a very precarious assurance on which to move into a recognized danger. Maxon did not know the passenger, the passenger was inside the bus and could not be aware of traffic conditions on all sides of the bus. The jury could well have concluded that Maxon's statement that he relied on so vague and unreliable a signal to hurl his 18,000-pound fire truck into an admittedly perilous situation was of itself enough to constitute the recklessness which would make the defendant liable in damages for what happened.

The Pennsylvania cases cited by appellant's counsel are not applicable to the facts at bar because in all those cases the collision occurred between an emergency vehicle and a nonemergency vehicle. Here we have an entirely novel situation where both vehicles stand in a favored class, and since neither can have an advantage over the other, they must in effect lose their priority rights and be treated as normal vehicles, insofar as rights against each other are concerned. It would be quite strange to hold, as appellant's counsel argues that if two fire trucks "meet at an intersection they both have an equal right to be there, and if they happen to collide that is simply one of the risks involved in getting to a fire as fast as possible." The law is not so fatalistic as that. The object of a fire truck's journey is not merely to make a show of rushing to a fire, but actually to get there. If the driver is to ignore all elements of safety driving at breakneck speed through obviously imperilling hazards, he may not only kill others en route, but he may frustrate the whole object of the mission and not get there at all! . . .

. . . Since we discover no abuse of discretion in the awarding of a new trial, that order will not be disturbed.

Affirmed.

STERN, C. J., and CHIDSEY, J., absent.

In the previous case, Justice Musmanno distinguishes cases of collision between two emergency vehicles and between an emergency vehicle and a nonemergency vehicle. However, as stated in the decision of the following case, the legal duty to respect the rights of others remains the same.

H. C. JOHNSON, Appellant,
v.
Oren BROWN, Respondent.
No. 4187.

Supreme Court of Nevada.
Oct. 29, 1959.
345 P.2d 754

Action against fireman, who drove fire engine, for injuries sustained by truck passenger when truck was struck by fire engine. The Second Judicial District Court, Washoe County, Clel Georgetta, J., Dept. No. 3, rendered judgment for passenger, and fireman appealed. The Supreme Court, McNamee, C. J., held that where fireman driving fire engine in response to emergency call proceeded through stop sign on heavily traveled through street at a speed in excess of speed limit, substantially cut corner and, while on wrong side of street, collided with truck, fireman was not driving with due regard for safety of all persons using street, and in absence of contributory negligence he was liable to guest passenger of truck for injuries sustained.

Affirmed.

McNAMEE, Chief Justice.

Respondent was a guest passenger in a truck as it was being driven north on Chestnut Street (now Arlington Avenue) in the City of Reno toward the intersection of that street and West Commercial Row. At this time appellant, a fireman for the city, was driving a fire engine in response to an emergency call proceeding west on West Commercial Row. After rounding the corner and while driving in excess of the statutory speed limit the fire engine came in contact with the truck on Chestnut Street just a few feet south of the intersection, causing to respondent the injuries complained of. Evidence showed that at the time of the collision, the truck driver was wholly within his own lane of traffic and was applying his brakes.

The jury's verdict was for respondent, and appeal is from the judgment based thereon. Three errors are assigned for our consideration.

1. Appellant maintains that the trial court erred in giving to the jury the following instruction relative to the degree of care required of the driver of an emergency vehicle while responding to an emergency call.

"It is the law of this state that a fire engine, responding to an emergency call and displaying a red light visible from the front, and sounding a siren, is exempt from and shall not be required to observe certain laws that generally apply to the drivers of vehicles on the public streets.

"This means that the driver of a fire engine, under such circumstances, need not observe laws regarding speed, the use of different lanes of the street, rights of way, stop signs, and turns at intersections. This exemption, however, will not relieve the driver of an emergency vehicle from the duty to drive with due regard for the safety of all persons using the street, nor shall it protect the driver of any such vehicle from the consequence of the reckless disregard of the safety of others."

This instruction merely restates the then existing law* pertinent to the rights and duties of drivers of emergency vehicles.

There is sufficient evidence in the record to show that appellant was entitled to those exemptions afforded by law to operators of emergency vehicles as are enumerated in said instruction.

It is appellant's contention that the quoted instruction contradicts itself in that while properly charging the jury that the fire engine was exempt from observing certain rules of the road it went on to state ". . . This exemption, however, will not relieve the driver of an emergency vehicle from the duty to drive with due regard for the safety of all persons using the street. . . ."

. . . Appellant argues that the requirement that he drive "with due regard for

---

*Nev. Stat. 1955, c. 118, § 1; NCL 1943-1949 Supp., § 4350.3; Ord. City of Reno, Sec. 10-60, Sec.10-74, Sec. 10-75.

the safety of others'' is met by his compliance with the conditions that entitle him to the exemptions.

There is substantial authority to sustain this view of appellant.

In the case of *Lucas* v. *City of Los Angeles,* 10 Cal.2d 476, 75 P.2d 599, 601, where a statute similar to the Reno ordinance provided that an authorized emergency vehicle is exempt by law from complying with specified statutes and ordinances regulating the operation of vehicles on public roads, the court held that if such vehicle is responding to an emergency call, is displaying a visible red light from the front, and is sounding a siren, the driver thereof cannot be negligent in violating such regulations in the absence of reckless disregard of the safety of others. With respect to that part of the statute which states that the provisions thereof "shall not, however, relieve the driver . . . from the duty to drive with due regard for the safety of all persons using the highway," the court said:

"A simple analysis of these statutory provisions discloses the clear intention of the Legislature to recognize the paramount necessity of providing a clear and speedy pathway for such vehicles when actually confronted with the emergency in which the entire public may be assumed to be concerned. The expression 'with due regard for the safety of all persons using the highway' was explained in the Balthasar Case [*Balthasar* v. *Pacific Electric R. Co.*], where the court said, 187 Cal. 302, at page 311, 202 P. 37, 41, 19 A.L.R. 452: 'It is evident that the right-of-way of fire apparatus over other vehicles is dependent upon "due regard to the safety of the public" only insofar as such "due regard" affects the person required to yield the right-of-way. Notice to the person required to yield the right-of-way is essential, and a reasonable opportunity to stop or otherwise yield the right-of-way necessary in order to charge a person with the obligation fixed by law to give precedence to the fire apparatus.' This is the only reasonable interpretation that the statute will bear. If the driver of an emergency vehicle is at all times required to drive with due regard for the safety of the public as all other drivers are required to do, then all the provisions of these statutes relating to emergency vehicles become meaningless and no privileges are granted to them. But if his 'due regard' for the safety of others means that he should, by suitable warning, give others a reasonable opportunity to yield the right-of-way, the statutes become workable for the purposes intended. . . .

". . . Our conclusions from the foregoing are that when the operator of an emergency vehicle responding to an emergency call gives the statutory notice of his approach the employer is not liable for injuries to another, unless the operator has made an arbitrary exercise of these privileges. In such cases speed, right-of-way, and all other 'rules of the road' are out of the picture; the only questions of fact, insofar as the public owner is concerned, are first, whether there was an emergency call within the terms of the statute; second, whether the statutory warning was given; and third, whether there was an arbitrary exercise of these privileges.

Here the emergency was conceded, the sounding of the siren was proved by the only substantial evidence offered, and an arbitrary exercise of the privileges has not been shown. . . .''

. . . It is clear to us that the majority and better rule is in opposition to the California rule as expressed in the *Lucas* case, *supra,* and requires the driver of an emergency vehicle answering an emergency call to exercise reasonable precautions against the extraordinary dangers of the situation which duty compels him to create, and he must keep in mind the speed at which his vehicle is traveling and the probable consequences of his disregard of traffic signals and other rules of the road.

In the case of *Montalto* v. *Fond Du Lac County,* 272 Wis. 552, 76 N.W.2d 279, 282, the Wisconsin Supreme Court construed its statute which contained a provision exempting authorized emergency vehicles from certain rules of the road and which stated that ''this provision shall not relieve the operator of an authorized . . . vehicle from the duty to operate with due regard for the safety of all persons using the highway.'' After noting the construction placed on such a statute by the California Courts, it opined as follows:

''Appellants argue that the same construction should be given to sec. 85.40(5), Stats. We cannot agree. We deem the better rule to be that expressed in the following cases:

'' 'The right-of-way given to public service vehicles and their exemption from traffic regulations, however, do not relieve their operators from the duty of exercising due care to prevent injury to themselves and others lawfully upon the ways. . . . while they have a right to assume in the first instance that the operators of other vehicles will respect their right-of-way at an intersection, they are warned by a red light flashing against them that other vehicles on the intersecting way are invited to proceed by a green light and may do so. Even if the driver of the other vehicle through negligence disregards their right-of-way, they must still use due care to avoid a collision. The measure of their responsibility is due care under all the circumstances.' *(Citing cases.) Russell* v. *Nadeau,* 1943, 139 Me. 286, 29 A.2d 916, 917.

'' 'As we understand appellant's position, it contends that since its ambulance was on an emergency call, the issue of negligence in the rate of speed could not arise against it. This contention is not sound. The law simply exempts it from the arbitrary speed of 20 miles per hour, but it was still under the duty of exercising ordinary care, and in the rate of speed the court convicted it of failing to exercise ordinary care.' *Grammier-Dismukes Co.* v. *Payton,* Tex.Civ.App., 1929, 22 S.W.2d 544, 546.

''In our opinion, the giving of visible and audible warnings may or may not afford a reasonable opportunity to others to yield the right-of-way, depending upon the particular circumstances that are present. And the failure to afford that opportunity may be ordinary negligence or reckless disregard, depending on those circumstances.

"To adopt the view of the appellants would mean that a lack of 'due regard' would have to amount to a 'reckless disregard' before an ambulance driver could be held negligent as to speed. That the legislature had no such intention is clear from the fact that sec. 85.40(5), Stats. both requires that a driver operate with due regard for the safety of others and prohibits the exercise of his privilege with a reckless disregard for their safety."

Similarly the Maryland Court in *City of Baltimore* v. *Fire Insurance Salvage Corps,* 219 Md. 75, 148 A.2d 444, 447, refused to follow the California rule, and said:

"The appellant earnestly urges us to adopt what may be called the California rule. In interpreting statutes somewhat similar to ours, that State has held that if the audible signal be given, 'speed, right-of-way, and all other "rules of the road" are out of the picture,' and the driver cannot be held responsible for ordinary negligence, but only for 'an arbitrary exercise of these privileges' (the words of the California statute corresponding to 'reckless disregard of the safety of others' in ours). *[Citations omitted.]* However, we think the correct rule is that adopted by the Supreme Court of Wisconsin in *Montalto* v. *Fond Du Lac County,* 1956, 272 Wis. 552, 76 N.W.2d 279, 283. . . . We, therefore, hold that under a proper construction of Section 214, the provision that requires the operator of an authorized emergency vehicle to do so 'with due regard for the safety of all persons using the street' renders him liable for ordinary negligence, namely, a failure to exercise reasonable care and diligence under the circumstances."

In *City of Kalamazoo* v. *Priest,* 331 Mich. 43, 49 N.W.2d 52,54, the court said:

"It will be noted that the quoted statute, according emergency vehicles the right-of-way, conditions the same upon the sounding of an audible signal by siren, *etc.* In addition, the statute expressly provides that the driver thereof shall not be relieved from the duty to drive with due regard for the safety of others. The statute exempting such driver from speed limits contains a like provision concerning due regard for the safety of others. Plaintiff cites cases (all California cases, the primary being *Lucas, supra*) for the proposition that the statutory requirement of due regard for the safety of others is met by the giving of suitable warning. Had such been the legislative intent in the enactment of the Michigan statute, which expressly requires the giving of an audible warning as a condition precedent to an emergency vehicle's acquiring the right-of-way, no purpose would have been served by the further express requirement of the statute that such vehicle be driven with due regard for the safety of others. . . .

". . . Driving a fire truck into an intersection in full reliance upon the right to exceed speed limits and the right to proceed without stopping for the stop sign or the through street, but without observing or giving any heed to oncoming traffic on the intersecting through street did not amount to driving with due regard for the safety of others as required by the statute. Such driving in reliance upon a statutory right-of-way has frequently been held to constitute contributory negligence as a matter of law on the part of plaintiff drivers of private vehicles. . . .

. . . While we are not unmindful that the Reno ordinances and state statutes are designed to give emergency vehicles extraordinary rights, they were not intended to absolve the drivers thereof from the duty to be on the lookout at all times for the safety of the public whose peril is increased by their exemptions from the rules of the road. We believe that sound public policy requires such a construction. As we stated in the case of *Springer* v. *Federated Church,* 71 Nev. 177, 283 P.2d 1071,1072, "It may well be true that the public conscience of today demands a more extensive acceptance of tort liability . . . ; that the general custom and practice of today is to accept such liability and insure against it." If such be true, the public is better able to carry this burden than the individual.

Under the facts of the present case where the evidence shows that appellant, while responding to an emergency call, approached a through street upon which there is a stop sign for entering vehicles, the through street being heavily traveled at that particular time of day, and proceeds through the stop sign without stopping at a speed in excess of the established speed limits, substantially cutting the corner, and is at the time of the collision on the wrong side of the street, and his vision of oncoming traffic is substantially obstructed, he is not driving with "due regard for the safety of all persons using the street," and, in the absence of contributory negligence, he is liable to any person injured as a proximate result of his conduct.

BADT and PIKE, JJ., concur.

VIOLATION OF LEGAL DUTY

In tort, the legal duty of one person to another must be breached or violated. The injury or damage thus caused may be the result of:
- **Nonfeasance** — or the failure to peform a duty.
- **Misfeasance** — or the improper performance of a duty.
- **Malfeasance** — the deliberate violation of a legal duty.

The fundamental test of whether one person has a cause of action in tort against another is whether the person sought to be held liable owed a legal duty to do something and did not do it, or owed a legal duty not to do something and did it.

Sir Frederick Pollock, in his *Treatise on Torts,* defines "tort" as an act or omission (not being merely a breach of duty arising out of a personal relation or undertaken by contract) that is related to harm suffered by a determinate person in the following ways: (1) it may be an act which, without lawful justification or excuse, is intended by the agent to cause harm, and does cause the harm complained of; (2) it may be an act in itself contrary to law, or omission of specific legal duty that causes harm not intended by the person so acting or omitting; (3) it may be an act or omission causing harm that the person so acting or omitting did not intend to cause, but might and should, with due diligence, have foreseen and prevented; (4) it may, in special cases, consist merely in not avoiding or preventing harm that the party was bound absolutely or within limits to avoid or prevent.

DAMAGE, INJURY, OR HARM

To sustain an action in tort, damage, injury, or harm must be proven to be the proximate and direct result of the breach or violation of duty, as described in the following excerpt from "Principles of Torts," by Warren A. Seavey:[5]

> In spite of the varied purposes served in actions of tort, harm is the tort signature. In general, the action is based upon the theory that one person has caused harm to another. Thus it is distinguished from criminal law, which directly vindicates the interests of the state; from the law of contracts, which gives sanction to promises; and from the law of restitution, which seeks primarily to prevent unjust enrichment. It is true that in such actions as those for trespass and defamation, the existence of harm is sometimes a legal fiction, and that a person may become responsible for harm done by things or persons under his control although his exercise of control has in fact caused no harm. [Note: This aspect of the law will be presented in the next chapter.] Nevertheless the causing of harm is predominantly the basis of tort actions.

In general, a party is subject to liability in tort if the party:
- Has caused direct and personal harm, damage, or injury to another.
- Has caused harm or injury by failure to perform a duty towards those who are dependent on that legal duty.
- Has allowed a possession or a person within the party's control to cause harm, damage, or injury to another.

REMEDY AT LAW

Compensation or legal remedy for the damage, harm, or injury caused can be:
- **Nominal** — a small amount of money awarded in intentional tort cases to provide a public record that the plaintiff was wronged by the defendant, but damage did not merit a substantial award.
- **Compensatory** — award that reflects the actual damage or injury sustained and suffered by the plaintiff.
- **Special damages** — awards given in addition to the compensatory damages awarded.

---

[5]Warren A. Seavey, "Principles of Torts," 56 *Harvard Law Review*, 72-98. Printed in *Selected Essays on the Law of Torts*, The Harvard Law Review Association, 1942.

- **Punitive** — awards to punish the defendant. These damages are generally quite large and are awarded if it is found that the defendant acted with malice and/or recklessness.

In "Principles of Torts," Seavey further explains the five purposes that are served by the entire group of remedial actions as follows:[6]

> First, to assign torts to a position in the legal scheme: The entire group of remedial actions serve five distinct purposes: (1) to give to a person what another has promised him (usually vindicated in an action upon a contract); (2) to restore to a person what another has unjustly obtained at his expense (usually the basis of a quasi-contractual action); (3) to punish for wrongs (the historical function of the criminal law) and to deter future wrong-doing; (4) to compensate for harm (the most important function of tort actions); and (5) to determine the rights in property (the basis of many different types of action).

KINDS OF TORTS

There are many kinds of torts. Torts may be divided into two general categories: (1) property, and (2) personal.

PROPERTY TORTS

Property torts are torts in which the possession and/or ownership of property is involved. This property may be real property (for example, land) or personal property. Torts involving property include:

Trespass: A violation of an owner's right to use the property owned without interference is called trespass. A trespass may occur through unlawful entry onto the land of another or the interference of one's use, possession, and enjoyment of the land. The subject of trespass will be further detailed later in this chapter.

Conversion: The unauthorized taking control of and assuming rights of the property of another is called conversion. For example, if an owner loses some personal property and this property is found by one who knows to whom the property belongs and keeps it anyway, this action is called conversion.

Nuisance: The unlawful and unreasonable use of property by an owner in a manner that could cause injury, harm, or damage to others is called nuisance.

A nuisance can be a public nuisance or a private nuisance. Public nuisance is defined as "an unreasonable interference with a right common to the general public."[7] A public nuisance is one that affects the community at large and must

---

[6]*Ibid.*

[7]§ 821B, *Restatement of Torts, Second.*

be controlled by the community in its governmental function of providing for the general health, safety, and welfare.

A private nuisance is defined as "a nontrespassory invasion of another's interest in the private use and enjoyment of land. It is not inconsistent with an accompanying trespass to the land."[8] Private nuisance causes harm, damage, or injury to an individual. The difference between trespass and private nuisance is primarily one of the right of the owner. An action in trespass deals mainly with the owner's right of exclusive possession of land, while an action in private nuisance centers on the owner's right to the use and enjoyment of the land.

An "attractive nuisance" is a dangerous condition that exists on land, is attractive to others (especially children), and may cause harm, damage, or injury.

PERSONAL TORTS

Personal torts include assault and battery, false arrest and/or imprisonment, defamation, slander, libel, invasion of privacy, emotional harm, malpractice, and personal negligence.

Assault and Battery: Assault, or a trespass against the person of another, is an unlawful attempt or threat to do violent and personal injury to another. Assault can result in an action in tort. Assault, which can be an offense against the peace and dignity of the state as well as an invasion of private rights, can also be an infraction of criminal law. Prosser's *Handbook of the Law of Torts* states:[9]

> The interest in freedom from apprehension of a harmful or offensive contact with the person, as distinguished from the contact itself, is protected by an action for the tort known as assault. No actual contact is necessary to it, and the plaintiff is protected against a purely mental disturbance of his personal integrity. This action, which developed very early as a form of trespass, is the first recognition of a mental, as distinct from a physical, injury. There is "a touching of the mind, if not of the body." The explanation of its early appearance lies in the obvious likelihood that assaults will result in breaches of the peace, against which the action of trespass was created to enforce the criminal law.

Whereas assault may be defined as the *threat* of violence or harm to another's body, battery may be defined as the actual *contact* with the body of another person in a rude, angry, or hostile manner. Battery can be inflicted on any part of the body or on anything that is attached to the body.

The term "assault and battery" is generally used because, once a battery is committed, the assault or threat to commit battery is generally implied.

False Arrest, False Imprisonment: False arrest is the unlawful detention of a citizen by someone acting in the name of governmental authority. False arrest can occur when a citizen is arrested by a person without legal authority to do so. If the

---

[8]§ 821D, *Restatement of Torts, Second.*
[9]Prosser, *Handbook of the Law of Torts*, p. 37.

detention continues, false arrest becomes false imprisonment. Both are considered trespasses against a person's right to personal liberty.

Defamation, Slander, Libel: A person's right to enjoy a good reputation is the central issue in the torts of defamation, libel, and slander. However, a person may lose this right to a good reputation by becoming involved in criminal activities. Defamation is the issuing of a statement or statements that hold a person up to "hatred, ridicule, or contempt." The defamatory statement or statements must be communicated to a third party since one's reputation is based on one's standing in the community. Proof of harm to the person's reputation is necessary in such actions in tort. Slander is defamation of a person's character or reputation, and is made orally or through gesture. Libel is defamation that is accomplished through some manner of publication for exposure to all in the community. Slander may also occur in publications through the use of pictures or effigies.

NEGLIGENCE

Negligence, the basis for most tort liability cases in the United States, is an unintentional breach of one's legal duty to another. Simply stated, negligence is the lack of ordinary or due care, or the failure to use ordinary care. Because harm or injury caused by negligence is not the result of a deliberate, malevolent act, malfeasance is not a consideration in negligence. Misfeasance, the doing of an act in an improper or careless manner, and nonfeasance, the failure to perform an act that is one's legal duty to perform, are both proper considerations in the tort of negligence. Negligence is often characterized by disregard, heedlessness, thoughtlessness, or inattention. In an article from the *Harvard Law Review*, Henry T. Terry defines negligence as follows:[10]

> Negligence is conduct, not a state of mind. It is most often caused by carelessness or heedlessness; the actor does not advert properly to the consequences that may follow his conduct, and therefore fails to realize that his conduct is unreasonably dangerous. But it may be due to other states of mind. Thus the actor may recognize the fact that his conduct is dangerous, but may not care whether he does the injury or not; or though he would prefer not to do harm, yet for some reason of his own he may choose to take a risk which he understands to be unreasonably great. This state of mind is recklessness, which is one kind of wilfulness, and negligent conduct due to recklessness is often called wilful negligence. . . . Negligent conduct may also be due to a mere error of judgment, where the actor gives due consideration to his conduct and its possible consequences, and mistakenly makes up his mind that the conduct does not involve any unreasonably great risk. He is not therefore excused, if his conduct is in fact unreasonably dangerous.

---

[10]Henry T. Terry, "Negligence," (1915) 29 *Harvard Law Review*, 40-41. ©1916 by The Harvard Law Review Association.

ELEMENTS OF NEGLIGENCE

Negligence, as has been previously stated, is a kind of conduct. However, negligent conduct is only one factor in assessing liability for negligence. Negligence has been defined as conduct that creates an undue risk of harm to the person or property of another. Four major elements are involved in assessing liability for negligence:

Legal Duty: A legal duty to protect others against unreasonable risks must exist. Danger is a factor in negligence. It has been held that negligence does not exist unless there is a reasonable likelihood of danger as a consequence of the act complained of, and it must be measured in the light of some danger that can be reasonably anticipated. The duty to use care is based on the knowledge of danger, and knowledge of danger on the part of the actor, or a reasonable opportunity to acquire such knowledge, is vital to the creation of a duty to exercise care. Negligence or want of ordinary care includes reasonable anticipation of harm. A test of negligence is whether the person or persons accused of negligence could have foreseen the likelihood that harm or injury could have resulted as a consequence of the act or omission. If the harm or injury was not reasonably foreseeable, *i.e.,* the possibility of harm or injury was a remote possibility, then no negligence is said to be involved in an injury. As stated by Judge Gaines in *Texas & P. Ry. Co.* v. *Bigham,* 38 S.W. 162, 163, 90 Tex. 223:

> As applied to the law of negligence, the rule is that a party should not be held responsible for the consequences of an act which ought not reasonably to have been foreseen. In other words, it ought not to be deemed negligence to do or fail to do an act, when it was not anticipated, and should not have been anticipated, that it would result in injury to any one. To require this is to demand of human nature a degree of care incompatible with the prosecution of the ordinary avocations of life. It would seem that there is neither a legal nor a moral obligation to guard against that which cannot be foreseen, and, under such circumstances, the duty of foresight should not be arbitrarily imputed.

Failure to Conform to a Required Standard: Negligence is defined by Dr. Thomas Cooley, in his *Treatise on the Law of Torts,* as:[11]

> . . . in a legal sense, no more or less than the failure to observe, for the protection of the interests of another person, that degree of care, precaution, and diligence which the circumstances justly demand, whereby such other person suffers injury.

In the absence of statutory definition, negligence is defined as the doing of what an *ordinarily* or *reasonably prudent* person would not have done, or the failure to do what an *ordinarily* or *reasonably prudent* person would have done, in like or similar circumstances.

---

[11]Thomas Cooley, *Treatise on the Law of Torts,* Callaghan and Company, Chicago, 1932.

This general standard of reasonable care is defined in §283 of the *Restatement of the Law of Torts, Second:*[12]

> . . . the standard of conduct to which (one) must conform to avoid being negligent is that of a reasonable man under like circumstances.

The "reasonable man" standard presents a model by which negligent actions are measured. This standard measures human actions against the yardstick of what an ordinary person using ordinary care and skill would do in like circumstances. In *Handbook of the Law of Torts,* Professor Prosser defines the "reasonable man" standard as follows:[13]

> The courts have gone to unusual pains to emphasize the abstract and hypothetical character of this mythical person. He is not to be identified with any ordinary individual, who might occasionally do unreasonable things; he is a prudent and careful man, who is always up to standard. Nor is it proper to identify him even with any member of the very jury who are to apply the standard; he is rather a personification of a community ideal of reasonable behavior, determined by the jury's social judgment.

Proximate Cause: The third element in a cause of action for negligence is that of proximate cause; the fact that the conduct or the action was the close or direct cause of the resulting harm or injury must be established. Without the action, the resultant harm or injury would not have occurred. In *The Law of Negligence,* Mirabel and Levy describe proximate cause as follows:[14]

> It is the closeness of the causal relation, and not of time or distance, which determines the existence of proximate cause. The cause, to be proximate, must also be one which is adequate to produce the result complained of. Where negligence for which recovery is sought was the efficient cause setting in motion a chain of circumstances which, uninterrupted by any intervening and independent cause, produced the injury, the negligence will be considered the proximate cause of the injury.

The proximate or legal cause aspect in negligence cases is the most difficult aspect to determine and has been the subject of much controversy. Seldom is the issue of proximate cause a clear one. Intervening circumstances, unforeseen circumstances, and a myriad of other circumstances often cloud the issue. These circumstances can affect the degree of liability and present juries with issues that even the most learned legal scholars have difficulty resolving. In *Rationale for Proximate Cause,* Leon Green states:[15]

---

[12]§283, *Restatement of Torts, Second.*
[13]Prosser, *Handbook of the Law of Torts,* p. 154.
[14]Joseph T. Mirabel and Herbert A. Levy, *The Law of Negligence,* Acme Book Company, Amityville, NY, 1962, p. 90.
[15]Leon Green, *Rationale for Proximate Cause,* Vernon Law Book Company, Kansas City, MO, 1927, p. 134.

A point of equal importance to be observed is that in any given case the inquiry is not directed toward discovery of *the* cause of the damage, but is whether the defendant's conduct was *a* cause of the damage. There are innumerable cause factors in every case. There is no such thing as a sole cause. The court is not set up as a scientific commission to ferret out all of these factors and apportion the respective shares of the losses to each. The court is only interested in determining what *part* the conduct of the *defendant* played in producing the results . . . nor is it quite this. The court wants to know whether the defendant's conduct played *such* part as to make *him* responsible. The complainant himself has by allegations limited the inquiry to the conduct of the person charged. The consideration of other cause factors is incidental and only material on two points: *First,* whether the part played by any other cause factor is a hazard for which the *defendant* should be held; and, *second,* whether in the light of all the other factors, the defendant's conduct played an *appreciable* part in the result.

The decision in the following case demonstrates some of the problems that may occur in determining proximate or legal cause.

GILBERT v. NEW MEXICO CONST. CO.
et al.

Supreme Court of New Mexico
Feb. 26, 1935.
Rehearing Denied May 13, 1935.
44 P. 2d 489

WATSON, Justice.

Plaintiff, J. C. Gilbert, sued the city of Roswell and the New Mexico Construction Company for a fire loss upon his residence, claiming that it was occasioned by low pressure in the city's water mains, insufficient to enable firemen to extinguish the flames. Trial to the court resulted in a judgment of $3,700 against defendant construction company, and it was appealed.

Appellant was in pursuit of a paving contract with the city. Though informed of the location of the water mains, it so operated its power shovel, in excavating, as to break a main. This occurred at 10 o'clock A.M. Appellant immediately notified the city, which undertook the necessary repairs. With ordinary diligence the break could have been repaired in two hours. It remained unrepaired until about 6 o'clock P.M.

At about 5:30 P.M. appellee's house caught fire. When discovered, if water pressure had been normal, a garden hose would have been sufficient to extinguish the flames. The firemen, with their equipment, were on the spot within two to five minutes after discovery of the fire. If the pressure had been normal it would have extinguished the fire without appreciable damage.

To facilitate the repairs, the city water superintendent directed the city engineer to reduce pressure to twenty pounds, the normal being sixty. The fire fighters, finding the pressure insufficient, requested the operator of the pumps for more. He declined to increase it until so ordered by the water superintendent. It was increased later, and the fire extinguished, after damage of $3,700 to appellee's house, shrubbery, trees, lawn, and flowers.

The court below found the foregoing facts, and held that appellant's negligence was the proximate cause of appellee's injury. It held that the city's negligence was a contributing cause, but that the city was not legally liable to appellee.

The first of appellant's contentions which we need notice is that its breaking of the main was not negligent. The argument is not impressive, and is based on some facts which the court refused to find. The theory is that the contract must have contemplated the use of machinery such as was used, and that, in excavating with such machinery, pipes near the required grade are likely to be broken. We find in this no justification for saying, as matter of law, that thus breaking a pipe, the location of which is unknown, is not negligence.

It is next contended that the breaking of the pipe was not, in any sense, a cause of the damage. It is claimed that the repairs were complete before the fire alarm was given; that the broken main had ceased to affect the situation; and that the low water pressure was attributable solely to the failure of the city to restore it immediately upon completion of repairs, or at least when the alarm was given. This depends on nice calculation as to the time of events. It attacks the findings that the fire occurred at about 5:30 o'clock, and that the repairs were completed at about 6. We doubt if the evidence brought to our attention would warrant disturbing these findings.

It is apparent that the repairs were completed and the fire discovered at about the same time. An attempt to reduce events to an exact and accurately timed sequence would no doubt fail, and does not seem warranted. We think that a few minutes' delay by the city in restoring pressure after appellant had made it necessary to reduce it, would not enable us to say, as matter of law, that the primary negligence had spent its force.

It is also contended that, even with the break unrepaired, sufficient pressure could have been furnished, at slight damage, to have enabled the firemen to accomplish their task. The trial court refused to find that, notwithstanding the break, "the water could have been turned on at the . . . plant when the . . . telephone communication was received from the scene of the fire . . . and sufficient pressure could have been forced to plaintiff's residence to have extinguished the fire." Appellant points to evidence which might have required the court to find that the mere existence of the break would not prevent putting the hydrant pressure to fifty pounds. It is not shown how long, the break existing, it would require to advance pressure from twenty pounds to fifty. After the fire broke out, time was important. Assuming that there is merit in the legal contention, we find no error in refusing the proposed finding.

Appellant's most important and interesting contention is that the breaking of the water main was not the proximate cause of the fire loss. The first aspect of its proposition is that such cause, regardless of any intervening negligence of the city, is but remote.

We find no contention between counsel as to the general rule of proximate cause. So far as definition goes, all seem content with the round statement. *[Citations omitted.]* "The 'cause which, in natural and continued sequence, unbroken by any efficient, intervening cause, produced the result complained of, and without which that result would not have occurred.' "

A general survey of appellant's argument discloses these grounds for asserting the remoteness of this cause: the fire itself was the proximate cause of appellant's loss; appellee had no right to this municipal fire extinguishment service capable of supporting an action for interference with it; appellant could not reasonably have anticipated that its act would produce the injury so long subsequently.

The first contention we deem too well settled against appellant to require more than citation of some of the decisions. In the leading case the court answered it thus: "The law regards practical distinctions, rather than those which are merely theoretical; and practically, when a man cuts off the hose through which firemen are throwing a stream upon a burning building, and thereupon the building is consumed for want of water to extinguish it, his act is to be regarded as the direct and efficient cause of the injury." *[Citation omitted.]*

Leaving now this most interesting phase of the case, we pass to the third of the above-stated grounds on which it is urged that the negligent act was but a remote cause of the loss; that appellant could not reasonably have anticipated such a result.

Here is stated a recognized test of proximate cause. Undoubtedly it is a proper inquiry whether appellant anticipated this result or whether it may reasonably be said that it should have been anticipated. . . .

". . . The other aspect of appellant's contention on proximate cause is that, even if its act could proximately have caused the injury, it did not in this case, because the chain of causation was broken by the intervening negligent delay of the city in making the repairs or in restoring pressure after completion of the repairs."

It is no doubt true that the injury to appellee might have been avoided by greater diligence or better directed effort on the part of the city. That does not in our opinion interrupt the natural and continued sequence of the original force. The city's negligence was passive merely. [Citation omitted.] It was not independent. The necessity for it to act at all arose wholly from appellant's act. In reason, one who sets a harmful force in motion should not be heard to defend against liability by the claim that another than the injured party might have prevented the injurious result by an active and timely intervention. The negligence of physician, surgeon, or nurse may intervene to aggravate an injury or to make it fatal. That will not sever the chain of casualty leading to him who made the treatment necessary. [Citations omitted.] And if it could be questioned whether the city's negligence, being passive, is a cause set in motion by the original wrongdoing, the intervening cause may as readily be given an active character by conceiving it the act of the city in reducing the water pressure. . . .

. . . We do not see how appellant can thus rid itself, as matter of law, of the primary liability. Having broken the main, damaged the city's property, and endangered appellee's, we cannot say that appellant has no concern or duty or liability with respect to the time or manner of restoring the main to service. If the city were not under a legal exemption, its negligence would be a concurring cause, we think. We do not consider that its nonliability changes the situation.

Aetna Insurance Co. v. Boon, 95 U.S. 117, 130, 24 L. Ed. 395, is a fruitful source in this connection. Mr. Justice Strong there says: "The proximate cause is the efficient cause, the one that necessarily sets the other causes in operation. The causes that are merely incidental or instruments of a superior or controlling agency are not the proximate causes and the responsible ones, though they may be nearer in time to the result. It is only when the causes are independent of each other that the nearest is, of course, to be charged with the disaster."

And he quotes thus from a leading case [citation omitted]: "The inquiry must always be whether there was any intermediate cause disconnected from the primary fault, and self-operating, which produced the injury. . . ."

. . . Before leaving proximate cause, another matter should perhaps be mentioned. The findings disclose that the city's system of mains was equipped with shut-off valves, and that on the occasion in question they were found useless from corrosion, due to the city's negligence. Except for this, we take it, the effects of the break would have been so localized that in other places, including the situs of the fire, water pressure would have been maintained.

Appellant makes no use of these facts in argument other than as partly explanatory of the delay in making repairs and of the necessity of reducing pressure throughout the city. We readily perceive, however, that these facts, in connection with others not shown, might have been of importance in the case, at least in the trial court. If it had been shown, either that appellant knew of the shut-off equipment, or that such equipment was in universal or general use, it could more forcefully have urged that the injury occurring eight hours later, and in a different part of the city, was not reasonably to have been anticipated.

The judgment will be affirmed, and the cause remanded. It is so ordered.

SADLER, C.J., and JOSEPH L. DAILEY, District Judge, concur.

On Motion for Rehearing.
WATSON, Justice.

The motion for rehearing is denied. We take advantage of the opportunity to add a word of explanation.

In concluding against the contention that the city's negligence broke the chain of causality leading to appellant's negligence so as to render the latter but the remote cause of the injury, we may have placed more emphasis than we intended upon the passive character of the city's negligence. We are aware of some situations in which nonaction is held to be an intervening cause.

The real basis of decision is that the city's negligence was not an independent cause. It was the primary fault, and it alone, that called for action on the part of the city. Its failure to act or to act in time was not a self-operating cause, and it was directly connected with the force previously set in motion.

The principle is quite different where a railroad company fails to perform its duty to inspect a car delivered to it for transportation, and an injury results from a defect which proper inspection would have discovered. There the passive negligence is independent of the original fault which occasioned the defect and serves as an intervening and the proximate cause. It is cases in this class upon which appellant has relied.

SADLER, C.J., and JOSEPH L. DAILEY, District Judge, concur.

Actual Loss, Harm, Injury, or Damage: The one element of a cause of action in negligence that is similar in all tort cases is that it must be shown that actual loss, harm, injury, or damage has occurred as a result of negligence.

DEGREES OF NEGLIGENCE

- *Slight negligence* is the failure to exercise great or extraordinary care, or the absence of that degree of care that persons of extraordinary prudence and foresight are accustomed to use — unless the circumstances were such that a duty to exercise more than ordinary care would be imposed upon the person sought to be charged.

- *Ordinary negligence* is the want of ordinary care and diligence.
- *Gross negligence* is far greater than ordinary negligence, and consists of an entire absence of care or an absence of even slight care or diligence.

DEFENSES AGAINST NEGLIGENCE

There are legal defenses against complete liability for negligence. These defenses involve proof that an injured party's actions have fallen below the standard of care necessary to ensure one's own protection from accidents or acts of negligence and, by so doing, have helped to bring about the party's own harm, damage, or injury. The five major defenses against complete liability for harm, damage, or injury that are of special concern to members of the fire service are: (1) assumption of risk, (2) the doctrine of rescue, (3) contributory negligence, (4) comparative negligence, and (5) the last clear chance doctrine.

Assumption of Risk: A party may be barred from recovering for harm, damage, or injury if it can be proved that the party (or plaintiff) knew that peril or danger was to be encountered and willingly encountered the peril in spite of this knowledge, as explained in the following:[16]

> One who knows of a danger from the negligence of another, and understands and appreciates the risk therefrom, and voluntarily exposes himself to it, is precluded from recovering for an injury which results from the exposure.
> It would be unjust that one who freely and voluntarily assumes a known risk for which another is, in a general sense, culpably responsible, should hold that other responsible in damages for the consequences of his own exposure.

Voluntary assumption of risk relieves the one who causes the negligent situation from all legal duty to the one who is harmed or injured. In situations where parties are willing to assume risk or "take a chance," the decision to do so may be a reasonable one based on many factors and may not be a negligent act in itself.

Two requirements are necessary in using the defense of assumption of risk: (1) the risk must be known, and (2) the choice to incur the risk must be voluntary.

Although members of the fire service are expected to exercise a higher degree of care in emergency situations because of the hazardous nature of their work, this higher standard of care does not require that fire fighting personnel abrogate all rights to the legal duties that are owed them by others.

As you read the following case study, determine how you would summarize the preceding theory based on the facts presented. In the following case, the defendants argued that fire fighters, by the very nature of their occupation, assume risks of danger. This defense is based on the contention that a plaintiff cannot recover

---

[16]Fitzgerald, Conn. R.P. CO., 155 Mass. 155.

from a negligent defendant if the plaintiff knows of the risk and willingly accepts it. However, in *Restatement of Torts, Second,* § 496e, the plaintiff does not assume a risk if, as a result of the defendant's negligence or negligent conduct, the plaintiff has no reasonable alternative to avoid the harm, or where the risk is to let the threatened harm occur.

Thomas E. BARTELS, Kathleen M. Bartels,
Judith A. Bartels and Georgine L. Bartels,
minors, by their mother and next friend,
Kathleen P. Bartels, and Kathleen P. Bartels,
Widow of George E. Bartels, Respondents,

v.

CONTINENTAL OIL COMPANY
a corporation, Appellant.

Supreme Court of Missouri,
Division No. 2.
Nov. 9, 1964.
Motion for Rehearing or for Transfer
to Court *En Banc* Denied
Dec. 14, 1964.
384 S.W.2d 667

Action for wrongful death of fireman caused when inadequately vented oil storage tank at scene of fire "rocketed" into air and engulfed him in a ball of fire. The Circuit Court, Jackson County, J. Donald Murphy, J., gave judgment to fireman's heirs and storage tank owner appealed. The Supreme Court, Barrett, C., held that finding of existence of hidden danger known to owner and failure to warn fireman of danger which he was not bound to accept as a usual peril of his profession was supported by the evidence.

Affirmed.

BARRETT, Commissioner.

George E. Bartels died as a result of injuries sustained while engaged in the performance of his duties as a captain in the Kansas City, Missouri, fire department. In this action to recover damages for his negligent injury and death his widow and four minor children have recovered a judgment of $25,000. The defendant-appellant, Continental Oil Company, operated a bulk storage plant and filling station on the northwest side of Southwest Boulevard at 31st Street in Kansas City, Kansas, abutting the Missouri-Kansas state line. On the front of the lot, facing Southwest Boulevard, there was a filling station and to the rear of the station on concrete saddles there were four 21,000-gallon capacity storage tanks. Tank number one contained 6,628 gallons of kerosine, tank number two contained 14,307 gallons of regular gasoline, tank number three contained 3,051 gallons of regular gasoline and tank number four contained 15,555 gallons of premium gasoline.

On August 18, 1959, about eight o'clock in the morning, Fred Berry, one of the appellant's tank-truck drivers was engaged in loading his truck from storage tanks two and four. Hoses from these two tanks were open at the same time, filling truck compartments three and five. Berry was on the truck's catwalk when Jim Mitchum, another tank-wagon driver on vacation, climbed up the side of the truck to show Berry his new cigarette lighter. Almost immediately (possibly when Mitchum flicked his lighter) flames flashed from the tank-truck's fifth compartment. Berry cut off the hose from one of the storage tanks but ran from the blazing flames and did not cut off the hose from the other storage tank and throughout the ensuing fire that open hose continued pouring gasoline into the flames. Firemen from both Kansas City, Missouri, and Kansas City, Kansas, responded to the alarm and with hand lines almost wholly from Southwest Boulevard, fought the fire with water. The streams of water were employed to confine the burning gasoline to the area of the storage tanks thereby preventing the gasoline from running down Southwest Boulevard and into the sewers of Kansas City, Missouri, endangering larger areas. About 9:15 storage tank number one ruptured at its rear and the area was engulfed in flaming kerosine intensifying the heat in the area of the other three tanks. In the next half hour tanks two and three ruptured, and about 9:40 tank number four left its concrete cradle and "rocketed" or was catapulted 75 to 100 feet over the filling station into Southwest Boulevard and "a ball of fire" engulfed the several crews of fire fighters in the street, killing one bystander and five firemen, including Captain Bartels, and injuring twenty-three people.

Since, as a general rule, "the duty of a possessor of land to firemen is the same as to licensees" (*Anderson v. Cinnamon,* 365 Mo. 304, 282 S.W.2d 445, 447, 55 A.L.R.2d 516), it is urged that upon this record there is no evidence of a violation of duty by Continental. It is said that the appellant's duty was not enlarged by the fact that Bartels was in the street when injured, that "(t)here was no unusual, hidden hazard that was unknown to the deceased and known to the defendant" and therefore there was no duty to warn that the tanks might rupture. In this connection it is urged that there was "no duty to warn deceased that the storage tank in question might rocket onto the street because (1) that fact was not known by the defendant (Continental); and (2) even had such a duty existed, there was no evidence that defendant had an opportunity to give such a warning." Further precisely pointing up the appellant's contention that its motions for a directed verdict should have been sustained is its claim that the respondents' instruction one was "reversibly erroneous" in several respects, but particularly in that it "was not supported by the evidence."

It is not necessary to a disposition of this appeal to enter upon an extended discussion of the general rules or to determine Bartels' status. The general rules are set forth and the applicable cases collected in *Anderson* v. *Cinnamon, supra,* and its annotation, 55 A.L.R.2d 525, "Duty of a possessor of land to warn adult licensees of danger," and see more recently 86 A.L.R.2d 1205, "Duty and liability of owner or occupant of premises to fireman or policeman coming thereon in discharge of his duty." And upon the merits of the cause there are no precisely applicable and governing Kansas cases. Nevertheless it is safe to say that Kansas applies the general rules both as to the status of firemen and the duties of landowners, particularly in the maintenance of gasoline storage facilities. *[Citations omitted.]* In the Buchholz case gasoline storage tanks at Hays, Kansas (an installation comparable to that of Continental's here), burned and exploded and when the larger tank exploded "one end of it was shot southward across the railroad tracks and against a flour mill which was about 200 feet away, the other end, like a torpedo, went toward the north, . . . and landed in the rear of a house which stood about 500 feet distant."

Under these general rules, admittedly, an experienced fire captain would of course accept the presence of kerosine and gasoline as a known hazard of a fire in a gasoline storage facility. *[Citations omitted.]* But the law does not compel firemen in fighting a fire to assume all possible lurking hazards and risks (*Gannon* v. *Laclede Gas Light Co.,* 145 Mo. 502, 46 S.W. 968, 43 L.R.A. 505); it may not be said that a "fireman has no protective rights whatever." *[Citations omitted.]* As indicated, the majority of cases, including *Anderson* v. *Cinnamon* and other cases relied on by the appellant, recognizes certain modifications or exceptions to the general rules relating to landowners and firemen and it is these principles upon which the respondents rely, — "that an owner or occupant of premises which firemen enter upon in the discharge of their duty may be held liable to a fireman injured by a hidden danger on the premises, where the owner or occupant knew of the danger and had an opportunity to warn the fireman of it." 86 A.L.R.2d 1. c. 1214, 55 A.L.R.2d 1. c. 526. And the essentially meritorious question upon this appeal is whether there was upon this record evidence to support the inference and finding of a hidden danger known to the appellant Continental Oil Company for "(c)ertainly, no meritorious reason can be advanced to justify the view that a property owner, with knowledge of a hidden peril, should be allowed to stand by in silence when a word of warning might save firemen from needless peril. . . . Although firemen assume the usual risks incident to their entry upon premises made dangerous by the destructive effect of fire, there is no valid reason why they should be required to assume the extraordinary risk of hidden perils of which they might easily be warned." *[Citations omitted.]*

Continental's four storage tanks, constructed of quarter-inch steel, were installed in 1924. They were 30 feet long, eleven feet in diameter, with flat ends, and when installed as well as 35 years later "were fitted with two-inch pressure valves." Originally the tanks were installed perpendicularly but in 1958 Continental Oil Company dismantled the storage facilities, removed the two-inch vents, and installed the four tanks horizontally on the concrete saddles. But in changing or replacing the installation Continental again placed two-inch vents on the tops of the storage tanks. And in response to interrogatories and "requests for admissions" Continental stated that the "four tanks was (sic) vented only by a single A. Y. McDonald 2-inch Plate 925 combination gauge hatch and P & R valve, known as a 'breather valve.' . . . that these said four tanks had no vent or mechanism designed to relieve pressure (such as might occur during a fire) other than whatever relief was provided by the single 2-inch diameter breather valve on each tank." The appellant has made unnecessary further recitation of the facts in connection with these valves because in its brief there are these further concessions: "(t)he four tanks in question were vented with 2-inch pressure and relief valves, which were admittedly (and obviously) not adequate to keep these tanks from rupturing"; and, "we will concede for the purpose of that question only (viewing the record favorably to the respondents), without admitting it, that by reason of the size of the vents, defendant did not maintain the premises in a reasonably safe condition, and that that was the cause of the ruptures."

In addition to these admissions the Chief of the Fire Prevention Division of the Kansas City, Missouri, Fire Department, who was present while the fire was in progress and when tank number four rocketed, and afterwards made an investigation, testified that the reason the tanks ruptured was that "(t)he vents were too small." He testified that the *Flammable Liquids Code* required "5½ inches of emergency venting." An expert witness, "a protection engineer specializing in petroleum hazards," referred to all the literature connected with the petroleum industry (with which Continental was familiar) and, in connection with his years of experience in the industry, testified that for petroleum storage tanks of 18,000, to 25,000 gallon capacity "a free circular opening of 5½ inches in diameter" was required. As he testified on both direct and cross-examination this witness gave the formulas and made the computations as to the "normal safety operating pressures of a tank" as well as the pressures built up in a storage tank during a fire. He testified that tank number four rocketed because the two-inch vent was not large enough to take care of the vapor generated by the fire. And, he said, "I can say without any worry at all that no tank that has been equipped with the vent of the size specified in these suggested standards (5½ inches) has ever rocketed." The appellant's expert, a professor of chemical engineering at Kansas University, was of the opinion that a 5½ inch valve would not have prevented the rocketing; but even this witness was of the opinion that a 7½ inch vent "would have been sufficient to vent all the vapor which was formed" in this fire and would have prevented the rocketing of tank number four.

And upon this particular phase of the case, in its answers to interrogatories and in counsel's opening statement, there were these additional admissions: ". . . that Continental Oil Company and Continental Oil Company officials, the officials of my client, knew at the time of this casualty that a two-inch vent was small enough that there was more apt to be a rupture in the event of an exposure fire. . . . And I want to tell you here at the outset that there is no real issue about that, there is no issue whatever. These were old-fashioned tanks. The two-inch vents in these tanks were smaller than was recommended by the literature, by the National Fire Protection Association and the Association of Petroleum Industries. . . ." And further of knowledge by the industry, he said, "Our company people . . . read the literature, . . . their safety director belonged to some of these organizations, and was on committees, and so they did know full well long before 1959 that a larger emergency vent would be safer."

While Bartels was an experienced fireman and was therefore necessarily aware of some of the hazards of petroleum products fires, there is no fact or circumstance indicative of any knowledge on his part that these particular storage tanks were equipped with inadequate safety vents. And as the appellant says in its brief "(a)lthough the men on the line knew the No. 4 tank would probably rupture (since the three other tanks had ruptured), they did not know the tank would rocket in the freakish manner in which it did. They had never seen a tank act like that. The action of this particular tank was entirely unpredictable." Also aerial photographs taken while the fire was in progress and before tank number four rocketed, as well as the testimony of witnesses, reveal that it was not possible through the enormous black smoke clouds and flames that firemen or anyone else could or did see the two-inch valves on the tanks, much less appreciate their danger. The fire prevention chief and other firemen testified that they did not know the size of the safety valves on the tanks. Six of Continental's employees, including its bulk plant supervisor, were present prior to and during the course . . . there were no warnings in any manner to the fire fighters that the tanks were equipped with the inadequate two-inch vents. . . .

. . . Further and explicit details could be recited and permissible inferences indicated; it is sufficient, however, to say that in these particularly detailed circumstances the evidence favorable to the plaintiffs supports the finding of a known hidden danger of which there was no warning whatever, a hazard that in any and all events Bartels as a fireman was not bound to accept as a usual peril of his profession. In these circumstances reasonable minds could differ and Continental was not for any of the reasons advanced here entitled to a directed verdict. . . .

STOCKARD and PRITCHARD, CC., concur.

PER CURIAM.

The foregoing opinion by BARRETT, C., is adopted as the opinion of the court.

All of the Judges concur.

Doctrine of Rescue: Another defense against negligence that can affect fire fighters in a special way is a defense arguing that the doctrine of rescue bars recovery by a plaintiff because of negligence or negligent conduct on the part of the defendant. As you read the following case, consider how the doctrine of rescue was used as a defense.

---

110 Ga. App. 620
WALKER HAULING COMPANY, Inc., et al.
v.
Helen P. JOHNSON, Executrix.

Court of Appeals of Georgia,
Division No. 1.
Oct. 26, 1964.
Rehearing Denied Nov. 12, 1964.
139 S.E.2d 496

Action by alleged skilled fire fighter for injuries received while helping to fight bulk petroleum storage plant fire. The Superior Court, Fulton County, Ralph H. Pharr, J., overruled renewed general demurrers of both defendants to the petition as amended, and exceptions were taken. The Court of Appeals, Felton, C. J., held that if skilled fire fighter was injured while voluntarily and without rashness or recklessness helping to fight bulk petroleum storage plant fire proximately caused by negligence of plant owner and common carrier, plant owner and carrier were liable.

Affirmed.

Syllabus by the Court.

The petition, seeking damages for injuries sustained by a "skilled fire fighter" while he was voluntarily and without rashness or recklessness helping to fight a bulk petroleum storage plant fire, the alleged sole proximate cause of which was the negligence of the defendant plant owner and defendant common carrier, stated a cause of action under the doctrine of rescue which was good as against the general demurrers.

Hayward Johnson filed suit in Fulton Superior Court against J. Ran Cooper, a resident of Taylor County, and Walker Hauling Co., Inc., a common carrier with principal office in Fulton County, to recover damages for injuries alleged to have been caused by the defendants' negligence. The petition as amended alleged substantially as follows: that on August 17, 1962, defendant Cooper owned and operated a bulk plant for the storage of petroleum products near Rupert in Taylor County and the defendant company was engaged as a common carrier in the delivery of gasoline by transport truck to Cooper's storage tanks; that said plant consisted of four 7,500 gallon tanks set upon brick pillars, together with accompanying equipment, at which he stored Class 1 and other inflammable [flammable] petroleum products as classified by the Georgia Safety Fire Regulations; that said plant was constructed in such a manner that the fumes from any spillage in filling operations would be concentrated upon the parking space allotted for transport supply trucks; that the plant was situated in a congested, built up area, with the tanks being within 30 feet of a heavily traveled highway; that at about 6:30 P.M. on the aforementioned day, the defendant company, through its agents and servants, undertook replenishing operations at the plant; that defendant Cooper ordered Walker's agent to pump 7,050 gallons of regular gasoline into one tank, assuring him it would hold that amount; that Walker's agent then connected an auxiliary gasoline pump, which he carried for that purpose, to an electric pump maintained as a permanent installation at the plant, utilizing both pumps simultaneously, the auxiliary pump being placed in a position where the fumes from any spillage would be directed upon it; that the agent connected the tank truck to the storage tank but failed to connect the terminals of the unloading hose securely and to establish an electrical bond connection between the truck and the unloading devices; that during the pumping operation spillage occurred from loose or defective fill pipe connections; that when 6,500 gallons of gasoline, including the spillage, had been pumped, the tank began to overflow from the top valve; that the fumes from the spilled gasoline under the storage tanks were concentrated upon the truck and the auxiliary pump and ignited in an explosion which enveloped the plant in a petroleum fire; that the only practicable method of extinguishing and controlling such a fire was by the use of foam equipment, the nearest of which to the fire was at Manchester, GA; that the fire was burning out of control and a message for help was relayed by the Georgia State Patrol to the Manchester Fire Department Chief, who asked for volunteers who would, over and beyond the call of any duty, be willing to go to said fire and try to control or extinguish the same with foam, in the interest of rescuing life and property imperiled by the fire; that the plaintiff, Hayward Johnson, was a "skilled fire fighter" and was among those who, under no duty or compulsion to do so, went to said fire for the purpose of assisting in extinguishment or control of the same before further explosions occurred; that the plaintiff drove his own automobile from Manchester to the scene of the fire and, together with others, had succeeded in partially extinguishing the flames by foam by about 10:30 P.M.; that the heat of the fire had caused the valves and pipes of the tanks to become warped, allowing more leakage to accumulate under the tanks; that the plaintiff was engaged, taking his regular turn with others, in playing a fire hose upon the hot and glowing wreckage surrounding the plant in order to prevent a resurgence of the fire, when an explosion of the spilled gasoline occurred, inflicting the various alleged injuries upon the plaintiff; that certain rules and regulations, set forth in the petition, promulgated by the State Fire Marshal, pursuant to law, governing the handling and storage of petroleum products, were in force at the time of the occurrence. The specifications of negligence alleged against defendant Walker Company are as follows: (a) In failing to ascertain what quantity of gasoline was already contained in the tank before engaging in an unloading operation. (b) In failing to use ordinary care or precaution in relying solely on information received as to how much gasoline the tank already contained before unloading. (c) In locating its auxiliary internal combustion pump and exhaust in close and hazardous proximity to its truck and the tanks, in violation of regulations duly promulgated by the State Safety Fire Commissioner pursuant to law, which act was negligence *per se*. (d) In failing to use ordinary care and precaution to shield the exhaust discharge pipe of its auxiliary internal combustion pump from gasoline vapors. (e) In failing to establish an electrical circuit bond connection between its truck and the unloading devices before unloading, in violation of regulations duly promulgated by the State Safety Fire Commissioner pursuant to law, which act was negligence *per se*. (f) In failing to use ordinary care or precaution to secure its hose terminals so as to prevent leakage of gasoline from the connections of its unloading hose. (g) In pumping gasoline into the tank in excess of its capacity, causing it to overflow. (h) In failing to devote adequate attention to its unloading operations so as to assure that no condition, such as overflow, developed which imperiled the life and property of others. (i) In causing, suffering and permitting a fire and consequent explosion to take place at the plant. (j) In utilizing and parking its truck at and upon the provided space during unloading operations when defendant knew, or in the exercise of ordinary care should have known, that said space was located in a spot whereby the fumes and vapors from any spillage from the tanks would be directed toward and channeled upon the space. The specifications of negligence against defendant Cooper were as follows: (a) In failing to use ordinary care to ascertain the amount of gasoline that his tank held before unloading operations were ordered. (b) In failing to use ordinary care to give the defendant company's agent correct information in regard to the capacity of his tank for gasoline, (c) In failing to assure that the defendant company's unloading operations were carried out safely and in the exercise of ordinary care. (d) In causing, suffering and permitting a fire and consequent explosion to take place at the plant as aforesaid. (e) In furnishing and providing a parking space for the use of tank trucks during unloading operations which was in hazardous proximity to his storage tanks. The concurring negligence of both defendants was alleged as the proximate cause of the various injuries and damages sustained. While the case was pending the plaintiff died and his executrix, Helen Pike Johnson, was substituted as party plaintiff. The court overruled the renewed general demurrers of both defendants to the petition as amended, to which judgments exceptions are taken.

Gambrell, Harlan, Russell & Moye, E. Smythe Gambrell, Edward W. Killorin, Donald O. Clark, Greene, Neely, Buckley & DeRieux, John D. Jones, Atlanta, for plaintiff in error.

George C. Kennedy, Manchester, Jack Turner, Atlanta, H. Briscoe Black, Manchester, for defendant in error.

FELTON, Chief Judge.

The allegation that the plaintiff was a "skilled fire fighter" is not reasonably susceptible to the construction that he was a "fireman," volunteer or otherwise. True, pleadings must be construed against the pleader on demurrer but the rule does not require strained or unreasonable or illogical constructions. Proof that one is a skilled fireman would not alone authorize a finding that he was a fireman. Likewise, an allegation that the Fire Chief of Manchester, GA, asked for volunteers to fight the fire is not reasonably susceptible to the construction that the plaintiff was a volunteer fireman of the City of Manchester. But even if we are wrong in the above conclusions a volunteer fireman who receives no remuneration is even more entitled to the benefits of the rescue doctrine than one who is less experienced and who would more likely assume an unreasonable or foolhardy risk. To reduce the ranks of rescuers to the less competent would be to contradict and weaken the application and consequences of one of the most advanced doctrines evolved by the conscience of mankind.

The allegations, that the defendants' negligence was the proximate cause of the explosion and fire which caused the plaintiff's injuries and that the plaintiff was free of contributory negligence, raised issues of negligence which a jury must resolve. There is no issue involved as to assumption of risk, since the doctrine of rescue necessarily contemplates an assumption of the risk inherent in the peril created by the defendants' negligence and allows recovery for injuries thereby incurred, for the reason that the defendants were charged with the foreseeability of their negligence's attracting rescuers to assume the risks. . . .

The court did not err in its judgment overruling the general demurrers to the petition as amended.

Judgment affirmed.

Contributory Negligence: Assumption of risk and contributory negligence are the two most common defenses in a negligence action. Of the two, contributory negligence presents more difficulty in terms of deciding liability relative to the actions of both the defendant and the plaintiff. Although contributory negligence is a defendant's argument in negligence cases (and therefore the burden of proof is on the defendant to prove that the plaintiff contributed to the liability in question), the plaintiff is often forced to provide evidence to prove that no contributory negligence was involved. This is difficult to do in many cases, and all too often inequitable judgments can result.

Contributory negligence is conduct on the part of the plaintiff that contributed to the legal harm or injury suffered by the plaintiff. The plaintiff's conduct or actions must fall below the standard of action required for one's own protection.

Simply stated, proof of contributory negligence relieves one party of liability for negligence if the other parties contributed to the harm, damage, or injury. In *A*

Treatise on the Law of Contributory Negligence, Charles Fisk Beach, Jr., states:[17]

> . . . Contributory negligence in its legal signification, is such an act or omission on the part of a plaintiff, amounting to a want of ordinary care, as, concurring or cooperating with the negligent act of the defendant, is a proximate cause or occasion of the injury complained of. To constitute contributory negligence there must be a want of ordinary care on the part of the plaintiff, and a proximate connection between that and the injury.

The defenses of contributory negligence and assumption of risk are major defenses against charges of negligence. One of the major distinctions between these two defenses is the time factor, as described in *Warner* v. *Markoe:*[18]

> The distinction between contributory negligence and voluntary assumption of risk is often difficult to draw in concrete cases, and under the law of this state usually without importance but it may be well to keep it in mind. Contributory negligence, of course, means negligence which contributes to cause a particular accident which occurs, while assumption of the risk of accident means voluntary incurring of an accident which may not occur, and which the person assuming the risk may be careful to avoid after starting. Contributory negligence defeats recovery because it is a proximate cause of the accident which happens, but assumption of the risk defeats recovery because it is a previous abandonment of the right to complain if an accident occurs.

When argued, the defense of assumption of risk rests on the issue that the defendant is relieved of any liability toward the plaintiff because the defendant was relieved of any duty to the plaintiff when the plaintiff voluntarily assumed the risk in question. In using the contributory negligence argument, the defendant argues that, although negligence was involved, duty was breached and liability would be owed to the plaintiff, the plaintiff's own conduct or actions so contributed to the harm or injury suffered that the plaintiff is denied recovery. (In many instances, although plaintiff's contributory negligence does not bar complete recovery, recovery is awarded based on degree of contributory negligence.)

As you read the following case study regarding fire fighters who suffered injuries from the negligence of the defendant, consider how the defense of contributory negligence is used.

CAMPBELL et al. <u>v.</u> PURE OIL CO.

Supreme Court of New Jersey, Atlantic
County
Nov. 12, 1937.
194 A. 873

---

[17]Charles Fisk Beach, Jr., *A Treatise on the Law of Contributory Negligence,* 3rd Ed., Baker, Voorhis & Co., New York, 1978, pp. 7-8.

[18]C.J. Bond, *Warner* v. *Markoe,* 171 Md. 351.

Action by Frederick Campbell and others against the Pure Oil Company. On motion to strike the complaint.

Motion denied.

JAYNE, Supreme Court Commissioner.

Concisely stated, the complaint alleges that on July 16, 1937, and for a long time prior thereto, the defendant maintained at the southeast corner of Virginia and Drexel avenues in the City of Atlantic City certain gas and oil tanks, pumps, pipes, and other apparatus for the storage of gasoline, kerosine, and lubricating oils. On the date mentioned a fire originated in and about the pumping plant on the premises of the defendant. The plaintiffs were firemen and were summoned to extinguish the fire or to arrest its progress. While the plaintiffs were endeavoring to protect adjacent properties from damage, a sudden and unexpected explosion occurred which caused them bodily injury. It is alleged that the defendant had negligently permitted large quantities of oily, greasy, and gaseous substances of a highly inflammable and combustible nature to accumulate on the premises in the vicinity of its pump house and tanks and had permitted its tanks and storage containers to become out of repair so that the contents could escape from them. Additionally it is alleged that the storage tanks were not encircled by an embankment or dike and were not equipped with fire protecting and safety devices as required by an existing ordinance of the city. It is averred that the plaintiffs were wholly unaware that oily and gaseous substances had been permitted to saturate the premises in the vicinity of the pump house and tanks, and were ignorant of the defective condition of the storage tanks and of the failure of the defendant to equip them with the required devices. It is to be inferred that these conditions which were unknown to the plaintiffs proximately occasioned the unanticipated explosion. The present motion can therefore be considered in the light of these recited facts.

The complaint also embodies a general allegation that the explosion and resultant injury to the plaintiffs were proximately caused by the negligence of the defendant. . . .

. . . At the argument of this motion counsel for the defendant particularly emphasized the contention that the only duty owed by the landowner to a fireman who in the pursuit of his duty enters the premises to abate a fire is to abstain from acts willfully and wantonly injurious to the fireman. . . .

. . . The factual circumstances alleged in the present complaint do not project for decision the question relating to the legal status of the plaintiffs upon the mere assumption that they were upon the premises of the defendant. . . . There is no allegation that the plaintiffs were on the property of the defendant when injured. The averment is that they were engaged in protecting adjacent properties. It was stated by counsel at the argument that the evidence to be submitted under the complaint would disclose that the plaintiffs, when injured, were on the public street. Therefore it is essential to here recognize the distinction between the more usual case of negligence in which the basic duty involved is the duty to abstain from acts of commission or omission which may be reasonably expected to wrongfully cause injury to another and the case in which a positive duty created by law or the situation of the parties is alleged to have been ignored.

It is elementary that one who utilizes one's property for the storage of relatively large quantities of highly inflammable or explosive substances or materials must exercise a degree of care commensurate with the risks and dangers to be reasonably anticipated in such an undertaking. This basic duty must be observed for the safety of others. The complaint alleges a failure on the part of the defendant company to observe this duty. . . . It may be inferred from the limited statement of facts in the complaint that the alleged negligent accumulation of inflammable substances around the closely located pump house and tanks and the defective state of the storage tanks not only created dangerous conditions which according to common knowledge and experience would be likely to cause instantaneous injury, but that it was such dangerous conditions that proximately caused the explosion.

It is argued that in the circumstances disclosed by the complaint it is apparent that the plaintiffs were guilty of contributory negligence and voluntarily assumed the very risks and dangers from which their injuries resulted. It will be noticed that in one of the counts of the complaint the conditions therein described are alleged to have constituted a nuisance. The availability of the defense of contributory negligence depends upon the character of the nuisance. [Citation omitted.] The contributory negligence of the injured person is not in general to be presumed. It is to be averred and established by proof as a matter of defense. It is an affirmative defense which the defendant must plead or it must be otherwise drawn in issue by the pleadings in the action. [Citations omitted.] Counsel for the defendant, however, has very earnestly contended that the complaint in the present action discloses circumstances which of themselves would impute to the plaintiffs contributory negligence or assumption of risk as a matter of law. It seems that at the common law, if the declaration definitely and specifically alleged acts which clearly constituted contributory negligence on the part of the plaintiff, a demurrer might be properly interposed. [Citation omitted.] To sustain such a demurrer it must clearly appear that the alleged act of the plaintiff did as a matter of law constitute contributory negligence. It is not therefore amiss to point out that one who exposes himself to a known danger is not, ex necessitate, guilty of negligence. The test is whether an ordinarily prudent person would, in the same or like circumstances, have incurred the risk which such conduct involved; and, where fair-minded men might honestly differ concerning whether the conduct was such as an ordinarily prudent person would have pursued in such existing circumstances, the problem must be solved by the trier of the facts. [Citations omitted.]

It cannot in reason be stated that a fireman has no protective rights whatever while engaged in the pursuit of his employment. It is contemplated that a fireman in the performance of his duty shall endeavor to extinguish fires however caused and encounter those risks and hazards which are ordinarily incidental to such an undertaking and which may be reasonably expected to exist in the situation in which he places himself. It does not follow that a fireman must be deemed as a matter of law to have voluntarily assumed all hidden, unknown, and extrahazardous dangers which in the existing conditions would not be reasonably anticipated or foreseen. It was the duty of the plaintiffs to exercise reasonable care and prudence for their own safety in the performance of their duty as firemen. The mere fact that it now appears that the plaintiffs imperiled their personal safety in rendering the service required by their employment does not conclusively establish their contributory negligence as a matter of law. Their conduct must be viewed in the light of the conditions and circumstances then existing. In the absence of some notice or knowledge, circumstantial or otherwise to the contrary, the plaintiffs had a right to presume that the defendant had complied with the applicable provisions of the ordinance of the city. *[Citation omitted.]* Although those who voluntarily pursue employments necessarily fraught with danger are taken to assume such risks and dangers as are normally and necessarily incident to the occupation, they are not treated as assuming obscure and unknown risks, which are not naturally incident to the occupation and which, in the existing conditions, would not be reasonably observed and appreciated. Recognizing the difference between assumption of risk and contributory negligence *[citations omitted]*, the bare facts alleged in the complaint do not justify the conclusion, as a matter of law, that the plaintiffs either knowingly assumed the risk resulting in their injury or were guilty of contributory negligence.

The allegations of the complaint are legally sufficient to withstand the present motion. The motion will therefore be denied to the end that all of the true and actual circumstances surrounding the occurrence of the explosion may be fully disclosed and the conduct and situation of the plaintiff may be more definitely indicated by testimony introduced at the trial.

Comparative Negligence: Under the doctrine of contributory negligence, the plaintiff is barred from recovery. This complete bar to recovery has been widely criticized. Under the doctrine of comparative negligence, which is in effect in many states either through legislation or judicial decisions, the relative degree of negligence on the part of both parties is assessed. For example, if a plaintiff is found to be 40 percent correspondingly negligent in contributing, through negligence, to the plaintiff's own harm, injury, or damage, then the plaintiff will be allowed to recover only 60 percent of the damages. Under the doctrine of contributory negligence, the plaintiff also would be allowed no recovery at all.

Last Clear Chance Doctrine: Under the last clear chance doctrine, the plaintiff's contributory negligence may be completely ignored. If it can be shown that the defendant was aware of the plaintiff's negligence, had an opportunity to avoid the danger and did nothing about it, then the plaintiff's contributory negligence will not bar recovery. In *The Complete Guide to Everyday Law,* Samuel Kling summarizes the application of the last clear chance doctrine as follows:[19]

> Under the doctrine of "the last clear chance" a defendant will still be liable if he was aware of the plaintiff's peril, or was unaware of it through carelessness, and had the opportunity to avert the accident. If, for example, a motorist is stalled on the railroad tracks through no fault of his own and a train smashes into the car when the engineer could have stopped the train, the railroad might still be held guilty of negligence, because it had "a last clear chance" to avoid the accident. The "last clear chance" doctrine does not apply, however, when the negligence is coincident and concurring or when both parties are actively and simultaneously negligent.

TRESPASS

Basically, there are two kinds of trespass in tort: (1) personal trespass, examples of which are assault and battery, false imprisonment, or infliction of mental and/or emotional distress, and (2) property trespass, which is the intentional interference with the land or property of another.

Trespass on the property of another is of vital concern to members of the fire service, whether during actual fire fighting conditions or in the many inspection tasks and obligations undertaken by members of the fire service.

A tort, it will be recalled, is a breach of legal duty by one person to another. When a person enters onto the property of another, the duty owed that person by the owner or possessor of the property depends on the legal capacity of the person who enters the property. For example, is that person an invitee, a licensee, or a trespasser? Different degrees of legal duty are owed to each, and harm and subsequent liability are judged on these terms.

The laws of trespass involve, among other things, landlord-tenant relationships, violation of air-space, and environmental rights and protection. It is those aspects of the law of trespass most relevant to the fire service with which the remainder of this chapter is concerned.

Although trespass can be unintentional, liability arises only if the unintentional intruder's behavior is reckless, negligent, or dangerous. An intentional trespasser is liable even if no harm has been caused and even if the trespass was a result of mistaken judgment, unless the mistake is the result of the owner's inducement.

The law of trespass has become more flexible with regard to nonconsensual trespass privileges. The range of circumstances to which such nonconsensual privileges apply is wide but is subject, of course, to the statutory provisions of the

---

[19]Samuel G. Kling, *The Complete Guide to Everyday Law,* Follett Publishing Company, Chicago, 1973, p. 348.

state. Some of the nonconsensual privileges that have been recognized are public necessity and entry by a public official to abate a public nuisance. The *Restatement of Torts, Second* defines public necessity as follows:[20]

> One is privileged to enter land in the possession of another if it is, or if the actor reasonably believes it to be, necessary for the purpose of averting an imminent public diasaster.

When one enters the land of another, the possessor of the land has responsibilities and legal duties to the person who enters the land, based on that person's legal status. The duty owed by an owner or possessor of the land to one who is on the land may be lessened, depending upon whether the person is an invitee, a licensee, or a trespasser.

TRESPASSER

The lowest duty of care is owed by the owner or possessor of the land to a trespasser. However, even to the trespasser the owner owes a duty. Generally, the owner is not free to inflict wanton or intentional injury on the trespasser. The exact nature of the legal duty owed to trespassers varies from state to state.

LICENSEE

The major point of difference between a trespasser and a licensee is that a licensee enters the land with the owner's consent. Generally, the licensee comes for the licensee's own purposes and not in the interests of the possessor or owner of the land. A licensee is not considered to have any right to demand that the land be made safe for this kind of "intrusion" and, in general, must assume responsibility for any risks encountered. Examples of the kinds of people who enter other's land with nothing more than the possessor's or owner's consent are: salespeople, people taking short cuts across property, servants, social visitors, those who come to borrow items from the owner or possessor, and so forth. There is no duty to a licensee on the part of the owner or possessor of the land except for the duty, as to a trespasser, not to inflict willful or wanton injury. No duty is required to assure the licensee that the premises have been made or are certified safe for the licensee. However, courts, concerned with giving priority to human safety, have promulgated the general tenet that an occupier of land has an obligation to exercise reasonable care to assure the safety and protection of the licensee. The *Restatement of Torts, Second* specifies these duties as follows:[21]

§ 343. Dangerous Conditions
Known to Possessor
A possessor of land is subject to liability for physical harm caused to licensees by a condition on the land if, but only if:

---

[20]§196, *Restatement of Torts, Second.*
[21]§342, *Restatement of Torts, Second.*

(a) the possessor knows or has reason to know of the condition and should realize that it involves an unreasonable risk of harm to such licensees, and should expect that they will not discover or realize the danger, and

(b) he fails to exercise reasonable care to make the condition safe, or to warn the licensees of the condition and the risk involved, and

(c) the licensees do not know or have reason to know of the condition and the risk involved.

INVITEE

Based on the distinctions between "trespasser" and "licensee," what distinctions are left to the invitee? The distinction between "trespasser" and "licensee" is one of consent. A licensee is not on the owner's property for business that will benefit the owner. An owner does not have to certify to a licensee that the property has been made safe for the licensee's entry. Basically, these two legal duties on the part of an owner distinguish a licensee from an invitee. An owner or possessor of property owes the highest duty of care to an invitee. Typically, invitees are customers in stores, theaters, and other businesses open to the public, and workers employed on the premises of the owner or possessor. Such persons as these, although a benefit will accrue to them as well, are functioning in the interests of the owner or possessor of the property. People who accompany customers into a store with no intention of purchasing anything themselves, those who go into a place open to the public to use the telephone, and those who go to see people off at railway stations have also been considered invitees by the courts.

The *Restatement of Torts, Second* has defined the duties of owners or possessors of property to an invitee as follows:[22]

§ 343. Dangerous Conditions
Known to or Discoverable by Possessor
A possessor of land is subject to liability for physical harm caused to his invitees by a condition on the land if, but only if, he:

(a) knows or by the exercise of reasonable care would discover the condition and should realize that it involves an unreasonable risk of harm to such invitees, and

(b) should expect that they will not discover or realize the danger or will fail to protect themselves against it, and

(c) fails to exercise reasonable care to protect them against the danger.

PUBLIC OFFICERS AND EMPLOYEES

Corpus Juris Secundum defines public convenience, necessity, or safety as follows:[23]

---

[22]§343, *Restatement of Torts, Second.*
[23]86 C.J.S. 930.

It is a maxim of the common law that, where public convenience and necessity come in conflict with private right, the latter must yield to the former; hence, the necessity or safety of the public or of some portion, or member, of the public at times will permit the doing of acts to prevent imminent loss or destruction or great general inconvenience, which without the existence of such necessity would be a wrongful invasion of right, but which, in view of the element of necessity, are not tortious. This doctrine of necessity applies with special force to the preservation of human life, but is not confined thereto. However, although conduct be nontortious by reason of necessity, a subsequent exceeding of one's authority may constitute mone a trespasser *ab initio.* . . .

The courts have had many problems resolving issues concerning those who enter another's property in performance of their public duty, but not with the owner's or occupier's consent. Such public officers and/or employees do not readily fit into the established legal classifications, and such public officers or employees are not trespassers since they are privileged to enter the premises. However, this privilege is not conferred through the owner's or occupier's consent.

The problem of whether public officers or employees are invitees or licensees revolves around the duty or care owed to such persons and the resultant liability of the owner to these visitors. The following case study involves a claim made by a fire fighter who brought action against a landowner for injuries sustained in the performance of duty. As you read the decision, consider how it demonstrates the problems involved in the classification of public officers or employees as invitees or licensees.

Ernest M. KRAUTH, Plaintiff-Appellant,

v.

**Israel GELLER and Buckingham Homes,
Inc., a corporation of the State
of New Jersey, also known as
Buckingham Builders, Defendants-
Respondents.**

Supreme Court of New Jersey
Argued Nov. 9, 1959.
Decided Jan. 11, 1960.
157 A.2d 129

Action by a fireman against a landowner for injuries sustained by fireman in fall from a balcony on which the railing had not yet been installed in house under construction when fire company was called to the house to extinguish an overheated salamander which was burning unattended in the house. The plaintiff obtained a judgment on the jury verdict and the Appellate Division, 54 N.J.Super. 442, 149 A.2d 271, reversed with directions that judgment be entered for the defendant and the fireman appealed. The Supreme Court, Weintraub, C.J., held the evidence did not establish wanton misconduct of the owner.

Judgment of the Appellate Division affirmed.

Jacobs. J., dissented.

The opinion of the court was delivered by

WEINTRAUB, C. J.

This case involves the liability of the owner or occupier of lands to a fireman injured while discharging his duty as a public employee. Plaintiff obtained a judgment on a jury verdict. The Appellate Division reversed with the direction that judgment be entered for defendant. One judge dissented, agreeing that a reversal was required by errors in the charge to the jury, but concluding that upon the facts plaintiff could succeed on a retrial. 54 N.J.Super. 442, 149 A.2d 271 (1959). Plaintiff's appeal comes to us as of right. Constitution of 1947, Article VI, § 5, par. 1(b); R.R. 1:2-1 (b). (Plaintiff has since passed away and his administratrix was substituted. "Plaintiff," as hereinafter used, refers to the deceased.)

We agree the judgment was properly reversed for the reasons upon which all members of the Appellate Division agreed. The sole matter that concerns us is the correctness of the holding of the majority that upon the most favorable view of the situation plaintiff must fail as a matter of law.

Much has been written with respect to the duty owed to and the status of a fireman who enters private property pursuant to his public employment. He is not a trespasser, for he enters pursuant to public right. Although it is frequently said he is a licensee rather than an invitee, it has been correctly observed that he falls within neither category, for his entry does not depend upon permission or invitation of the owner or occupier, nor may they deny him admittance. Hence his situation does not fit comfortably within the traditional concepts. *[Citations omitted.]* . . .

In what circumstances should the owner or occupier respond to the injured fireman? That the misfortune here experienced by a fireman was well within the range of foreseeability cannot be disputed. But liability is not always coextensive with foreseeability of harm. The question is ultimately one of public policy, and the answer must be distilled from the relevant factors involved upon an inquiry into what is fair and just. *[Citations omitted.]*

It is quite generally agreed the owner or occupier is not liable to a paid fireman for negligence with respect to the creation of a fire. *[Citations omitted.]* The rationale of the prevailing rule is sometimes stated in terms of "assumption of risk," used doubtless in the so-called "primary" sense of the term and meaning that the defendant did not breach a duty owed, rather than that the fireman was guilty of contributory fault in responding to his public duty. *[Citations omitted.]*

Stated affirmatively, what is meant is that it is the fireman's business to deal with that very hazard and hence, perhaps by analogy to the contractor engaged as an expert to remedy dangerous situations, he cannot complain of negligence in the creation of the very occasion for his engagement. In terms of duty, it may be said there is none owed the fireman to exercise care so as not to require the special services for which he is trained and paid. Probably most fires are attributable to negligence, and in the final analysis the policy decision is that it would be too burdensome to charge all who carelessly cause or fail to prevent fires with the injuries suffered by the expert retained with public funds to deal with those inevitable, although negligently created, occurrences. Hence, for that risk, the fireman should receive appropriate compensation from the public he serves, both in pay which reflects the hazard and in workmen's compensation benefits for the consequences of the inherent risks of the calling. *[Citations omitted.]*

Although there is virtual unanimity with respect to nonliability for negligence as to the creation of fire, there is appreciable authority which would impose liability upon the land occupier for negligence with respect to conditions creating undue risks of injury beyond those inevitably involved in fire fighting. Thus it has been held that a fireman may recover if the injurious hazard was created in violation of statute or ordinance. *[Citations omitted.]* So also, he has prevailed if the occupier failed to utilize an available opportunity to warn him of a hidden peril. *[Citations omitted.]* And the land occupier has been held where he failed to exercise due care with respect to the condition of places intended as a means of access by contemplated visitors. *[Citations omitted.]* There are, of course, decisions the other way, but where liability is found the emphasis is not upon culpability with respect to the inception of the fire but rather with respect to the other risks of injury we have described.

The present case does not fall within any of the exceptions, if such they may be called, outlined in the paragraph above. Defendant was the owner and builder of a one family home under construction at the time of the occurrence. While proceeding along an interior balcony and meaning to descend the stairs, plaintiff mistook layers of smoke for them and fell. Neither the balcony nor the stairs were protected by a railing — the construction simply had not reached that stage. No breach of duty could be found with respect to the condition of the premises. And defendant not being on the scene, there was no culpable failure to seize an opportunity to warn of a hidden danger. In fact, plaintiff, who had been on the premises a few days before the accident, admitted he was fully aware of the precise state of construction of the balcony and stairs.

What plaintiff seeks is an inroad upon the basic rule that the occupier is not liable to a fireman for the creation of a fire. Placing himself in the category of a licensee, or at least by analogy thereto, he urges defendant was guilty of "wanton" conduct. The allegedly wanton conduct occurred before plaintiff entered the premises and at most can consist of precipitating unnecessarily a call upon the fire force. Defendant loaned a salamander to the plasterer (defendant claims he was an independent contractor, but we here assume he was an employee) who used it to provide heat and thus to dry the plaster. The device consists of a container for the fuel and a pipe or stack of some four feet in height, with a movable cap or cover. If the controls are set properly, the flame remains within the pipe, causing it, as we read the record, to become cherry red, a condition which from the outside could suggest an uncontrolled fire. On the evenings of March 1, 2, and 5, as the jury could find, the flame actually rose above the top of the pipe resulting in danger of conflagration of the premises. The fire department answered the calls of the neighbors on each of those days. The plaintiff was a member of the team that responded on March 1 and 5, the accident occurring on the latter day.

On March 2 (prior to the incident of that day), the assistant fire chief talked with defendant. He said,

> "I asked him if he would please leave a man there in attendance with it or if he would check it himself." . . .

and

> ". . . It was just a matter of cooperation between him and the fire department. We already had one run there and he knew what happened. He was building a house, and if he didn't want to have another run we were asking his cooperation to leave a man or to see that it was taken care of. That's it."

Defendant's version was that the assistant chief "requested that I have placed above the top of the salamander a protective covering, either metal or sheet rock," and this advice he passed on to the plasterer, a man who, he said, was experienced in the use of a salamander.

Plaintiff of course is entitled to the most favorable view of the testimony. But there was no evidence that prudent practice required constant attendance of a salamander. There was no testimony that a salamander inevitably or likely will behave as this one did. Defendant testified without contradiction that salamanders were in regular use in the area. The assistant chief said a salamander could go out of control for several reasons but was unable to tell why this one did.

We are not concerned with the liability to a fireman of an arsonist or one who deliberately induces a false alarm. Rather we are asked to hold that "wanton" conduct resulting in a fire and consequent alarm will suffice. Wantonness is not too precise a concept. It is something less than intentional hurt and, so viewed, it is an advanced degree of negligent misconduct. In the context of the policy considerations which underlie the rule of nonliability for negligence with respect to the origination of a fire, it is debatable whether or not degrees of culpability are at all pertinent.

At any rate, we need not decide the question since we can see no basis for a claim of wanton misconduct as the term is ordinarily defined. To warrant that characterization, the act or the omission to discharge a duty must be intentional, and coupled with a consciousness, actual or imputed, of a high degree of probability that harm, here to a fireman, will ensue. *[Citations omitted.]* Plaintiff does not suggest that the incidents of March 1 or March 2 fall within this area. Rather he claims a third episode crosses the line. Three such occurrences within five days evoke exasperation, but, however indicative of a lack of care, they do not reveal a purpose to cause a conflagration or to attract the fire department. And although injury to a fireman is surely foreseeable, and this despite the rather freak circumstance that layers of smoke simulated a stairway, yet there was no evidence of a consciousness, actual or imputable, of a high degree of probability that harm would befall a fireman. In fact on the first two occasions none resulted although the physical scene was identical.

We think it would be artificial to squeeze the situation within the category of wanton conduct. Rather, in proper perspective, the question is whether the land occupier should be held liable for a second or third act of carelessness. If the time interval had been greater, it would be more apparent that the issue is the one just stated, and the sense of exasperation to which we referred would be less pronounced or absent. But if the prevailing rule of nonliability to a paid fireman with respect to the origination of the occasion for his work is sound, and we are not persuaded otherwise, we think it unwise to attempt to substitute, or to qualify it with, a formula in which the factor of numbers and time intervals will somehow be decisive.

There being no suggestion that plaintiff could fare better upon a retrial, the judgment of the Appellate Division directing the entry of judgment is affirmed. No costs.

For affirmance: Chief Justice WEINTRAUB and Justices BURLING, FRANCIS and PROCTOR — 4.

For reversal: Justice JACOBS — 1.

In 1960, the Supreme Court of Illinois reviewed the entire question of the legal classifications of visitors on premises, the status of fire fighters with regard to these classifications, and the resultant obligations on the part of the owner or occupier of the premises to fire fighters. The following decision summarizes the law up to modern times and demonstrates one aspect of "the living law" in giving a new dimension and interpretation to "American common law."

Gino DINI et al., Appellants,

v.

Irving NAIDITCH et al., Appellees

Supreme Court of Illinois.
Sept. 30, 1960.
As Modified on Denial of Rehearing
Nov. 30, 1960.
170 N.E.2d 881

Actions for injury to and death of firemen in a fire on the defendants' property.
The Superior Court, Cook County, John J. Lupe and Thomas E. Kluczynski, JJ.,
entered judgments for defendants, and appeals were taken. The Supreme Court,
Bristow, J., held that a wife could maintain an action for loss of consortium due to
the negligent injury of her husband, and also that a fireman performing his duty
on private property is an invitee, not a licensee, so that recovery from a landowner
may be based on a failure to exercise reasonable care.

Reversed and remanded with directions.

Schaefer, C. J., and Klingbiel and House, JJ., dissented.

BRISTOW, Justice.

This is a combined appeal by plaintiff Elizabeth Dini from a summary judgment dismissing her action for loss of consortium, and by plaintiff Gino Dini and
plaintiff Lillian M. Duller, as administratrix of the estate of Edward J. Duller,
from a judgment notwithstanding the verdict entered in their actions for the injury and death of city firemen, allegedly caused by defendants' negligence and
statutory violations in the maintenance of their premises. The superior court of
Cook County set aside jury verdicts awarding damages of $235,000 for personal in-
juries sustained by fireman Gino Dini, and $20,000 for the wrongful death of fire
captain Edward Duller, on the ground that there was no legal basis for liability.

Our jurisdiction to review the cause on this direct appeal having been determined on motion, we must now consider the two major issues presented by this
appeal: first, whether landowners and operators are liable to city firemen for the
negligent maintenance of their premises in violation of certain fire ordinances;
and secondly, whether a wife is entitled to damages for loss of consortium due to
the negligent injury of her husband.

The operative facts discernible from the controverted testimony are that since
1946 defendants Albert and Rae Naiditch have been the owners of a four-story
brick building erected in 1896 at the intersection of Milwaukee Avenue and North
Green Street in Chicago. Each floor contained about 6,000 square feet of space.

Most of the ground floor was occupied by Naiditch for selling store and restaurant fixtures; the basement was used for storage for that business and for the boiler; and the second, third, and fourth floors of the building were operated as the Green Mill Hotel. There were 27 rooms, most of which were single, on each of the three hotel floors, and five or six rooms on each floor had kitchens. These premises were occupied by some 84 persons. Access to the three hotel floors from the vestibule of the Green Street entrance was by means of a wooden stairway, approximately six feet wide, supported by stringers that were nailed to the walls rather than recessed.

Adjacent to the stairway landing on the second floor was an office maintained by defendant Kenneth Oda and Thomas Sato, who, as lessees of Naiditch since 1951, operated the hotel as partners at the time of the fire. Next to that office, and some 17 or 20 feet from the stairwell was a storage room in which, according to the uncontroverted testimony, paint and benzene were kept, including a four or five gallon aluminum can of benzene at the time of the fire. There was also testimony that there were numerous paint cans, brushes and rags in the office, despite a provision in the lease that no naphtha, benzene or other enumerated flammable products were to be kept on the premises without the written permission of Naiditch.

The record respecting the condition of the premises prior to and at the time of the fire is extensive. Apparently there was no compliance with the lease provision that the lessee would spend an average of $1,500 annually for maintenance and improvement of the premises and give Naiditch monthly itemizations of such expenditures. Naiditch had his attorney write to Oda and Sato demanding that repairs be made, and later filed a lawsuit, which culminated with $1,500 put in escrow for repairs. Neither Naiditch nor Oda, however, admitted having any records whatsoever respecting the maintenance of the property. While the lease required Naiditch to inspect the hotel once a month, he admitted on trial that he had not made such regular inspections, and that he had not seen the hotel some six weeks prior to the fire, having been on vacation.

It also appears from the testimony of residents of the hotel that there were oil drums converted into open garbage cans in the hotel corridors, which were emptied only two or three times a week, that paper and other waste was piled in the corridors beside the cans, that the walls were cracked and rain had leaked through the roof into the fourth floor hallway, and that the janitor, Jimmy Sato, was "always drunk," but retained despite complaints. It also appears from the record that prior to the fire, defendants' attention had been called to nine separate violations of city ordinances within the building, although there is some controversy as to the findings of a former building inspector who testified for defendant.

With reference to the condition of the premises on the night of the fire, one of the residents testified that two garbage cans on the first floor of the hotel were full and overflowing, and that paper was piled about a foot high on the floor. Another resident, who returned home about 11 P.M., said that three or four garbage cans in his section of the fourth floor were full, with paper piled around the cans so that he had to "cross around." According to firemen who fought the blaze, they could see rubbish in the hotel corridor when they reached the second floor of the building, as well as trash and litter on the stairs. Moreover, there were no fire doors, according to the testimony of the deputy fire marshal who was on the premises during the fire and made a minute inspection after the fire, and that of the chief building inspector, and of the division fire marshal who was also inside the building during the fire and directed the fighting of the blaze. Nor were there any fire extinguishers of any kind in the hotel, according to the original admission of Naiditch and the testimony of residents who had lived there for three or four years, and that of the police detective who examined the premises after the fire.

With reference to the occurrence, it appears that at about 12:50 A.M. on April 28, 1955, after the fire had apparently been burning for at least thirty-five minutes, a police officer on duty some blocks away was attracted by the flames. He drove to the scene, where he found the Green Mill Hotel burning, and he radioed a report. Within minutes fire equipment arrived, but the flames were then shooting through the hotel roof and people were hanging out of the windows yelling and screaming.

According to the fire battalion chief, the fire was located in the stairway at the Green Street entrance, blocking the exit. He therefore ordered an engine company up the inside of the stairway to cool the fire off in order to effect rescue operations. Fire captain Duller, and firemen Smith, Collins and Dini, who was carrying a hose on his shoulder, entered the building through the Green Street entrance, and proceeded up the stairs to the second floor landing, where they could hear the roar of the fire above. Collins was sent for a smaller hose, and Dini was left on the landing to couple the smaller hose into a shutoff pipe, while Duller and Smith started on up to the third floor where they could see the fire raging above them. At that moment, and without any warning, the entire stairway collapsed and fell into a heap at the first floor level. Captain Duller was buried in the burning debris, and his body was not recovered until the following day. Smith, who escaped, testified that something hit him on the head and drove him through the stairs onto the area below. Dini was pinned in a pile of burning wood, but extricated himself with great difficulty, and made his way out in flames which he extinguished by jumping into a puddle of water at the curbing.

Dini was so severely burned that his recovery was in doubt for two months. He suffered third degree burns on his scalp, face, neck, chest, arms, left leg and knee. Both outer ears were almost completely burned off, as were his nose, lips and eyelids. He also suffered severe burns inside his mouth and throat, which not only made breathing difficult, and swallowing and eating impossible, but interfered with the administration of anesthesia. When the burned skin sloughed off, leaving raw areas, it was necessary, in order to prevent infection, to make some thirteen skin grafts from other areas of his body. That phase of his hospitalization treatment lasted until August 13, 1955; and from October 13, 1955, to February 25, 1959, Dini underwent some fifty-nine additional operations for skin grafting and for the reconstruction of eyelids, and ears, and the removal of scar tissue.

His injuries are permanent insofar as loss of motion and flexion in the affected members is concerned, and insofar as they affect his appearance. Moreover, since March 1956, except for periods of hospitalization, Dini has worked only approximately three hours a day in the Fire Prevention Bureau operating a typewriter, with the resulting loss of income.

Captain Duller was 54 years of age at the time of his death in the fire. He had been married some 21 years, and left surviving a widow, an 18-year-old son, and a 16-year-old daughter.

On the basis of substantially the foregoing facts, the jury returned the verdicts for plaintiffs Dini and Duller, as hereinbefore noted, and the superior court entered judgment for defendants notwithstanding the verdict on the ground that there was no basis of liability, since the fire ordinances violated by defendants were not enacted for the benefit of firemen. Moreover, the court entered a summary judgment dismissing the complaint of Elizabeth Dini on the ground that a wife has no cause of action for loss of consortium resulting from the negligent injury of her husband.

In reviewing this cause, we shall consider first the claims of Gino Dini and administratrix Lillian Duller, which involve the issue of the landowners' liability to firemen for the negligent maintenance of the premises. We believe that this question, considered last by this court in 1892, should properly be re-examined in its entirety.

It must be recognized at the outset that the English common law, from which our law is derived, was part and parcel of a social system in which the landowners were the backbone, and that it was inevitable that in such a legal climate supreme importance would be attached to proprietary interests. Bohlen, *Fifty Years of Torts*, 50 Harv.L.Rev. 725, *et seq.* It was the feudal conception that the landowner was sovereign within his own boundaries that gave birth to the rule that the only duty a landowner owed a licensee was not to willfully or wantonly injure him. *[Citation omitted.]* It was, then, hardly a "giant step" to give the label of "licensee" to a member of the fire department who, in an emergency, enters the premises in the discharge of his duty, and to hold, as the early cases did, that the owner or occupant owed the fireman no greater duty than to refrain from infliction of willful or intentional injury. *[Citations omitted.]*

However, the history of the law on the subject of landowners and "licensees" shows a tendency to whittle away a rule which no longer conforms to public opinion. As Bohlen points out, "Like so many cases in which a barbaric formula has been retained, its content has been so modified by interpretation as to remove much of its inhumanity." 50 Harv.L.Rev. 725, 735. Thus, to avoid extending what has been deemed a "harsh rule" [citations omitted], courts have held that firemen were entitled to be warned of "hidden dangers" or "unusual hazards" known to the landowner or occupant. [Citations omitted.]

Other courts have avoided the harsh rule by finding from slight variations of circumstances that the injured fireman was an "invitee" [citation omitted]; or a "business visitor" to whom the landowner owed a duty of reasonable care to keep the premises safe. [Citation omitted.]

Still other courts, as well as legal scholars, have forthrightly rejected the label of "licensee," with its concomitant set of rights and duties for firemen. [Citations omitted.]

In the Meiers case the court allowed a fireman to recover for injuries caused when he stepped into a hole while fighting a fire on defendant's premises. While the court did not clearly define the status of a fireman, it refused to categorize him as a "licensee." In a closely circumscribed opinion, the court allowed recovery to "one not a licensee, entering business property as of right over a way prepared as a means of access for those entitled to enter, who is injured by the negligence of the owner in failing to keep that way in a reasonably safe condition for those using it as it was intended to be used." [229 N.Y. 10, 127 N.E. 493.]

The Minnesota court, in Shypulski v. Waldorf Paper Products Co., 232 Minn. 394, 45 N.W.2d 549, observed that since firemen have a unique status, it follows that the duties owed to them may properly be unique; and that same approach was followed by our Illinois Appellate Court in Ryan v. Chicago & Northwestern R. Co., 315 Ill.App. 65, 75, 42 N.E.2d 128. After reviewing the status of public officers, including firemen, who come on the land in the exercise of a legal privilege, the Illinois court allowed damages for the death of a police officer while on defendant's right of way, caused by the negligent operation of defendant's trains, on the ground that the owner was obligated to use reasonable care not to injure the police officer.

In reviewing the law on this issue, we note further that this legal fiction that firemen are licensees to whom no duty of reasonable care is owed is without any logical foundation. [Citations omitted.] It is highly illogical to say that a fireman who enters the premises quite independently of either invitation or consent cannot be an invitee because there has been no invitation, but can be a licensee even though there has been no permission. The lack of logic is even more patent when we realize that the courts have not applied the term "licensee" to other types of public employees required to come on another's premises in the performance of their duties, and to whom the duty of reasonable care is owed. If benefit to the landowner is the decisive factor, it is difficult to perceive why a fireman is not entitled to that duty of care, or how the landowner derives a greater benefit from the visit of other public officials, such as postmen, water meter readers and revenue inspectors, than from the fireman who comes to prevent the destruction of his property. *[Citation omitted.]*

Consequently, it is our opinion that since the common-law rule labelling firemen as licensees is but an illogical anachronism, originating in a vastly different social order, and pock-marked by judicial refinements, it should not be perpetuated in the name of *"stare decisis."* That doctrine does not confine our courts to the "Calf Path," nor to any rule currently enjoying a numerical superiority of adherents. *"Stare decisis"* ought not to be the excuse for decision where reason is lacking. *[Citation omitted.]* As aptly stated by Lord Goddard in *Best* v. *Samuel Fox & Co.* (1952) Appeal cases, 716, 731, "English law is free neither of some anomalies, nor of everything illogical, but this is no reason for extending them."

Inasmuch as firemen obviously confer on landowners economic and other benefits which are a recognized basis for imposing the common-law duty of reasonable care (*Restatement, Torts,* § 343a; Harper, Torts § 96; Prosser, 26 Minn.L.Rev. 573, 574; 35 Mich.L.Rev. 1161), we would agree with the court in the Meiers case [*Meiers* v. *Fred Kock Brewery,* 1920, 229 N.Y. 10, 127 N.E. 491, 13 A.L.R. 663] and with its adherents, that an action should lie against a landowner for failure to exercise reasonable care in the maintenance of his property resulting in the injury or death of a firemen rightfully on the premises, fighting the fire at a place where he might reasonably be expected to be.

This interpretation does not run counter to any imposing body of precedent in this jurisdiction. As hereinbefore noted, this court has only considered the problem once, and that was in 1892 in *Gibson* v. *Leonard,* 143 Ill. 182, 32 N.E. 182, 17 L.R.A. 588. A careful reading of that case, however, indicates that the court did not analyze the common-law status of firemen, but was concerned primarily with whether a volunteer member of a fire insurance patrol, injured by a defective elevator counterweight in the basement of a building, was entitled to the protection of a particular safety ordinance. Nevertheless, for the clarification of the law, insofar as any language contained therein might be inconsistent with our interpretation of the common law in this case, it must be deemed to be overruled, along with any Appellate Court cases following the archaic licensee concept. *[Citations omitted.]*

In the instant case, from the evidence previously noted that defendants failed to provide fire doors or fire extinguishers, permitted the accumulation of trash and litter in the corridors, and had benzene stored in close proximity to the inadequately constructed wooden stairway where the fire was located, the jury could have found that defendants failed to keep the premises in a reasonably safe condition and that the hazard of fire, and loss of life fighting it, was reasonably foreseeable. Hence, since there was legal basis of liability in this case, it was error to set aside the jury verdict and enter judgment notwithstanding the verdict.

Plaintiffs have alleged, as a second basis of liability, defendants' violation of certain safety ordinances. We have recognized the rule that the violation of a statute or ordinance designed for the protection of human life or property is *prima facie* evidence of negligence, and that the party injured thereby has a cause of action, provided he comes within the purview of the particular ordinance or statute, and the injury has a direct and proximate connection with the violation. *[Citations omitted.]*

In determining whether the violation of safety ordinances gives an injured fireman a right of action against the guilty party, we find no unanimity in the case law. Apparently the crucial question in each case is whether it was intended that a fireman should come within the scope of the protection afforded by the statute or ordinance. [141 A.L.R. 584, 592.]

Plaintiffs have called our attention to defendants' violation of certain provisions of the municipal code of Chicago requiring, for structures such as the Green Mill Hotel, enclosed stairwells (sec. 62 — 3.2); fire doors (63 — 3.6); fire extinguishers (64 — 4.1); and specifying that oil rags and waste shall be kept during the day in approved waste cans of heavy galvanized iron with self-closing covers, and shall be removed at night, and that rubbish shall not be allowed to accumulate in any part of any building (90 — 25). These ordinances further provide that, "It shall be unlawful to continue the use of or occupy any structure or place which does not comply with these provisions of this code *which are intended to prevent a disastrous fire or loss of life in case of fire,* until the changes, alterations, repairs or requirements found necessary to place the building in a safe condition have been made." (Sec. 90 — 3.) *(Italics ours.)*

Defendants, however, argue that firemen are not within the purview of these ordinances, and that, therefore, plaintiffs cannot predicate liability thereon. In support of that contention, defendants cite the aforementioned Gibson case, 143 Ill. 182, 32 N.E. 182. Recovery was denied in that case on the ground that the ordinance on which the fireman based his claim specified that it was designed to insure against injury to a specified class — employees of the building; hence, the court reasoned that a volunteer fireman was not within that class for whose protection the ordinance was passed. The court distinguished the leading Massachusetts case of *Parker* v. *Barnard,* 1882, 135 Mass. 116, also cited by plaintiffs herein, on the ground that recovery by the injured policeman in that case was predicated on the landowners' violation of a safety statute respecting elevator shafts, which was general in its terms, and stated at page 195 of 143 Ill., at page 185 of 32 N.E.: "But here, in section 1074 of the ordinances of Chicago, *instead of general language, such as was used in the statute considered in Parker v. Barnard,* is found language which shows, in express terms that the ordinance was intended only for the protection and benefit of employees in factories, workshops, and other places or structures where machinery is employed." *(Italics ours.)*

In the light of that precise statement in the Gibson case, differentiating it, in effect, from the case at bar where the safety ordinance is general in its terms and has the avowed purpose of preventing loss of life in case of fire, regardless of whose life it may be, we can hardly find the Gibson case to be a determinative precedent for the denial of this action. On the contrary, the entire implication of that case is that the ordinances herein, being general in terms, might properly include firemen within their protection.

Such an interpretation, moreover, would be consistent with the case law, both in Illinois and in other states. *Bandosz v. Daigger & Co.,* 255 Ill.App. 494; 141 A.L.R. 584, 592, and cases cited. Thus, in *Drake v. Fenton,* 1912, 237 Pa. 8, at page 12, 85 A. 14, at page 14, where the injured fireman was allowed to recover from a landowner who violated a statute requiring elevator shafts to be kept closed, the Pennsylvania court stated "It [the Act of 1903] is not restricted to a specific class, but is general in its terms; and it is a reasonable construction to hold that it was passed for the benefit of all persons lawfully on the premises." Similarly, in *Maloney v. Hearst Hotels Corp.,* 1937, 274 N.Y. 106, 8 N.E.2d 296, at page 297, the court stated ". . . we have in this case a direct violation of ordinances which were enacted for the benefit of firemen as well as guests in the hotels; at least firemen entering into the premises had a right to assume that the law in this particular had been complied with." *[Citations omitted.]* In contrast, where the fireman's claims were rejected, the statutes or ordinances on which the claims were predicated were found to be expressly for the protection of "employees"; *[Citations omitted]* or for persons lodged or residing on the premises. *[Citations omitted.]*

In accordance with this approach our Appellate Court in *Bandosz v. Daigger & Co.* allowed recovery in a wrongful death action where a fireman was killed by the explosion of inflammable liquid stored in a basement in violation of an ordinance and statute. The statute provided, in substance, that it shall be unlawful to store benzene or any combustibles in such manner or *under such circumstances as will jeopardize life and property,* and the ordinances prohibited the storage of more than 10 gallons of benzol or ether in any building within a certain area. The Court rejected the Gibson case as determinative, and relied instead on *Parker v. Barnard,* on the ground that "the ordinance and statute were general in their terms, and not limited in their operation to any particular class of persons." Consequently the court held that firemen were entitled to the protection of such laws, and recognized a claim based on their violation.

There is a similarity between the italicized general language of the statute involved in the Bandosz case, and the ordinances herein, which are intended "to prevent a disastrous fire or loss of life in case of fire." This analogy is not marred by defendants' attempt to distinguish that case and the others allowing such claims on the ground that the statutes in those cases related to explosives or combustible materials, whereas the safety ordinances herein deal only with the maintenance of the premises. The cases, however, warrant no such refinement. The criterion they impose is based not upon the subject matter of the safety statutes, but rather upon the intended coverage, *i.e.*, whether the law is general in its application, or restricted by its terms to a particular class. This is evident from the language in both the cases recognizing and those denying liability. In no case was the fireman's action rejected merely because the statute on which he based his claim did not deal with combustibles or with extra-hazardous materials, as the defendants suggest.

We must, therefore, recognize, as an additional basis of liability, defendants' alleged violations of the particular firesafety ordinances and that the owners, defendants Naiditch, can not avoid such liability on the ground that the premises were leased. *[Citations omitted.]* We have already noted that there was some evidence from which the jury could reasonably find that defendants were in fact guilty of violating these ordinances, and that such violations proximately caused the injuries of plaintiff Dini and the death of Captain Duller. Therefore, it was reversible error for the trial court to enter judgment notwithstanding the verdict with respect to these claims.

In considering next the issue of whether plaintiff Elizabeth Dini may assert an action for loss of consortium due to the negligent injury of her husband, we note that this question has been considered in this jurisdiction in only a single Appellate Court case, decided in 1913 (*Patelski* v. *Snyder*, 179 Ill.App. 24), which the federal court deemed declarative of Illinois law in the absence of a Supreme Court decision overruling it. *Seymour* v. *Union News Co.*, 7 Cir., 1954, 217 F.2d 168, 169. In the Patelski case, our Appellate Court denied the action on the ground that it was neither recognized as common law prior to the act of 1874 (Ill.Rev.Stat. 1874, chap. 68, par. 1,) removing the civil disabilities of married women, nor specifically conferred by that act.

Therefore, inasmuch as this question is not overladen with Illinois precedents and is essentially one of first impression for this court, we shall review the origin of the common-law rule, examine its application by the courts of other jurisdictions and the reasons advanced for its retention or rejection.

The common-law rule adhered to in the Patelski case denying the wife an action for loss of consortium due to the negligent injury of her husband was promulgated at a period in history when all the wife's personal property, money and chattels of every description became her husband's upon marriage. She could neither contract, nor bring any action of any kind. Husband and wife were one, and "he was that one." *[Citations omitted.]* Since the husband was entitled to his wife's services in the home, as he was to those of any servant in his employ, if he lost those services through the acts of another, that person had to respond in damages. 8 Holsworth, *A History of English Law* (2d Ed. 1937), 427, 430. He had a right of action for injury to her grounded on the theory that she was his servant. However, should the husband be injured, the wife, being a legal nonentity (1 Blackstone, *Commentaries,* 442), could bring no action. A servant could hardly sue for the loss of services of the master. 3 Blackstone, *Commentaries,* 142, 143. As vividly explained in the Montgomery case, 101 N.W.2d at page 230: "This, then, is the soil in which the doctrine took root; the abject subservience of wife to husband, her legal nonexistence, her degraded position as a combination vessel, chattel, and household drudge whose obedience might be enforced by personal chastisement."

Notwithstanding the obvious changes in the social, economic and legal status of married women during the ensuing centuries, these common-law rules allowing the husband a right of action for loss of consortium due to the negligent injury of his wife, but denying the wife a reciprocal action, were uniformly adhered to by the courts until 1950. *[Citations omitted.]*

In 1950, however, the issue was re-evaluated by the federal court in the well-known Hitaffer case (*Hitaffer* v. *Argonne Co.*), 87 U.S.App.D.C. 57, 183 F.2d 811, 23 A.L.R.2d 1366. In that case, after the husband recovered under the Longshoreman's Compensation Act for negligently caused injuries, his wife asserted a claim for loss of her husband's aid, assistance, enjoyment and sexual intercourse, resulting from the injuries. The trial court, as in the instant case, granted defendant's motion for summary judgment, but, on appeal the federal court for the District of Columbia held that the complaint stated a cause of action. After recognizing the unaminity of authority denying the wife recovery for such a loss, the court pierced what it called "the thin veil of reasoning employed to sustain the rule," and found no substantial rationale on which it could predicate a denial of the action. On the contrary, the court found that the husband and wife have equal rights in the marriage relation which should receive equal protection of the law. It could not justify denying the wife protection of an interest while allowing the husband protection of the same interest. Nor could it justify denying her action for the loss of consortium negligently incurred when she is allowed protection for the same interest in cases of intentional invasion. After buttressing the opinion with the critical comments of legal scholars and dissenting jurists condemning the common-law rule, the court stated at page 819 of 183 F.2d: "The medieval concepts of the marriage relation to which other jurisdictions have reverted in order to reach the results which have been handed to us as evidence of the law have long since ceased to have any meaning."

The cases adjudicating this issue after the Hitaffer decision have been conflicting. The *ratio decidendi* of the Hitaffer case itself has been circumscribed by the court which decided it, so that no right of action for loss of consortium may be asserted by the wife where the husband's injuries are compensable under a workmen's compensation act, in view of the exclusiveness of that remedy. *[Citations omitted.]*

Some courts have expressly rejected the rule of the Hitaffer case, and have adhered to the old doctrine denying the wife's action for loss of consortium in all cases where her husband has been negligently injured. *[Citations omitted.]*

In some of these cases, however, one perceives that the adherence to the old rule was based on a compulsion to follow precedent rather than upon a deep conviction of the wisdom or applicability of the rule. *[Citations omitted.]*

Other courts, however, have rejected the antiquated precedents, and have followed the Hitaffer case in recognizing the action. *[Citations omitted.]*

In view of this maze of conflicting precedents, it is not feasible to review the cases individually; we can only evaluate the reasons they reiterate in support of the recognition or denial of this action.

One of the principal and more popular grounds for denying the action is that the wife's injury is too remote and indirect to warrant protection. *[Citations omitted.]* It has been pointed out, however, that injury to the same interest of the husband has never been regarded too remote or indirect when the husband sues for his reciprocal loss. *[Citations omitted.]* This inconsistency would seem to detract from the cogency of this argument for denying the action.

Another reason frequently advanced in denying the right of action is that it may entail double recovery for the same injury, since the husband could recover in his action for his diminished ability to support his family. *[Citations omitted.]* This argument emphasizes only one element of consortium — the loss of support. Consortium, however, includes, in addition to material services, elements of companionship, felicity and sexual intercourse, all welded into a conceptualistic unity. *[Citation omitted.]* Consequently, in this action the wife is not suing for merely loss of support, but for other elements as well. Any conceivable double recovery, however, can be obviated by deducting from the computation of damages in the consortium action any compensation given her husband in his action for the impairment of his ability to support. *[Citation omitted.]* Hence, since the possibility of double recovery can be eliminated by this simple adjustment of damages, it should not constitute a basis for denying her action, which includes many elements which are in no way compensable in the husband's action. The "double recovery" bogey is merely a convenient cliche for denying the wife's action for loss of consortium.

The same emphasis on the material aspect of consortium and the arbitrary separation of the various elements of consortium is evident in the argument favored in some of the cases that since the wife has no right to her husband's services, she can have no action for loss of consortium, for the law does not allow recovery for sentimental services. *[Citations omitted.]* This argument not only gratuitously assumes that the concept of consortium is capable of dismemberment into material services and sentimental services — which is but a theoretician's boast — but also overlooks the case law allowing the husband recovery for loss of consortium even where there is no loss of his wife's services, as well as the actions for criminal conversation and alienation of affections where damages are given for the so-called "sentimental services." *[Citations omitted.]* Hence, the denial of the wife's action cannot logically be predicated on her inability to allege a loss of services according to medieval pleading practices.

It is further argued by courts which bar the action that the Married Women's Acts removing the common-law disabilities not only conferred no cause of action for loss of consortium on the wife, but must be construed as abolishing the husband's action, as an archaic legal concept. Inasmuch as we are not called upon to adjudicate the husband's claim, we need not consider that novel theory which would remedy the arbitrary denial of a cause of action to one partner by denying it also to the other one; nor can we predicate our interpretation of the common law on the supposition that one day a court will be able to equalize the situation by striking down the husband's action. Nor do we find that in Illinois the concept of "consortium" is ready for the discard pile. On the contrary, its vitality was reaffirmed in *Heck* v. *Schupp*, 394 Ill. 296, 68 N.E.2d 464, 167 A.L.R. 232, where we held unconstitutional a statute abolishing the action for alienation of affection, which involves this precise concept.

Other courts concede either expressly or impliedly the inadequacy of the common-law rule denying the wife an action for loss of consortium for the negligent injury of her husband, but insist that the remedy lies with the legislature. *[Citations omitted.]* We disagree. Inasmuch as the obstacles to the wife's action were "judge invented," there is no conceivable reason why they cannot be "judge destroyed." *[Citation omitted.]* We find no wisdom in abdicating to the legislature our essential function of re-evaluating common-law concepts in the light of present day realities. Nor do we find judicial sagacity in continually looking backward and parroting the words and analyses of other courts so as to embalm for posterity the legal concepts of the past. On the contrary, we are mindful of the caveat of the Georgia court in *Brown* v. *Georgia-Tennessee Coaches, Inc.*, 88 Ga.App. 519, 77 S.E.2d 24, 32, in adjudicating this precise issue: ". . . we do indeed have a 'charge to keep,' but that charge is not to perpetuate error or to allow our reasoning or conscience to decay or to turn deaf ears to new light and new life."

Consequently, we must agree with those jurists and critics who find that the reasons advanced in the cases for denying the wife's right of action for loss of consortium are without substance, and apparently have been added to support a predetermined conclusion dictated by history and the fear of extending liability. *[Citations omitted.]* Obviously the historical milieu in which the rule originated has been completely changed. Today a wife is no longer her husband's chattel, but stands as his equal in the eyes of the law. *[Citations omitted.]* Therefore, precedents predicated on a medieval society are out of harmony with the conditions of modern society, and cannot in good conscience be deemed determinative. As Justice Cardozo aptly stated: "Social, political and legal reforms have changed the relations between the sexes and put woman and man upon a plane of equality. Decisions founded upon the assumption of a bygone inequality are unrelated to present-day realities, and ought not to be permitted to prescribe a rule of life." Cardozo, *The Growth of the Law,* pp. 105, 106.

Inasmuch as we prefer cogent reasoning to numerical superiority of authorities, we must follow those courts which hold that since the husband's right to the conjugal society of his wife is no greater than hers, an invasion of the wife's conjugal interest merits the same protection of the law as an invasion of the husband's conjugal interest. Furthermore, if the law protects the wife's conjugal interest from so-called intentional invasions, as in the alienation-of-affections cases, it cannot deny protection to the same interest [conjugal] where it has been injured by a negligent [rather than intentional] invasion. The same basic reason for granting relief exists in both cases, namely, the protection of the family, as the unit upon which our society is founded.

This approach, moreover, is in accord with the entire movement of the law toward protecting familial interests, and recognizing the changing obligations of its members. *[Citations omitted.]* Therefore, the complaint of plaintiff Elizabeth Dini seeking damages for the loss of consortium of her husband due to the injuries he sustained as a result of defendants' negligence, properly sets forth a basis of liability which the law must recognize. It was error for the trial court to enter summary judgment dismissing her complaint.

Although initially on this appeal defendants argued essentially the issues of liability, in their petition for rehearing they urge the propriety of a new trial. Defendants claim the verdict is not supported by the evidence, and refer to certain conflicts in the evidence. It was the province of the jury, however, to determine whether to give greater credence to the testimony of the defendant building owner and lessee, or to the testimony of the tenants and the fire and police department officials as to the condition of the premises. As previously noted in this opinion, the evidence was overwhelming, even including some admissions of the defendant owner, that the premises were poorly maintained in violation of numerous fire ordinances, and that these conditions contributed to the intensity of the fire. Under those circumstances, it can hardly be seriously contended that the jury verdict was unsupported and against the manifest weight of the evidence so as to warrant a new trial.

Defendants argue further that a new trial should be granted because of the rejection of certain alleged building inspection reports made sometime prior to the date of the fire. While the trial court probably erred in admitting only one of these exhibits, this ruling can hardly be deemed reversible error in view of the overwhelming testimony and exhibits revealing that the premises were a virtual fire hazard at the time of the fire. Moreover, those rejected inspection reports, made over a period of years, were of limited relevancy, and of questionable authenticity, since the person who prepared them did not testify, and the testimony of the building inspector who was called by defendants was thoroughly discredited on cross-examination. We find no justification, therefore, for authorizing a new trial merely on the basis of the improper rejection of these exhibits. We are of the opinion that the result would have been no different had the alleged trial errors not intervened.

In accordance with our analysis, the judgments entered in these claims should be reversed, with directions to adjudicate the complaint of plaintiff Elizabeth Dini, and to reinstate the jury verdicts in favor of plaintiffs Gino Dini and Lillian Duller.

Reversed and remanded with directions.

The following legal problem, adapted from Henderson and Pearson's text titled *The Torts Process,* concerns injuries sustained by a fire fighter while in the performance of duty.[24] Read the problem as described under *Facts.* Use the *Preliminary Comments* for help in deciding whether you would consider the plaintiff (P) an invitee or a licensee. Be prepared to give reasons for your decision. Then read the section *How the Court is Likely to React* and compare it with your conclusions.

The Visiting Fire Fighter

Facts: P, a fire fighter, sues D for injuries sustained on the latter's premises. At the trial, the evidence was that P responded to a telephone alarm reporting a fire at D's icy walkway. D requested the trial judge to instruct the jury that P was a licensee to whom only the limited duty to warn of hidden defects was owed. Should the trial judge grant the instruction?

Preliminary Comments: Does P fit within either of the invitee categories of § 332 of the *Restatement of Torts, Second*? One argument that P might make is that the walk on which he fell was open to guests of the tenants and thus was open to the public, which would make P a public invitee. P might also argue that he was in the business of putting out fires and that he was on D's premises for

[24]James A. Henderson and Richard N. Pearson, *The Torts Process,* Little, Brown & Co., Boston, 1975.

business purposes, and thus would be a business visitor. In any event, would it make any difference if D himself had called in the alarm?

Keeping your decision in mind, read the following section, titled *How the Court is Likely to React.* Compare your decision with the following rulings.

How the Court Is Likely to React: In *Roberts* v. *Rosenblatt,* 146 Conn. 110, 113, 148, A.2d 142, 144 (1959), the Supreme Court of Errors of Connecticut ruled, in a similar case, that the trial judge should have granted the requested instruction: "Upon these facts, the court should have instructed the jury as a matter of law that the plaintiff entered upon the premises in the performance of a public duty under a permission created by law and that his status was akin to that of a licensee and the defendants owed him no greater duty than that of a licensee. *[Citations omitted.]* The status of the plaintiff was not dependent in any respect upon the identity of the person who sounded the alarm, be he an occupant or owner of the premises or a passerby."

Police officers, like fire fighters, are also generally considered to be licensees. See *Scheuer* v. *Trustee of the Open Bible Church,* 175 Ohio St. 163, 192 N.E.2d 38 (1963); *Cook* v. *Demetrakas,* 108 R.I. 397, 275 A.2d 915 (1971).

SUMMARY

It is everyone's legal duty to respect the legal rights of others. A breach or violation of this legal duty is called a tort. If a tort occurs, the injured party (plaintiff) may sue the party who caused the harm, damage, or injury (defendant) for recovery.

All social infractions are not true torts. Certain elements must be present to comprise an actionable tort — that is, a breach of legal duty for which one party can sue another. Those elements that comprise an actionable tort are: (1) there must be a *legal* duty of one party to another, (2) this legal duty must be violated, (3) harm, damage, or injury must result from this breach of legal duty, and (4) a remedy at law must be available.

In general, torts are divided into two categories: (1) property torts or trespass against another's land or property, and (2) personal torts or trespass against another's personal rights, such as the right to safety and the right to enjoy a good reputation. Some actions in tort with regard to violation of personal rights are: assault and battery, false arrest and imprisonment, slander and libel, and the major category of negligence.

Trespass on the property of another is of vital concern to members of the fire service whether during actual fire fighting conditions or in the many inspection tasks and obligations that they undertake.

The elements that comprise negligence are similar to those of torts in general. However, the standard by which a breach of legal duty is measured, called the reasonable person standard, requires judgment based on what a reasonable person under the same or similar circumstances would do. This standard is of particular concern to members of the fire service. Often, members of the fire service function in emergency or dangerous situations. Therefore, the standard of care required of members of the fire service must be carefully weighed and judged. In addition, the issue of negligence to fire service personnel is of special concern, given the fact that fire fighting is considered the most dangerous occupation in America.

ACTIVITIES

1. Using the four elements of a tort, describe a tortious act that you know of. Point out the four elements in the situation you describe.
2. Define and describe property torts. Give examples with your definitions.
3. Define and describe personal torts. Give examples with your definitions.
4. Describe three of the five defenses against negligence presented in this chapter, and tell how these defenses have particular implications for the fire service.
5. With a classmate, role play the parts of an owner of a burned-out building and a fire fighter who has been injured while attempting to control the fire. The fire fighter is now attempting to sue for negligence. Defend yourself either as the owner or as the fire fighter. Have a classmate play the part of the other party and offer defense from that point of view.
6. Why is the distinction of trespasser, licensee, and invitee important to members of the fire service?
7. Describe the accepted legal status of members of the fire service for your community in terms of their rights with regard to trespass.
8. Using the "reasonable person" standard, convince a jury of the standards that you consider reasonable "fire fighter standards."
9. Explain the difference between contributory negligence and assumption of risk. Describe situations in which each may be used as a defense.
10. Describe the differences between nonfeasance, misfeasance, and malfeasance, and explain how each can relate to the fire service.

BIBLIOGRAPHY

Bahme, Charles W., *Fire Service and the Law*, National Fire Protection Association, Boston, 1976.

Cooley, Thomas, *Treatise on the Law of Torts*, Callaghan and Co., Chicago, 1932.

Holmes, Oliver Wendell, *The Common Law*, Harvard University Press, Cambridge, 1963

Ross, Martin J., *Handbook of Everyday Law*, Harper & Row, Publishers, New York, 1959.

For detailed, in-depth study of the topics presented in this chapter, the following are recommended:

Henderson, James A., Jr. and Pearson, Richard N., *The Torts Process*, Little Brown & Company, Boston, 1975.

Prosser, William L., *Handbook of the Law of Torts*, 4th Ed., West Publishing Co., St. Paul, MN, 1971.

IV

EMPLOYER AND EMPLOYEE RELATIONSHIPS

OVERVIEW

Because the fire service is involved in public service, it is administered by the government through various governmental agencies. This chapter presents another dimension of the relationship of the fire service to the government with regard to tort liability. This relationship, commonly called "the master and servant doctrine," asks the question, "When is the master (or employer) responsible for the tortious acts of the servant (or employee)?" This area of legal consideration is known as "vicarious liability." A further topic of discussion is compensation systems that are alternatives to tort liability under the master and servant doctrine and the relationship based on this doctrine. Also included is a brief summary of civil rights involved in an employer/employee situation.

THE MASTER AND SERVANT DOCTRINE

Under English common law, the legal relationship of one person working for another came to be known as "The Master and Servant Doctrine." In America, this same relationship is often termed the "employer-employee relationship." A New Jersey court decision described this relationship as one in which there is

". . . a hiring for a fixed period of time, for fixed wages, and in which the employee's work should be *subject to the control and direction of the employer.*"

Defining the nature of the employer-employee relationship (and, in fact, whether or not such a relationship exists) is vital because an employer may be held responsible if an accident or injury occurs on the job, or if tortious conduct is committed by an employee while on the job. This area of vicarious liability is governed by the principle of *respondeat superior,* which means "let the master answer." There are several major ideas behind this principle of holding an employer financially responsible for an employee's acts or injuries, including the following:

- The employer is responsible because the employer has some measure of control over the employee's acts.
- The employer is responsible for selecting and delegating responsibilities to the employee.
- Because monetary damages are often involved as a result of injuries, the employer is more able to meet any damages assessed so that the plaintiff or injured party has a more financially responsible defendant and can be compensated in a more equitable manner.

In the following excerpt from his book titled *Studies in the Law of Torts,* Clarence Morris describes some of the reasons which justify the use of the rule of *respondeat superior:*[1]

> The proponents of the "entrepreneur theory" have offered a rationale of the workings of the rule of *respondeat superior* which throws light on its utility. In sum, their position is this: The employee who commits a tort is usually judgment proof; if the injured person must look to the employee for reimbursement he will, as a practical matter, have no recourse at all. The employer, on the other hand, is more usually financially responsible; and since he knows of the liability for the torts of his servants, he can and should consider this liability as a cost of his business. He may avoid this cost by staying out of business entirely, or he can plan to carry such losses by insuring against them and adjusting his prices so that his patrons must bear part, if not all, of the burden of insurance. In this way losses are spread and the shock of accident is dispersed. The theory is bottomed on the reparative office of the law of torts. Those who advance it see the desirability of compensating injured plaintiffs and are willing to do so at the expense of a certain class of defendants, because they believe that these defendants are in a position to spread losses so that the ultimate incidence is diversified and none of the bearers will suffer appreciably.

The following case reflects the importance of the doctrine of *respondeat superior* in liability cases. In this case, damages in the amount of $10,000 were awarded to the plaintiff whose wife was killed by an automobile belonging to the fire department and driven by a member of the fire department in response to an alarm. The City of New York was held liable for the $10,000 and appealed, claiming the doctrine of *respondeat superior* did not apply.

---

[1]Clarence Morris, *Studies in the Law of Torts,* The Foundation Press, Brooklyn, 1952, p. 277.

As you read the case, consider: What might have been the result if the City of New York had won on appeal?

MILLER v. CITY OF NEW YORK.

Supreme Court, Appellate Division, Second Department.
April 29, 1932.
257 N.Y.S. 33

Appeal from Supreme Court, Kings County.

Action by Max Miller, as administrator of the goods, chattels, and credits of Rebecca Miller, deceased, against the City of New York. From a judgment in plaintiff's favor upon a verdict of the jury for $10,000 damages, defendant appeals.

Judgment unanimously affirmed.

Argued before LAZANSKY, P. J., and YOUNG, SCUDDER, TOMPKINS, and DAVIS, JJ.

TOMPKINS, J.

Plaintiff sued for the death of his wife, who was killed on the 27th day of May, 1930, by reason of the alleged negligence of firemen of the city of New York in the operation of an automobile belonging to the fire department of the defendant and carrying the chief of battalion in response to a fire alarm. The jury rendered a verdict for $10,000.

The findings of the jury in plaintiff's favor on the issues of negligence and contributory negligence are amply supported by the evidence. The only other question is whether the defendant is liable for the negligence of its servant in the operation of its automobile. The defendant claims that it is not liable, because the man operating the fire department automobile was engaged in the discharge of a governmental duty, and that therefore the *respondeat superior* rule does not apply, citing *Gaetjens* v. *City of New York,* 132 App. Div. 394, 116 N.Y.S. 759. That decision was in 1909, and the rule therein restated continued to be the law until the enactment by the Legislature of Chapter 466 of the Laws of 1929, by which the Highway Law was amended by the addition of Section 282-g, which reads as follows: "Every city, town and village shall be liable for the negligence of a person duly appointed by the governing board or body of the municipality, or by any board, body, commission or other officer thereof, to operate a municipally owned vehicle upon the public streets and highways of the municipality in the discharge of a statutory duty imposed upon the municipality, provided the appointee at the time of the accident or injury was acting in the discharge of his duties and within the scope of his employment. Every such appointee shall, for the purpose of this section, be deemed an employee of the municipality, notwithstanding the vehicle was being operated in the discharge of a public duty for the benefit of all citizens of the community and the municipality derived no special benefit in its corporate capacity.'' The defendant claims that this act is invalid because it violates Section 2 of Article 12 of the State Constitution, which is as follows: ''The legislature shall not pass any law relating to the property, affairs or government of cities, which shall be special or local either in its terms or in its effect, but shall act in relation to the property, affairs or government of any city only by general laws which shall in terms and in effect apply alike to all cities except on message from the governor declaring that an emergency exists and the concurrent action of two-thirds of the members of each house of the legislature.''

Defendant contends that Section 282-g of the Highway Law, above quoted, is a special or local act, and, not having been enacted on an emergency message from the governor, is unconstitutional. The act applies to every city, town, and village where the negligence is that of a person in the employ of the municipality engaged in operating a municipally owned vehicle in the discharge of a statutory duty and within the scope of his employment. The defendant's position is that this provision discriminates between municipalities which provide police and fire protection by reason of a statutory duty imposed upon them, and those which provide such protection, not because of statutory requirement, but of their own volition. We think this claim is untenable. There may be some municipalities in the state where police and fire protection are not required by law, but it is a matter of common knowledge that most, if not all, of the cities of the state have police and fire departments established by legislative enactment. Besides, by statutory provisions, cities, villages, and towns have their highway and health departments operating vehicles on the public streets; likewise water, lighting, park, and sewer departments and districts are established by law in cities, towns, and villages, in the maintenance of which vehicles are used on the public highways. With the many uses to which municipally owned vehicles are employed in the discharge of the statutory functions of government, it cannot be said that the statute in question is special or local. . . .

. . . Our conclusion is that Section 282-g of the Highway Law does not violate Section 2, Article 12 of the State Constitution and is valid, and that the judgment should be affirmed, with costs.

Judgment unanimously affirmed, with costs.

All concur.

THE SCOPE OF EMPLOYMENT

The basis of the relationship between the employer and the employee is, of course, employment. But what constitutes ''employment''? The establishment of this relationship as a matter of ''control'' is described in the following excerpt from the case of *Glover* v. *Richardson & Elmer Co.*:[2]

---

[2] *Glover* v. *Richardson & Elmer Co.*, 64 Wash. 403.

To create the relationship of employer and employee or master and servant, there must be an express or implied contract or acts such as will show that the parties recognize one as the employer and the other as the employee. Whether a person performing work for another is performing it as an independent contractor or as a servant or employee of that other is a question not always easy of solution but all of the authorities agree that the test of the relationship is the right of control on the part of the employer.

An employer is responsible for the acts of an employee only when the employee is acting within the scope of employment. Determining scope of employment is, however, difficult with reference to litigation. The following definition from *Restatement of the Law of Agency, Second* provides some guidelines for this difficult area:[3]

SCOPE OF EMPLOYMENT
§ 228. *General Statement*

(1) Conduct of a servant is within the scope of employment if, but only if:
   (a) it is of the kind he is employed to perform;
   (b) it occurs substantially within the authorized time and space limits;
   (c) it is actuated, at least in part, by a purpose to serve the master; and
   (d) if force is intentionally used by the servant against another, the use of force is not unexpectable by the master.
(2) Conduct of a servant is not within the scope of employment if it is different in kind from that authorized, far beyond the authorized time or space limits, or too little actuated by a purpose to serve the master.

In addition to determining whether actions fall within the scope of employment, *Restatement of the Law of Agency, Second* also provides guidelines for kinds of conduct that can be considered within the scope of employment:[4]

§ 229. *Kind of Conduct within Scope of Employment*

(1) To be within the scope of the employment, conduct must be of the same general nature as that authorized, or incidental to the conduct authorized.
(2) In determining whether or not the conduct, although not authorized, is nevertheless so similar to or incidental to the conduct authorized as to be within the scope of employment, the following matters of fact are to be considered:
   (a) whether or not the act is one commonly done by such servants;
   (b) the time, place and purpose of the act;
   (c) the previous relations between the master and the servant;
   (d) the extent to which the business of the master is apportioned between different servants;
   (e) whether or not the act is outside the enterprise of the master or, if within the enterprise, has not been entrusted to any servant;
   (f) whether or not the master has reason to expect that such an act will be done;

---

[3]*Restatement of the Law of Agency, Second,* Vol. 1, American Law Institute Publishers, St. Paul, MN, 1958, § 228.

[4]*Ibid.,* § 229.

(g) the similarity in quality of the act done to the act authorized;
(h) whether or not the instrumentality by which the harm is done has been furnished by the master to the servant;
(i) the extent of departure from the normal method of accomplishing an authorized result; and
(j) whether or not the act is seriously criminal.

EMPLOYMENT RELATIONSHIPS

The doctrine of *respondeat superior* as it applies to the employer-employee relationship is based on two major questions, each of which must be considered when determining liability. These two questions are:

- Was the conduct within the scope of employment?
- Was there a true employer-employee relationship?

All people who work in an employment capacity do not fall within the range of the master-servant doctrine or the corresponding principle of *respondeat superior*. One major exception is the independent contractor. Whether a person is an employee or an independent contractor is determined by the degree of control an employer exercises over the person in question. This is a vital distinction, because an independent contractor is generally not protected by an employer's insurance. *Restatement of the Law of Agency, Second* distinguishes these relationships in the following way:[5]

§2. *Master; Servant; Independent Contractor*
(1) A master is a principal who employs an agent to perform service in his affairs and who controls or has the right to control the physical conduct of the other in the performance of the service.
(2) A servant is an agent employed by a master to perform service in his affairs whose physical conduct in the performance of the service is controlled or is subject to the right to control by the master.
(3) An independent contractor is a person who contracts with another to do something for him but who is not controlled by the other nor subject to the other's right to control with respect to his physical conduct in the performance of the undertaking. He may or may not be an agent.

Under common law, an employer was liable for any injuries sustained by an employee during the course of employment if the injuries resulted from the employer's direct negligence. An employer, however, was not considered liable for any injuries to an employee if the employee was guilty of contributory negligence or assumption of risk, or if the injury resulted from another employee's negligence. As will be seen later in this chapter, state and federal statutes such as workmen's compensation laws have virtually eliminated these defenses against employer liability. However, under common law, proof of an employer's negligence was essential to recovery by an employee.

Although compensating statutes have altered considerations with regard to an

---

[5]*Ibid.*, § 2.

employer's liability, an employer still has very definite duties and obligations to employees. For example, an employer must use the reasonable care of an ordinarily prudent person (under similar circumstances) to protect employees from unnecessary risks during the course of employment. Although an employer is not expected to anticipate every possible danger, the "due care" standard has been interpreted to mean that an employer must:

- Provide a safe place to work. This also includes areas in which employees await work.
- Provide safe entrances, exits, materials, and appliances with which to work.
- Correct any unsafe working condition that is detected. (This duty applies only to permanent work areas, and not to temporary areas made unsafe because of the nature of the work performed there.)
- Warn employees of any dangers that may exist of which they may be unaware.
- Warn employees of any unusual or peculiar risks associated with a particular job.
- Establish rules and regulations to ensure employee safety. (This is especially important if employment conditions are hazardous.)
- Use reasonable care to select employees of sufficient skill and adequate judgment to ensure that other employees will not be jeopardized on account of the incompetence or negligent actions of their coworkers.

The employer's (master's) duty in employer-employee relationships is summed up in the following excerpt from the case of *Kreigh* v. *Westinghouse*:[6]

> The master's duty has been summed up as requiring: (1) the use of reasonable diligence to provide a safe place for the men in his employ to work, (2) that no obligation is imposed upon the servant to examine into the safety of the master's method of doing business, and he may assume, in the absence of notice, that reasonable care is used, (3) that the master cannot delegate this duty to others, although he is not responsible for the place of work becoming unsafe through the negligence of fellow servants in carrying on the work if he has discharged his primary duty; nor is he obliged to maintain safety at every moment, insofar as it depends on the due performance of the work by the servant and his fellow-servants, and (4) the duty is a continuing one.

Employer: In addition to making employers financially responsible for the torts of employees, the master and servant doctrine performs another vital social function. This function is explained by Clarence Morris in his book, *Studies in the Law of Torts*:[7]

> . . . there is another end to be accomplished by holding a master for the torts of his servants. The same fact — the servant's financial irresponsibility — which

---

[6]*Kreigh* v. *Westinghouse, etc., Co.*, 214 U.S. 249.
[7]Morris, *Studies in the Law of Torts*, pp. 227-228.

practically deprives an injured person of compensation if his remedy is limited to an action against the servant, also puts the servant beyond the reach of the law of torts. Unless the servant's tort is also a crime, it is impossible to do anything in a law court to discourage such wrongs directly. But employers have a considerable measure of control over the lives of their servants; employers are in a good position to discourage the commission of torts by their servants. The employee who commits a tort may be punished by discharge, refusal of letters of recommendation, withholding of advancement, and so on. Yet, there would be little incentive for employers to punish wrong-doing servants if there were no responsibility for their wrongs. The liability afforded by the application of the rule of *respondeat superior* supplies this incentive. Servants, who cannot be reached by the law directly, are deterred from committing torts through threatened and actual punishment by their masters. If these surmises have counterparts in the actual workings of the law, there are two reasons for holding the master liable for the torts of his servants: first, the injured person is compensated by one who is able to distribute the loss; and secondly, the employer is given an incentive to punish wrong-doing employees who might otherwise suffer no sufficient deterrent.

Employee: The term "servant" identifies those persons for whose physical conduct and actions the employer is responsible to a third party, whether it be in the form of financial responsibility to hospitals, doctors, and the employee's family, or legal responsibility to plaintiffs for tortious actions performed by the employee. *Restatement of the Law of Agency, Second* defines "servant" as follows:[8]

§ 220. *Definition of Servant*
(1) A servant is a person employed to perform services in the affairs of another and who with respect to the physical conduct in the performance of the service is subject to the other's control or right to control.
(2) In determining whether one acting for another is a servant or an independent contractor, the following matters of fact, among others, are considered:
   (a) the extent of control which, by the agreement, the master may exercise over the details of the work;
   (b) whether or not the one employed is engaged in a distinct occupation or business;
   (c) the kind of occupation, with reference to whether, in the locality, the work is usually done under the direction of the employer or by a specialist without supervision;
   (d) the skill required in the particular occupation;
   (e) whether the employer or the workmen supplies the instrumentalities, tools, and the place of work for the person doing the work;
   (f) the length of time for which the person is employed;
   (g) the method of payment, whether by the time or by the job;
   (h) whether or not the work is a part of the regular business of the employer;

---

[8]*Restatement of the Law of Agency, Second*, § 220.

(i) whether or not the parties believe they are creating the relation of
master and servant; and

(j) whether the principal is or is not in business.

The scope of an employee's job is determined by the functions and duties performed by the employee with the employer's knowledge and consent. The test of
whether an employee is performing actions within the scope of employment is,
again, the "reasonable person" standard; *i.e.,* would an ordinary and reasonable
person assume that a given action or actions would be expected to be performed
as part of one's employment? An employee may reasonably alternate or vary
means and methods of performing a task and still be functioning within the
scope of the employee's employment, as the following excerpt from *Pittsburgh
C. & St. L. R. Co.* v. *Kirk* illustrates:[9]

> Whether a servant in a given case was acting within the scope of his employ-
> ment, in pursuance of his line of duty, or his own responsibility, in pursuit of his
> own pleasure or convenience, must usually depend upon the facts in such case.
> To undertake to lay down a general rule applicable to all cases would not only be
> difficult but impossible. But we think this much may be said: Where a servant is
> engaged in accomplishing an end which is within the scope of his employment,
> and while so engaged adopts means reasonably intended and directed to the
> end, which result in injury to another, the master is answerable for the conse-
> quences, regardless of the motives which induced the adoption of the means.
> And this, too, even though the means employed were outside of his authority,
> and against the express orders of the master. . . .
>
> Where a servant steps aside from the master's business and does an act not
> connected with the business which is hurtful to another, manifestly the master is
> not liable for such an act, for the reason that, having left his employer's
> business, the relation of master and servant did not exist as to the wrongful act;
> but if the servant, continuing about the business of the employer, adopts
> methods which he deems necessary, expedient, or convenient, and the methods
> adopted prove hurtful to others, the master is liable.

Independent Contractor: The relationship between an employer and an independent contractor is quite different from the relationship between an
employer and an employee. The most important distinction between an
employee and an independent contractor is that an independent contractor has
exclusive and total control over all aspects of the work and the way it is performed or conducted. The employer of an independent contractor controls only
the results of the work.

Although the independent contractor is liable for any negligence on the part
of the contractor and the contractor's employees or agents, the employer has certain duties and obligations that could result in liability if these duties and obligations are breached.

An employer has a duty to provide safe working conditions for an independent
contractor. As with employees, it is the employer's duty to hire a competent and

---

[9]*Pittsburgh C. & St. L. R. Co.* v. *Kirk,* 102 Ind. 399.

qualified independent contractor. The employer is also liable if the employer's duty to protect the public from danger or dangerous products — usually a duty mandated by statute — is violated. In *Mallinger* v. *Webster City Oil Co.*, the duties of an independent contractor are defined and described as follows:[10]

> An independent contractor under the quite universal rule may be defined as one who carries on an independent business and contracts to do a piece of work according to his own method, subject to the employer's control only as to results. The commonly recognized tests of such a relationship are, although not necessarily concurrent or each in itself controlling: (1) The existence of a contract for the performance by a person of a certain piece of work at a fixed price; (2) independent nature of his business or of his distinct calling; (3) his employment of assistants with the right to supervise their activities; (4) his obligation to furnish necessary tools, supplies, and materials; (5) his right to control the progress of the work, except as to final results; (6) the time for which the workman is employed; (7) the method of payment, whether by time or by the job; (8) whether the work is part of the regular business of the employer.

Agent: In determining vicarious liability, the following two questions must be answered:
- Who actually caused the injury or harm to the plaintiff?
- Who is legally responsible for the actions of the individual who caused the harm or injury?

The answers are based on the legal concept of agency, which is described in *Restatement of the Law of Agency, Second* as follows:[11]

> Agency is a legal concept which depends upon the existence of required factual elements: the manifestation by the principal that the agent shall act for him, the agent's acceptance of the undertaking, and the understanding of the parties that the principal is to be in control of the undertaking. The relation which the law calls agency does not depend upon the intent of the parties to create it, nor their belief that they have done so. To constitute the relation, there must be an agreement, but not necessarily a contract, between the parties; if the agreement results in the factual relation between them to which are attached the legal consequences of agency, an agency exists although the parties did not call it agency and did not intend the legal consequences of the relation to follow. Thus, when one who asks a friend to do a slight service for him, such as to return for credit goods recently purchased from a store, neither one may have any realization that they are creating an agency relation or be aware of the legal obligations which would result from performance of the service. On the other hand, one may believe that he has created an agency when in fact the relation is that of seller and buyer.

Restatement of the Law of Agency, Second defines agents and agency as follows:[12]

---

[10] *Mallinger* v. *Webster City Oil Co.*, 211 Iowa, 847.
[11] *Restatement of the Law of Agency, Second*, p. 8.
[12] *Ibid.*

§ 1. *Agency; Principal; Agent*
  (1) Agency is the fiduciary relation which results from the manifestation of consent by one person to another that the other shall act on his behalf and subject to his control, and consent by the other so to act.
  (2) The one for whom action is to be taken is the principal.
  (3) The one who is to act is the agent.

The following excerpt from the decision in *Marks, J., Orfanos* v. *California Insurance Co.* further defines the relationship between an agent and his or her principal:[13]

> An agent for hire is bound to exercise a greater degree of care and diligence than an agent acting gratuitously, but the latter has no greater license to indulge in misrepresentations, concealments, or other breaches of good faith than the former. . . . Even a gratuitous agent is therefore liable to his principal for damages resulting from a gross neglect of duty which he owes to his superior.
>
> In entering upon the relation of principal and agent, the principal bargains for the exercise of the skill, ability, and industry of his agent, and the agent agrees that he has reasonable skill and will do the work with reasonable care.
>
> Loyalty to his principal's interest requires that an agent shall make known to his principal every material fact concerning his transactions and the subject matter of his agency that comes to his knowledge or is in his memory in the course of his agency, and if he fails to do so, he is liable in damages to his principal for any injury or loss suffered in consequence of such failure.

WORKMEN'S COMPENSATION AND RELATED LAWS

Workmen's compensation laws, both state and federal, have virtually eliminated the consideration of tort or fault in cases dealing with injuries to employees. The social desirability of ensuring compensation for employment-related injuries has removed from the courts many situations that formerly required litigation to settle. However, workmen's compensation laws and statutes are highly specific and, if neither the employer-employee relationship nor the definition of scope of employment applies under the law, then the only recourse available is through the courts. Some of the major compensatory statutes and laws are: Workmen's Compensation, Federal Employers' Liability Act, Federal Labor Standards Act, and Social Security.

WORKMEN'S COMPENSATION

During the last half of the 19th century, the Industrial Revolution brought a rapid expansion of industrial activity to the United States. As a consequence, the industrial work force increased dramatically. Unfortunately, this expanding work force very often worked under conditions that were unsafe, resulting in a high rate

---

[13]*Marks, J., Orfanos* v. *California Ins. Co.*, 84 F.2d 233, Cal. 1938.

of industrial accidents. Many injured employees were no longer able to continue working and they and their families became dependent on the state to provide assistance. When an employee died as a result of an industrial accident, the employee's family became dependent on the state for continuing economic assistance. If a disabled employee tried to recover compensation for the disabling injury, the only recourse was to recover damages in tort from the employer. But in order to recover damages from the employer in an action in tort, the employee was required to prove that someone else's negligence was the proximate cause of the injury. And even if negligence was proved — if this negligence was the result of carelessness on the part of a fellow worker — the employer was often not held liable. Moreover, even if the employer was proved negligent, the injured employee was required to prove that the doctrines of contributory negligence and/or assumption of risk were not present in order to recover damages from the employer. As a result, disabled employees often received little or no compensation for themselves or for their dependents. If an employee did recover damages for the disabling injury, the long and tedious litigation that often resulted, having placed severe economic burdens on the employee and the employee's dependents, lessened the benefits derived from such compensation. Thus, because the standard tort's procedure resulted in rare compensation for injuries sustained during the course of employment, employers were not under any duress to improve working conditions, as explained by William L. Prosser in the following excerpt from *Handbook of the Law of Torts*:[14]

> Under the common law system, by far the greater proportion of industrial accidents remained uncompensated, and the burden fell upon the workman, who was least able to support it. Furthermore, the litigation which usually was necessary to any recovery meant delay, pressure upon the injured man to settle his claim in order to live, and heavy attorneys' fees and other expenses which frequently left him only a small part of the money paid. Coupled with this were working conditions of an extreme inhumanity in many industries, which the employer was under no particular incentive to improve.

The overriding social interest on behalf of disabled employees and the dependents of those employees who were killed as a result of employment-related injuries led to workmen's compensation statutes. This legislation replaced the common law tort action with a system of absolute liability on the part of employers for injury or death to employees in work-related activities. The rallying cry for the passage of workmen's compensation statutes became: "The cost of the product should bear the blood of the workman."

Workmen's compensation statutes were first legislated for federal employees in 1908. The state of New York passed the first state statute in 1910. Many other states followed very quickly with similar legislation that guaranteed compensation to employees for at least a part of their economic losses from work-related injuries or death.

---

[14]William L. Prosser, *Handbook of the Law of Torts*, 3rd Ed., West Publishing Co., St. Paul, MN, 1964, p. 554.

Workmen's Compensation and the Fire Service: Workmen's compensation laws are statutory laws. They vary according to their provisions and to whom they apply. Some statutes exclude fire fighters altogether; others include only members of the fire service who are public employees and not public officers; still others exclude all members of the fire service whose employment is based on civil service appointment.

Compensation acts pertaining to volunteer fire fighters are a special issue which is dependent on the employment status of the volunteers. The employment status of volunteer fire fighters is usually specified in the statute under which a company was organized. Generally, however, compensation is allowed for volunteer fire fighters who are members of a fire company subject to municipal control and jurisdiction.

The following decision with regard to workmen's compensation addresses the question of whether fire fighter William Edenfield functioned in the status of an employee of the City of Brunswick so that compensation to his widow could be allowed.

CITY OF BRUNSWICK v. EDENFIELD

Court of Appeals of Georgia, Division No. 2.
Jan. 17, 1953.
74 S.E. 2d 133

Proceedings on claim for workmen's compensation for death of claimant's husband while serving as fireman of municipality. The Superior Court, Glynn County, Douglas F. Thomas, J., affirmed award of compensation of single director, and the city excepted. The Court of Appeals, Gardner, P.J., held that where it was customary for chief of fire department to hire and discharge firemen at will, without confirmation or approval of such action by any other official, and firemen were not subject to civil service, fireman, third class, who was so hired by chief was an "employee" of the city, and not an "officer" thereof, and compensation was payable upon his death while in performance of his duties.

Judgment affirmed.

Syllabus by the Court.

It appearing that the claimant's husband was employed by the chief of the fire department of the City of Brunswick as a third-class fireman, that it was customary for the chief to hire and discharge firemen at will, that the claimant's husband was employed in this manner, and that such contract of hiring did not have to be confirmed or approved by any other municipal officer or department, the single director of the State Board of Workmen's Compensation, was authorized to find that the claimant's husband was an "employee" of the City of Brunswick within the meaning of the compensation act of this State, and to award compensation to her for his death.

On November 21, 1951, William Wilbur Edenfield, a fireman, third class, with the City of Brunswick, was acting in his capacity as such, in answering a fire alarm. He received an injury when the fire truck and an automobile collided, from which injury he died on November 22, 1951. On May 7, 1952, application was made by Mrs. Hazel B. Edenfield, the widow of the deceased, for compensation under the Georgia Workmen's Compensation Act, Ga. Code Ann. §§ 114-101 *et seq.*, and a hearing was had pursuant thereto on July 2, 1952, before hearing director Lawton W. Griffin, a member of the State Board of Workmen's Compensation. There was no dispute as to the salient facts, and it was conceded by both parties that the sole question for determination by the director was whether the deceased was an employee of the City of Brunswick, within the meaning of the compensation act. The deceased was hired by the Chief of the Brunswick Fire Department. There is no civil service act or merit system act applicable to any employee or officer of the City of Brunswick. The charter of the city provides: That the city manager "shall have the appointment, subject to confirmation by the Commission (Brunswick having a city manager-commission form of government), of all heads of departments of said city, and all employees thereof, except the Secretary of the Commission and the Recorder. His appointment of employees below the grade of heads of departments shall not be subject to confirmation by the Commission. He shall have the right to remove heads of departments and other employees (except the Secretary of the Commission and the Recorder) without the consent of the Commission and without assigning any reason therefor, except that in case he removes the head of any department he shall state to the Commission in writing the cause of such removal." Ga.L.1920, pp. 757, 769. The charter further provides: "The following shall constitute the various departments of the city government referred to in Section 14 (act 1920), and the titles of the heads thereof. . . . Fire Department (including inspection of buildings and issuance of building permits), the head of which shall be designated chief. . . . The Commission may from time to time on the recommendation of the city manager create other departments and may define the duties pertaining to each department." Ga.L.1920, pp. 757, 790. In actual practice, the Chief of the Brunswick Fire Department hires and discharges firemen at will. The deceased was hired or employed by the Chief of the Fire Department in this manner.

The single director found as a matter of fact and ruled as a matter of law that the deceased fireman was an employee of the City of Brunswick, and that, his death having been the result of injuries sustained by an accident arising out of and during the course of his employment, the widow was entitled to compensation therefor under the compensation laws of this State. From this judgment and award the City of Brunswick appealed to the Superior Court of Glynn County, on the ground that the deceased was an officer thereof and not an employee and that under the law his death was not compensable under the compensation statute of this State. The Superior Court affirmed the award of the director, and to this judgment the City of Brunswick excepts to this court. . . .

. . . In *Liberty Mutual Ins. Co.* v. *Kinsey,* 65 Ga.App. 433, 441, 16 S.E.2d 179, 184, this court stated that, in determining whether a person was the servant of another, in order to hold the latter liable for torts committed during the course of employment, "the criterion by which to determine whether the relation existed as alleged is to ascertain whether, at the time of the injury, the alleged servant was subject to the defendant's orders and control and was liable to be discharged by him for disobedience to orders or for misconduct." The record here discloses that the Chief of the Fire Department of Brunswick, who was the head of his department, employed the deceased fireman and that it was customary for him to "hire" all firemen, and to "discharge" them at will, and that there was no civil service act as to firemen as in some cities. It is agreed by both parties here that the hiring and firing of those situated as was the deceased at the time of his death were exclusively at the will and discretion of the chief of the fire department. It is our opinion that insofar as the record here discloses the deceased fireman, Edenfield, was subject to the immediate control and supervision of the Chief of the Fire Department of said city and, having been hired at the instance of the chief, could be discharged by him as could any ordinary servant and employee in any department of the city. The firemen did not have to measure up to any required standard of qualifications as is required by a civil service or merit system. They are hired by the chief as are any ordinary employees and servants hired in any department of the city. Nothing to the contrary appears. The chief enters into a contract with the firemen and employs them as he sees fit, and they are hired at will and may be discharged at will. The firemen are not appointed by the city government, the commission, or the manager, as is the chief of the department.

The present case is not like that of *City of Macon* v. *Whittington [citation omitted]*, and the others, so strongly relied on by the city to show that this fireman was not an employee of the city but was an officer. The ruling of the case of *McDonald* v. *City of New Haven [citation omitted]*, that "A regularly appointed member of a city fire department is not an employee of the city within the meaning of a workmen's compensation act which requires a contract relationship of employer and employee in order to bring an injured employee within its provisions," is not applicable under the agreed facts before the director here. That case as do the other cases from Georgia relied on by the city seems to turn on the fact that the city firemen or policemen involved were not employed at will by the head of the department, but were appointed under civil service or other similar method, and could not, therefore, be hired and fired at the will of the municipality involved. In the case at bar, however, a man employed by the Chief of the City of Brunswick Fire Department — such chief having the right to discharge him, with or without cause, at any time he pleases, and not under any civil service laws and regulations whatever or subject to the city commission, etc., is not as a matter of law an officer. In such a case, we think that the finding of fact made by the director is supported by the approved facts in the record and is not contrary to law for any of the reasons assigned, and it will be upheld by this court. . . .

. . . However, it does not appear here that the fireman for whose death compensation is claimed by his widow was not an employee within the meaning of the act but an officer, a public officer of said City of Brunswick. The fact that his services were engaged by being hired by the Chief of the Fire Department of the City of Brunswick to work for the City as a fireman, and he could be discharged by the chief at will, indicates that his status was that of a servant and employee and not of a municipal officer. The fact that he is designated as a "fireman" and that the courts have held that certain "firemen" selected or appointed under the civil-service rules under which the department operated or policemen appointed by the mayor or other governmental department and subject to approval by council and who exercise a portion of the sovereign power in performing their duties, are officers of the city and not entitled to compensation, does not necessarily and as a matter of law mean that every fireman is an officer and not an employee of the city, simply because he is a fireman and is called a fireman. The fact that he is engaged in performing governmental functions in the performance of his duties as a fireman, and for which the city is not liable does not as a matter of law constitute him an officer of the city. So far as the record discloses, this fireman was hired by the chief and could have been discharged by him at will. He was subject to the supervision and direction of the chief in performance of his duties. He was engaged by the chief under a contract of employment and not chosen or appointed. The question is whether there was an employment of the claimant's husband under a contract of hiring. Because in the performance of the duties of his employment as a fireman the claimant's husband might engage in the performance of duties and the city might not be liable in damages for his improper performance on the ground that the same were governmental functions, does not prevent him from being an employee under the statute.

The facts in the record authorized the single director to find that the claimant's husband was an "employee" of the City of Brunswick under the provisions of the Workmen's Compensation Act of Georgia, and came under the compensation laws. It follows that the judge of the superior court did not err in approving and affirming such finding of the director, awarding compensation to the claimant for the death of her husband.

Judgment affirmed.

TOWNSEND and CARLISLE, JJ.,

Concur.

In the previous case, William Edenfeld's employee status was, in part, determined by the fact that the chief of the fire department of the City of Brunswick could hire and discharge fire fighters at will without approval of any other municipal officer or department. The following decision addresses the status of a fire fighter, Isaac Fisher, as one which is *not* described by statute and as one which is involved in a governmental function.

CITY OF HUNTINGTON v. FISHER

Supreme Court of Indiana.
April 6, 1942.
40 N.E.2d 699

FANSLER, Judge.

The appellee, widow and sole dependent of Isaac Fisher, a member of the Fire Department of the City of Huntington, who was killed while in the discharge of his duty as a city fireman, was granted compensation by the Industrial Board under the Workmen's Compensation Law, Burns' Ann.St. 40-1201 *et seq.* The city brought this action in the Appellate Court questioning the lawfulness of the award. The Appellate Court held, on authority of *City of Fort Wayne* v. *Hazelett*, 1939, 107 Ind.App. 184, 23 N.E.2d 610, that the appellee was not entitled to compensation. The case comes to this court upon petition to transfer, which questions the correctness of the decision of the Appellate Court. *[Citation omitted.]*

In the Hazelett case it was pointed out that a city acts in a governmental capacity in maintaining a fire department, and is controlled by statute in hiring and discharging firemen, and that there is a statute providing for a pension system for members of the fire department, and the court concluded that, in view of these facts, firemen are not in the service of a city under a contract of hire within the meaning of the Workmen's Compensation Law. It is said that (page 189 of 107 Ind.App., page 612 of 23 N.E.2d): "Considering these statutes together, it would seem there was no intent on the part of the Legislature to include members of a city fire department among the persons entitled to the benefits of our Workmen's Compensation Law." The opinion was by a divided court. See dissenting opinion of Stevenson, J.

In the Compensation Law, "an employer" is defined to include any municipal corporation using the services of another for pay, and an "employee" is defined to include every person in the service of another under contract of hire. . . .

. . . Nor can we agree that the fact that the city acts in a governmental capacity, and that the manner of hiring and discharging firemen is regulated by statute, and that a pension system is provided for firemen, furnishes a basis for concluding that it was not intended that they should come within the terms of the Compensation Law. The state is included within the definition of employer, and it acts in a governmental capacity, and the manner of its employment and discharge of employees is regulated by statute, and it may be noted that, under federal statutes, private employers are regulated in the employment and discharge of men where labor union membership is involved. The statute providing for pensions for municipal utility employees expressly excludes them from the provisions of the Workmen's Compensation Act. Section 48-6607, Burns' Ind. St. 1933, Section 12321, Baldwin's Ind. St. 1934. But we find no similar provisions in the Firemen's Pension Law, Burns', Ann. St. 48-6501 *et seq.,* and must assume, in the light of the express exclusion of municipal utility employees, that if an exception was intended because of the Firemen's Pension Law, it would have been expressed.

The judgment is reversed, and the cause remanded to the Appellate Court for further proceedings not inconsistent with this opinion.

Volunteer Fire Departments: If a person is a member of an authorized volunteer or regularly organized fire department, compensation is generally allowed. Even though a volunteer fire fighter may be expressly brought within the coverage of workmen's compensation laws, the relationship may not be considered a strict employer-employee relationship and, as such, coverage may be allowed only during the actual performance of duty while acting in the capacity of volunteer fire fighter. This is explained in the following excerpt from 99 C.J.S. 117:[15]

> It has been held, however, that since it would be contrary to public policy to impose on a volunteer fireman the burden of determining in each instance whether the service he is called on to perform by an apparent lawful command of a superior officer is one which is within the scope of a volunteer fireman's duties, it generally will be considered that a volunteer fireman is performing services which will bring him within the protection of the compensation act when he is following the directions of his superior officer in the performance of duties which are within the apparent line of duty.

FIRE FIGHTERS' RELIEF ASSOCIATIONS

Many volunteer fire departments form relief associations, organized and incorporated to manage and disburse fire fighters' disability and/or pension funds. By law, these organizations must be incorporated. The statutes relating to a fire fighter's disability fund are generally to protect active members injured during the performance of their duties; however, benefits must fall within those conferred by the association. In order to qualify under the statutes relating to disability funds, a person must be a confirmed member of an organized department that is recognized by the municipality.

---

[15] 99 C.J.S. 117, p. 418.

FEDERAL EMPLOYERS' LIABILITY ACT

The passage of the first Federal Employers' Liability Act in 1906 signalled the beginning of laws that compensated for work-related injuries. This act was passed by Congress with the specific intention of providing work-related compensation to railroad workers. The Constitution empowers Congress to act in matters regarding interstate commerce, and it was on this premise that Congress enacted the Federal Employers' Liability Act. However, because it also applied to employees involved in intrastate activities at the time of injury, this act was declared unconstitutional. A second act, passed in 1908, limited compensation to those employees who were themselves engaged in interstate or foreign commerce. This act was amended in 1939 to include compensation for injuries received while engaging in activities furthering interstate commerce, or directly or substantially affecting interstate commerce.

FAIR LABOR STANDARDS ACT OF 1938

The Fair Labor Standards Act of 1938 provided the first federal regulations for working conditions in America. It set guidelines for conditions of employment such as child labor restrictions, minimum wages, hours of employment, *etc.* The federal government was able to include the states in this act by mandating that it was applicable to all those engaged in interstate commerce. Most employers were (and are) affected by this act, and even those who technically are not must still abide by the guidelines and regulations established in the act. [*N.B.:* The Constitution is very specific regarding the powers delegated to the Federal Government. The preceding concept is vital in terms of federal encroachment of state power. The Constitution delegates most power to the states.]

The decision in the following case, which concerns compensation for the "sleeping time" of fire fighters while on duty, describes and defines the guidelines that bring employees within a state under federal jurisdiction. As you read the case, consider the court's decision and how it applies to the following questions: (1) Should a fire fighter be paid for sleeping during on-duty hours if all other responsibilities are completed and there is a lull in fire department calls? and (2) What might happen if a fire fighter chose not to respond to a call during "sleeping time"?

BELL et al. v. PORTER et al.

Circuit Court of Appeals, Seventh Circuit.
Dec. 10, 1946.
Writ of Certiorari Denied April 7, 1947.
See 67 S.Ct 1092,
159 F.2d 117

Appeal from the District Court of the United States for the Northern District of Illinois, Eastern Division; John P. Barnes, Judge.

Action under the Fair Labor Standards Act by Virgil D. Bell and others against Seton Porter and another, copartners, to recover overtime compensation, liquidated damages and attorney's fees. From a judgment for plaintiffs, 66 F.Supp. 49, defendants appeal.

Reversed and remanded with instructions.

Before KERNER and MINTON, Circuit Judges, and LINDLEY, District Judge.

KERNER, Circuit Judge.

In this appeal 53 employees and former employees of Sanderson and Porter, a copartnership, sued to recover for overtime, liquidated damages, and attorneys' fees, under the Fair Labor Standards Act, 29 U.S.C.A. § 201 *et seq.* The case was tried by the court without a jury. The court, after making findings of fact and conclusions thereon, entered judgment in favor of plaintiffs-appellees totalling $276,619.16, including therein the sum of $27,600 as fees for plaintiffs' attorneys. 66 F.Supp. 49.

On this appeal, the questions raised are whether plaintiffs-appellees were engaged in the production of goods for commerce, and whether the employees' sleeping time under the two-platoon system was compensable under the Act.

We shall discuss first the question whether appellees were engaged in production of goods for commerce. As a necessary prerequisite to an understanding of the case, we state the facts insofar as they relate to the point raised. . . .

. . . Appellants operated the Elwood Ordnance plant at Elwood, Illinois, where they were engaged in the manufacture of shells, explosives, and munitions for the armed forces, under a cost-plus-fixed-fee contract with the United States Government. This plant, including all buildings, machinery and equipment, was owned by the Government, but all of it was maintained and operated by appellants as independent contractors. The Government procured, owned, and furnished appellants all powder and other component parts used in the manufacture of the munitions. Appellants, however, as consignees, procured and received certain other materials and supplies used in the assembling and loading of the munitions from various consignors without the State of Illinois. Title to these supplies vested in the Government at the time of delivery. All munitions produced at the plant were shipped from the plant to various army installations throughout the United States upon orders received by the Commanding Officer from the War Department in Washington, D.C. Appellants had complete supervision of all employees, including the hiring and discharging of all employees, and maintained their own fire department in which appellees were employed as fire fighters. Upon these facts the court concluded that appellees were engaged in the production of goods for commerce within the meaning of the Act. . . .

. . . As to the second point, we note that the Act is made applicable to any employee "who is engaged in commerce or in the production of goods for commerce." Section 7(a), 29 U.S.C.A. § 207(a). By § 3(b) of the Act, commerce is defined as "trade, commerce, transportation . . . from any State to any place outside thereof." But nowhere in the Act is it suggested that Congress intended that transportation effected by the Government or of Government goods be treated differently from all other transportation; hence we believe, as the court did in the case of *Atlantic Co.* v. *Walling,* 5 Cir., 131 F.2d 518, that when Congress defined "commerce" in the Act, it intended to give the term the broadest possible meaning, so as to include all transactions, conditions and relationships as have been heretofore known and acknowledged as constituting commerce in the constitutional sense. . . .

. . . The Constitution confers upon Congress the power to regulate commerce among the several States. U.S. Const. Art. 1, § 8, cl. 3. This power to regulate commerce is not confined to commercial or business transactions. From an early date such commerce has been held to include the transportation of persons and property, no less than the purchase, sale, and exchange of commodities, *United States* v. *Hill,* 248 U.S. 420, 423, 39 S.Ct. 143, 63 L.Ed. 337, and goods may move in commerce though they never enter the field of commercial competition. For example, the movement of people across State lines and the unrestricted ranging of cattle across the boundary between two States is commerce. The interstate transportation of whiskey for personal consumption, of a woman from one State to another for an immoral purpose without any element of commerce, of a kidnapped person or a stolen automobile — all constitute interstate commerce in the constitutional sense. *[Citations omitted.]* These cases, we think, make it clear that interstate commerce is not limited to interstate trade. [*N.B.*: The remainder of the decision does not address the question of federal jurisdiction and is reprinted here for information only.]

We now consider whether the employees' eight-hour sleeping time was compensable working time under the Act.

Appellees' duties were to fight fires, scrub the premises, drill, make inspection and trial runs, pick up and deliver reports, fill extinguishers, attend school after supper, and clean equipment. These duties ordinarily took about five hours per day of their time, the remainder of the time between 7:30 a.m. and 11:30 p.m. being spent in playing cards, reading, listening to radio programs, eating, and other personal activities. They were permitted to retire at 10 p.m. except that one fireman in each station was required to remain awake until 11:30 p.m. During the period in question, appellees spent 338,265 hours at the plant, of which 229,545 were hours worked between 7:30 a.m. and 11:30 p.m., for which they were paid at overtime rates for hours worked in excess of 40 hours per week. The remaining 108,720 hours were sleeping period hours between 11:30 p.m. and 7:30 a.m., for which no compensation was paid.

The court found that the two-platoon system was proposed to the men in December, 1943, at which time they voted unanimously in favor of the plan; that the plan was put into effect on February 27, 1944, and continued in effect until December 1, 1945; that under the plan, on the first day of a work week, one platoon reported for duty at 7:30 a.m. and remained on the premises until 7:30 a.m. the following day, at which time they were relieved by the other platoon; that on alternate days each platoon was on the premises for 24 consecutive hours followed by 24 consecutive hours off duty; that appellants furnished the men in each station sleeping facilities, bedding, clothes lockers, showers, toilet facilities and laundry service for bedding; that if the men had been free to sleep elsewhere, they would not have chosen sleeping quarters similar to those furnished by appellants; and that under the three-shift system, 108 men were necessary at a weekly wage of $44.20, while under the two-platoon system but 68 men were necessary, the average weekly wage of each man being $54.40, and that thereby a saving in manpower and money accrued to appellants.

The court also found that during the period in question there were six occasions on which the sleeping periods of the men were interrupted by fire alarms, three occasions when certain of the men were required to remain during a portion of their sleeping period at the scene of a fire which had occurred prior to 11:30 p.m., and one occasion on which two of the men were detailed for a portion of their sleeping period to stand by at the scene of hazardous operations. The total number of hours spent by all appellees on such night duty amounted to 136 out of a total of 108,720 sleeping period hours. Four of the appellees were never called out for such duty and the remainder spent from 30 minutes to 8 hours on such duty.

The court concluded that the time spent by appellees at the plant during the period between 11:30 p.m. and 7:30 a.m. was predominantly for appellants' benefit as an incident of their employment and as the only practical means for getting instant service in case of fire, and that the eight-hour period between 11:30 p.m. and 7:30 a.m. constituted working time.

Appellees argue that the sleeping quarters were uncomfortable and unsanitary. They make the point that whether in a concrete case sleeping time falls within or without the Act is a question of fact to be resolved by appropriate findings of the trial court, *Skidmore* v. *Swift & Co.*, 323 U.S. 134, 136, 65 S.Ct. 161, 89 L.Ed. 124, and since the court found that if the men had been free to sleep elsewhere, they would not have chosen quarters similar to those furnished, the findings may not be disturbed. . . .

. . . In this case, as in the *Bowers* v. *Remington Rand* case, 7 Cir., 159 F.2d 114 appellees rely principally on *Armour & Co.* v. *Wantock*, 323 U.S. 126, 65 S.Ct. 165, 90 L.Ed. 118, and *Skidmore* v. *Swift & Co.*, *supra.* In the Armour case the court had no occasion to pass upon the question of whether "sleeping time" was "working time." True, the Swift case involved the question of whether the firemen were entitled to be paid for waiting, but the case was reversed because of the application of an erroneous principle of law. They also cite the case of *Anderson* v. *Mt. Clemens Pottery Co.,* 328 U.S. 680, 66 S.Ct. 1187, 1194, in which the court said: "the statutory workweek includes all time during which an employee is necessarily . . . on duty or at a prescribed work place . . . ." This case involved time spent in walking upon the employer's premises after punching a time clock and time spent in preparatory work. We do not think these cases are controlling and decisive of our case. They are cited because of the quoted language used in the Anderson case and certain statements of the court in the Armour and Swift cases, but the words of "opinions are to be read in the light of the facts of the case under discussion." *Armour & Co.* v. *Wantock, supra,* 323 U.S. 133, 65 S.Ct. 165, 168, 89 L.Ed. 118.

The court in the instant case did not find that the sleeping quarters were inadequate or of such a character that the men were unable to obtain normal rest. But be that as it may, the ultimate finding was clearly a mixed question of fact and law, *United States* v. *Anderson,* 7 Cir., 108 F.2d 475, and if based upon a misapplication of the law, it is not binding upon the reviewing court. [Citation omitted.] Here, under the facts and circumstances, as in the *Bowers* v. *Remington Rand* case, it is clear that appellees, in consideration of their employment as firemen, were willing to sleep on the premises and to keep themselves available for duty if called upon during their rest period; their contract was to wait to be engaged; hence the time spent in sleeping is not compensable.

The judgment is reversed and the cause is remanded to the District Court with instructions to dismiss the complaint.

Reversed and remanded.

The following case of *Bridgeman et al.* v. *Ford, Bacon, Davis, Inc.,* which concerns the recovery of overtime pay for "sleep time" by members of an industrial fire brigade, was heard in district court (a federal court) because federal rights under the "commerce clause" of the Constitution were involved. The rulings and interpretations of the federal Administration of Wages and Hour Division were used as guidelines in this decision.

BRIDGEMAN et al. v. FORD, BACON & DAVIS, Inc.

District Court, E. D. Arkansas, W. D.
Feb. 25, 1946.
64 F. Supp. 1006

Action by Jesse C. Bridgeman and others against Ford, Bacon, & Davis, Inc., to recover overtime, liquidated damages, and attorneys' fees under the Fair Labor Standards Act of 1938, 29 U.S.C.A. § 201 *et seq.*

Judgment in accordance with opinion.

LEMLEY, District Judge.

This is an action under the Fair Labor Standards Act of 1938, 29 U.S.C.A. § 201 *et seq.*, for overtime, liquidated damages, and attorneys' fees, brought by the plaintiff Jesse C. Bridgeman and thirty-eight other men employed as firemen by the defendant, Ford, Bacon & Davis, Inc., in the operation of the Arkansas Ordnance Plant.

The Arkansas Ordnance Plant was one of approximately ninety similar plants, engaged in the loading and assembling of ammunition for the United States Army in the late war, supervised by the Ordnance Division of the War Department of the United States and under the management of various and sundry corporations, in the present instance, the defendant, Ford, Bacon & Davis, Inc., which operated under a cost plus fixed fee contract, all costs expended being reimbursed by the United States Government.

The plant, located at Jacksonville, about fiteen miles from Little Rock, Arkansas, was constructed in 1941-42. The loading of ammunition began on March 16, 1942, and continued until the end of the war. Since that time the plant has been taken over by the Corps of Engineers of the United States Army.

Jacksonville is a village without fire protection, the nearest well-equipped fire department being located at North Little Rock, about fourteen miles away. Therefore, in the course of the building of the ordnance plant it became necessary to install such a department. Two fire stations were built for this purpose, and about seventy men employed. The stations, known as No. 1 and No. 2, were of practically the same construction. They were two-story buildings with a fire engine room, inspectors' office, wash room and toilet downstairs, and a dormitory for the firemen, a kitchen, a toilet and shower room, and a room for the occupancy and use of an assistant fire chief and a captain upstairs. The dormitory, which was occupied by the firemen and used as their sleeping quarters at night, was separated by a narrow hall from the room of the assistant chief and captain.

It was necessary, of course, that the plant have fire protection at all times, so the workdays of the men were so shifted that there was an adequate force in the fire stations, day and night, to man the equipment. Prior to November 29, 1943, about seventy men, each working six days a week, in three eight-hour shifts, were employed.

On November 29, 1943, the eight-hour shift arrangement was abandoned, and what is commonly called the two-platoon system was inaugurated. Under this system each fireman undertakes a twenty-four hour tour of duty three times a week. Sixteen hours of such tour are supposed to be devoted primarily to the duties of the fire department, and eight hours to sleep and recreation. Compensation is paid for the sixteen hours, but no general compensation for the eight hours.

The inauguration of the two-platoon system was brought about in the following manner:

During the summer of 1943 the War Department had under consideration the adoption of the two-platoon system in order to conserve manpower. In this connection, considerably fewer men are required in the operation of the two-platoon system than in the eight-hour shift arrangement. It seems that the War Department was considering the inauguration of the two-platoon system in practically all of the ordnance plants under its supervision. At least, the plan was ultimately adopted in approximately 80 percent of such plants.

Before any definite steps were taken to install the system, the plan was first submitted to the U.S. Department of Labor and approved by it in a letter from Honorable William R. McComb, Deputy Administrator, Wage and Hour Division, to Lt. Col. William J. Brennan, Jr., Chief, Labor Section, Office of the Chief of Ordnance, War Department, and thereafter submitted to the firemen for their approval by means of a secret, unsigned ballot, upon which they would express their approval or disapproval. All of the firemen except two voted in favor of the adoption of the system. These two remained in the employ of the company.

The plan as submitted by the War Department to the Department of Labor, and approved by the latter, was substantially complied with during all of the period of time involved in this action.

In operation the two-platoon system gave the firemen many more consecutive hours of freedom from duty than they had previously enjoyed; and on their days off, numbers of the men tended their farms and certain others worked at times at other jobs, or went hunting and fishing.

The men served in shifts of twenty-four hours each, followed by a period of twenty-four hours off duty. Each man normally served three such shifts in each week of employment, and was off duty four twenty-four hour periods each week. About every two months each man would have four consecutive periods of twenty-four hours each off duty. The firemen were required to remain in or within call of their respective fire stations at all times. When the system was first installed, the twenty-four hour shifts began at 11 p.m. and ended at 11 p.m. the following night, the first eight hours of each shift being considered rest or sleeping time, and the other sixteen hours as working time or time on active duty. This arrangement, insofar as it required the men to check in and check out at 11 p.m., was a rather awkward one, and after approximately two months of such operation, namely, between November 29, 1943, and January 31, 1944, the checking in and checking out times were changed from 11 p.m. to 7 a.m., and so continued throughout the entire period of time under consideration. Under the latter arrangement the sixteen-hour period between 7 a.m. and 11 p.m. was regarded as work time, and the eight hours from 11 p.m. to 7 a.m. as sleep or rest time, but in practice the men were permitted to go to bed at 9:30 p.m. instead of 11 p.m.

The men were paid at the rate of $35 for the first forty hours worked in each week, and time and one-half for the remaining hours worked, but received no compensation for the eight hours of rest or sleeping time, unless summoned to answer a fire call or make an ambulance run, in which event they were paid, for fire calls, time and one-half with a minimum of thirty minutes for any time spent less than that period, and a minimum of one hour for any period of time worked between thirty minutes and one hour, and for ambulance runs, time and one-half. In practice the men earned about $45.50 per week, exclusive of the special compensation allowed them for fire and ambulance runs during the eight-hour rest period.

The general duties of the firemen during the sixteen-hour work period consisted of cleaning of the fire stations and of the sleeping and living quarters, checking equipment, and washing and polishing of the fire trucks and accessories, with occasional drills and practice runs, and of course the answering of fire alarms. Ordinarily, two hours out of the sixteen-hour work period would be all that was required for the routine duties. There were comparatively few fire alarms, and most of the sixteen-hour period was occupied by the men in reading, playing games, listening to the radio, and eating. The facilities provided by the company for the use of the firemen included shower bath and toilets, steam heat, electric fan, cards, checkers and chess for games, a well-equipped kitchen, with range, sink and utensils, an electric refrigerator, beds, blankets and bed linen, with laundry service therefor.

The company pursued a liberal policy with respect to tardiness and brief absences in the course of the shift. Many of the men lived at considerable distances from the plant and it was the practice of the fire chief not to have the men docked for brief absences during the work period if they had a good excuse, and during the rest period it was the policy of the company not to dock the men at all, and they were not so docked except in a few instances, which have been explained to the Court's satisfaction as being errors of timekeepers.

From the time the two-platoon system was inaugurated up until May 14, 1944, the men were called out during the rest period for fire runs only, and were specially compensated therefor as above set forth. On said date two ambulances, which had theretofore been stationed at the plant hospital, were moved, one to each fire station, and one or two firemen were called from time to time to drive the ambulances and answer ambulance calls. This work was divided around among the men, and usually assigned to those who preferred to drive rather than to remain at the station. Ambulance calls made during the rest period were specially compensated for as above indicated. Fire and ambulance runs made during the rest period were termed by the witnesses in this case "emergency runs," and from now on will be referred to under that term.

The allegations of the plaintiffs' complaint were general and not specific. It was charged that between the dates of November 29, 1943, and May 27, 1945, the defendant had failed to pay the complainants certain overtime compensation, and as a result was due them an aggregate amount of approximately $130,000 for overtime and liquidated damages. Two specific issues were developed at pretrial conferences and in the excellent pretrial briefs which were filed at the request of the Court: (1) Whether under the Fair Labor Standards Act the firemen were entitled to compensation for the eight-hour rest and sleeping period, and (2) whether, as was contended at the conferences and in plaintiffs' brief, the men had reported for work a substantial period in advance of the regular clocking-in time, had been immediately put to work and had not been compensated therefor. No proof was offered on the second contention and it has been abandoned.

The case was submitted to the Court on an agreed statement of facts and oral testimony, in the course of which proof was offered by the plaintiffs to the effect that during the first two months of service under the two-platoon system, that is, during the period when the firemen checked in at 11 p.m. when their rest period was supposed to begin, the men who drove the fire trucks inspected their trucks, and all of the men, including the truck drivers, made up their beds and did certain other chores, all of which it was contended amounted to work for which no compensation had been paid. Counsel for the defendant objected to the introduction of this evidence on the ground that it conflicted with certain provisions of the agreed statement of facts, but finally agreed that, while he was unwilling to waive any of the provisions of the stipulation, he would concede that if the proof showed that the men were permitted to do actual work for any definite length of time during the rest period, they would be entitled to recover therefor.

We will first take up the question as to whether the men were permitted to work any part of their rest period during the two months in question, and if so, for how long. On this point, while there is a conflict in the evidence as to just what chores or services were performed immediately upon checking in, a preponderance of the evidence discloses that between the time the men checked in and the time that they retired, the drivers of the fire trucks, as routine, went to the clothing rack in the engine room, got their helmets and fire coats and placed them on their trucks; that thereafter they gave their trucks a careful inspection, including warming up the motors, testing their booster pumps, and checking of the oil, gas and water and all material and appliances on the trucks; that after this inspection they went upstairs to the dormitory, secured their combination fire suits and boots, placed them beside their beds, procured their bed clothes from their lockers and made up their beds. The proof with respect to what was done by the men other than the drivers, between the time that they checked in and the time that they retired, is that they did all of the things that were done by the drivers, except the inspecting of the fire trucks. In this connection, there was some testimony to the effect that the men other than the drivers assisted in inspecting the trucks and cleaned out the dormitory before going to bed, but we do not so find. We believe that at the time the men other than the drivers placed their helmets and fire coats on the trucks to which they were assigned, they glanced at their stations on the trucks — the plugmen to see that their plug wrenches were at hand, and the hosemen that their nozzles were in place — but did not make an actual inspection of the trucks. As to the cleaning out of the dormitory, we find that it was done before 11 p.m. by the men of the preceding shift.

We find that it took the truck drivers who served during the two months period in question twenty (20) minutes each night to inspect their fire trucks, and in our opinion this constitutes employment time within the meaning of the Fair Labor Standards Act. Each driver was responsible for his truck, and it was necessary for the protection of the ordnance plant that each truck be ready for action at any moment. The inspection in question was a substantial one, involving the checking not only of the gasoline, oil, and water in the fire trucks, but also of the pumps, ladders, tools, and all fire fighting equipment carried thereon; and this service was performed in furtherance of the company's business, with no benefit to the employees other than to aid them in the performance of their work. The service, moreover, was not an inconsequential or elusive one, and in our judgment is compensable.

With respect to the time spent by the employees, both truck drivers and others, in getting their helmets and fire coats from the clothing rack in the engine room and placing them on the trucks to which they were assigned, securing their combination fire suits and boots and their bed clothes from their lockers in the dormitory, and making up their beds, the situation is different, and one which we think comes within the rule announced by the Circuit Court of Appeals of the Fifth Circuit in *Tennessee Coal, Iron & R. Co.* v. *Muscoda Local No. 123 et al.*, 135 F.2d 320, 323, in which that Court said:

"With respect to time spent by employees checking in and out and procuring and returning tools, lamps, and carbide, ... In addition to the practical difficulties incident to the computation of isolated moments so elusive in character, we think these pursuits should not be computed as work-time, since they fall within the category of duties incident to qualifying the employee to perform his work rather than within the scope of his actual employment."

On the main issue in this case, namely, whether the plaintiffs are entitled to recover for the entire eight-hour rest period provided for under the two-platoon system, their contention, as we understand it, is two-fold: First, that since it is admitted that the men were required to remain at their fire stations and away from their homes during the entire rest period, as a protection to the plant, they were at all times "employed" within the meaning of the statute, and are entitled to recover as a matter of law; and secondly, if such is not the law, that they are still entitled to recover on the facts, which they insist disclose that the rest period was interrupted to such an extent as to deprive them of the enjoyment thereof and of a normal night's sleep.

We cannot agree with the plaintiffs in either of these contentions.

In *Johnson et al.* v. *Dierks Lumber & Coal Co.*, 8 Cir., 130 F.2d 115, 120, the Circuit Court of Appeals of this Circuit had under consideration an appeal involving among other questions the one as to whether sleeping time should be excluded from employment hours. The plaintiffs, two in number, had been employed by the defendant company to grease and service trucks and tractors and to keep watch, seven days a week, over motor equipment, mules and other property at certain stations in the woods, termed "grease camps." Under their agreement they lived in trailers provided by the company and spent their entire time at the camps. The manual labor performed by them, consisting mainly of servicing trucks and caring for mules, consumed about six hours per day, for which, under their agreement, they were paid. The rest of each twenty-four hour period was spent in caring for the company's property at the camp and in sleeping. For this, under the agreement, they were not paid. They brought suit under the Fair Labor Standards Act to recover compensation for the eighteen hours of each twenty-four hours spent as caretaker and in sleep. The District Judge directed a verdict for the defendant. The Circuit Court of Appeals reversed and remanded the case, holding that upon the record it could not be said as a matter of law that the plaintiffs should be allowed compensation only for the time devoted to manual labor. A concurring opinion was written by Judge Kimbrough Stone, in the course of which he stated that since there was to be a new trial it seemed advisable to make an additional statement, which might be helpful to the trial judge on retrial. Whereupon, he said:

"My view is that the time when these plaintiffs could be sleeping is not to be treated as hours of employment *in the factual situation shown in this present record.* I do not mean that, in every case nor under all circumstances, sleeping hours should be excluded from employment hours. Whether they should or not must depend upon the factual situation of each case as it appears. That such hours may be excluded under certain fact situations is not a novel idea (see *Muldowney* v. *Seaberg Elevator Co.*, D.C., 39 F.Supp. 275, 282, and Interpretative Bulletin No. 13, Department of Labor, Wage and Hour Division, Paragraphs 1, 2, 6 and

7). Nor do I mean to exclude the time when plaintiffs might be aroused at night to care for the property. I cover solely the time when they were or could have been sleeping at night.''

In *Muldowney* v. *Seaberg, supra* [39 F. Supp. 282], cited by Judge Stone, the District Court of the Eastern District of New York had the following to say with respect to whether sleeping hours should be excluded from or included in employment hours:

"In computing the hours of overtime I will be guided by Interpretative Bulletin No. 13, Paragaphs 6 and 7, United States Department of Labor, Wage and Hour Division, Office of Administrator, November 1940, from which it is clearly apparent that where the employee stays at the employer's place of business at night, as in the case at bar, and in the ordinary course of events has a normal night's sleep, and ample time for meals, and time for relaxation, and entirely private pursuits, such time must be excluded as this seems to me to be clearly the intent of the law.''

In both of the cases just cited, reference is made to Interpretative Bulletin No. 13 of the Department of Labor, Wage and Hour Division. This Bulletin will be referred to later.

In *Harris* v. *Crossett Lumber Co., D.C., W.D. of Ark.*, 62 F.Supp. 856, 858, an action to recover overtime compensation under the Fair Labor Standards Act, the plaintiff was employed as a pumper or tender of five water wells, each equipped with an electric motor, and was required to keep them in operation for a period of twenty-four hours per day, seven days a week, and to give whatever time was necessary therefor. No definite hours of employment were considered in fixing his monthly salary. His residence was equipped with a telephone, and he was required to be on the property at all times or to have someone else present who could be notified if the company desired any of the wells to be cut off, or put on if they had previously been cut off for any reason. This was necessary in order to maintain a uniform and adequate supply of water at all times. He was subject to call any time day or night, and often received calls during the nighttime. He brought suit for compensation for the eighteen-hour period above mentioned. In the findings of fact filed by Judge John E. Miller in the case is the following:

"A reasonable computation of the working hours of the plaintiff per week during the time from May 20, 1940, to October 3, 1942, is 84 hours. It appears reasonable to the court that 12 hours of every 24 was available to the plaintiff for uninterrupted sleep, the conduct of his own personal affairs and the ordinary normal routine of living and, therefore, the court finds that the plaintiff did not work 24 hour per day as contended by him, but that he did work 12 hours per day, or 84 hours per week." . . .

. . . We conclude from the foregoing authorities that the rulings and interpretations of the Administrator of the Wage and Hour Division, United States Department of Labor, hereinabove referred to, namely, Wage and Hour Division Interpretative Bulletin No. 13, and the informal ruling made in the letter of the Deputy Administrator, Wage and Hour Division, above mentioned (wherein he stated that the two-platoon system as operated in the instant case appeared to be unobjectionable and that payment to the employees in accordance with the plan would not appear to be violative of the Fair Labor Standards Act), while not controlling upon us, constitute a body of experience and informed judgment to which we may properly resort for guidance in this action.

These interpretations are fair, just, and reasonable, and should be applied here.

We further conclude that time spent by an employee in a rest or sleeping period — the employee being subject to call for emergencies, and being compensated specially therefor — may, but need not necessarily, be considered as employment hours under the provisions of the Fair Labor Standards Act. Whether or not such sleeping time constitutes employment hours depends upon the facts in each particular case. In this connection, it is important to determine the degree with which the employee is free to obtain normal rest and the degree of interference with his normal mode of living and freedom of action is that the employee is required to sleep away from home, the time devoted to rest and sleep should not be considered as employment hours. But if the rest period be interrupted to such an extent as to deprive the employee of a normal period of consecutive hours for sleeping, then such period should be considered as hours of employment under the provisions of said Act.

In the case at bar, plaintiffs, during the first two months of operation under the two-platoon system, were given from 11 p.m. to 7 a.m. within which to sleep and engage in activities of their own choosing. After the first two months they were allowed, under their agreement with the company, a similar eight-hour period of rest extending from 11 p.m. to 7 a.m., but, in practice, were permitted to retire at 9:30 p.m. Their contention is that the rest periods referred to were so interfered with by outside noises and emergency calls that they were unable to enjoy the same or obtain a normal night's sleep. The plaintiffs have failed to sustain the burden of proof in this regard.

In the course of the first two months, the plaintiffs were required, during the rest period, to answer fire calls only, and the proof shows that only sixteen fire runs were made during the rest period between November 29, 1943, and May 27, 1945, a period of eighteen months, and of average duration of slightly less than one hour. After the first two months, as stated, the men checked in at 7 a.m. instead of 11 p.m. Some three and one-half months after the change was made, the ambulances were moved from the hospital to the fire stations, and emergency calls were more frequent. A compilation from defendant's records, filed as an exhibit herein, and which we accept, discloses however, that, taking into consideration both fire and ambulance runs made during the eight-hour rest period, over the entire period from November 29, 1943, to May 27, 1945, thirty-eight plaintiffs averaged forty-one minutes per one hundred hours rest or sleeping time on both fire and ambulance calls, or, conversely, ninety-nine hours and nineteen minutes of each one-hundred hour's rest or sleeping time were uninterrupted by reason of emergency calls of any nature.*

Some of the firemen testified that they were disturbed by the fire inspectors who had an office on the first floor of the fire station, but the fire chief, Mr. Lawrence A. Pluche, testified that he had received only one complaint in this regard during the entire period under consideration, which was on account of the inspectors storing their clothes in lockers on the second floor; and that he immediately moved the lockers downstairs; also, that the inspectors spent practically all of their nighttime in inspecting various parts of the ordnance area, which comprised about 7,400 acres. Some of the men complained that the beds were not comfortable. These were small iron beds, fitted with mattresses and springs, such as are frequently used in dormitories occupied by numbers of men. Chief Pluche stated that he and his wife sleep on identical beds every night. Others complained of snoring. Doubtless it took time for the men to adjust themselves to sleeping in a dormitory along with a number of others, but, on the whole, the Court finds that their rest period was not materially interrupted, and that they obtained a normal night's rest.

In our opinion the only substantial interference with the men's normal mode of living and freedom of action during the rest period was that they were required to sleep away from their homes.

FEDERAL EMPLOYMENT STANDARDS

Federal influence is increasingly involved in state and local fire service decisions and operations. A list of federal agencies involved in fire protection and fire prevention appears in Appendix D at the end of this text. In the following excerpt from *Urban Politics*, Murray S. Stedman, Jr., explains the state-city-federal government relationship:[16]

---

*The exhibit also shows that thirty-eight plaintiffs spent a total of 6,153 nights in the dormitory, or an average of 162 nights each, and that based on the average time of emergency runs during the rest or sleep period, each plaintiff on an average suffered 15 seconds interruption during each eight hours of sleeping time. For no apparent reason this compilation was made as to thirty-eight instead of thirty-nine of the plaintiffs. The case of the omitted plaintiff, however, is not atypical.

[16]Murray S. Stedman, Jr., *Urban Politics*, 2nd Ed., Winthrop Publishers, Inc., Cambridge, MA, 1975, p. 64.

Under the American constitutional pattern, the cities are tied to the states, not to the federal government, and the state-city relationships are the basic ones. But in practice these relationships have often been restrictive and negative, and city officials have found it more rewarding to turn toward Washington for assistance. In fact, the federal government is more urban-oriented than the state governments, and has recognized the special needs of large cities for the last three or four decades. Where the federal government has responded to such needs, the response in recent years has mostly been expressed through grants-in-aid, which is to say, money.

The Occupational Safety and Health Act (OSHA) of 1970 requires that employers covered by this act provide safe and healthy working conditions for their employees. This act applies wherever there is an employer-employee relationship. For the fire service, this includes even private or volunteer departments where no pay is received. Charles W. Bahme, in his book titled *Fire Service and the Law,* describes some of these:[17]

One of the most common violations has been the failure to keep a record of all injuries and fatalities on prescribed forms (obtainable from regional OSHA offices). Pilot inspections have disclosed other violations such as a lack of first aid kits, insufficient exit doors, doors that swing the wrong way out of assembly rooms, lack of signs for exits and for fire extinguishers, lack of "No Smoking" signs (in battery recharging areas), lack of labels on electric panels, ungrounded electric tools, unshielded grindstones, lack of handles on files, lack of inspection tags on fire extinguishers, *etc.*

PUBLIC SERVICE EMPLOYMENT

Members of the fire service, whether in the employ of state, local, or federal agencies, are still considered to be employed in a public service capacity. As such, their employment status may be determined and defined by statute, while policy decisions with regard to public employees may be submitted to the voters. Although both public and private employment can be said to have generally similar characteristics, the employment status and statutory restrictions concerning public employment are often much more complex than private employment considerations, as described in the following excerpt from *Governing Urban America,* by Adrian and Press:[18]

The Jacksonian frontiersman taught us to be skeptical of those who would "feed at the public trough." The lesson was taught so well that Americans often

---

[17]Charles W. Bahme, *Fire Service and the Law,* NFPA, Boston, 1976, p. 38.

[18]Charles R. Adrian and Charles Press, *Governing Urban America,* McGraw-Hill Book Company, New York, 1968, pp. 334-35.

still accept uncritically the notion that public employees are loafers, far less efficient than those in private employment. Some citizens believe that the nature of public employment itself makes people lazy and inefficient.

While students and practitioners of municipal government are in virtually universal agreement that this viewpoint is false, it is difficult to present conclusive evidence to that effect because there is no certain basis for comparing public and private employment. In the field of private endeavor, the test of profit exists. Cities, however, operate few services that are expected to be revenue-producing. Even when they do — such as in public transportation — the service is often one in which profit opportunities are too doubtful for most businessmen to be interested. The effectiveness of most city services — its recreation, welfare, highways, fire fighting, and other functions — can be given only a subjective evaluation. Where one person may think the city recreation program, for example, is inefficient and a waste of the taxpayers' money, another may think it excellent. No clear standard of measurement exists, and the recreation program certainly cannot be compared in efficiency with the method by which some local manufacturer makes motorboats.

GOVERNMENT EMPLOYMENT

Every unit of state government and virtually every unit of local government is involved in employment. As such, questions regarding the employment status and rights of public employees are questions that are addressed by almost every government agency. Questions involving the right of public employees to join a union, job security, hiring and firing practices, salaries, and the right of public employees to strike all require special consideration.

Government employment is usually referred to as civil service employment, and the term "civil servant" means a civilian who is employed by the government. In *Governing Urban America,* Adrian and Press comment on the degree of professionalization that has evolved in the municipal civil service in recent times:[19]

> It seems likely that most urban citizens do not realize the degree of professionalization that has taken place in the municipal civil service in recent decades. Over two million people are employed in local government service in the United States (not counting school employees). Most of them are employed by municipalities, and most of them are performing specialized jobs that require training of one degree or another. While the number of Federal employees has had a general downward trend since 1945, municipalities continue to expand their services, the total number of employees, and the total size of the payroll.

CIVIL SERVICE AND OTHER MERIT SYSTEMS

Generally, personnel administration in government employment is the function of an independent agency developed to keep politics out of government

[19]*Ibid.,* p. 328.

employment. Included in personnel administration is recruitment; for example, paid fire fighters are generally not recruited directly by a fire department but rather by a municipal personnel or civil service agency. Percy Bugbee, in *Principles of Fire Protection,* explains the independent agency recruitment process in a fire department:[20]

> In many states the Civil Service Commission maintains a roster for individuals who wish to join a paid fire department. To become eligible, a candidate usually must pass a written examination, possibly a practical one, and a physical checkup. Those who qualify are usually entered on the roster in order of their test results. Several names from the top of the list are submitted to a fire department for specific selection. Either a personnel officer, the company officer, or a small selection team must then choose from these candidates the one who is best qualified to fill the position.

The most common type of independent agency for personnel administration is the Civil Service Commission. The uniformity in the legal basis for local personnel systems is discussed in the following quotation from George S. Blair's *American Local Government*:[21]

> There is little uniformity in the legal basis for local personnel systems. In some states, mandatory state laws apply to local employees. More commonly, states enact general "civil service" provisions applicable to various classes of local governments, often depending upon their size; and in some communities, the local personnel systems are provided by special acts of the state legislature. Home rule charters, where authorized, normally include authorization for local merit systems. Although less binding than the above legal bases, personnel arrangements may be governed by local ordinance or by local executive or administrative orders.

Generally, a civil service commission makes rules with regard to examinations, classifies positions, conducts examinations, and develops a list of eligible potential employees. Civil service commissions can also make rules regarding wages, hours, employment conditions, transfers, and promotions.

One of the major responsibilities of a civil service commission is the classification of employment positions or categories. Classification of employment positions is explained by Adrian and Press as follows:[22]

> Although a position-classification plan may be greeted as "bureaucratic red tape" by the general public and municipal employees alike, it performs many useful functions. Classification raises employee morale by standardizing job titles. It permits the use of a uniform pay plan so that persons doing approximately the same work receive the same pay. If salaries are standardized, classification permits

---

[20]Percy Bugbee, *Principles of Fire Protection*, NFPA, Boston, 1978, p. 246.

[21]George S. Blair, *American Local Government*, Harper and Row, Publishers, New York, 1964, p. 346; from Kenneth O. Warner, "Municipal Personnel Administration in the United States," *Local Government in the United States of America*, International Union of Local Authorities, 1961, p. 93.

[22]Adrian and Press, *Governing Urban America*, pp. 331-32.

the use of a system of periodic pay raises. These may be based on the employee's increased value and faithful service over a period of time or on his receiving additional training. Often, however, they are based almost purely on the length of employment. Since the opportunities for promotion are necessarily limited, it is important to the morale of employees that they may be advanced in pay up to a maximum limit while continuing to perform the same type of duties. Classification allows for the transfer of employees among departments on the basis of their job descriptions. It simplifies recruitment, since a single standard examination may be given for a class, even though many positions in that class are to be filled.

FIRE SERVICE EMPLOYEE ORGANIZATIONS

As employees in public service capacities, fire fighters are free to join unions. Many fire fighters are members of the International Association of Fire Fighters (IAFF) — a division of the AFL-CIO. Adrian and Press summarize the pros and cons of unionization of municipal employees as follows:[23]

> Some writers have argued — from a theoretical standpoint — that there should be no organization of municipal employees because "the civil service commission is their union." But to the rank-and-file worker, the civil service commission and its executive secretary or the personnel director are the *employer* — and hence the opposition when it comes to bargaining. The commission cannot serve simultaneously as employer and as employee representative. At least it cannot do so in the mind of the employee.

Although fire fighters are not required to join a union, the majority of paid fire fighters are union members. However, the smaller the city, the less likely it will be that fire fighters are unionized. Even though many fire fighters are members of a union, they are generally barred from striking because of their unique position in terms of public service, as explained by Adrian and Press:[24]

> The semimilitary services are barred from striking by one rule or another. This, plus the fact that the public will tolerate less "labor agitation" from them, has sometimes resulted in their salaries falling behind the trend of other employees. On the other hand, they often enjoy a pension plan that is more liberal than that of other municipal employees. Firemen are commonly on duty for many hours consecutively, after which they have a long period off duty. In many cities, therefore, firemen take on outside employment on their off days. Some police officers do, too, although this is specifically prohibited in some cities because it might interfere with the policeman's duty to enforce the law impartially, or to be available on emergency notice.

FIRE SERVICE EMPLOYMENT PRACTICES

Statutes and charters often place members of the fire service under civil service rules or comparable merit system regulations, and usually affect municipalities of a

---

[23]*Ibid.*, p. 244.
[24]*Ibid.*, pp. 337-38.

certain size. Such provisions are enacted to ensure that members of the fire service meet proper qualifications, and to protect members of the fire service, especially from unjust or arbitrary removal. In *State* v. *Morris,* the purpose of such provisions is explained as follows:[25]

> The purpose of the statute . . . is to promote efficiency of the personnel of paid fire departments . . . ; to preclude discrimination in appointments and promotions therein based on any consideration other than fitness for the performance of the duties required; and to prevent discharge without cause.

Appointment and Promotion: The manner of appointment, selection, and promotion of many members of the fire service is prescribed by law — usually by statute, charter, or some form of local ordinance. In assessing the right to a promotion, credits are usually granted for service or seniority. Sometimes these are prescribed by statute; in other cases an administrator or administrative body (such as a civil service commission) is allowed to determine the credits awarded for seniority or length of service. Seniority within a rank, as opposed to total time of employment, is also a factor that affects promotion decisions. In different cases, either may be the determining factor. The decision in the following case, concerning an appointment to the position of second assistant fire chief, is based on an appeal regarding these two points of view.

STATE ex rel. O'CONNELL v. ROARK et al.

Supreme Court of Florida, Division B.
March 15, 1946
25 So. 2d 275

Appeal from Court of Record, Escambia County; Ernest E. Mason, Judge.

Mandamus proceeding by the State of Florida, on the relation of Charles B. O'Connell, against George J. Roark and others to secure appointment to position of second assistant fire chief of the City of Pensacola. From judgment denying application of peremptory writ, the relator appeals.

Affirmed.

BROWN, Justice.

This was a proceeding in mandamus instituted by the appellant in the Court of Record of Escambia County whereby the relator sought to have the court require the members of the Civil Service Board of Pensacola to certify the name of relator to the City Manager as the only person eligible for appointment as Second Assis-

---

[25]*State* v. *Morris,* 37 S.E.2d 87.

tant Fire Chief of said city, and to require the City Manager to appoint relator to this position. . . . The relator's application for peremptory writ was denied by the court and relator took this appeal.

Axel W. Largergreen and the appellant Charles B. O'Connell were both Captains of the Fire Department of said city on July 20, 1945. On said date there was a vacancy in the next higher rank, to-wit: Second Assistant Chief of the Fire Department; and the Civil Service Board, on the basis of seniority within the rank of Captains, named Axel W. Largergreen as the one eligible for appointment to said vacancy.

Appellant propounds the question as to whether or not the court erred in construing the intention of the legislature to be that in filling this vacancy, the principle of seniority within rank, and not seniority of employment by the city, should control.

The relator had been continuously an employee of the Fire Department of Pensacola since October 1, 1919, and became a Captain in the Fire Department on July 16, 1922, but was demoted from the rank of Captain to that of driver, the next lower grade, on September 1, 1923, and remained in said lower grade until June 16, 1925, at which time he was reinstated to the rank of Captain. Largergreen became a captain of the Fire Department on July 7, 1922, and had maintained his rank of Captain ever since his employment up to July 20, 1945, at which time he was promoted and made Second Assistant Chief of said Department. It is clear therefore that Largergreen had seniority within the rank of Captain (there being in all eight captains), but appellant claims that as he had seniority of service for the city, dating back to 1919, he should have been given the appointment to fill the vacancy.

Both relator and Largergreen were members of the Civil Service of the City of Pensacola. Chapter 15425, Acts of 1931, which is the Charter of the City of Pensacola, as extended and amended by Chapter 16867, Acts of 1935, designates certain officers and employees and classes of officers and employees, as civil servants and grants them protection against suspension, removal, demotion or reduction in pay, except for just cause and after an opportunity to be heard. These acts also confirm such officers and employees in the respective ranks held by them on April 1, 1931. . . .

The last clause of section 90, subd. C, of Chapter 15425 reads as follows:

"And said persons' names and their respective rank shall be entered upon the Civil Service Register as aforesaid, taking precedence by rank and those persons of same rank shall take precedence by seniority."

The last clause of section 10 of Chapter 16867 reads as follows:

"Each employee's name and rank shall be entered upon the Civil Service Register as aforesaid, taking precedence by rank, and those employees of the same rank shall take precedence by seniority."

Here we are dealing with a promotion made from an intermediate rank — that of Captain — to the next higher rank, Second Assistant Chief of the Fire Department. Section 75 of Chapter 15425, which regulates promotions, specifically provides that whenever practicable vacancies in the classified services shall be filled by promotion and charges the Civil Service Board with the duty of indicating the lines of promotion from each lower to higher grade "whenever experience derived in the lower grade tends to qualify for the higher," and provides that any advancement in rank shall constitute a promotion. Construing this section 75 in connection with section 90, subd. C, it would indicate that the legislature intended that seniority within the rank rather than seniority of service in city employment should govern promotions from an intermediate rank to the next higher rank. As was well said by the learned Judge of the trial court in his very able opinion accompanying his decision in this case:

"It stands to reason that the employee with the longest period of service, that is to say the greater experience, in the rank from which the promotion is to be made should be more qualified for the next rank of employment. And in this case the Civil Service Board, which is the agency charged under Section 75 with the duty of determining if experience derived in the lower grade tends to qualify for the higher, has in the exercise of the discretionary power granted it by said section determined that experience — the longer service — of Largergreen in the rank of captain has qualified him over the other captains for promotion to Assistant Chief. In the absence of a showing of abuse of this discretion by the Board this court would not be justified in submitting its judgment for that of the members of that Board. No such abuse is shown."

Section 90, subd. E, of said Chapter 15425 is also strongly persuasive of this conclusion. The Judge of the trial court ended his opinion with these words:

"Therefore, it is evident to the court from a close scrutiny of all of the provisions of the Civil Service laws which are applicable to the principle that seniority within the rank should govern and not seniority of employment with the city. . . ."

. . . The judgment of the court below denying the application for the peremptory writ and quashing the alternative writ is accordingly

Affirmed.

CHAPMAN, C. J., and THOMAS and SEBRING, JJ.,

Concur.

Removal, Suspension, Demotion: Because of the early political overtones of public service, employees were often subjected to the political whims of those in charge of employment. Today, administrative procedures in most governmental areas provide that a public employee can be discharged only for cause. They give the employee the right to appeal a decision for dismissal, removal, or demotion, and the right to engage an attorney to defend the employee's position with regard to the allegations for dismissal.

The following decision in the case of *Robertson* v. *City of Rome* highlights the right of an employee to fair representation at a hearing.

ROBERTSON v. CITY OF ROME

Court of Appeals of Georgia, Division No. 2.
May 15, 1943.
25 S.E. 2d 925

Syllabus by the Court.

The amendment to the charter of the City of Rome, providing that one against whom charges had been filed would have the right to employ and be represented by counsel at the hearing before the Civil Service Board, Ga.L. 1941, p. 1690, contemplated that counsel should have the right to argue the case before the board, and the judge erred in overruling the *certiorari* assigning error on the refusal of the board to allow counsel for one accused to argue the case on the hearing provided for in said amendment. . . .

Reversed.

FELTON, Judge.

. . . The assignments in the instant case are that the evidence did not warrant the finding that the plaintiff in error was guilty of conduct unbecoming an officer; that the Civil Service Board committed reversible error through its chairman when he stated at the opening of the hearing, "Of course we don't know anything about the legal technicalities and all we want to know is what he knows about it. This is not a trial, this is just a hearing here before the board"; and that the board erred in refusing counsel for plaintiff in error the right to argue the law and the facts to the board upon the completion of the evidence. Because of the fact that the case will be reversed on other grounds we shall not decide whether the evidence authorized the finding of the board, or whether the statement by the chairman of the board was error, since it will not likely occur on another trial. The case will be decided on the question whether the board erred in refusing counsel the right to argue the law and the facts to the board.

The hearing before the Civil Service Board of the City of Rome was held in accordance with the amendment to the charter of the City of Rome, Ga.L. 1941, p.

1690, which provides in part as follows: "No member of the fire or police department shall be removed or discharged . . . except for cause upon written charges or
complaint and after an opportunity for an open public hearing in his own defense
by the Civil Service Board. . . . The person against whom charges are preferred
shall have the right to employ counsel to represent him on the hearing before said
board." It is evident that the hearing contemplated by the act is not a common-
law or a criminal proceeding. But while it is not a common-law or criminal proceeding it is of judicial character and must be so construed. *[Citations omitted.]*
The power of the Civil Service Board of the City of Rome is derived from the act
creating it, and it has no power not granted by the act, and in performing its functions it must do so in terms of the act. "The full performance of all conditions
established by the civil service laws is an essential prerequisite to the jurisdiction of
the removing body over the subject matter of the removal of an officer . . . , and
where there is no substantial compliance with the statutory procedure, an order of
removal is a nullity." 43 C.J. 679.

It is earnestly contended by counsel for the city that the right to have the case
argued is not in the act creating the Civil Service Board or in any general statute,
and that all of the technicalities of the law are not to be observed on such a hearing, but that the hearing is in the nature of an investigation and not a trial, and
that it is within the discretion of the board to refuse to hear argument if it sees fit.
We cannot agree with this contention because we are of the opinion that the hearing is in the nature of a trial. "The right to a hearing, with notice of charges,
especially where, as here, the right to be represented at such hearing by counsel is
especially secured, contemplates a proceeding in the nature of a judicial investigation, although it is one in which the attainment of substantial justice rather than
the observance of any particular formalities is aimed at." *[Citation omitted.]*

Being of the opinion that the hearing contemplated by the amendment to the
charter of the city is in the nature of a judicial proceeding, an administrative act to
be performed judicially, we look to see whether depriving counsel for plaintiff in
error of the right to argue the law and the facts on the hearing deprived the plaintiff in error of such a substantial right, under the act empowering the board to
hold such a hearing, as to require a reversal of the judgment overruling the *certiorari* from the decision of the board. . . .

. . . The Supreme Court said: "It is a most valuable right to be represented by
learned and eloquent counsel, not only before the Court, as to the law, but also
before the Jury, as to the facts. It means something — it is a guarantee against the
encroachments of power upon the personal rights of a citizen. It is, in this country,
no mean privilege. . . . The true view of the position of counsel, before the Jury,
is that of aids or helps. They are officers of the Court — amenable to its authority,
subject to its correction, and restrained by usages of honor and courtesy, which,
however, in some instances disregaded, are as ancient in their origin and as potent
for good, and as generally respected, as any usages which belong to any class of the

highest grade of civilized man. The duties of the advocate are among the most elevated functions of humanity. . . . His business is to comment on the evidence — to sift, compare and collate the facts — to draw his illustrations from the whole circle of the sciences — to reason with the accuracy and power of the trained logician, and to enforce his cause with all the inspiration of genius, and adorn it with all the attributes of eloquence. . . .''

In view of the authorities just cited, we are of the opinion that the right given by the act to one against whom charges had been filed to employ and be represented by counsel carried with it all the rights connoted by the word "represent"; *i.e.*, to stand in his place; to act as his substitute; to exercise his right; and that, by virtue of the provision securing him the right of representation by counsel, he had the right to have his counsel conduct his case in the manner which is generally accepted by our courts — examination of witness, cross-examination, argument of the law and the facts.

The refusal on the part of the board to allow such argument was error, and the judge erred in overruling *certiorari.*

Judgment reversed.

STEPHENS, P. J., and SUTTON, J.,

Concur.

––––––––––––

Removal "for cause" is determined by a body empowered by law to make such decisions. This removal cannot be made "at pleasure," but only within prescribed boundaries. The reasons for removal for cause may vary and, because of the wide variety of circumstances and interpretations, may be subject to judicial litigation as in the following case, which concerns the reinstatement of the plaintiff as a member of a fire department following the plaintiff's removal for "conduct unbecoming an officer of the city."

––––––––––––

KENNETT et al. v. BARBER

Supreme Court of Florida, en Banc.
June 10, 1947.
31 So.2d 44
Rehearing Denied July 7, 1947

THOMAS, C. J., and CHAPMAN, J., dissenting.

Proceeding in mandamus by Harry C. Barber against D. C. Kennett, as Chief of the Fire Department of the City of Miami Beach, a municipal corporation, and others to compel reinstatement of plaintiff as a member of the fire department. From a final judgment granting peremptory writ directing reinstatement, defendants appeal.

Judgment reversed. . . .

TERRELL, Justice.

Pursuant to Chapter 18696, Sp. Acts of 1937, the City of Miami Beach promulgated regulations governing the behavior of civil service employees of the city, one of which authorized the removal of any employee "guilty of conduct unbecoming an officer of the City."

Appellee was a member of the fire department. The City Manager and the Chief of the Fire Department, being authorized to remove members of the department for infractions of the quoted regulations, removed appellee on the charge that "on April 21, 1946, at about the hour of 6 o'clock p.m. at or near 411 Michigan Avenue, in the City of Miami Beach, Florida, you were guilty of conduct unbecoming an employee of the City in that you were drunk and you did then and there grab and twist the arm of your pregnant wife and threw her to the ground and you did then and there beat, bruise, and batter Adele Betancourt with your fists and your feet."

The Personnel Board of the city sustained the removal. On petition of appellee alternative writ of mandamus was directed to the city commanding it to reinstate him. A motion to quash the alternative writ was overruled. An answer was filed and a motion for peremptory writ, notwithstanding the answer, was granted. Appellants complied with the peremptory writ and have appealed from the final judgment. . . .

The point for determination is whether or not a city fireman who gets drunk and seizes the arm of his pregnant wife and throws her to the ground and then beats, bruises, and batters a second woman with his fists and feet is guilty of "conduct that is unbecoming an employee of the City," subjecting him to removal or discharge from the service, as contemplated by the applicable city regulation.

There is no dispute about the facts recited. Appellee's defense is that he was not on duty at the time he committed the acts with which he is charged; that they do not relate to the duties of his employment; that they in no way affect his qualification to be employed as a member of the fire department or to perform the duties thereof; and that the mere fact that his superiors consider his "conduct unbecoming an employee of the City" does not make it so.

The defense so tendered requires an answer with a double aspect. (1) May the city inquire into and discipline or discharge one of its employees for unbecoming conduct committed while he was off duty? And (2) did the acts committed amount to "conduct unbecoming a city employee," subjecting him to discharge from the service of the city?

The defense that an employee's conduct while off duty is immunized from challenge by the city stems from a philosophy that had considerable support in this country prior to the day of the direct primary election. It was implicit in the doctrine, "The public be damned" which we thought had gone the way of the little red school house, the country doctor, and the muzzle loading shot gun. We think the rule is now generally accepted that anyone who seeks public employment or public office or who makes his living by dealing with the public or otherwise seeks public patronage submits his private character to the scrutiny of those whose patronage he implores, and that they may determine whether it squares with such a standard of integrity and correct morals as warrants their approval. [Citation omitted.]

By this yardstick the City of Miami Beach would certainly be authorized to inquire into the private character and conduct of a member of its fire department, and determine whether it met the standard required of those who seek employment at its hands. After they are employed, their character is subject to inspection at all times. No business is required to keep one in its employ who is known to be stealing, cheating, gambling, or who is guilty of other conduct embarrassing to or inimical to the interest of his employer. The first aspect of the answer disposed of, we are next confronted with the second aspect. Did the acts with which the appellee was charged amount to conduct unbecoming a city employee? The charge must of course be specific and one that can be defended against.

We see no escape from an affirmative answer to this question.

In reaching its determination of whether or not one has been "guilty of conduct unbecoming a City employee" the city cannot act on an arbitrary or capricious judgment; neither must it act on a judgment shrouded in the fog of bias and skepticism. Such judgment must above all be one illuminated by the light of truth and justice. So far as the record discloses, the judgment under review meets this test. In fact, it is not claimed by appellee that it fails to do so.

We think it was competent for the City Manager, the Chief of the Fire Department, and the Personnel Board to determine what constitutes "conduct unbecoming a City employee." Chapter 18696 clothes them with this power and they proceeded to make an unequivocal charge against him for which he was convicted. After all is said, it is a question of what standard of social ethics the city sees fit to exact of those who work for it, and if a member of the fire department becomes so enamoured of John Barleycorn that he beats up his wife and a second woman who appears to have been a mere bystander, the city should not be forced to keep him against its regulations. We are aware of no right of an employee that rises above the right of a city to promulgate and enforce as high a standard of social and cultural conduct as its people want. Certainly this Court should not say nay when the record reveals nothing more than an orderly attempt to enforce a reasonable standard of conduct on the part of city employees.

There is a well-recognized general effort over the country at the present to conduct municipalities as business institutions. Any informed person knows this so we may assume that the Court knows it. There is not a bank, store, railroad office or business of any kind that does not look first to the moral equipment of every applicant it has for employment, and whether or not he is employed depends on that equipment. Not only that, but whether or not he keeps the job after he gets it depends on how well he keeps his private conduct squared with an approved standard of morals. It is the very bedrock on which sound democratic institutions rest. One cannot get and hold a job in a gambling enterprise without a certificate of good moral character.

The fire department is a very important arm of the city government. It should be staffed by men who are dependable and feel their responsibility, else the city may become liable for heavy penalties. It is utterly foolish to contend that a man who gets under the influence of liquor and beats women in the manner charged meets all the requirements of a city fireman. It is equally as foolish to contend that such conduct has no relation to the duties of his employment or in no way affects his qualifications as a city fireman.

In the category of deeds and attributes that make a city a good place in which to live we do not find drunkenness and wife beaters listed. If the city wants to proscribe them it has a right to do so, and no rule of the civil service may be invoked as a shield against them. That the city's thesis is supported by a social fact rather than a physical or a literal one is no less convincing.

"Conduct unbecoming an employee" is not susceptible to a definition that would serve the purpose every time a court is confronted with the charge. There must be some stability to the rule by which it is determined. At the same time it must be fluid enough to accommodate itself to the social aspirations of those who impose it. Perhaps the most delicate task of the juristic craftsman is that of reshaping old rules of law to constant changes in the social structure. A rule grown stale and inept should not be permitted to paralyze the present. "The Earth in usufruct belongs to the living" said Jefferson. Neither should the law of today be overburdened by what some of the historical school of jurisconsults have chosen to call, "ancient law lumber."

From what has been said, it follows that the judgment appealed from must be and is hereby reversed.

Reversed.

BUFORD, ADAMS, and BARNES, JJ., concur.

THOMAS, C. J., and CHAPMAN, J., dissent.

The preceding decision was presented with three judges agreeing and two judges disagreeing. On what grounds do you think the dissenting judges disagreed? This decision was handed down in 1947. Do you think the same facts would result in a different decision today? How would the employee rules in your own jurisdiction affect a decision based on the facts of this case? Do you think the statements made in the preceding decision draw a distinction between private and public employment? If so, how?

Retirement: Generally, the retirement pensions for employees of small units of government are provided for by state or federal retirement systems. Larger units of government provide retirement plans that parallel the pension plans of private industry. In *Governing Urban America,* Adrian and Press discuss retirement systems as partial remuneration for public employment:[26]

> As a part of the total remuneration for public employment, municipal retirement systems are commonly provided. Nearly all cities of over 10,000 and many smaller municipalities provide retirement plans for all of their employees. Many cities operate their own local systems, often with the police and fire services under a separate plan, but about one-half of them are members of state-administered systems. Smaller cities, in particular, are aided in maintaining solvency by being banded together with other cities in a state system. Since 1951, there has been a sharp trend toward the covering of municipal employees in the federal old-age and survivors' insurance program ("social security"). Amendents to the Social Security Act in 1954 aided in bringing more local governments into the federal program, which often provides more liberal benefits than do local systems.

EMPLOYEES' RIGHTS

Many rights of employees are constitutional rights. The Constitution of the United States carefully and specifically defines the rights and powers of the federal government. Those powers not specifically defined as federal powers are delegated to the individual states. However, both state and federal constitutions guarantee and protect each and every citizen's rights. The employment rights of individuals have become more and more important in the field of constitutional law, with the Civil Rights Act of 1964 providing major impetus in this direction.

---

[26]Adrian and Press, *Governing Urban America,* p. 235.

INDIVIDUAL RIGHTS

Constitutional law, very simply, is concerned with the relationship between the individual and government, whether the government is federal, state, or local. Both federal and state constitutions have been drawn to define the rights of the individual, to ensure that no governmental action deprives an individual of the right to life, liberty, and/or property without due process of law, and to provide equal protection under the law for all. This means that any government action — federal, state, or local — that violates a person's individual rights is subject to challenge by that individual. Because such challenges are constitutional challenges, they may be taken as high as the Supreme Court of the United States.

The following excerpt from the decision in *People* v. *Hurlburt* (24 Mich. 44) describes the rights of an individual under the Constitution:

> Mr. Justice Story has well shown that constitutional freedom means something more than liberty permitted; it consists in the civil and political rights which are absolutely guaranteed, assured, and guarded, in one's liberties as a man and a citizen — his right to vote, his right to hold office, his right to worship God according to the dictates of his own conscience, his equality with all others who are his fellow-citizens; all these guarded and protected, and not held at the mercy and discretion of any one man or any popular majority.

The following case, as described in *American Law Reports, Annotated*, involves a fire hazard and individual rights as guaranteed by the Constitution:[27]

> And a somewhat similar situation confronted the court in *Tebbetts* v. *McElroy* (1932; DC) 56 F(2d) 621, where the court enjoined the arbitrary exercise of the closing of a hall devoted to a "walkathon" contest by the director of the municipal fire department. In this case it appeared that the contest was initiated in a building which was at least semifireproof in construction, with the knowledge and consent of the municipal authorities; that extraordinary precautions were taken against possible dangers of any kind; that the promoters of the contest did everything that they were asked to do by the municipal authorities; that they received no intimation from anyone that anything was wrong with the contest or with the arrangements under which it was conducted; and that after the contest had been underway for almost three weeks the director of the municipal fire department, without any previous intimation that anything was wrong, appeared at a time when the grandstands were filled with people, and peremptorily ordered the promoters to get the audience then in attendance out of the building within five minutes, to close the place to the public and keep it closed. Granting a temporary injunction, the court said: "1. It goes without saying that no individual as such lawfully may destroy the business of another by threat or use of force. 2. To anyone who has the slightest knowledge of the provisions of the Constitution it need not be said that neither the government of the United States nor that of any state, nor that of any subdivision of any state may deprive any person of liberty or

[27]140 ALR 1048-49.

property without due process of a law. No servant of the people in public office arbitrarily may say to any person 'Your business is closed.' Not all the forces of government may lock the door of the humblest shop save in accordance with the law of the land, for 'this is a government of laws and not of men.' These principles, which are founded in natural justice, already were old when Magna Charta was wrested from a tyrant king. They are written in the Bill of Rights of every nation of freemen. 3. Learned counsel for defendants do not question these principles, but they say there is a statute somewhere, some provision of the city charter, some ordinance, that vests the director of the fire department in Kansas City with the power immediately to close any place of business, any hall or building where people may assemble, if, in his judgment, danger from fire exists in that place, that hall, that building. In effect they say that by virtue of this law this official may go into a mercantile establishment, look about awhile, and seeing goods upon the shelves which, if fired will burn, seeing counters and floors of wood, seeing the aisles crowded with customers, as at the Christmas season, conceiving in his mind there is a fire hazard present, may send for the proprietor and tell him to get his customers out at once and to lock his doors. He may go into a church, crowded with worshipers, perhaps beyond the seating capacity of the church, and if he thinks there is a fire hazard there, may interrupt the clergyman in the middle of his sermon, drive the congregation into the street, and decree that the edifice may not be used again for religious worship. . . . What a deadly weapon of oppression such a legislative act, if there were one, would be! If such a law could anywhere be found it would be no law, so plainly would it contravene the supreme law of the Constitution.''

Generally, in other cases, the right to close places of public assemblage deemed to be fire hazards has been upheld.

CIVIL LIBERTIES

Civil liberties are the natural rights that each person has as a human being:[28]

> The theory upon which our political institutions rest is, that all men have certain inalienable rights — that among these are life, liberty, and the pursuit of happiness; and that in the pursuit of happiness all avocations, all honors, all positions, are alike open to everyone, and that in the protection of these rights, all are equal before the law.

Though civil liberties are protected by the Constitution, they are not specifically defined and mandated by law as civil rights are. Civil liberties are considered to be immunities or restraints on government.

CIVIL RIGHTS

A civil right is a right that is defined by law. As such, a civil right is a legally enforceable claim, and a deprivation of a civil right may be contested in a court of law. The right to vote, for example, is a civil right. The right to own and enjoy property is also a civil right. In the following excerpt, civil rights are described as specific economic and social rights which are defined and mandated through

[28]*Cummings* v. *Missouri*, 71 U.S. 277.

legislation and are specific claims that legislation makes available to an individual. Civil liberties, on the other hand, are immunities or restraints on government mandated by the Constitution which refer to the natural rights each person has as a human being.

SOWERS v. OHIO CIVIL RIGHTS COMMISSION

252 N.E. 2d 463.

WINTER, Judge (of Medina County, by assignment).

At the July 31, 1969, hearing before this court, rulings on two motions by counsel for the respondent-plaintiff herein were deferred pending the filing of briefs by the respective parties on or before the 11th day of September, 1969. Accordingly, upon consideration of said briefs, now filed by respondent-plaintiff and the defendant Ohio Civil Rights Commission, and upon review of the transcripts and the exhibits in question the court finds:

Did the defendant-commission's order of September 26, 1968, violate the respondent-plaintiff's constitutional civil liberties of freedom of speech and freedom of religion?

Before attempting to answer these questions this court is of the opinion that some discussion is in order concerning what appears to be a confused conception of the term "civil rights." The facts of this case seemingly are illustrative of this observation. As stated in 15 American Jurisprudence 2d 406: "'Civil rights' have been defined simply as such rights as the law will enforce, or as all those rights which the law gives a person." *I.e.*, a civil right is a legally enforceable claim of one person against another.

"Natural rights" are those rights which appertain originally and essentially to each person as a human being and are inherent in his nature, as contrasted to civil rights, which are given, defined, and circumscribed by such positive laws, enacted by civilized communities, as are necessary to the maintenance of organized government, *Byers* v. *Sun Savings Bank,* 41 Okl. 728, 139 P. 948, 52 L.R.A., N.S., 320.

Professor Pollack, 27 Ohio State Law Journal 567, points out that indiscriminate use of the term "rights" to describe an immunity, or privilege, has fostered confusion in the law. Our constitutional government was originally framed on a system of limited powers, reserving rights in individuals. "However," says Pollack:

". . . what were reserved in this context were not actually rights but immunities — restraints on the government."

Accordingly, the distinction between claims and immunities becomes increasingly important in relation to our consideration of problems arising from alleged violations of our Ohio Civil Rights Act.

President Kennedy carefully drew this distinction, identifying immunities as civil liberties, and claims as civil rights. He said:

"The Bill of Rights, in the eyes of its framers, was a catalogue of immunities, not a schedule of claims. It was, in other words, a Bill of Liberties . . . . When civil rights are seen as claims and civil liberties as immunities, the government's differing responsibilities become clear. For the security of rights the energy of government is essential. For the security of liberty, restraint is indispensable."

In line with this distinction, civil, or economic rights, functioning as claims, are structured in legislation, while civil liberties are constitutionally protected.

Pollack goes on to say that this distinction and separation wisely suggests a constitutional definition of historic rights and a developmental pattern of economic and social activities through legislation. He further states:

"Liberty is preserved through constitutional assurances that prevent governmental encroachments, and economic and social change is re-enforced by legislative action."

Civil rights then, within the meaning of Sections 4112.01 to 4112.08, inclusive, and 4112.99, Revised Code, are economic rights functioning as legally enforceable claims which are structured in legislation. On the other hand civil liberties are natural rights which appertain originally and essentially to each person as a human being and are inherent in his nature; such rights, which are constitutionally protected, are not actually rights but are immunities, or restraints on government.

Thus, the Ohio General Assembly, by legislative enactment effective July 27, 1959, gave to those persons whose civil rights were violated, a legally enforceable claim against the violator. This enactment, however, was not expressive of any legislative intent to interfere with the preservation of the constitutional assurances that prevent governmental encroachments. And it is this balance between civil rights as claims, and civil liberties as immunities, that the courts must weigh carefully and judiciously in all controversies such as the case at bar.

CIVIL RIGHTS LEGISLATION

The Constitution of the United States contains the principal federal guarantees of civil rights. The first provision of the Fourteenth Amendment to the Constitution specifies three civil rights that have been the basis for many of the civil rights claims the courts have been asked to enforce. First, the Fourteenth Amendment forbids states from making and enforcing any law abridging the privileges (civil rights) and immunities (civil liberties) of any citizen of the United States. Second, it prohibits states from depriving any citizen of the United States of life, liberty, or property without due process of law. Finally, it forbids states to deny any person the equal protection of the laws. The "due process" clause and the "equal protection" clause have been of great importance to civil rights decisions.

The Civil Rights Act of 1957 was the first major civil rights legislation passed in almost one hundred years. One of the major provisions of this act was the creation of a Commission on Civil Rights. The Commission on Civil Rights is a fact-finding agency and thus its claims and judgments often must be enforced through the courts, as in the following decision concerning a volunteer fire department's violation of the Law Against Discrimination.

James H. BLAIR, Director, Division on Civil Rights, Department of Law and Public Safety, State of New Jersey, Complainant-Respondent,

v.

The MAYOR AND COUNCIL, BOROUGH OF FREEHOLD, et al., Respondents-Appellants.

Superior Court of New Jersey.
Appellate Division.
Argued Oct. 4, 1971.
Decided Dec. 13, 1971.
285 A.2d 46

Proceeding on appeal from order of Director of Division on Civil Rights determining that membership and admission procedures for entry into borough volunteer fire department violated Law Against Discrimination. The Superior Court, Appellate Division, held that where only rational reason for requirement of vote of membership of volunteer fire department for admission of a new member was exclusion, and such exclusion was motivated at least in part by race, admission procedures constituted unlawful employment practice because of establishment of requirements irrelevant to proper performance of duties of firemen.

Affirmed.

Before Judges CONFORD, MATTHEWS and FRITZ.

PER CURIAM.

This is an appeal from an order of the Director of the Division on Civil Rights which determined that the membership and admission procedures for entry into the volunteer fire department of the Borough of Freehold, as contained in a certain ordinance of the borough, violated the provisions of the Law Against Discrimination, N.J.S.A. 10:5-1 *et seq.* The order of the director has been appealed by both the mayor and council of the borough and the fire department.

Except insofar as he concludes that the facilities of the fire department constitute a public accommodation under the provisions of the Law Against Discrimination, the findings, determination and order of the director are affirmed, essentially for the reasons stated by hearing examiner Pressler.

We are not persuaded that the facilities maintained for the pleasure and sociability of members of the volunteer fire department are the equivalent of facilities maintained for the use of the general public of a personal nature. The facilities of the fire department, as shown by the record here, are maintained for the use of its members and not for the general public. Such facilities are therefore not an accommodation within the meaning of the act.

Our reading of the record satisfies us that no overt act of discrimination with respect to either Jews or blacks was established at the hearing below. This in itself does not establish the lack of discrimination. *[Citation omitted.]* It is our conclusion that the admission procedures established under the various borough ordinances, including the latest, constitute an unlawful employment practice because of the establishment of requirements irrelevant to the proper performance of the duties of firemen. We cannot conceive of any lawful reason for the requirement of a vote of the membership of a volunteer fire department for admission of a new member thereto. The only rational reason for such a requirement is exclusion. The overall record contains substantial credible evidence to warrant the conclusion that such exclusion was motivated at least in part by race. The determination of the director must therefore be sustained. *[Citation omitted.]*

Finally, we have no difficulty in affirming the director's conclusion that he has jurisdiction over municipalities and volunteer fire companies. To hold otherwise would undermine the very substance of the Law Against Discrimination. As Mr. Justice Jacobs pointed out in his opinion for the court in *Jenkins* v. *Morris Tp. School Dist. and Bd. of Ed.*, 58 N.J. 483, 279 A.2d 619 (1971):

> [Political subdivisions of the states whether they be "counties, cities or whatever" are not "sovereign entities" and may readily be bridged when necessary to vindicate federal constitutional rights and policies. . . . It seems clear to us that, similarly, governmental subdivisions of the state may readily be bridged when necessary to vindicate state constitutional rights and policies. This does not entail any general departure from the historic home rule principles and practices in our State in the field of

education or elsewhere; but it does entail suitable measures of power in our State authorities for fulfillment of the educational and racial policies embodied in our State Constitution and in its implementing legislation.] . . . [at 500, 279 A.2d at 628].

Affirmed.

Although there are many legal sources of protection from discrimination in employment, Title VII of the Civil Rights Act of 1964 is the comprehensive federal guarantee against discrimination in employment. Title VII prohibits employment discrimination on the basis of race, color, religion, sex, and national origin. Title VII also created the Equal Employment Opportunity Commission (EEOC), empowered to investigate charges of employment discrimination, but not to initiate court action. In 1972, however, the Title was amended to allow the Equal Employment Opportunity Commission to institute court action. Originally, Title VII prohibited discrimination by employers, labor organizations, joint apprenticeship committees, and employment agencies; state and local government employers were excluded from its provisions. But in 1972, the Title was amended to include state and local government employers. In the following excerpt from the book *Equal Employment Opportunity — Responsibilities, Rights, Remedies,* John Pemberton, Jr., presents some of the many ways that employment discrimination can come about:[29]

> Employment discrimination may arise from an isolated act, or from a series of acts. It may arise as a result of overt acts, or subjective judgments. It may exist because of positive action, because of failure to take action, or because of failure to take enough action. It may occur with or without intent. It is by nature a class phenomenon, but it may involve a single individual. Discrimination may arise from treating employees unequally, or from treating them equally.

An example of employment discrimination that may arise when people are treated "equally" is in the area of testing. In many instances, employers rely heavily or exclusively on tests as a basis for making employment decisions. The following two cases involving the New York City and City of Boston fire departments demonstrate how testing procedures can be discriminatory.

The VULCAN SOCIETY OF the NEW YORK CITY FIRE DEPARTMENT, INC., et al., Plaintiffs-Appellees-Appellants,

v.

CIVIL SERVICE COMMISSION OF the CITY OF NEW YORK et al., Defendants-Appellants-Appellees,

---

[29]John de J. Pemberton, Jr., *Equal Employment Opportunity — Responsibilities, Rights, Remedies,* Practicing Law Institute, New York, 1975, pp. 76-77.

Nicholas M. Cianciotto et al., Interve-
nors-Defendants-Appellants-
Appellees.

United States Court of Appeals.
Second Circuit.
Argued Oct. 15, 1973.
Decided Nov. 21, 1973.
490 F.2d 387

Minority individuals who had applied for employment with New York City fire department and organizations representing minority fire fighters brought class action against civil service commission, city's department of personnel and others based on claim that procedures used to select New York City firemen discriminated against blacks and Hispanics in violation of equal protection clause of the Fourteenth Amendment. The District Court for the Southern District of New York, Edward Weinfeld, J., 353 F. Supp. 1092, 360 F. Supp. 1265, entered judgment from which defendants appealed and intervenors and plaintiffs cross-appealed. The Court of Appeals, Friendly, Circuit Judge, held that finding that written and physical examinations for positions of New York City firemen had had a racially disproportionate impact was not clearly erroneous, that satisfactory examinations for positions of city firemen were not limited to those which have been subjected to "predictive validation" or "concurrent validation," that in combination with defects in preparation and content of written examination, the use of a merely qualifying physical examination rendered fire department's selection procedures insufficiently job-related to withstand constitutional attack, and that it was altogether appropriate for trial judge to seek agreement of parties that he might rule on minimum height issue as on final hearings and, having obtained this, to make a disposition of this issue, but, in absence of an agreement by plaintiffs' counsf to abandon minimum height issue, which was squarely raised as to one of plaintiffs, trial judge was obligated to, on question either one way or the other, and case would be remanded for this purpose.

Affirmed in part, and remanded with respect to issues not decided on merits.

I.

Plaintiffs, five minority individuals who had applied for employment with the Fire Department and two organizations representing minority fire fighters, brought this suit as a class action in the District Court for the Southern District of New York. Their complaint alleged that the procedures used to select New York City firemen discriminated against blacks and Hispanics in violation of the equal protection clause of the Fourteenth Amendment. The defendants were the Civil Service Commission of the City of New York, the City's Department of Personnel, the chairman of the Commission and director of the Department, two members of the Commission and then Fire Commissioner Lowery, hereafter referred to as the municipal defendants. Attacking on a broad front, plaintiffs first sought injunctive relief to prevent the Fire Department from making any more appointments based on an eligibility list reflecting performance on Exam 0159, a written civil service examination given on September 18, 1971. In addition, they sought to block further use of various other screening measures, including the Department's minimum height requirement, a requirement that every fireman have a high school or high school equivalency diploma, and the bar, arising from the combined effect of § 487a-3.0(b) of Chapter 19 of the Administrative Code of New York City and § III-4.3.2(b) of the Rules and Regulations of its Civil Service Commission, against any applicant who had been convicted of a felony or of petty larceny.*

At a hearing on plaintiffs' application for a preliminary injunction, Judge Weinfeld took evidence on the alleged discriminatory impact of the written examination and on the question whether the test was sufficiently related to a fireman's job to survive constitutional attack. At the conclusion of the seven-day hearing, he suggested that the parties agree to treat the hearing as a final trial on the merits of the case under F.R.Civ.P. 65(a)(2). Both parties agreed, but the plaintiffs requested that several points, including the automatic disqualification issues, be left open. Judge Weinfeld subsequently ruled that any further evidence on these points would have to be submitted before the court's decision was handed down, and that the decision would be final as to all issues in the case except for the challenge to the Fire Department's promotional examination, which would be left open for later consideration.

On June 12, in a comprehensive opinion, the district judge, 360 F. Supp. 1265, ruled that the written examination had a discriminatory impact and that it was not sufficiently job-related to justify its use. He enjoined further reliance on the challenged eligibility list, without prejudice to the parties' applying for interim relief which would permit appointments from the list on a quota basis until a new examination could be given and a new eligibility list established. Because the injunctive relief would benefit all persons similarly situated, Judge Weinfeld declined plaintiffs' request that he designate a class. As to the automatic disqualification issues, he ruled that it was unnecessary to consider those questions, stating:

> This disposition makes it unnecessary to consider the other grounds
> urged by plaintiffs in support of their claim, particularly since little

---

*Plaintiffs also challenged the defendants' recruitment program, the scheduling and location of qualifying examinations, the character review process, the absence of a competitive physical examination, the absence of a city residency requirement, the Fire Department's promotional examination and the promotional merit point system. Each of these practices, plaintiffs charged in their complaint, is discriminatory in impact and not sufficiently job-related to survive constitutional attack. None of these points is directly raised on appeal, although, as will appear below, the district court relied in part on the absence of a competitive physical examination in finding that the qualifying examination process was not adequately job-related. Plaintiffs apparently did not challenge the bar created by the Rules and Regulations against persons dishonorably discharged from the armed forces of the United States, § III-4.3.2(b).

evidence was adduced with respect thereto upon the hearing. The submissions as to these matters were included in post trial briefs or affidavits and in some instances raise issues of fact, the resolution of which would require reopening of the trial.

He added there was serious doubt whether the plaintiffs had standing to raise either the criminal conviction bar or the high school diploma requirement, since it appeared from the complaint that none of the named plaintiffs was subject to exclusion for those reasons.

Two months later, after having allowed intervention by nonminority candidates who had qualified under Exam 0159 but had not yet been appointed, the district court issued an order granting interim relief. The order instructed that in making future appointments from the challenged eligibility list, defendants would be required to hire one minority applicant for every three nonminority applicants hired. In an accompanying memorandum, the court further directed the municipal defendants "to exert every good faith effort to accelerate the establishment of a new list."[*]

The defendants, intervenors, and plaintiffs all appealed from various portions of the district court's decision and order. We denied an application by the defendants and intervenors for a stay but brought the appeal here on an expedited basis. As is usual in cases of this sort, we have had a number of *amicus* briefs filed on behalf of various individuals and organizations. However useful *amicus* briefs may be on an issue of first impression in this circuit, see *Chance, supra,* 458 F.2d at 1169 & n.5, they only add to our burdens when the controlling principles have been established and the parties are so capably represented as here.

II.

Under *Chance* and *Bridgeport Guardians* our analysis must be three-pronged. Was Judge Weinfeld "clearly erroneous" in finding that Exam 0159 had had a "racially disproportionate impact?"[**] If not, did he err in concluding that the City had not made the requisite showing that Exam 0159 was sufficiently job-related; that is, did the City fail to prove that the disproportionate impact was simply the result of a proper test demonstrating lesser ability of black and Hispanic candidates to perform the job satisfactorily? If the district court was correct on that point also, we reach the third issue, the propriety of the relief.

---

[*]The order permitted the Fire Department to make a maximum of 152 appointments from List 0159 every two months, until June 30, 1974. If any further relief were needed, the defendants were instructed to apply to the court at that time.

[**]This phrase, used by Judge Coffin in *Castro* v. *Beecher,* 459 F.2d 725, 732 (1 Cir. 1972), seems preferable to "discrimination" or even "*de facto* discrimination," since these terms have acquired a pejorative connotation not warranted in the initial phase of the inquiry. The first endeavor is to ascertain whether the challenged procedure has had disparate effects on particular racial groups. Discrimination in the invidious sense exists only if such effects are not the result of job-related tests.

The municipal defendants do not here challenge the findings of racially disproportionate impact, but the intervenors do. The basic facts are these: Roughly 11.5 percent of the 14,168 applicants who entered the examination halls were black or Hispanic. Yet minority members comprised only 5.6 percent of those who had passed the written, physical and medical examinations at the time of the hearing. Nonminority candidates thus survived the screening process at a rate more than twice that of minority candidates. Perhaps even more important, 18.4 percent of the whites who took the examination ranked in the top 4,000 and survived the physical* while the comparable figure for minority candidates was 6.6 percent, a disparity of 2.8 to 1.

In challenging these statistics, the intervenors attack the reliability of the two ethnicity surveys by which the figures were gathered. The first was conducted by the Vulcan Society, an organization representing black firemen. The Society posted in front of the examination halls observers who counted the minority candidates as they entered. The Fire Department itself conducted the second survey, a "sight survey" of those candidates who passed all stages of the selection process and were deemed "finally qualified." While the rather crude procedures of physical observation used in the surveys doubtless led to error in some cases, it is hard to believe that survey errors could have accounted for the striking racial imbalance that the results indicated. Indeed, it is arguable that the statistics probably underestimate the disparity between the percentage of minority candidates who took the examination and the percentage who qualified for appointment, since in the initial survey the Society instructed its observers to register candidates as minority members only when they were certain the candidates were black or Hispanic, but there is no indication that the Fire Department was similarly conservative in its survey of the candidates who qualified.

The intervenors also criticize the plaintiffs' comparison of the number of minority candidates who took the written examination with the number who finally qualified for appointment as a means of proving the racially disproportionate impact of the written test. The comparison was invalid, they claim, because minority members might have been disproportionately eliminated by either the physical or medical examinations, rather than by the written test. In theory this is true, and since there is no claim that the physical and medical examinations were biased, such a result would vitiate plaintiffs' constitutional claim. But plaintiffs' statistical expert produced an analysis showing that on the hypothesis that the written examination did *not* discriminate against minority applicants, the nonminority candidates must have passed the physical and medical examinations at a rate almost three times that of the minority candidates, a result he properly regarded as extremely unlikely.

---

*There was evidence that those not ranking in the top 4,000 had only a marginal chance of appointment.

The intervenors point to several other factors which they claim may undercut the validity of plaintiffs' statistics, but none of them casts serious doubt on the court's finding. They contend first that a substantial number of minority candidates may have done well on the written examination but been prevented from becoming "finally qualified" by the diploma or height requirement or the criminal conviction bar. Yet since the factors requiring automatic disqualification were publicized prior to the administration of the examination, it seems unlikely that a substantial number of candidates took the written test in the face of certain rejection. Second, the intervenors point to the large number of candidates who passed the written test but did not appear for their medical or physical examinations; however, there is no reason to suppose that minority candidates were significantly overrepresented among the "no-shows." Finally, the intervenors claim that the "character review" procedure may have been responsible for eliminating a large number of minority candidates. This claim is frivolous, as it appears that only four candidates had been eliminated by the character review screening process at the time of the hearing.

It may well be that the cited figures and other more peripheral data relied on by the district judge did not prove a racially disproportionate impact with complete mathematical certainty. But there is no requirement that they should. "Certainty generally is illusion, and repose is not the destiny of man."[*] We must not forget the limited office of the finding that black and Hispanic candidates did significantly worse in the examination than others. That does not at all decide the case; it simply places on the defendants a burden of justification which they should not be unwilling to assume.

III.

In *Castro* v. *Beecher, supra*, 459 F.2d at 732, the First Circuit stated that

> The public employer must, we think, in order to justify the use of a means of selection shown to have a racially disproportionate impact, demonstrate that the means is in fact substantially related to job performance.

Judge Coffin later referred to the defendants' obligation to "come forward with convincing facts establishing a fit between the qualification and the job." *Id.* We do not consider that this court's references to "a heavy burden" in *Chance, supra*, 458 F.2d at 1176, and *Bridgeport Guardians, supra*, 482 F.2d at 1337, meant anything more. The point we were endeavoring to underscore is that a showing of a racially disproportionate impact puts on the municipal or state defendants not simply a burden of going forward but a burden of persuasion. This may indeed prove to be "heavy" because plaintiffs are likely to produce experts who will find numerous grounds for criticizing the examinations, whether legitimately or not.

---

[*]Holmes, *The Path of the Law*, 10 Harv. L. Rev. 457, 461 (1897).

But if the public employer succeeds in convincing the court that the examination was "substantially related to job performance," an injunction should not issue simply because he has not proved this to the hilt.

We shall first consider the written examination. No one now challenges Judge Weinfeld's conclusion that Questions 81-100, entitled "City Government and Current Events," were not job related. * Taking this as a given, appellants advance two major contentions: One is that the judge found nothing else wrong with the written examination, with the consequence that if he erred in insisting on a competitive physical, the only relief required would be to regrade the examination on the basis of the first 80 questions. The alternative contention is that if he found more to be wrong, he was in error. We reject both.

Although the judge placed particular emphasis on the unrelatedness of the civic affairs questions, this was not the limit of his criticism of the written examination. He sustained plaintiffs' contention that defendants failed to perform an adequate job analysis in preparing the examination [footnote omitted] and said that "The record compels the conclusion that the procedures employed by defendants to construct Exam 0159 did not measure up to professionally accepted standards concerning content validity." He added that "Even if defendants were not required to conform precisely to all the requirements of a professional job analysis, it is clear that the methods actually employed were below those found unsatisfactory in Chance. . . ." Turning to defendants' contention that a faulty method of developing the examination should not be fatal if the result is satisfactory, he said that "Even if this contention is accepted, under these circumstances only the most convincing testimony as to job-relatedness could succeed in discharging their [defendants'] burden," but that "The testimony of defendants' expert . . . not only failed to meet this burden, but even acknowledged the presence of a major flaw in the examination which is in itself fatal." The "major flaw" that the court pointed to was the inclusion of the twenty civics questions, but we do not think the mention of this serious defect indicated that, in the judge's view, the resitting of the examination was without substantial flaws. . . .

IV.

The defendants and the intervenors ask us to set aside Judge Weinfeld's finding that Exam 0159 was insufficiently job-related because of the absence of a com-

---

*In addition to the example cited in Judge Weinfeld's opinion, another good illustration is Question 99:

The New York State Legislature recently passed a welfare requirement that has been challenged in the courts. This requirement provides that:
    (a) people who seek welfare must be residents of the State for a year.
    (b) everyone on welfare must take a job.
    (c) drug addicts must give up their habit before they can receive any money.
    (d) people who receive welfare must promise to vote in state elections.
All four answers are plausible; we fail to understand the relevance of picking the right one to the act of becoming a good fireman.

petitive, as distinguished from a merely qualifying, physical examination. We decline to do so.

We can speedily reject the first ground of attack, namely, the absence of evidence that the minority group candidates would do better than whites on a competitive physical examination. This misinterprets Judge Weinfeld's opinion. He did not hold that the use of a merely qualifying physical in itself necessarily or even probably worked against the minorities; what he held was that the absence of a competitive physical in the selection process for a largely physical vocation was additional evidence of the lack of job-relatedness of the selection procedure considered as a whole.

We likewise reject the claim that there was insufficient evidence to support this finding. Several witnesses testified to the high physical demands of a fireman's job. The Department had conducted competitive physical examinations from 1919 to 1968, and Fire Commissioner Lowery and Fire Chief O'Hagen expressed a strong and well reasoned preference for the practice. While Captain Meyers supported the use of a qualifying physical, he conceded what indeed must be obvious even to judges, namely, that many of a fireman's duties are very strenuous and that the work requires substantial dexterity. The only truly contrary opinion was McCann's, and the court was warranted in considering his reasons to be unpersuasive. It is true that some of a fireman's duties, *e.g.*, inspection, may require little or no physical prowess and that, as was urged by counsel for the intervenors, the intellectual content of a fireman's work may have increased far beyond that familiar in our youth. But that does not mean that no significant physical content remains.

We stress the limited nature of our holding. We do not read Judge Weinfeld as having said that if a written examination were sufficiently job-related, a competitive physical would *always* be constitutionally required, although he obviously would view such a physical with favor. There are considerations of cost and convenience that militate against giving a competitive physical to an extremely large group, including some who will rank so low on a proper written examination that even an Olympic score on a competitive physical would not put them within hiring range. Plaintiffs say these difficulties can be readily overcome, but they do not tell us how. In any event, there is no need to decide the question at this time. All that we regard the judge as having held, and all that we now approve, is that, in combination with the defects in preparation and content of Exam 0159 which we have described, the use of a merely qualifying physical examination rendered the Fire Department's selection procedures insufficiently job-related to withstand constitutional attack.

V.

All parties object to the interim relief directed by the trial judge. What we have said sufficiently disposes of the objections of the municipal defendants and the intervenors based on the contention that the only defect in Exam 0159 was the 20 civics questions and that therefore the court should have done no more than to order that the examination be regraded. There remain the challenges to the court's use of an interim quota system and to the ratio of minority to nonminority candidates that the court selected. As in *Bridgeport Guardians, supra,* 482 F.2d at 1340, we approach the use of a quota system "somewhat gingerly" and approve this course only because no other method was available for affording appropriate relief without impairing essential city services. As to the ratio chosen, the intervenors argue that the proper ratio should be seven majority candidates to one minority candidate, and the plaintiffs contend that the ratio should be 1:1.*

The argument for the 7:1 figure is that the ratio of whites to minority members who took the examination was 8:1 and that only a slight adjustment in that figure is needed to take account of improper disparities in appointments already made. Arguing against this and in support of a 1:1 ratio, plaintiffs urge that the nature of Exam 0159 and its predecessors had discouraged minority members from taking the examination; that although there was no specific evidence as to the discriminatory effect of previous examinations, on which Exam 0159 was patterned in considerable measure, this could be gleaned, sufficiently for the purpose of framing relief, from the fact that only 5 percent of the New York City Fire Department consisted of minority members as against a 32 percent city-wide minority population in the eligible age group; and that at most the district court's 3:1 ratio up to June 30, 1974 would bring the percentage of minority firemen up to only 6.7 percent.

The judgment is affirmed except that the cause is remanded for further proceedings on the issues discussed in Part VI of this opinion. Plaintiffs may recover costs against the municipal defendants.

BOSTON CHAPTER, N.A.A.C.P., INC.,
et al., Plaintiffs-Appellees,

v.

Nancy B. BEECHER et al., Defendants-
Appellees,
Director and Commissioners of Civil
Service, Defendants-Appellants.

United States Court of Appeals,
First Circuit.
Argued May 7, 1974.
Decided Sept. 18, 1974.
Costs Allocated Nov. 8, 1974.
504 F.2d 1017

---

*The municipal defendants had suggested that if there was to be a quota system, the ratio should be 2:1. However, they do not appeal from the judge's choice of a 3:1 ratio.

Actions instituted by United States and by NAACP based on violation of civil rights with respect to the hiring of fire fighters were consolidated. The United States District Court for the District of Massachusetts, Frank H. Freedman, J., entered judgment adverse to defendants, 371 F.Supp. 507, and an appeal was taken. The Court of Appeals, Levin H. Campbell, Circuit Judge, held that uncontroverted expert testimony that black and Spanish-surnamed candidates typically performed more poorly on paper-and-pencil tests of multiple-choice type as used in test for fire fighters could be found in conjunction with other evidence to establish a *prima facie* case that the test had a racially discriminatory impact thus shifting burden to defendant employers to justify test by showing that it was job-related; that record supported court's conclusion that defendants had not met burden of showing that written multiple-choice test for fire fighters was sufficiently job-related; and that where written multiple-choice test for fire fighters was found not job-related, court properly enjoined use of eligibles list from most recent test and any further administration of similar test until validated, and court also, in order to remedy past discrimination, properly ordered that a preference be given members of minorities in future hiring.

Affirmed.

LEVIN H. CAMPBELL, Circuit Judge.

For many years applicants for the position of fire fighter in the cities and towns of Massachusetts have had to pass a written multiple-choice test (the "test"), administered by the Massachusetts Division of Civil Service. The Division appeals from a district court decision holding the test insufficiently related to a fire fighter's duties to justify its disproportionate impact upon black and Spanish-surnamed applicants and ordering a preference to be given members of those minorities, in future hiring, to remedy past discrimination. 371 F.Supp. 507 (D.Mass. 1974).

Two actions brought against Boston, its Fire Commissioner and Massachusetts Civil Service officials were consolidated in the district court. The first was brought late in 1972 by the Boston Chapter, N.A.A.C.P., Inc., and by black and Spanish-surnamed individuals under 42 U.S.C. §§ 1981, 1983, and the Fourteenth Amendment.* Plaintiffs alleged that standards and procedures for recruiting and hiring fire fighters had the foreseeable effect of discouraging minority employment. The test, a swim requirement, and the disqualification of those with felony records were all challenged. A second action was brought early in 1973 by the Attorney General of the United States under Title VII of the Civil Rights Act of

---

*This was brought and certified as a class action. The first certified class was "All black or Spanish-surnamed persons who have applied for the position of fire fighter in any fire department . . . subject to Massachusetts Civil Service law, but have not become eligible for appointment under existing requirements." The other class included all who never applied because they were deprived of information concerning fire fighter employment opportunities.

1964, 42 U.S.C. § 2000e *et seq.*, as amended by the Equal Employment Opportunity Act of 1972. Both suits sought not only orders forbidding the challenged practices but also remedial hiring of enough minority individuals to offset past discrimination.

The district court held a hearing at which evidence was introduced concerning the alleged discriminatory hiring practices and the disproportionate racial impact of the test. After the hearing the parties stipulated that it would be treated as one on the merits of the testing issue, but would cover only the "preliminary injunction stage" of the recruiting challenge. Objections to the felony disqualification and the swim test were not pressed at the hearing, but have not been abandoned. The district court's opinion and judgment enjoined use of the test in its current form, ordered Boston and its Fire Commissioner to engage in additional recruiting of minorities, and awarded minorities a preference in hiring to ameliorate the effects of past discrimination. Boston and its Fire Commissioner took no appeal from the court's adverse rulings.

I.

In *Castro* v. *Beecher [citation omitted]*, we held that an employer may use a means of selection having a "racially disproportionate impact" only if he can show "that the means is in fact substantially related to job performance." . . . The approach is thus two-pronged: those challenging an employment test must establish its disproportionate impact by demonstrating that, for whatever reason, it is more of a hurdle for minority members than for others; once this is shown, the test's proponents acquire a burden of justification and must "prove that the disproportionate impact was simply the result of a proper test demonstrating lesser ability of black and Hispanic candidates to perform the job satisfactorily." *[Citation omitted.]*

Some courts, including the court below, describe the showing plaintiffs must make as a *"prima facie* case" of "racial discrimination." We use "racially disproportionate impact" because it is a neutral and seemingly more accurate description. A means of selection may disqualify proportionally more minority candidates than others and thus have a racially disproportionate impact, yet not be discriminatory in the constitutional sense. In *Castro,* for example, we approved a high school diploma requirement for police even while recognizing a disparity between blacks and Spanish-surnamed candidates and others in respect to a high school education. *[Footnote omitted.]* We thought a high school education was a "bare minimum for successful performance of the policeman's responsibilities." *Castro, supra* 459 F.2d at 735. But we disapproved a paper-and-pencil test which also bore more heavily on blacks and Spanish than others because it was not proven "convincingly" that there was a "fit between the qualification and the job." *Id.* at 732.

Plaintiffs usually meet their initial burden by demonstrating that minority candidates have a higher test failure rate; defendants are then put to their proof of job-relatedness. Here, however, the district court found inadequate the only available sampling showing how blacks and Spanish have fared on the test *[footnote omitted]*, although it found much evidence that blacks and Spanish have held disproportionately few jobs in the fire departments of the major Massachusetts cities where most of them reside.* Until recently relatively few minority members applied for fire fighting jobs, resulting in a very small sample from which to draw conclusions about their comparative test performance.

The district court concluded that the census figures, especially those for Boston and Springfield, when used ''in support of the meager exam statistics,'' established a *prima facie* case of the test's discriminatory effect. The court correctly noted that:

> . . . ''such a finding is not determinative of the issue but merely shifts the burden to the defendant to justify the use of the exam. This is a burden a public employer should not be unwilling to assume.'' 371 F.Supp. at 514.

We need not decide whether census figures showing a gross disproportionality in the employment of black and Spanish-surnamed fire fighters and others are enough, standing alone, to shift the burden of justification to defendants. In *Castro,* when dealing with a relatively innocuous height requirement, we declined to impose a burden of justification upon defendants in the absence of any evidence that the height requirement adversely affected minority candidates. On the other hand, the present test, given for more than half a century, is a far more salient selection device, and it can be argued that a showing of significant disproportionality in minority employment, coupled with even minimal proof of a higher minority failure rate, is enough to shift to the Division of Civil Service the burden of justification. *[Citations omitted.]* Disproportionate impact or *prima facie* discrimination are simply labels that aid in singling out qualifications which it is reasonable to ask an employer to justify; ''complete mathematical certainty'' is not required. *[Citation omitted.]* When widespread minority underemployment is shown to exist in a given occupation, primary selection devices should not

---

*1970 census figures show that in Boston, with a 1970 black population of over 16 percent, black fire fighters made up less than 1 percent of the force; Springfield's black population exceeded 12 percent but blacks made up less than 0.2 percent of the force. And, as the district court pointed out, the combined black and Spanish minority in Boston is today probably closer to 23 percent than to 16 percent. Cambridge shows a lesser discrepancy, although still a very sizable one. New Bedford and Worcester, with smaller minority populations, reflect the least disparity. Census data as to Spanish-surnamed individuals show that nearly 3 percent of Boston's population, but only 0.1 percent (two individuals) of the fire department, fit within that category. Springfield has no Spanish-surnamed fire fighters, although Spanish comprise 3.4 percent of its population. Cambridge, New Bedford and Worcester had substantial disparities. . . . It would, of course, be unreasonable to expect perfect correlation between ethnic groupings and the holders of a particular job, but when there is only one minority fireman out of 475 in a city like Springfield with a large black or Spanish-surnamed population, the imbalance is obvious.

be immunized from study by placing an unrealistically high threshold burden upon those with least access to relevant data. This seems especially so when the small size of the sample may be traceable to the test's discouraging effect as well as to unequal recruitment practices.

But we do not decide on this issue alone. What in our view conclusively tips the scale in plaintiffs' favor is the uncontroverted testimony, from experts called by both sides, that black and Spanish-surnamed candidates typically perform more poorly on paper-and-pencil tests of this type. *See* Cooper & Sobol, *Seniority & Testing Under Fair Employment Laws: A General Approach to Objective Criteria of Hiring & Promotions,* 82 Harv. L. Rev. 1589, 1640 (1969). (*Cf. Castro, supra,* where black and Spanish police candidates were shown to have performed more poorly than did whites on a test of similar design.) In light of the expert testimony, we cannot say that the district court was clearly erroneous in its ultimate fact finding that plaintiffs had established a *prima facie* case. The burden thus shifted to the defendants to justify the test by showing that it was job-related.

That Massachusetts did not intentionally discriminate is immaterial. . . . The question is whether the test denied applicants equal protection of the laws by creating "built-in headwinds" for those who, although qualified to perform the job, cannot pass the test. If it did, the inequality may be remedied without regard to official malice, specific intent, or actionable neglect.

II.

The district court found that defendants had not met their burden. The defendants' major task in doing so was to show a substantial relationship between the test results and job performance. A test fashioned from materials pertaining to the job (here, from a preliminary fire fighters' manual) superficially may seem job-related. But what is at issue is whether it demonstrably selects people who will perform better the required on-the-job behaviors after they have been hired and trained. The crucial fit is not between test and job lexicon, but between the test and job performance.

In fairness to the state, we must not forget that civil service tests were instituted to replace the evils of a subjective hiring process. Little will be gained by minorities if courts so discourage the use of tests that the doors to political selection are reopened. Moreover, a test, even one the cutoff of which does not demonstrably predict job performance, may serve worthwhile goals in gross by sifting from the pool of potential applicants those without enough motivation even to try to acquire the skills the test demands, and by discarding some few candidates who take the test but whose mental ability is so low that they are obviously unsuitable. Finally, it is virtually impossible for an employer to justify to a mathematical certainty every selection device. Controlled experiments in which applicants are hired without regard to their test scores may be impractical in many cases, and so those who fail the test often are not available to be evaluated on the job. If they are not, it is impossible to prove that they would have performed less competently.

Nonetheless, we think that the judgment of the district court is amply supported by the record, and we agree with it. Too many doubts persist concerning the validity of this test, the format of which has persisted for years, to make a convincing case for its unaltered use in fire departments notable for the absence of minority employees. Although perfect tests are goals as illusory as perfect schools or perfect courts, the evidence justifies compelling defendants to attempt to fashion a more sensitive test, one that will not needlessly serve as a "built-in headwind" to competent minority members, depriving both them and the Commonwealth of an opportunity for which they are qualified.

The test in recent years has had two parts, one of twenty-five questions covering current events, spelling, vocabulary, and arithmetic, and the other of seventy-five questions taken verbatim from the "Red Book," a fire fighter's preliminary manual available from state officials. A score of 70 is required for qualification. Some questions call for considerable verbal skill: an applicant is asked whether "condense" means "(a) conduct (b) expand (c) evaporate (d) contract" and to recognize that "presurrised" rather than "buoyancy" was misspelled. The second part asks questions such as: "the name given to the fire fighter who carries the play pipe end of the hose up the ladder is (a) engineman (b) pumpman (c) nozzleman (d) hoseman." In some instances, questions on the second part required knowledge of apparently obsolete equipment.

The test was not professionally developed. The civil service examiner who wrote the latest versions was without professional training in employment or psychological testing, and did not consult with anyone who had such training. No analysis of required job skills was conducted, and the author (who was not skilled in fire fighting) relied on the civil service "poster" setting forth the principal "duties" of a fire fighter, and on the "Red Book." After a test was given, civil service personnel would analyze 50 to 100 answer sheets and, in later versions, eliminate questions that were too easy (because most people got them right) or too hard (because most answers were wrong). The "passing" score of 70 was selected arbitrarily, without reference to the expected or actual distribution of scores.

The general questions in the first part of the test have nothing to do with fire fighting, and plaintiffs' expert was of the opinion that they were not useful even as an index of general intelligence. Defendants may plan to delete such questions in the future; in any event, unless the correlations discussed below between the overall test scores and the performance of certain groups of tasks by selected fire fighters may be said to validate the first-part questions, there is no evidence at all that they are job-related.

The second-part questions deal with fire fighting, yet there is a difference between memorizing (or absorbing through past experience) fire fighting terminology and being a good fire fighter. If the Boston Red Sox recruited players on the basis of their knowledge of baseball history and vocabulary, the team might acquire authorities like John Kieran but who could not bat, pitch, or catch. The test does not examine traits seemingly more relevant to a fire fighter's performance such as agility, stamina, quick thinking under pressure, poise, mechanical aptitude, and the ability to work with others. Experts for both sides agreed that verbal memory is not a very important attribute for the job. And unlike the motor vehicle rules covered in a driver's test, it seems unessential whether the candidate absorbs the tested vocabulary before or after acceptance. Nomenclature and similar matters can be mastered during training and on the job. Testing them before acceptance puts a premium on ability to memorize terms that, at the time, contain only abstract meaning.

Because the test measures a type of achievement not especially relevant to the job, one would not expect it to predict job success. It may be, however, that neither the test designer nor the court can understand how the score on the test successfully predicts. Perhaps, for example, the questions about machines are indirect indicators of mechanical aptitude. Recognizing these possibilities, the district court felt that careful scrutiny of any claimed predictive value was in order. We agree, and now seek to answer whether, against all odds, the test has been shown to be an actual predictor of on-the-job success. . . .

. . . The district court's conclusion was not erroneous. Substantial doubt is cast on the test's validity by its failure to correlate at all with any of the subjective ratings or with the overall objective ratings. Only two of the tasks on the objective portion of the study correlated with the test, and the correlation there was not impressive. These rather meager signs of validity do not convince us, nor did they convince the district court, that the test is, as *Castro* demanded, "substantially related to job performance." The state has not come forward with "convincing facts establishing a fit between the qualification and the job."*Id.* 459 F.2d at 732. We do not think the district court was bound to find either compliance with the Guidelines or "job-relatedness" in a more general sense. The Guidelines are not satisfied by just any correlation to any facet of job performance. They require close scrutiny of "the use of a single test as a sole selection device . . . when that test is valid against only one component of job performance." § 1607.5(c). The test here, and the validation study performed by defendants' expert do not survive close scrutiny. . . .

. . . There are, in sum, too many problems with the test for us to approve it here. The district court was correct in ruling that defendants had not discharged their burden.

The other arguments raised by defendants have been examined and found to be with merit. The judgment is

Affirmed.

In addition to various federal statutes, many states and local governments have fair employment practice statutes. Increasingly, state and municipal agencies are relieving federal agencies of some of the heavy volume of cases and complaints, as explained in the following excerpt from *American Jurisprudence*:[30]

> State fair employment practices statutes typically provide that one who is discriminated against in violation of the statute may file a complaint with an administrative agency which, after investigation of the complaint, may, if the facts warrant, conduct a hearing. Under the "voluntary" or "educational" type of statute, which relies on moral pressure for enforcement, the agency, if it finds that the employer has discriminated against the complainant, may make recommendations to the parties and may give publicity to its findings, but it cannot issue an enforceable order and the complaining party has no cause of action against the employer. On the other hand, under the "compulsory" type of statute, the administrative agency, if it finds that the employer has engaged in a practice proscribed by the statute, may issue a cease and desist order which is enforceable in the courts. But even if the proper administrative agency finds that a prospective employer has discriminated against a job applicant because of race, it cannot order the employer to cease and desist from refusing to employ such person; it can only order that if the applicant presents himself for employment, the employer shall not refuse, because of the applicant's race, to employ him. Under a statute authorizing the enforcement agency to order the hiring of a person unlawfully discriminated against, the agency may require an employer to offer the person discriminated against the first vacancy of a job for which he has applied, provided he meets the standard qualifications of other applicants for employment, without regard to race, color, or national origin.

SUMMARY

The role of a member of the fire service as an employee is a complex one. Questions concerning who is responsible for the acts of an employee, what employment relationship one person has to another, who is responsible for an employee's injuries, and what safeguards an employee can expect in terms of employment conditions are all areas of vital concern.

The fact that members of the fire service are considered "public service employees" distinguishes them in some respects from those engaged in private employment. A public employee is allowed safeguards, often by statute, not as readily available to an employee in private industry. On the other hand, members

---

[30]*American Jurisprudence*, 2nd Ed., Vol. 15, pp. 505-506.

of the public service may be subject to more restrictions than those employed in private industry. Both private employees and members of the public service, however, are protected by virtue of their civil rights; their civil rights are guaranteed under the Constitution of the United States as well as under state constitutions.

A civil right is a legally enforceable claim based, usually, on economic or social factors. Beginning with the Civil Rights Act of 1871 and continuing through the acts of the 1960s and 1970s, federal legislation provides a body of law supplementary to state remedies for civil rights violations. Thus, exhausting all state remedies before a federal remedy is sought may not be necessary if the state remedy is inadequate or if the state remedy, though adequate in theory, is not adequate in practice.

ACTIVITIES

1. Explain why the Master and Servant Doctrine is an important legal concept.
2. (a) How would the principle of *respondeat superior* affect a member of the fire service who is injured or incapacitated?
   (b) Under what circumstances would this principle have no relevance to an injured or incapacitated member of the fire service?
3. What effect does the legal concept of agency have on the Master and Servant Doctrine?
4. (a) Cite at least five activities that are within the scope of employment in the fire service. (Try to avoid obvious activities.)
   (b) For each of the five activities cited, describe: (1) a circumstance when the employer would be responsible for the activity of the employee, and (2) a circumstance when the employer probably would not be responsible for the act or acts of the employee.
5. Give examples of civil rights violations that could affect the fire service.
6. (a) Describe the different employment relationships of an employer, an employee, an independent contractor, and an agent.
   (b) Give examples of each of these relationships in the fire service.
7. Explain how federal regulations influence the fire service in your community.
8. Describe to your classmates what, in your opinion, are some of the major differences between public and private employment.
9. With your classmates, debate the advantages and disadvantages of a civil service system as related to the fire service.
10. Describe how unions affect the fire service. Include in your description some of the ways in which belonging to a fire service union is different from belonging to a union in private industry.
11. Describe how constitutional rights affect discrimination in employment.
12. Discuss some recent instances from the media, or that you have heard of, or with which you are familiar concerning discriminatory employment practices in public service. Include in the discussion your feelings concerning the handling of these situations.

BIBLIOGRAPHY

Bahme, Charles W., *Fire Service and the Law*, NFPA, Boston, 1976.

Kling, Samuel G., *The Complete Guide to Everyday Law*, Follet Publishing Company, Chicago, 1969.

Miller, Vernon X., *Selected Essays on Torts*, Dennis & Col., Inc., Buffalo, 1960.

Shapo, Marshall S., *Cases and Materials on Tort and Compensation Law*, West Publishing Co., St. Paul, MN, 1976.

For further study of the topics covered in this chapter, the following are recommended:

American Law Institute, *Restatement of the Law of Agency, Second* (3 Vols.), American Law Institute Publishers, St. Paul, MN, 1958.

Pemberton, John de J., Jr., ed., *Equal Employment Opportunity — Responsibilities, Rights, Remedies*, Practicing Law Institute, New York, 1975.

Prosser, William L., *Handbook of the Law of Torts*, 4th Ed., West Publishing Co., St. Paul, MN, 1971.

V

CRIMINAL LAW

OVERVIEW

The field of criminal law is a vast and often complex area of the law that has been increasingly affected by constitutional law. Increasing also is the involvement of the fire service with criminal law and the need for members of the fire service to become familiar with the basic tenets and aspects of this field of law. Over the years, the field of criminal law has become so broad that consideration of it now requires concentrated study of its two major aspects: (1) ascertaining whether a crime has indeed been committed (the area of substantive criminal law), and (2) ascertaining who was responsible for the commission of the crime (the area of criminal procedure). Substantive criminal law is the subject of this chapter; criminal procedure will be discussed in Chapter VI.

CRIMINAL LAW AND THE FIRE SERVICE

In earlier times police and fire personnel were considered one and the same with regard to the responsibility of guarding the public peace, harmony, and welfare. Although the complex needs of our society resulted in the police and fire services branching into specialized activities and responsibilities, the two remained "distant relatives." Today, however, this distance is being closed; problems and complexities in the very society that once required specialization now demand that fire service members become involved with various aspects of criminal law in the course of fulfilling their duties.

Although increasing numbers of fire service organizations have been, or may become, involved with crimes of malicious mischief, criminal assault and battery, riot, and drunkenness, this chapter addresses itself primarily to the crime of arson.

This is because arson has become the crime of greatest importance to members of the fire service.

Criminal law deals with public wrongs; *i.e.,* injury to the order and peace of a society. This is in contrast to tort law which, as we have seen, concerns actions between citizens or groups in a private or civil capacity. Under an offense in tort, one citizen may sue another for redress. Under criminal law, however, the state prosecutes an individual on behalf of the society that the state represents because the offense is considered a violation of the peace and order of society in general. This is an important point, as the Supreme Court has stated,[1]

> Punishment is spoken of sometimes as the purpose of criminal law, but this is quite erroneous. The purpose of criminal law is to define socially intolerable conduct, and to hold conduct within limits which are reasonably acceptable from the social point of view.

Today the scope of criminal law is very much broader than in former times. As Rollin M. Perkins states in his book titled *Criminal Law:*[2]

> Within recent times there has been a tremendous expansion of the area of human conduct regulated by the criminal law. In early days this law concerned itself only with conduct seriously antisocial in its character. Other agencies such as the church and the home were relied upon to regulate conduct in other respects. The present tendency is to place the entire burden upon the criminal law, and while this shift has been taking place, changes in the social and economic structure of the community have created many new conflicts and have added ingenious modes of infringing recognized rights. Out of the combination have come countless regulations of trades, occupations, monopolies, banking and finance, sale of securities, foods, drugs, liquor, and the use of automobiles — a legion of restrictions scattered throughout the statutes and frequently not in the criminal code itself.

WHAT IS A CRIME?

The distinction between statutory law and common law is an important one with regard to criminal law. Additions to the field of common law are constantly being made because the courts are constantly being asked to rule on matters that reflect and affect the changing economic and social conditions of this country. The body of common law often has the effect of enabling the courts to make law through its decisions. However, this process is not possible in the field of criminal law. In the United States, a crime *must* be designated a crime by law. In other words, there is no "unwritten" criminal law. The state, local, or federal government must decide what acts are criminal acts, and must designate them as such. The courts can punish only those offenses that have been declared crimes by law.

---

[1] *Sauer* v. *United States,* 241 F.2d 640, 648, 1957.
[2] Rollin M. Perkins, *Criminal Law,* 2nd Ed., The Foundation Press, Inc., Mineola, NY, 1969, p. 4.

As Oliver Wendell Holmes stated, "A man may have as bad a heart as he chooses, if his conduct is within the rules."[3]

In the following excerpt from the dissent in *Commonwealth* v. *Mochan*, Judge Woodside said:[4]

> . . . There is no doubt that the common law is a part of the law of this Commonwealth, and we punish many acts under the common law. But after nearly two-hundred years of constitutional government in which the legislature and not the courts have been charged by the people with the responsibility of deciding which acts do and which do not injure the public to the extent which requires punishment, it seems to me we are making an unwarranted invasion of the legislative field when we arrogate that responsibility to ourselves by declaring . . . that certain acts are a crime.

In addition, Chief Justice Warren added a constitutional safeguard, as indicated in the following excerpt from *United States* v. *Harris:*[5]

> The constitutional requirement of definiteness is violated by a criminal statute that fails to give a person of ordinary intelligence fair notice that his contemplated conduct is forbidden by the statute. The underlying principle is that no man shall be held criminally responsible for conduct which he could not reasonably understand to be proscribed.

However, the common law does influence criminal statutes in certain states. Many states consider as crimes only those crimes that are specifically designated as such by statute. Often, however, statutes will refer to common-law crimes by name only (such as "murder"). Because the various interpretations and common-law descriptions are omitted in the wording of the statutes, the common law can be followed by the courts in deciding cases.

Owing to its origins as a French colony, Louisiana's judicial system is based on the Napoleonic Code. Louisiana is, therefore, not a common-law state, but a civil-law state: each legal case must be argued on its own merits without reference to previous judicial decisions. In 1805, however, the state adopted statutes making English common law the basis of criminal jurisprudence. Thus murder (and all other felonies) are classified as crimes by law, and that point does not have to be argued in each case.

In federal courts no crime is punishable merely because there is precedent for it in common law. Some act of Congress must determine it to be a crime, and must set a corresponding penalty. In addition, the statute must specify it to be a federal crime (such as murder of a federal agent acting in the line of duty) in order for it to be tried in federal court. Common law may be used to argue legal points and decisions. Usually, the common law used in these arguments is taken from previous federal court cases.

---

[3]Oliver Wendell Holmes, Jr., *The Common Law*, Harvard Ed., Little, Brown, and Company, Boston, 1963.

[4]*Commonwealth* v. *Mochan*, 110 A.2d 788.

[5]*United States* v. *Harris*, 347 U.S. 612.

The Constitution imposes limits on both state and federal courts to safeguard the rights of citizens in criminal matters. Martin Ross, in the *Handbook of Everyday Law,* explains what some of these limits are:[6]

Each state is sovereign in its power to determine those acts which are defined as crimes and to prescribe adequate punishments. This power is, however, restricted since the states are prohibited from passing laws which are contrary to the Bill of Rights of the federal constitution and to those limitations in their own constitutions which are intended to protect the liberty of its citizens.

"Bills of attainder" are prohibited. No state legislature may pass any law which, in effect, will cause a person convicted of a felony to forfeit his property.

A person finally sentenced to a term of life imprisonment, though deprived of all his civil rights and considered legally and civilly dead, has the right to own property, make a will, and devise his property.

"*Ex post facto*" laws are prohibited. No state may pass a law which will, in effect, make an act, which was legal and permissible when committed, punishable as a crime. This includes those laws which increase the penalty of crimes committed prior to their enactment.

Illustrations:
A law which prohibits the sale of liquor cannot affect those sales which were legal before the law was passed.

A law which increases the punishment of a crime from 5 years to 10 years in a state prison cannot affect a crime which was committed before the penalty was raised. The punishment for the crime previously committed must be 5 years.

No law shall be passed which violates the right of the people against unreasonable searches and seizures.

No person shall twice be put in jeopardy of his life for the same offense. This is commonly known as double jeopardy. In effect, it means that once a person has been tried for a crime and found not guilty, he cannot again be tried for the same crime.

No person shall be compelled to be a witness against himself in a criminal prosecution. This is the well-known Fifth Amendment. Actually, it is only one portion of the Fifth Amendment of the United States Constitution, a part of the Bill of Rights. It guarantees a person protection from testifying in any proceeding which may tend to expose him as guilty of a crime. This testimony may not be used against him. However, this claim of self incrimination may not protect a person from testifying if he has been given immunity and a guaranty that he will not be prosecuted for any crime revealed by his testimony.

No person may be deprived of life, liberty, or property without due process of law. This guarantees to all the right to a hearing, the right to submit a defense, the right to cross examine and to be confronted by all witnesses. No law shall be passed by any state which will abridge and violate these constitutional safeguards.

---

[6]Martin J. Ross, *Handbook of Everyday Law,* Harper & Row, Publishers, New York, 1975, pp. 267-8.

DEGREES OF CRIMES

English common law classified crimes as felonies, misdemeanors, or treason. Under English common law, all felonies were punishable by death except the crime of mayhem,* for which the punishment of mutilation was substituted.

Statutes in this country generally categorize crimes as felonies or misdemeanors.

Felonies: English common law felonies are: murder, manslaughter, rape, robbery, larceny, arson, burglary, sodomy, and mayhem. In this country, statutes have created additional felonies such as kidnapping, extortion, or dealing in certain narcotics. The designation "felony" is generally based on one or both of the following conditions with regard to punishment: the type of institution in which an offender may be imprisoned (such as a state prison), and/or the length of time that may be imposed. In some states a felony is a crime punishable by death or by imprisonment for more than one year, while in other states a felony requires punishment by death or imprisonment in a *state* prison.

Misdemeanors: Any crime that is not a felony is a misdemeanor. Misdemeanors are "lesser" crimes than felonies. Punishment for a misdemeanor may result in imprisonment in a city or county jail rather than in a state prison. Although the punishment for all crimes includes imprisonment, in cases of "lesser" misdemeanors the sentence may be merely a suspended sentence and a fine.

Misdemeanors are classified as *mala in se,* or wrong in themselves, or *mala prohibita,* or crimes which, although not necessarily wrong in themselves, are prohibited by statute, such as regulations involving the sale of food or drugs. English common law punished no act that was not wrong in itself — all English common law crimes were *mala in se.* Statutory felonies, because they are considered major crimes against the public, are also considered "wrong in themselves."

Many states also recognize a violation of a lesser category of misdemeanor that does not carry a criminal conviction. Although not considered crimes, such violations are punishable — usually by fine. Terms can vary for this type of offense. Minor traffic infractions are often classed as "violations." In California such violations are called "infractions" and are expressly not punishable by imprisonment.

The United States Code divides offenses into felonies, misdemeanors, and petty offenses. A petty offense is an offense for which the penalty does not exceed imprisonment for six months, or a fine of $500, or both.

ELEMENTS OF A CRIME

Every crime has two basic elements: (1) the act must actually have been committed, and (2) the act committed must have been intended. Very simply, to be a

---

*Under common law, mayhem was the malicious deprivation of a bodily member causing an individual to lose permanently the ability to fight. Now it is chiefly used to mean any willful and permanent crippling, disfigurement, or mutilation.

crime, there must be a criminal act and a criminal intent. In addition, as in the law of torts, the act must have been the legal or direct cause of the injury inflicted. For example, if a burglar enters a house with the intent of robbing it and inadvertently kicks over a kerosine lamp, which then burns the house down, the burglar would not be guilty of arson because burglary — not arson — was the criminal intent.

CRIMINAL ACT

No matter what degree of criminal intent an individual possesses, criminal liability cannot be imposed until the criminal intent results in a criminal act:[7]

> One of the distinguishing elements in deciding whether a crime exists or not is the presence or absence of what is called an "overt act." By this is simply meant that the law does not, perhaps because it could not, punish a criminal frame of mind. There must be some open, palpable, *overt* act which goes to show the crime. Hence a man may be mentally desirous of committing crime, and have wicked designs in his heart, but until he commits some overt act he is not punishable; and when the crime has been committed, it is the *act* and not the criminal frame of mind that is punished. And it is claimed to be a principle of natural justice that the "intent" and the "act" must both concur to constitute a crime.

Criminal acts may be committed by commission (actually doing something) or omission (failing to do something). If an individual is under legal duty to act and fails to do so, the failure to act can be deemed criminally negligent. Criminal negligence is generally considered to be a *gross* lack of care.

CRIMINAL INTENT

Crimes that are *mala in se*, or "wrong in themselves," require a criminal intent. Criminal violations that are *mala prohibita* — prohibited by statute because, although not inherently wrong in themselves, they endanger public welfare, health, or safety — do not require criminal intent or proof of criminal intent.

There may be varying degrees of intent with regard to crimes. Many statutes use the word "malice" or a similar term to describe the level of intent that must be present. The following statements by Chief Justice Traynor, from the case of *People* v. *Conley,* refer to the level of intent required by the term "malice aforethought" in distinguishing between murder and manslaughter. The Chief Justice's comments also refer to level of intent in general:[8]

> We have previously noted the difficulty of formulating a comprehensive definition of malice aforethought that will serve to distinguish murder and

---

[7]Term 514; 16 Engl. L. & Eq. 480, quoted in *A Treatise on Criminal Law and Criminal Procedure,* Vol. XI, AMS Press, Inc., New York, 1976, p. 5.

[8]*People* v. *Conley,* 411 P.2d 911.

manslaughter. Penal Code, section 188 provides that malice "may be express or implied. It is express when there is manifested a deliberate intention unlawfully to take away the life of a fellow creature. It is implied when no considerable provocation appears, or when the circumstances attending the killing show an abandoned and malignant heart." These provisions create a presumption of malice when the commission of a homicide by the defendant has been proved, and place the burden on him to raise a reasonable doubt in the minds of the jurors that malice was present. The "conclusive presumption" of a malicious and guilty intent set forth in section 1962 of the Code of Civil Procedure offers no help to the jury. To bring it into operation the jury must find "the deliberate commission of an unlawful act for the purpose of injuring another," which involves subjective factors on which evidence of diminished capacity is also relevant. . . .

The mental state constituting malice aforethought does not presuppose or require any ill will or hatred of the particular victim. . . . When a defendant "with wanton disregard for human life, does an act that involves such a high degree of probability that it will result in death," he acts with malice aforethought. . . . This mental state must be distinguished from that state of mind described as "willful, deliberate, and premeditated," however. The latter phrase encompasses the mental state of one who carefully weighs the course of action he is about to take and chooses to kill his victim after considering the reasons for and against it. . . . A person capable of achieving such a mental state is normally capable also of comprehending the duty society places on all persons to act within the law. If, despite such awareness, he does an act that is likely to cause serious injury or death to another, he exhibits that wanton disregard for human life or antisocial motivation that constitutes malice aforethought. An intentional act that is highly dangerous to human life, done in disregard of the actor's awareness that society requires him to conform his conduct to the law is done with malice regardless of the fact that the actor acts without ill will toward his victim or believes that his conduct is justified. In this respect it is immaterial that he does not know that his specific conduct is unlawful, for all persons are presumed to know the law including that which prohibits causing injury or death to another. An awareness of the obligation to act within the general body of laws regulating society, however, is included in the statutory definition of implied malice in terms of an abandoned and malignant heart and in the definition of express malice as the deliberate intention unlawfully to take life.

In the previous statement by Chief Justice Traynor, the element of malice is related not to ill will toward a victim, but to an intentional act that displays a wanton disregard for human life. In other words, proof of malice of intent to commit a crime does not require any personal ill will on the part of the perpetrator toward the person or persons against whom the crime is committed. Malice, in terms of criminal law, can mean simply the wanton disregard of the rules and laws of society in general. The following decision reinforces this concept that "malice" in criminal law is different from its literal meaning and also relates "malice" to "intent."

As you read the following case study concerning attempted arson, consider which particular factors determined that the element of malice could be inferred from the willful act of setting a fire.

COMMONWEALTH
v.
LOUIS L. LAMOTHE, JUNIOR

Supreme Judicial Court of Massachusetts.
Middlesex.
Argued Dec. 4, 1961.
Decided Dec. 29, 1961.
179 N.E.2d 245.

Prosecution for attempting to burn a building. Defendant excepted to the denial of a motion for directed verdict, and to an instruction given, and the Superior Court, Vallely, J., reported the case with the consent of the defendant. The Supreme Judicial Court, Spalding, J., held that necessary element of malice could be inferred from willful act of setting fire.

Judgment affirmed.

SPALDING, Justice.

The defendant was tried and convicted under an indictment which charged that on December 17, 1960, he "did attempt to willfully and maliciously set fire to and burn a certain dwelling house of the property of Eugene Berube."

The material evidence was as follows: "On December 17, 1960, the defendant was drinking in the Moody Gardens on Moody Street, Lowell. About 8 p.m. he left by the street exit, walked to the rear of the building, and went up the rear stairway to a porch on the second floor of the same building. The defendant then set fire to some papers that were on the porch, intending to burn the building. He then left the premises and walked down the street. The fire scorched the floor and the wall of the porch before it was extinguished. The building was a dwelling house and was the property of Eugene Berube. The defendant was later questioned by the police and when asked by . . . [them] why he set the fire, . . . [he] said that he had no reason for doing it."

At the close of all the evidence, the defendant presented a motion for a directed verdict of not guilty. The motion was denied, subject to the defendant's exception. In his charge, subject to the defendant's exception, the judge instructed the jury "that they could infer malice from the willful act of setting the fire." The judge, being of the opinion that his rulings presented questions of law of such importance and doubt as to require the decision of this court, reported the case with the consent of the defendant. (G.L. c. 278, § 30.)

Both exceptions present the same question, namely, whether malice can be inferred from the willful attempt to burn the property.

The offense charged in the indictment is defined in G.L. c. 266, § 5A (inserted by St.1932, c. 192, § 5) which, so far as material, reads, "Whoever willfully and maliciously attempts to set fire to, or attempts to burn . . . any of the buildings, structures, or property mentioned in the foregoing sections . . . shall be punished," etc. The defendant concedes that the Commonwealth has proved all that is necessary to sustain a conviction except the element of malice. Proof of willfulness, he contends, is not enough; there must, he asserts, be proof also that the act was done out of a motive of cruelty, hostility, or revenge.

To ascertain the meaning of the word "maliciously" in the statute we must turn to the common law, for the statute was undoubtedly drawn against that background. At common law the offense of arson consisted of the willful and malicious burning of the house of another. [4 *Blackstone,* Commentaries (21st Ed.), p. 220.] But the meaning given to the word "malicious" when used in defining the crime of arson is quite different from its literal meaning. Sir Matthew Hale in his Pleas of the Crown (Vol. 1 at page 569) gives the following illustration, "But if A has a malicious intent to burn the house of B and in setting fire to it burns the house of B and C or the house of B escapes by some accident, and the fire takes in the house of C and burneth it, tho A did not intend to burn the house of C yet in law it shall be said the malicious and willful burning of the house of C, and he may be indicted for the malicious and willful burning of the house of C." Modern authorities are to the same effect. *[Citations omitted.]* "The malice which is a necessary element in the crime of arson need not be express, but may be implied; it need not take the form of malevolence of will, but it is sufficient if one deliberately and without justification or excuse sets out to burn the dwelling house of another."

We are mindful that the defendant is not charged with common law arson or with its statutory counterpart, defined by G.L. c. 266, § 1; he is charged with having attempted to burn a building. But what we have said with reference to the meaning of the word "maliciously" in the common law crime of arson is none the less applicable, for the statutory offense defined by § 5A is so closely related to arson that it is very unlikely that the Legislature intended the word to be used in a different sense. Support for this view may be found in *Commonwealth* v. *Mehales,* 284 Mass. 412, at page 415, 188 N.E. 261, at page 262 (case construing § 5A), where it was said, "The malice now essential under that statute is not necessarily against the owner of the building, but that malice which 'characterizes all acts done with an evil disposition, a wrong and unlawful motive or purpose; the wilful doing of an injurious act without lawful excuse.' Shaw, C. J., in *Commonwealth* v. *York,* 9 Metc. 93, 104."

The defendant relies heavily on a line of cases involving prosecutions for malicious mischief. *[Citations omitted.]* But these cases are not in point. In offenses of that sort it has generally been recognized that malice has a meaning quite different from its ordinary meaning in the criminal law. *[Citation omitted.]* This distinction has been well stated in *Commonwealth* v. *Goodwin,* 122 Mass. 19. There it was said at page 35, "The willful doing of an unlawful act without excuse is ordinarily sufficient to support the allegation that it was done maliciously and with criminal intent."

It follows that the judge did not err in submitting the case to the jury and in instructing the jury as he did.

The entry must be

Judgment affirmed.

In addition to the two basic elements of a crime described earlier in this chapter, other conditions must be present to constitute criminal liability. Among these conditions are:

- The individual must be of competent age.
- The individual must act voluntarily.
- The individual must possess sufficient mental capacity to be held liable.

ATTEMPT

An attempt to commit a *malum in se* crime can be a crime in itself. In *Martin* v. *Commonwealth,* the court stated:[9]

> It is well settled in this jurisdiction that in criminal law an attempt is an unfinished crime, and is compounded of two elements, the intent to commit the crime and the doing of some direct act toward its consummation, but falling short of the execution of the ultimate design; that it need not be the last proximate act towards the consummation of the crime in contemplation, but is sufficient if it be an act apparently adopted to produce the result intended. . . .

The basic elements of a crime must, of course, be present. However, in order to prove "attempt," the acts committed must be ones performed in the *execution* of the crime, not merely acts performed in *preparation* to commit a crime, as explained in *A Treatise on Criminal Law and Criminal Procedure:*[10]

> A criminal attempt is defined to be an act done in part execution of a crime. This attempt must be such an act as is proximately connected with the final illegal object. That is, not every overt act which may lead up to the commission of a crime is to be regarded as a criminal attempt; and preparation for a crime must be

---

[9]*Martin* v. *Commonwealth,* 81 S.E.2d 574.
[10]Hon. Charles E. Chadman, Ed.-in-Chief, *A Treatise on Criminal Law and Criminal Procedure,* AMS Press, Inc., New York, 1976.

distinguished from a criminal attempt, since a good deal of preparation might be made and yet no crime result. Thus, buying matches to set fire to a building was not an attempt to commit arson, but was only preparation for it, and not punishable.

The following two case studies involve attempted arson. The decisions in both cases address the distinction between preparation for a crime and actual attempt to commit the crime. When you have read the cases, compare the decisions and determine the factors that constitute the distinction between the preparation for the crime of arson and the actual attempt to commit the crime of arson.

MANER v. STATE
No. 21601.

Court of Appeals of Georgia, Division No. 1.
Sept. 4, 1931.
159 S.E. 902

Syllabus by the Court.

Strewing excelsior along and upon barrels and boxes, pouring alcohol and gasoline over and upon the same, putting croker sacks under a roller-top desk, and saturating them with alcohol and gasoline, are overt acts, "inexplicable as lawful acts," "tending to the commission of the crime" of arson. The indictment sufficiently alleged an attempt to commit such crime, and the court did not err in overruling the general demurrer thereto.

LUKE, J.

The indictment in this case charges "W. A. Maner with the offense of attempt to commit arson for that said accused . . . did willfully and maliciously attempt to set fire to and burn the offices and building known as No. 160½ Hunter Street, Southwest, in Atlanta, Fulton County, Georgia, then and there occupied by W. A. Maner, doing business under the trade name of Durham Medical Institute, said City of Atlanta being a municipal corporation, and said attempt to set fire to and burn the above designated offices and building having been made in the following manner, to wit: Accused strewed excelsior along and upon barrels and boxes and poured alcohol and gasoline over and upon the same, and accused stuck croker sacks under a roller-top desk and saturated the same with alcohol and gasoline, in said offices in said building, with intent thereby to set fire to and burn said building." Defendant demurred to the indictment, on the ground that "it sets out no overt act on the part of the defendant to commit the crime of attempt to commit arson, the allegations therein contained merely alleging preparatory acts." The demurrer was overruled, and on this ruling the defendant assigns error.

We recognize, of course, that one may intend to commit a crime and do certain acts towards its consummation, and repent of his intention and refrain from his original purpose before the commission of the act; and that "mere preparatory acts for the commission of a crime, and not proximately leading to its consummation, do not constitute an attempt to commit the crime." *Groves* v. *State*, 116 Ga. 516, 42 S.E. 755, 59 L. R. A. 598. The question for determination is: Does the indictment allege such preparatory acts as "proximately lead to the consummation of the crime?" We think it does. How much further could the defendant have gone and still be guilty only of an attempt? If he had actually set fire to the building, he would have been guilty of the crime of arson rather than an attempt to commit this crime, because the Acts of 1924 (Ga. Laws 1924, p. 193, § 2) provide that one is guilty of the offense who willfully or maliciously or with intent to defraud "sets fire to *or* burns *or* causes to be burned, *or* who aids, counsels, *or* procures the burning" of the buildings named in this section. Any one of these acts would constitute the crime of arson. Since setting fire to the building would constitute the crime of arson, and the law makes it an offense to attempt to commit this crime, at what stage of defendant's acts would he be guilty of the latter offense? Where is the dividing line between preparation and attempt? Mr. Clark in his work on Criminal Law (2d Ed.) p. 126, says, "An attempt to commit a crime is an act done with intent to commit that crime, and tending to, but falling short of, its commission," and that two of the essential elements of the offense are: "(a) The act must be such as would be proximately connected with the completed crime, and (b) There must be an apparent possibility to commit the crime in the manner proposed." And on page 127 he says: "To constitute an attempt there must be an act done in pursuance of the intent, and more or less directly tending to the commission of the crime. In general, the act must be inexplicable as a lawful act, and must be more than mere preparation. Yet *it cannot accurately be said that no preparations can amount to an attempt.* It is a question of degree and depends upon the circumstances of each case." Bishop says that, "An attempt is an intent to do a particular criminal thing, with an act towards it falling short of the thing intended." See 1 Bish. New Crim. L. § 728. A reasonable construction of the acts alleged in the indictment force one to the conclusion that such acts were committed for no purpose other than to set fire to the building, and thereby commit arson. Whether the defendant repented before the consummation of the crime intended would be a question of fact for the jury after the trial had reached that stage of the proceedings.

In the case of *Weaver* v. *State*, 116 Ga. 550, 42 S.E. 745, wherein the defendant was convicted of an attempt to commit arson, though he did not set fire to the building, the Supreme Court held: "Where police officers saw the accused throw oil upon a house for the purpose *(afterwards admitted)* of burning it, and at this juncture the officers came out from their hiding place, and could have been seen by the accused, who then started away from the house without attempting to ignite the oil or the house, the judge properly submitted to the jury the question as to whether the accused desisted on account of having repented, or because he had seen the officers, and was afraid of apprehension." In *Griffin* v. *State*, 26 Ga. 493 (1), the Supreme Court held: "The taking the impression of the key which unlocks the door of a storehouse, for the purpose of making or procuring a false key, with the intent of entering the house and stealing therefrom, is an attempt to commit larceny from the house, by the person taking the impression of the key, whether he intended to enter and steal, himself, or to procure another to do it." The indictment in the instant case alleges that the acts set forth were done "maliciously" and "with intent thereby to set fire to and burn said building." Strewing excelsior along and upon barrels and boxes, pouring alcohol and gasoline over and upon the same, putting croker sacks under a roller-top desk and saturating them with alcohol and gasoline, are overt acts, "inexplicable as lawful acts," "tending to the commission of the crime" of arson. The indictment sufficiently alleged an attempt to commit such crime, and the court did not err in overruling the general demurrer thereto.

Judgment affirmed.

BROYLES, C. J., concurs.

BLOODWORTH, J., absent on account of illness.

STATE v. TAYLOR

(Supreme Court of Oregon. Jan. 2, 1906. On
Rehearing, Jan. 30, 1906.)
84 P. 82

BEAN, C. J. Section 2159, B. & C. Comp., reads: "If any person attempts to commit any crime, and in such attempt does any act towards the commission of such crime, but fails or is prevented or intercepted in the perpetration thereof, such person, when no other provision is made by law for the punishment of such attempt, upon conviction thereof shall be punished," *etc.* Under the provisions of this section the defendant, Taylor, was indicted and convicted of an attempt to commit the crime of arson. The proof was that he entertained an enmity against John Bannister, one of his neighbors, because of some testimony Bannister had given in a divorce suit. Apparently for revenge, he desired to burn and destroy Bannister's barn and wheat. He solicited one McGrath to do the burning, who, in turn, asked one Palmer to assist in the commission of the crime. Palmer informed his employer, a friend of Bannister's of the proposed plan, and was advised to allow the matter to proceed, and that arrangements would be made to apprehend the parties. After some preliminary negotiations Taylor, McGrath, and Palmer met in the back room of a saloon in Athena on July 30, 1904, and Taylor there engaged McGrath and Palmer to burn Bannister's barn and wheat, agreeing to pay them $100 for so doing, and at the same time showed them how to start a slow burning fire with a pair of overalls, saying he had tested it. After the conference at the saloon the parties separated, agreeing to meet that night about 12 o'clock at Taylor's place, from which McGrath and Palmer were to start to Bannister's for the purpose of consummating the crime. Palmer advised Bannister's friends of what was about to take place, and they made arrangements to lie in wait and intercept the parties. At the appointed time Palmer went to Taylor's place and there met Taylor and McGrath, who were waiting for him. Taylor had his own horse saddled and ready for McGrath to ride. He produced a pair of overalls, and after again showing McGrath and Palmer how to use them in starting a fire tied them on the saddle of his horse and paid McGrath $100 in money. McGrath and Palmer then started towards Bannister's with the purpose, so Taylor supposed and believed, of setting the fire, with a parting expression from him of "Good luck go with you." Taylor "laid awake two hours to see the fire," but as McGrath and Palmer were going towards Bannister's they noticed fresh tracks in the road, and when they approached within 20 feet of the barn observed two or three buggies in the barnyard, which frightened McGrath, who was afraid to go on with the enterprise for fear they were being watched, and so it was abandoned. McGrath and Palmer were both witnesses for the prosecution. Palmer testified that he never had any intention of committing the crime, and McGrath said that he did not intend to set the fire, but that the arrangement was that it would be started by Palmer. Upon these facts the question for decision is whether the defendant was legally convicted of an attempt to commit the crime of arson.

The question as to what constitutes an attempt to commit a crime is often intricate and difficult to determine, and no general rule has or can be laid down which can be applied as a test in all cases. Each case must be determined upon its own facts, in the light of certain principles which appear to be well settled. An attempt is defined as an "intent to do a particular criminal thing, with an act toward it falling short of the thing intended." 1 Bishop, New Crim. Law § 728. Or, according to Wharton: "An attempt is an intended apparent unfinished crime." 1 Wharton, Crim. Law (9th Ed.) § 173. Another author says: "An attempt to commit a crime is an act done in part execution of a criminal design, amounting to more than mere preparation, but falling short of actual consummation, and possessing, except for failure to consummate, all the elements of the substantive crime." 3 Am. & Eng. Enc. Law (3d Ed.) 250. An indictable attempt, therefore, consists of two important elements: First, an intent to commit the crime; and, second, a direct, ineffectual act done towards its commission. To constitute an attempt, there must be something more than a mere intention to commit the offense, and preparation for its commission is not sufficient. Some overt act must be done toward its commission, but which falls short of the completed crime. It need not be the last proximate act before the consummation of the offense, but it must be some act directed toward the commission of the offense after the preparations are made. It is often difficult to determine the difference between preparation for the commission of a crime and an act towards its commission. There is a class of acts which may be done in pursuance of an intention to commit a crime, but not, in legal sense, a part of it, and do not constitute an indictable attempt; such as the purchase of a gun with the design of committing murder, or the procuring of poison with the same intent. These and like acts are considered in the nature of mere preliminary preparation, and not as acts toward the consummation of the crime. It is upon this principle that most of the cases cited by the defendant rest although some of them seem to have carried the doctrine to the utmost limit. *[Citations omitted.]*

In the case at bar we have something more than mere intention or preparation, so far as the defendant is concerned. His part in the transaction was fully consummated when he employed McGrath and Palmer to commit the offense, gave them the materials with which to do it, showed them how to start a slow burning fire, paid them a compensation for their services, furnished a horse for one of them to ride, and started them on their way. He had thus done all that he was expected to do, and his felonious design and action was then just as complete as if the crime had been consummated, and the punishment of such an offender is just as essential to the safety of society. The failure to commit the crime was not due to any act of his, but to the insufficiency of the agencies employed for carrying out his criminal design. One may commit a crime by his own hand or that of another, employed, aided, or encouraged by him. If he endeavors or attempts to commit it himself, and is interrupted or frustrated, he would clearly be guilty of an indictable attempt, and, if he uses another person to accomplish the same purpose, and the other fails to carry out his design, whether purposely or otherwise, the result is the same. *[Citation omitted.]* The statute under which the defendant was indicted was probably taken from that of the state of New York. It had received a judicial construction in that state long before it was enacted here. In *People* v. *Bush,* 4 Hill, 133, decided in 1843, the defendant was indicted for an attempt to commit the crime of arson. The proof was that he requested one Kinney to set fire to a barn, gave him a match for that purpose, and promised to reward him. The court held the conviction legal, although the defendant never intended to be present at the commission of the offense and Kinney never intended to commit; Mr. Justice Cowen saying: "The act imputed to Bush was no doubt an attempt to commit an offense. It is admitted that he endeavored to make himself an accessory before the fact; and to become an accessory is, in itself, an offense. A mere solicitation to commit a felony is an offense, whether it be actually committed or not. This was held in the *King* v. *Higgins,* 2 East, 5. In the case before us there was more. The solicitation was followed by furnishing the instrument of mischief. The question of principal and accessory does not arise, as it would have done provided the crime had actually been committed. Had it been committed, the attempt would have been merged in an actual felony — a crime of another species. There would have been a principal arson by Kinney and an accessorial offense by Bush. The attempt of the latter was to have both crimes committed, and, the question of principal and accessory not being in the case, I see nothing against considering the matter in the light of the ordinary rule that what a man does by another he does by himself; in other words, the course taken to commit the arson by the hand of Kinney was the same thing, in legal effect, as if Bush had intended to set the fire personally, and had taken steps preparatory to that end.''

The same principle was again applied in *McDermott* v. *People,* 5 Parker, Cr. R. 102. In that case the defendant solicited another to commit the crime of arson, offering in consideration thereof to deed and assign over to him certain property, and said he had camphene and other combustibles in his room. The court held the defendant properly convicted of an attempt to commit arson, saying: ''The two important and essential facts to be established to convict a person of an offense are, first, an intent to commit the offense; and, second, some overt act consequent upon that intent towards its commission. So long as the act rests in bare intention, it is not punishable. *'Cogitationis poenam nemo patitur.'* It is only when the thought manifests itself by an outward act in or toward the commission of an offense that the law intervenes to punish. As we cannot look into the mind and see the intent, it must, of necessity, be inferred from the nature of the act done, and, if that be unlawful, a wicked intent will be presumed. These are fundamental legal principles. Now applied to the facts of this case, what do we find? We find that the defendant intended to commit the crime of arson. Indeed, he had committed the offense 'already in his heart.' What were the overt acts toward the commission? He had prepared camphene and other combustibles, and had them in his room, and then he went a step further and solicited McDonnell to use those combustibles to burn the building, promising him, if he would do so, to 'give him the deeds of the place, and assign to him his right in the same.' We have, then, the fixed design of the defendant to burn this barn, and overt acts toward the commission of the offense, and a failure in the perpetration of it. The offense, then, is fully made out, for the intent to do the wrongful act, coupled with the overt acts toward its commission, constituted the attempt spoken of by the statute.'' These cases and the doctrine upon which they are grounded have been recently reaffirmed in *People* v. *Gardner [citation omitted],* and *People* v. *Sullivan. [Citation omitted.]*

Missouri has a similar statute. In *State* v. *Hayes,* 78 Mo. 307, the defendant solicited one McMahan to set fire to a building, furnished him a can of oil for the purpose, and gave him instructions for the burning. The court held that he was properly convicted of an attempt, although McMahan was acting under the advice of the police, and did not himself intend to commit arson. The court said: ''The evil intent which imparts to the act its criminality must exist in the mind of the procurer. And how the fact that the party solicited does not acquiesce or share in the wicked intent, exonerates the solicitor, baffles reason.''

The state of Georgia has a statute likewise taken from New York. In *Griffin* v. *State,* 26 Ga. 493, the New York cases are approved. The defendant intended to commit the crime of larceny by abstracting goods from a storehouse through the agency of one Jones. He took an impression of the key to the door of the building, and made a key for the purpose of opening it, which he sent in a box of fruit to

Jones, who feigningly agreed to become a participant in the accomplishment of the contemplated crime. It was held that the defendant was guilty of an attempt to commit the crime, and that Jones' intent had nothing to do with his offense.

The statute of Massachusetts provides that "whoever attempts to commit an offense prohibited by law and in such attempt does any act toward the commission of such offense" shall be punished as therein provided. In *Commonwealth* v. *Peaslee,* 177 Mass. 267, 59 N.E. 55, the evidence was that the defendant had arranged combustibles in a building in such a way that they were ready to be lighted, and, if lighted, would have set fire to the building and contents. The plan, however, required a candle which was standing on a shelf about six feet away from the combustibles to be placed on a piece of wood in a pan of turpentine and lighted. The defendant offered to pay a young man in his employment if he would go to the building, seemingly some miles from the place of the dialogue, and carry out the plan. This was refused. Later the defendant and the young man drove toward the building, but, when within about a quarter of a mile of the place, defendant said he had changed his mind and drove away. This was the only act he ever did toward accomplishing what he had in contemplation, and yet the court held that it was sufficient to convict him of an attempt to burn the building and its contents with intent to injure the insurers of the same.

We conclude, therefore, that the conviction of the defendant was right, and the judgment will be affirmed.

HAILEY, J., took no part in this decision.

On Rehearing.

BEAN, C. J. The doctrine of *State* v. *Hull,* 33 Or. 56, 54 Pac. 159, 72 Am. St. Rep. 694, and similar cases, has no application to the facts of this case. That was an indictment for larceny. The representative of the owner of the property alleged to have been stolen solicited the defendants to commit the offense. The property was taken by them by the express direction of the owner and with his assent. There was, therefore, no trespass in the taking and no crime committed. Here, however, the defendant, Taylor, planned the alleged arson and solicited McGrath and Palmer to assist him in its commission. Palmer informed his employer of the proposed plan and was advised to join Taylor and McGrath in appearance. This did not excuse Taylor for what he did personally. 1 Bishop, Crim. Law (5th Ed.) § 262.

The petition is denied.

TYPES OF CRIMES

Although an increasing number of crimes and corresponding punishments are being defined in clauses added to statutes on subjects other than criminal ones, the majority of crimes defined and made punishable by law can be grouped into the following general categories:

- Offenses against the person.
- Offenses against property.
- Offenses against public morals, health, safety, welfare, and public peace.
- Offenses against the administration of governmental functions.

OFFENSES AGAINST THE PERSON

Homicide: Homicide is the killing of one person by another. Not all homicides, however, are considered crimes. Some homicides are classed as *justifiable*. For example, if one person kills another in self-defense (*i.e.*, to keep from being killed), the homicide is considered justifiable. Similarly, if a police officer kills a person to prevent a serious crime such as burglary, the homicide is also considered justifiable and is not classified as a crime. Some other types of homicide are considered excusable. Although a person who commits excusable homicide is not necessarily without fault, the degree of fault does not merit the label of criminal intent. Excusable homicides are usually the result of accidents in which no gross or criminal negligence was involved.

Criminal Homicides: Criminal homicides are homicides neither legally justifiable nor excusable. A criminal homicide may be either murder or manslaughter. At common law, there were no degrees of murder or manslaughter — any unlawful homicide was considered either murder or manslaughter, and all murders were punishable by death. In this country, however, virtually all states have enacted statutes that define degrees of murder and manslaughter and the resulting punishment based on the perpetrator's state of mind and the circumstances of the crime.

First-degree Murders: First degree murders are those murders judged to be willful, deliberate, and premeditated. Killing by poisoning or in the commission of arson is considered murder in the first degree. Murder in the second degree is generally judged to lack the premeditation of first-degree murder. Homicide in the heat of passion or as a result of gross negligence is usually classified as second-degree murder. In *all* murders, however, the element of malice must be present, as explained in the following excerpt from the case of *Jacobs* v. *Commonwealth:*[11]

> The test of murder is malice. Every malicious killing is murder, either in the
> first or second degree — the former if deliberate and premeditated, and the latter

---

[11]*Jacobs* v. *Commonwealth*, 132 Va. 681.

if not. Furthermore, there is a *prima facie* presumption of malice arising from the mere fact of a homicide, but there is no presumption therefrom of deliberation and premeditation. This is merely another way of stating the familiar rule of law that every homicide is *prima facia* murder in the second degree, and that the burden is on the accused to reduce, and on the commonwealth to elevate, the grade of the offense. This, of course, does not mean that the accused may not rely upon circumstances of extenuation appearing in the evidence produced by the commonwealth with the same effect as if brought out in evidence offered by him.

Degrees of Manslaughter: Degrees of manslaughter are also mandated by statute. Generally, manslaughter is considered to be the unlawful killing of another human being without malice or premeditation, usually in the heat of passion or through gross negligence. A drunken driver who kills another person is guilty of manslaughter. Manslaughter is generally considered to have two degrees: voluntary and involuntary. Voluntary manslaughter results when, although the homicide is intentional, mitigating circumstances in which the homicide is committed (resulting from *sudden* heat of passion and *reasonable* provocation) reduce the degree of the crime from murder to voluntary manslaughter. Involuntary manslaughter is an unintentional homicide resulting from dangerous or unlawful acts or from criminal negligence. Negligence is further defined in the following excerpt from the case of *Chapman, J., Russ* v. *State:*[12]

> The killing of a human being by culpable negligence by the statute (of Florida and most states) is made manslaughter. Culpable negligence is the omission to do something which a reasonable, prudent, and cautious man would do, or the doing of something which such a man would not do under the circumstances surrounding the particular case. Negligence is the failure to observe for the protection of another's interest such care, precaution, and vigilance as the circumstances justly demand whereby injury is done to such a person, or, in another form, negligence is the failure to do what a reasonable and prudent person would ordinarily have done, or the doing of what such a person would have done under the situation whereby injury is done to another.

Other offenses against the person include assault and battery, kidnapping, rape, and false imprisonment.

OFFENSES AGAINST PROPERTY

Crimes against property include burglary, robbery, larceny, forgery, embezzlement, and arson. The crime of arson will be considered separately later in this chapter. Generally, burglary, robbery, and larceny are distinguished as follows: burglary at common law is the breaking and entering of the dwelling of another in the nighttime with the intent to commit a felony. Breaking is the use of force or fraudulent means to enter a building against the occupant's will. Entering is defined as the actual entrance of the burglar into the building. If any part of a

---

[12] *Chapman, J., Russ* v. *State*, 191 So. 296, 1939.

burglar's body enters the building — even if only a hand or foot — entry is considered to have been made.

Burglary: Most states have enacted statutes extending the common-law definition of burglary to include buildings owned by another; also, most states have eliminated the requirements of breaking *and* entering: only one of these two elements needs be present, as is illustrated in the following excerpt from the case of *Hayward* v. *State:*[13]

> It is fundamental that, in order to prove burglary, there must be evidence tending to prove the unlawful breaking and entry, and that such unlawful acts were with felonious intent *[intent to commit a felony]*. . . . The act of breaking may be shown by proof, either direct or circumstantial. . . . The breaking must be accompanied by an unlawful intent, if not, it is not burglary. This intent may also be proved by direct or circumstantial evidence. . . . These rules are fundamental and well understood, and are included in the statute.

Robbery: Robbery, another common-law felony, is defined as the unlawful taking and carrying away of the personal property of another from that person or in that person's presence through the use of violence, or the threat of violence, with the intention of permanently depriving the owner of the owner's property. The difference between robbery and theft is distinguished in the following excerpt from the case of *Krueger, J., Barfield* v. *State:*[14]

> The distinguishing feature between theft and robbery is that in theft the property is fraudulently taken from the possession of the owner or from some person holding the same for him; it is not necessary that the property be in the actual possession of the owner or the person holding the same for him; it is sufficient if he be in constructive possession thereof. In robbery the property must be taken by force and violence, not necessarily from the owner, but from any person in possession thereof whose right of possession is superior to that of the robber. The very fact that the property is taken from a person by the use of firearms, violence or threatened violence, is, within itself, sufficient to show that the person from whom it was taken was in possession thereof.

Larceny: Larceny is the taking and carrying away of another's personal property with the intent to steal the property. Robbery is larceny through violence and in the presence of the victim. At common law, larceny was a felony punishable by death; statutes in this country assign degrees of larceny depending on the value of the property stolen. Embezzlement and stealing the personal property of another through fraud are forms of larceny. The following excerpt from *State* v. *Holder* further defines elements of the crime of larceny:[15]

> To constitute the crime of larceny, there must be an original, felonious intent, general and special at the time of the taking. If such intent be present, no subse-

---

[13]*Hayward* v. *State*, 97 Nebr. 9.
[14]*Krueger, J., Barfield* v. *State*, 129 S.W.2d 310.
[15]*State* v. *Holder*, 188 N.C. 561.

quent act or explanation can change the felonious character of the original act. But if the requisite intent be not present, the taking is only a trespass, and it cannot be made a felony by any subsequent misconduct or bad faith on the part of the taker.

OFFENSES AGAINST PUBLIC MORALS, HEALTH, SAFETY, WELFARE, AND PUBLIC PEACE

This category reflects a moral code, as defined in the various statutes of the various states. Many crimes in this category are subject to varying degrees and descriptions by statute, and many of them are punishable as misdemeanors rather than as felonies. Subjects in this category include: adultery, bigamy, prostitution, obsenity, riot, libel, vagrancy, gambling, narcotic drugs, income tax evasion, and possession of deadly weapons.

OFFENSES AGAINST THE ADMINISTRATION OF GOVERNMENTAL FUNCTIONS

This category of crimes addresses interference with law enforcement processes and judicial processes. Under this category, perjury, bribery of law enforcement or judicial personnel, misconduct in office, tampering with evidence, obstructing police efforts, and contempt of legislative bodies or courts are all punishable crimes.

ARSON

Virtually from earliest times arson was considered a felony and, at one time, was punishable by death. Today, many modern statutes recognize arson as tantamount to murder in the first degree and punishable as such when homicide results from an act of arson.

Arson is generally considered to be committed by persons who fall into three general categories: mentally deranged persons, juveniles, and those mentally competent adults who for a variety of reasons — usually revenge or intent to defraud insurance companies — commit arson or cause arson to be committed. The following excerpt from Percy Bugbee's *Principles of Fire Protection* describes the wide variety of types of arsonists:[16]

> Firesetters come from all walks of life; neither sex, age, education, nor economic status in any way limits the possibility of becoming a potential arsonist. Professional arsonists are not as prevalent as they were in the past, but they still are in business.

Nearly anyone who deliberately sets a fire, however, can usually be classified

---

[16]Percy Bugbee, *Principles of Fire Protection*, NFPA, Boston, 1978, p. 134.

Table 5.1   Incendiary and Suspicious Fires and Losses, 1964-1974*

Year	Number	Property Loss
1974	114,400	$563,000,000
1973	94,300	$320,000,000
1972	84,200	$285,600,000
1971	72,100	$232,947,000
1970	65,300	$206,400,000
1969	56,300	$179,400,000
1968	49,900	$131,100,000
1967	44,100	$141,700,000
1966	37,400	$94,600,000
1965	33,900	$74,000,000
1964	30,900	$68,200,000

*From *Fire Protection Handbook,* 14th Ed., NFPA, Boston, 1976, p. 1-28.

under one or more of three headings, according to Paul L. Kirk's book titled *Fire Investigation:*[17]

Arsonists for profit. This includes perhaps the largest single group of persons who burn their own property for the sake of collecting insurance, or who will burn another person's property for hire. In this context, arson is a calculated act not basically different from burglary or armed robbery, although its perpetrators tend to feel less criminal and more justified by the unfortunate aspects of their individual financial situation, . . . Such persons will certainly meet all the legal qualifications of felons, although to place them formally in this category is often very difficult.

Arsonists for spite. Persons in this category are "getting even" or seeking revenge. Someone has wronged them, the wrong being either real or imagined, and the most obvious recourse to these warped minds is to burn the property of their persecutors. Arson, in this context, is undoubtedly more a rural than an urban crime, but no less a crime, and a serious one.

Arsonists for "kicks." This includes two different categories of personality, the first being the "firebug" who has a pathological attraction to fires and is happiest when witnessing a fire and its destructive effects. This person is distinctly abnormal. . . .

The second category of arsonists for "kicks" is the rather casual but malicious prankster who sets fires merely for the momentary excitement or as a general retaliation against society. This person, generally youthful and delinquent in more ways than one, has no special attraction to fires as such. . . .

Crimes of arson have increased dramatically over the past few decades. This fact, of course, is of vital interest to members of the fire service both from the point of view of fire protection and the possibility of major conflagrations as a result of acts of arson, and from the point of view of fire prevention. The successful prosecution

[17]Paul L. Kirk, *Fire Investigation,* John Wiley & Sons, Inc., New York, 1969, p. 159.

of arson cases has proven to be an effective deterrent to contemplated acts of arson. The steady growth in the number of incendiary fires and property losses resulting from such fires for the years 1964 to 1974 are shown in Table 5.1.

At the second national conference of the National Fire Prevention and Control Administration, Richard L. Best, Fire Analysis Specialist in the Fire Investigations Department of the NFPA, described in detail the dramatic rise in property loss and life loss from arson:[18]

> The increase in the number of incendiary and suspicious fires in the United States in the last ten years has been staggering. In 1966, the number of incendiary and suspicious fires, according to NFPA records, was approximately 37,000. In 1975, that number had soared to 144,000, an increase of about 285 percent.
>
> Aside from the dollar loss, other more important costs are the human lives lost in arson and incendiary incidents. That this is all too real is demonstrated by the fact that an estimated 1,000 people, including forty-five fire fighters, die each year in arson fires.

DEFINITION OF ARSON

Although most states have incorporated the common-law definition of arson in their statutes, they have also extended and broadened the common-law definition to include other forms and kinds of arson.

COMMON LAW ARSON

At common law, arson was defined as the willful and malicious burning of the dwelling house of another. Arson, at common law, was an offense against habitation and the resultant danger to the life of the person or persons in the dwelling. Arson was not considered an offense against property, nor was it considered an offense against the owner of the property. Rather, it was held to be an offense against the occupant of the property.

At common law, there was no question of the value of the property with regard to the charge of arson. Common law sought to protect the safety of the inhabitants. It was not necessary that the dwelling be inhabited when the act of arson was committed. At common law, it was sufficient that the structure burned be capable of being a dwelling for human beings.

Furthermore, under common law, the crime of arson did not include burning one's own dwelling or hiring someone to burn one's own dwelling. The crime of arson could only be committed against the dwelling of another. The only mitigating circumstance would be if a fire was set by an owner in order for the fire to spread to the dwelling of another.

---

[18]From an address given by Richard L. Best, fire analysis specialist of the National Fire Protection Association, at the second national conference of the National Fire Prevention and Control Administration, U.S. Department of Commerce, October 18-20, 1976.

STATUTORY ARSON

In recent times, arson has been extended by statute to include many acts of burning of structures not intended for habitation and not involving special danger to human beings. Statutes have expanded the common-law definition to make arson an offense against property as well as an offense against habitation. Many of these statutes include phrases such as "the property of any other person," or "burning of property." Statutory phrases can thus cover all forms of property, including cars, boats, crops, and so forth. Many statutes also define burning by means of explosive elements as arson.

Statutes relating to arson are not only legislated by states: there are also federal statutes relating to arson. Congress has defined and regulated punishment for crimes of arson committed in places and areas under federal jurisdiction. Lands and buildings reserved for the use of or under the protection and regulation of the United States, as well as seas and waters under federal jurisdiction or regulation, are subjects for federal arson statutes.

Arson-related Laws: The Statute of Limitations is important to fire investigators because it states that prosecution for a crime must be started within a certain period of time. Some of the states have a Statute of Limitations of five years for arson. Thus, arson prosecutions in these particular states must begin within five years of the date of the fire. The law, however, varies from state to state, with some states having no Statute of Limitations for arson. Fire service personnel acting as fire investigators should therefore be aware of the Statute of Limitations in the jurisdiction where the fire occurred.

The Federal Fugitive Felon Act is another important arson-related law with which fire investigators should be familiar. This Act prohibits flight from one state to another or to a foreign country to avoid prosecution, custody, or confinement after conviction for arson and certain other crimes. It also pertains to attempts to commit these crimes, and prohibits interstate flight to avoid giving testimony in any criminal proceeding involving punishment by imprisonment.

ELEMENTS OF ARSON

The *corpus delicti* of a crime consists of those elements necessary to prove that a crime has been committed. Two basic elements are necessary in order to demonstrate *corpus delicti:* (1) it must be shown that an injury or harm has actually occurred, and (2) it must be shown that the injury or harm has, in fact, been caused by a criminal agency.

In arson, the *corpus delicti* consists of demonstrating and proving that burning has occurred and that the burning was the result of a criminal act.

Burning: The crime of arson is not consummated unless some form of burning occurs. This does not mean that an entire building must be consumed by fire. If any part of the material involved in the fire is materially changed, that is, if wood is charred or fiber is destroyed, burning is considered to have occurred, even though no actual blaze results. Discoloration from heat or smoke is not considered burning.

The decision in the following case refers to both elements necessary to prove the *corpus delicti* of arson, and demonstrates the need to prove that the fire was not accidental.

STATE of Hawaii, Plaintiff-Appellee,
v.
Duprie K. DUDOIT, also known as Dupree K. Dudoit, Defendant-Appellant

Supreme Court of Hawaii.
Sept. 17, 1973.
514 P.2d 373

Defendant was convicted before the First Circuit Court, City and County of Honolulu, Nelson K. Doi, J., of arson, and he appealed. The Supreme Court, Richardson, C. J., held that state had burden of overcoming presumption that fire had natural or accidental origin, and that evidence did not support conviction.

Reversed.

LEVINSON, J., filed a concurring opinion; MARUMOTO, J., filed a dissenting opinion in which ABE, J., joined.

Syllabus by the Court

1. An essential element of the state's case is the establishment of a *corpus delicti* beyond a reasonable doubt.

2. The *corpus delicti* of arson consists of a burning that is the result of a criminal rather than accidental or natural cause.

3. The burden is on the state to overcome the presumption that a fire has natural or accidental origins.

4. Evidence that is equivocal in nature is insufficient to overcome the presumption that fires have natural or accidental origins.

5. The test on appeal in criminal cases is whether the verdict is supported by substantial evidence.

RICHARDSON, Chief Justice.

This is an appeal in *forma pauperis* from a conviction for the offense of arson.

The facts of the case are simple, if not sparse. Defendant Dudoit was seen leaving a rooming house lavatory. At the same time Dudoit was observed leaving, a small fire was discovered in that lavatory. Some testimony was presented as to Dudoit's suspicious actions and reactions regarding the fire. There was also testimony as to the presence of newspapers, cardboard, and pieces of rug inside the washroom in question. There was expert testimony that the fire was not electrical in origin. No evidence was introduced as to the origin of this fire other than the physical presence of newspapers, cardboard, and pieces of rug.

One of the most fundamental principles of the common law is that the occurrence of a crime must be proved before anyone can be convicted of the offense. The establishment of this *corpus delicti,* the body of the offense, is an essential element of the state's case.

Proof of the commission of a crime consists of three elements, each of which must be proved beyond a reasonable doubt: (1) the basic injury, such as . . . , the burning in arson, . . . , (2) the fact that the basic injury was the result of a criminal, rather than a natural or accidental cause, and (3) the identification of the defendant as the perpetrator of the crime. The first two of these elements constitute the *corpus delicti* or body of the crime, which is proved *when the prosecution has shown that a crime has been committed by someone. [Citations omitted.] (Emphasis added.)*

The concept of criminality thus requires that the state prove beyond a reasonable doubt a criminal, rather than a natural cause. This has been the law in Hawaii, that "[i]n reality the term *corpus delicti* means the actual commission of a crime by someone." *[Citation omitted.]*

In order to convict defendant Dudoit of the crime of arson, an essential element that the State of Hawaii had the burden of proving beyond a reasonable doubt was that the fire was kindled by other than natural or accidental means. The only proof submitted by the state to meet that burden was evidence of the presence of debris in the washroom. The presence of these newspapers, pieces of rug, and a cardboard box are all that the state relies on to raise the inference of a criminal agency in kindling the fire.

The state's main witness testified that he was employed for the purpose of cleaning such debris out of the washroom. It further appears from the testimony that it was not unusual for the items in question to be present in the washroom. State witness Kahalekomo testified that he sometimes saw newspapers in the washroom. We find the presence of this debris equivocal proof at best.

This court has adhered in the past to the "substantial evidence" test in reviewing criminal appeals. *[Citation omitted.]* We do not find the state's proof of *corpus delicti* to be supported by substantial evidence in this case.

The presence of material which has incendiary properties only incidental to their primary purpose is not substantial evidence. We could distinguish this case from *Republic of Hawaii v. Tokuji,* 9 Haw. 548 (1894), where the facts showed the fire to have started in three distinct places and where oily wicking and matches were found nearby. In the case at hand, only the presence of debris is shown, a presence that is susceptible of more than one reasonable interpretation.

The state did not establish nor attempt to show that the materials present were somehow related to the fire's origin. The state has resources not available to defendants, especially indigent defendants. City fire inspectors or other experts could have been used to tie the debris to the cause of the fire. The burden is on the state to overcome the presumption that a fire has natural or accidental origins. *[Citation omitted.]* We feel that the equivocal evidence presented did not overcome this presumption nor was it substantial evidence of a criminal agency.

When the state undertakes to overcome the presumption of natural origin through circumstantial evidence it must meet the further burden of disposing of any reasonable theory other than the guilt of the accused.

> The *corpus delicti* may be established by circumstantial evidence, but, as is the rule in all cases where circumstantial evidence is relied upon, the state of facts established must be inconsistent with any theory other than the guilt of the accused and the facts and circumstances disclosed and relied upon must be irreconcilable with the innocence of the accused in order to justify his conviction. *[Citation omitted.]*

The state may rely on the circumstantial presence of incendiary material as in *Republic of Hawaii v. Tokuji, supra,* to establish a *corpus delicti*. However, where the material present is of an ambiguous nature reconcilable with the innocence of the accused, as in this case, the state has not met its burden. The state by the expert testimony available to it must at least establish a causal relationship between the debris and the burning in order to show a criminal agency.

We reverse for failure of the *corpus delicti* and award attorney's fees on appeal to counsel for the appellant.

LEVINSON, Justice.

I concur in the opinion of Richardson, C. J., but add this note as a reminder of the view I have expressed previously in my concurring opinion in *State* v. *Butler,* 51 Haw. 180, 186, 455 P.2d 4, 8 (1969) and in my dissenting opinion in *State* v. *Rocker,* 52 Haw. 336, 348, 475 P.2d 684, 691 (1970) and to which I still adhere, that we should make it clear that "substantial evidence" is that evidence which would justify a reasonable trier of fact in finding that the defendant was guilty beyond any reasonable doubt.

MARUMOTO, Justice, with whom Justice ABE, joins.

I dissent. This court reverses defendant's conviction in this case on the ground that the state had the burden of proving beyond a reasonable doubt that the fire was kindled by other than natural or accidental means, and that "[t]he only proof submitted by the state to meet the burden, was evidence of the presence of debris in the washroom."

The record shows more than that. The fire took place in a small four feet by seven feet room in a tenement building. It was one of the two rooms in the building containing toilet facilities.

The police officer, who investigated the fire, testified that the areas in the room where the electric wiring ran were not burned at all. From that testimony, the trial court concluded that the fire was not caused by any short-circuiting in the electric wiring system.

The testimony of the officer showed the following: five pieces of three-inch wide boards were burned from floor to ceiling; the toilet seat was also burned; the top of the porcelain water tank was blackened; items which could have been used to start the fire were spread on the floor in front of the toilet. The trial court concluded from that testimony that the fire was of incendiary origin.

The caretaker of the building testified that he saw the fire in the room and defendant walking out of the room, "Just like nothing business, just keep on walking."

A tenant of the building testified to the same effect, his testimony being as follows:
"Q. Now, while you were putting out the fire, pioing* the fire, did you see Duprie around?
"A. Yeah; this man stop.
"Q. Was he helping you folks putting out the fire?
"A. No."

In finding defendant guilty, the trial court stated: "Defendant, when seen, certainly didn't express any concern for the fire; expressing concern for the fire would be normal. It is grossly abnormal that he did not express concern. I think it is also grossly inordinate that he would step into the bathroom when the bathroom is

---

*Pio is an Hawaiian word that means extinguishing a fire.

that small, when a fire that size is going on at that time that it is not necessary for him to step into the bathroom to see that fire.''

I think that the evidence in the record is sufficient for affirmance of the conviction under the substantial evidence test set forth in *State* v. *Kekaualua,* 50 Haw. 130, 433 P.2d 131 (1967), and *State* v. *Rocker,* 52 Haw. 336, 475 P.2d 684 (1970).

Willful and Malicious Intent: The second element necessary to prove the *corpus delicti* of arson requires proof that the burning was not accidental or the act of nature, but was a result of criminal agency; that is, that the burning was the result of malicious and willful intent, as stated in the following excerpt from *State* v. *Pisano:*[19]

> Although malice is a necessary ingredient of the offense, it need not be specially proved, as it may be presumed from evidence that accused deliberately, willfully, and without justification or excuse burned the property . . .

The varying degrees of willfulness and intent will be described in the next section of this text. Proof of willfullness will be considered in the section on ''Evidence'' in the next chapter.

The following commentary from Arthur Curtis' *A Treatise on the Law of Arson* describes the distinction between malice and intent *(footnotes and citations have been omitted):*[20]

> **Malice — Against other than owner.**
> One shadowy distinction between malice and intent is that intent is directed against property, while the term ''malice'' conveys the sense of ill feeling toward a person. This differentiation caused no difficulty to the common law courts, for the reason that the owner of the property was the one in occupation. If one intentionally set fire to the house in the possession of another, malice against the occupier was easily inferred. But, when the element of malice is continued with an extension of the crime to unoccupied buildings or those owned by the accused, a slightly different conception of malice is required.
> When it was sought to prosecute, under an indictment alleging common-law arson, one who at the owner's request burned property to aid the owner to defraud an insurance company, the element of malice interfered, for he could not have malice against the owner, and malice against the insurer was not an element of the charge. It is now recognized that an owner may have malice against his insurance company, and that a third person who sets the fire may be actuated by malice against both the owner and the company. Yet it has been thought that malice against an insurance company is not to be implied solely from the fact that the owner burned his own property while it was subject to an insurance policy.

---

[19]*State* v. *Pisano,* 141 A. 660.
[20]Arthur F. Curtis, *A Treatise on the Law of Arson,* Dennis & Co., Inc., Buffalo, 1936, pp. 86-87.

Although the rule may be adopted that malice must be aimed at some person, it is not necessary that the offender should actually know the person who will be injured, but it may be sufficient to show that he was bent on mischief against the owner whosoever he might be.

It has been difficult to apply the element of malice as against one who has burned a public building. This can be done by construing malice as the willful, rather than the negligent or accidental, burning of a building. Or a statute relative to such buildings may be interpreted as dispensing with the element of malice. When a statute extends the crime of arson to such buildings, it must impliedly abrogate the personal theory of malice, for one can rarely, if ever, be shown to have malice against the public.

The decision in the following case demonstrates the necessity of proving willful and malicious intent (*mens rea*) as a necessary element in a prosecution for statutory arson. As you read the case, consider the importance of this element in arson cases.

Peter J. BORZA, III
v.
STATE of Maryland

Court of Special Appeals of Maryland.
March 21, 1975.
Certiorari Denied June 25, 1975.
335 A.2d 142

Defendant was convicted in the Criminal Court of Baltimore, Albert L. Sklar, J., of statutory arson, and he appealed. The Court of Special Appeals, Moylan, J., held, *inter alia,* that testimony of asserted accomplice was sufficiently corroborated by evidence of defendant's presence and evidence that burned store was being operated at a loss by defendant; and that testimony of FBI agent that, on the day before the fire, alleged accomplice telephoned agent and confessed his involvement up to that point was admissible over hearsay objection to show abandonment of the conspiracy by the alleged accomplice and as a prior consistent statement, when credibility of accomplice was attacked because he remained silent for over two years after the fire.

Affirmed.

LOWE, J., filed a dissenting opinion.

MOYLAN, Judge.

The appellant, Peter J. Borza, III, was convicted in the Criminal Court of Baltimore by a jury, presided over by Judge Albert L. Sklar, of statutory arson. Upon this appeal, he raises six contentions:

1. That the trial judge should have ruled that Joseph Credge was an accomplice as a matter of law and that there was insufficient evidence to corroborate his testimony;

2. That there was insufficient testimony from which the jury could find that the accomplice's testimony was corroborated;

3. That the trial judge erred in refusing to give appellant's requested jury instructions;

4. That the trial judge committed reversible error in admitting prejudicial hearsay evidence;

5. That the trial judge erred in refusing to grant a mistrial; and

6. That the evidence was legally insufficient to sustain the convictions.

Because of our holding that the corroborative testimony was sufficient, even if Joseph Credge is assumed to be an accomplice, contentions one and two merge. Contention six may be dealt with along with them.

Late in the afternoon of Saturday, October 2, 1971, a fire occurred at the Castro Convertible furniture store at 315 North Howard Street in Baltimore. It was during business hours. The site of the fire was on the fifth floor, an area used for the storage of furniture. The business was one of three furniture stores in Maryland and New Jersey operated by the appellant, as franchisee of the Castro Convertible Company. The Hartford Mutual Insurance Company paid $13,917.52 for damages to the store and $24,655.57 for damages to the contents. An accountant testified that as of June, 1971, three months earlier, the store was operating at a net loss of $104,000.

Captain John Richter of the Baltimore City Fire Department's Fire Investigation Bureau arrived at the fire scene at 5:05 p.m. He estimated that the fire had been burning about one hour when he arrived. The first alarm had been turned in at 4:36 p.m. by a parking lot attendant next door who saw smoke and flames coming from the building. Captain Richter could not fully ascertain the cause of the fire. He found a large pile of trash burning on the fifth floor. The fire had also spread to the sixth floor. He effectively eliminated electrical or heating fixtures as a cause of the fire. He could not eliminate spontaneous combustion, noting that the source of possible combustion would have been consumed in the fire. He surmised that the fire probably resulted from careless smoking. He did interview the employee who had worked on the fifth floor that day, however. That employee was not a smoker, and he had not been in the store after 12:30 p.m. that day.

The testimony of the assumed accomplice now comes into play. Joseph Credge first became acquainted with the appellant in 1967. He was later employed by the appellant as a salesman in his Trenton, New Jersey, store. Credge later worked as general manager of both the Trenton store and the Baltimore store. In 1971, Credge and the appellant became partners in the Trenton store.

Credge testified that the Baltimore store was losing money and that in April or May of 1971, he and the appellant began discussions as to how to dispose of it. They finally determined to burn the Baltimore store. It was agreed that one or the other of them would go to the fifth floor of the Baltimore store and set fire to the packing materials and rubbish which were usually piled there awaiting disposal. On two occasions, on two successive Saturdays in September, 1971, Credge traveled from Trenton to Baltimore to set the fire. On both occasions, he abandoned the attempt because of fear. On the night of Friday, October 1, 1971, the appellant told Credge that he was going to Baltimore the next day and would set the fire if the opportunity presented itself. As a signal to the appellant upon his return to Trenton that the mission had succeeded, Credge was to leave the lights on in the Trenton store if word came in from Baltimore that the store had caught fire. Late on the afternoon of October 2, Credge, in Trenton, received a telephone call from the manager of the Baltimore store informing him that the store had caught fire.

The corroboration of Credge's testimony came largely from three employees of the Baltimore store. John Martin, the manager of appellant's Towson store, received a call from the appellant on the morning of October 2. The appellant informed Martin that he was taking a train in from Trenton and asked to be picked up at Pennsylvania Station. Martin picked him up at the station at between 3:30 and 4:00 p.m. and drove him immediately to the Howard Street store. The appellant spent a few minutes on the first floor talking to various employees. A customer came in, and the appellant took him upstairs. The employees assumed that the two were going to a second floor showroom. They were not certain whether the appellant returned to the first floor simultaneously with the customer. All that was certain was that the appellant was out of their sight on some upper floor. The appellant returned to the first floor about fifteen minutes after he had left it and within several minutes told Martin that he had to catch a train. They immediately departed. Martin dropped the appellant at Pennsylvania Station "at most one hour" from the time he picked him up. Martin was just three blocks away from the station when he heard on his car radio about the store fire on Howard Street. He returned there immediately. The appellant took the stand and acknowledged the account of his visit to Baltimore as given by the three employees.

We have no difficulty whatsoever in holding that the testimony of the assumed accomplice was amply corroborated. The test with respect to corroboration was well stated by Chief Judge Orth in *Early* v. *State,* 13 Md.App. 182, at 191-192, 282 A.2d 154, 160:

"'"Corroborate" means to strengthen not necessarily the proof of any particular fact to which an accomplice has testified, but the probative, criminating force of his testimony.' *Wright* v. *State,* 219 Md. 643, 649, 150 A.2d 733, 737. Quoting with approval 1 Underhill Criminal Evidence (5th Ed. 1956) § 185. It is settled that *the corroborative evidence must tend* either: (1) to identify the defendant with the perpetrators of the crime, or (2) *to show the defendant's participation in the crime itself.* [*Citations omitted.*] If with some degree of cogency it tends to establish either of these matters it would be sufficient, authorizing the trier of fact to credit the accomplice's testimony even with respect to matters as to which there had been no corroboration. [*Citation omitted.*] *Corroboration need not extend to every detail.* [*Citation omitted.*] *So it is not necessary that in and of itself the corroborating evidence be sufficient to convict,* and *not much in the way of corroboration is required; only slight corroboration is necessary. The corroborating evidence may be circumstantial.* [*Citation omitted.*] 'Whether the testimony of an accomplice has in fact been sufficiently corroborated must, of course, depend upon the facts and circumstances, and the inference deducible therefrom, in each case.' '' [*Citation omitted.*] *(Emphasis added.)*

The foregoing would also be dispositive of the appellant's sixth contention that the evidence was legally insufficient to permit the case to go to the jury, but for an added fillip in that regard. The fourth count charged specifically that certain merchandise, to the detriment of the defrauded insurance company, was "set fire to and burn[ed]." The fire investigator testified that the literal damage to the furniture was from its being "smudged from smoke and heat" and "water damage." Art 27, Sec. 9 reads, in pertinent part:

> "Any person who willfully and with intent to *injure or* defraud the insurer sets fire to or burns *or causes to be burned or who aids, counsels or procures the burning of* any goods, wares, merchandise or other chattels or personal property of any kind, *or of the property of himself or of another,* which shall at the time be insured by any person or corporation against loss or damage by fire; *shall upon conviction thereof, be sentenced to the penitentiary for not more than five (5) years." (Emphasis added.)*

It is clear to us that the gravamen of the offense is defrauding the insurance company by damaging goods through the agency of fire. That the literal damage comes from the heat of the fire, from the smoke of the fire, or from the water of the firemen's hoses, rather than through the chemical process of combustion, is not controlling. To hold otherwise is to make an absurdity of the law. The employment of the phrase "sets fire to" as an alternative *actus reus* to "burns" frees this law from ancient rigidity of the arson laws when it comes to the term "burn." *Cochrane* v. *State,* 6 Md. 400 (1854), holds that there is a difference between the two terms and cites with approval the Virginia case of *Howel* v. *Commonwealth,* 5 Grattan 664. That case states, at 670-671:

"We are not satisfied that *setting fire to* and *burning,* have been established by any legal authority to be synonyms, so as to justify, in an indictment upon the statute, the substitution of the former words in the place of the other. East, (2 Cr.L. 1020, § 2) remarking upon the statute 9 Geo. i. ch. 22, says, that that statute does, indeed, in enacting the felony, make use of the words 'set fire to,' but he was not aware of any decision which had put a larger construction on those words than prevails by the rule of the common law; and the contrary opinion may be collected (he says) from what is said in Spalding's Case (1 Leach 218), and Breeme's Case, (1 Leach 220) and in the case of Sarah Taylor, (1 Leach 49). With all the respect which is justly due to this writer, upon an attentive inspection of the authorities which he has referred to, it will be discovered that there is nothing in these cases which decides that these expressions are identical in their meaning. The same author, in a subsequent section, (§ 11) says, that at common law, it was necessary to state an actual burning, but that the statute 9 Geo. i. ch. 22, using the term *'set fire to'* the house, it is now become common to state both, though (as he says) in effect meaning the same thing. (2 East 1033). But yet in Salmon's Case, Russ. & Ry.C.C. 26 (which was a prosecution under this state, for setting fire to a hay stack), it was moved to arrest the judgment, on the ground that it was not averred in the indictment, that by reason of setting on fire, the stack of hay was burnt and consumed; and the point being reserved, the Judges were of opinion that the conviction was right — *that it was not necessary the stack should be burned, the words of the act being 'set fire to.'* (2 Bagg.Cr.L. 79.) This authority seems clearly to decide that *setting fire to* and *burning* are not legal synonyms. If they were, there would be no reason for that redundancy of language which is usual in the indictments under the stat. 9 Geo. i., according to the forms, laying burning as well as setting fire to. (See 3 Chit.Cr.L. 1109-1115.) In the statutory offences of arson, enacted in the 2d and 3d sections of our statute, 1 Rev. Code, ch. 160, the offences are described 'burn or set fire to'; but in the 4th section, under which it is that the present indictment is laid, the expression is 'burn' alone. The change in the phraseology would seem to intimate on the part of the Legislature that there was a difference in the meaning. It is not, therefore, without reason, that *Mr. Davis,* in his work on Criminal Law, regarded the difference in the language as distinguishing between *burning* and *setting fire to.* (Davis' Cr.L. 117, see note q.)"

The dissent raises an issue not raised by the appellant, unless arguably done so under the broad and undifferentiated umbrella of charging that the evidence was legally insufficient to permit the case to go to the jury. Out of chance phrases in *McDowell* v. *State,* 231 Md. 205, 189 A.2d 611, and *Bollinger* v. *State,* 208 Md. 298, 117 A.2d 913, the dissent spins the theory that the *corpus delicti* of arson may never be established by evidence which goes to show criminal agency. Because an uncritical reading of dicta in *Bollinger* might give rise to such a misperception in others, it may be well to try to lay the ghost to rest.

In *McDowell,* the critical question was whether the testimony of an accomplice had been adequately corroborated. The court held that it had, and the conviction for arson was affirmed. Preliminarily, the court did recite the facts of the case and did set forth the uncontested fact that the *corpus delicti* of arson had there been established.* It then quoted Wharton [1 Wharton's Criminal Evidence (12th Ed., Anderson, 1955) § 17, at 48] to the effect that ''[p]roof of the defendant's connection with the crime as the operative agent, although essential for conviction, is not part of the *corpus delicti.''* The court in *McDowell* was not remotely considering the question for which the dissent cites *McDowell* as authority — that of whether evidence of criminal agency may be used to prove the *corpus delicti.* The *McDowell* court was simply clearing the arena of the undisputed issues before coming to grips with the issue in that case — whether the testimony of the accomplice establishing the criminal agency of McDowell was adequately corroborated. That court was in effect saying, ''The *corpus delicti* is not in dispute, but the mere establishment of the *corpus delicti* does not establish the criminal agency of the defendant. We now turn to the question of criminal agency generally and to the corroboration of the accomplice specifically.'' Indeed, a quick glance at the quoted Wharton makes it clear that that authority was not there considering any question dealing either with arson law or with whether an item that may go to show criminal agency may also go to show a particular element of a particular crime. In the general introduction, rather, to the concepts dealt with in the criminal law, Wharton was attempting to define and to distinguish the notion of *corpus delicti* from the notion of criminal agency.

Indeed, a moment's reflection on the proposition urged by the dissent — that each and every element of the *corpus delicti* must be established totally independently of proof tending to show criminal agency — will supply its own refutation. Whereas the *actus reus* may conceivably be established by independent proof not going to the criminal agency of anyone, a *mens rea* — general or special — is also a necessary element of the body of a crime. *Mens rea,* by definition, exists in the head of the perpetrator and not in a vacuum. Arson, moreover, is a crime having as one of its elements the special *mens rea* that the act of burning shall have been ''willful and malicious.'' This specific intent of willfulness and malice must have existed in the brain of him who burned or him who procured the burning. Proof, therefore, of this particular mental element of the crime is, *ipso facto,* proof of criminal agency. It cannot be otherwise.

McDowell is also instructive on the further point of what is independent evidence tending to corroborate the testimony of an accomplice. In *McDowell,* an apartment was burned. The defendant was a subtenant enjoying residential

---

*It is interesting to note for present purposes that the Court found relevant in proving the *corpus* *delicti* of arson the fact that a captain of the Fire Investigation Bureau was able to eliminate spontaneous combustion or faulty wiring as possible causes of the fire.

privileges through January 3, 1962, the day of the fire. His presence in the apartment house, like the presence of the present appellant in the Howard Street store, was, therefore, not remarkable. Notwithstanding the relative innocuousness of such presence, standing alone, that presence was nevertheless held to be independent evidence corroborating the accomplice's testimony.

The *McDowell* court also considered as independent, corroborative evidence, the inference of a motive arising out of the fact that the defendant was being ejected from the apartment. Further evidence of motive came from the mouth of the accomplice, but the inference was held to be corroborative.* In the present case also, direct testimony as to motive came from the accomplice Credge. The fact of the insurance being carried on the appellant's business and the independently established fact that that business had been operating at a net loss of over $100,000 during the months immediately preceding the fire certainly give rise to an inference of motive as forcefully as did the ejectment by a landlord in *McDowell*. In referring to these two independent items of evidence, *McDowell* said, *passim*, at 231 Md. 212-214, 189 A.2d at 615:

> "Here the evidence of McDowell's presence, of his having to leave the apartment, and of his being the last person shown to have been in the immediate vicinity of the scene of the crime with both the opportunity and a possible motive to commit the offense serves as corroboration of the identification of the defendant with the commission of the crime, which is the other branch of the Polansky [*Polansky* v. *State*, 205 Md. 362, 109 A.2d 52] rule. . . .

> In an arson case the presence of the defendant in the vicinity of the fire, whether before or after its occurrence, is always relevant. *Bollinger* v. *State*, *supra*, at 208 Md. 307, 117 A.2d at 917. . . . As to both presence and motive the defendant's testimony corroborates that of the accomplice. He was in the immediate vicinity of the fire shortly before it was set, and he was being put out of the apartment along with Aldrich.

> The defendant's presence near the scene of the fire was also corroborated by the independent witness, the plumber. . . .

---

*The origin of the rule requiring corroboration of an accomplice's testimony is well traced in *Wright* v. *State*, 219 Md. 643, 150 A.2d 733. That case also illustrated how corroborative evidence need not itself be evidence of unlawful conduct. This aspect of *Wright* was summarized in *McDowell*, at 231 Md. 212, 189 A.2d at 615:

"The *Wright* case like the instant case, involved a conviction for arson based upon the testimony of accomplices. The opinion enumerates various items of evidence corroborative of the accomplices' testimony. A number of them concerned matters not in themselves unlawful and some were described as corroborating 'only the fact that the defendant had engaged in several lawful pastimes — such as going to the movies, eating watermelon and roaming the highways and byways late at night.' yet these were held (citing *Polansky* v. *State*, *supra*) to afford corroboration as showing 'that the defendant was identified with the admitted arsonists.' "

The most important corroborative evidence we think consists of that which (1) shows McDowell's presence, (2) suggests a motive for setting the fire, and (3) shows that Aldrich departed from the scene when the fire occurred, leaving McDowell at the scene. . . . At that time the appellant had ready access to the place where the fire soon afterwards appeared.''

Nor will *Bollinger,* under close examination, support the theory erected upon it. It is important to put *Bollinger* in perspective.

The barn burning, on which the conviction in *Bollinger* was had, occurred on December 16, 1954. Both defendants gave formal confessions to the State Police. The heart of the defense contention was that ''the court erred in admitting the confessions before the State had overcome the legal presumption that the fire was not a criminal fire and before the State had proved the *corpus delicti* which they claim the State had never been able to prove.'' 208 Md. at 303, 117 A.2d at 916. The central issue was the order of proof — the question of whether the State could offer the defendants' confessions before the *corpus delicti* had been shown. The Court of Appeals there held that there was no error in the *Bollinger* case and they affirmed the conviction. The Court adhered to the general truism that a confession of crime alone will not sustain a conviction and that, ordinarily, the confession should not be the first item of the State's proof. In retaining flexibility in application of the principle, however, the *Bollinger* court cited *Weller* v. *State,* 150 Md. 278, 132 A. 624, to the effect that an extrajudicial confession is not admissible ''unless there then exists, *or there is a proffer of proof later''* of the *corpus delicti. (Emphasis supplied.)* It went on to cite *Harris* v. *State,* 182 Md. 27, 31 A.2d 609, for the proposition that the order in which the evidence should be produced ''is largely a matter within the discretion of the trial court.''

The court set out the defense thesis and the applicable law, at 208 Md. 304, 117 A.2d at 916:

''Appellants further argue that without proof of the *corpus delicti* the confessions were not admissible. It is stated in Wharton's Criminal Law, 12th Ed., Vol. II, Section 1063, as follows: 'The burden is on the state to show that the burning was with a criminal design, — this is the *corpus delicti.* The *corpus delicti* cannot be established by proof of the burning alone, or by the naked confession of the accused. Where nothing except the burning appears, the law presumes it to have been accidental, and not by criminal design; and the state must overcome this presumption of law, and prove a criminal design beyond a reasonable doubt.''

It is easy to see where the dissent was logically ensnared, because the passage in Wharton, cited with apparent approval by *Bollinger,* is a model of poor exposition. It joins in a single sentence — ''The *corpus delicti* cannot be established by proof of the burning alone, or by the naked confession of the accused.'' — two very disparate propositions, without so much as a break of paragraph or even a period to make them twain.'' That freakish accident of murky exposition cannot give birth to the notion that the *corpus delicti* of an arson cannot be proved by *the combination of* proof of burning *and* the confession of the accused.

As to the first proposition, Wharton reiterates the universally accepted truth that the confession of a crime cannot, standing alone, warrant a conviction, absent some independent evidence of the *corpus delicti*. The thrust of the principle is to prevent mentally unstable persons from confessing to, and being convicted of, crimes that never occurred. *Bollinger* makes clear that it is this general principle — and not some rule peculiar to arson cases — that is being discussed. ''Apparently there is no difference in this rule as applied to arson than in other criminal cases.'' 208 Md. at 304, 117 A.2d at 916. ''There is no question that in this State an extra judicial confession of guilt by a person accused of crime uncorroborated by other evidence is not sufficient to warrant a conviction.'' *Id.* at 304-305, 117 A.2d at 916. ''Generally an uncorroborated confession does not establish as a matter of law the commission of crime beyond a reasonable doubt. The purpose of the rule requiring corroboration of confessions is to guard against convictions based upon untrue confessions alone.'' *Id.* at 305, 117 A.2d at 916. All of the cases cited by *Bollinger* make it clear that it is the general truism being discussed, and not some peculiarity of the arson law. *Weller* v. *State, supra* (where the crime was the manufacture of bootleg whiskey). *[Citations omitted.]*

Bollinger makes it very clear, in discussing specifically the proof of the *corpus delicti* itself, that a confession may well be one of the factors entering into that equation. It said at 208 Md. 305-306, 117 A.2d at 917:

> ''It is also not necessary that the evidence independent of the confession must be full and positive. . . . In addition, the evidence necessary to corroborate a confession need not establish the *corpus delicti* beyond a reasonable doubt. *It is sufficient if, when considered in connection with the confession, it satisfies the jury beyond a reasonable doubt that the crime was committed and that the defendant committed it.* As Judge Learned Hand said in *Daeche* v. *United States,* 2 Cir., 250 F. 566, 571, circumstances corroborating a confession need not independently establish the truth of the *corpus delicti* at all, either beyond a reasonable doubt or by a preponderance of proof, *but any such circumstances will serve which in the judge's opinion go to fortify the truth of the confession. [Citation omitted.] (Emphasis added.)*

> ''Of course, proof of the *corpus delicti* in an arson case is usually a difficult matter, as the burning is almost invariably done in a most secretive manner. The prosecution usually has to depend on circumstantial evidence. *[Citation omitted.]* 'It is sufficient if there is substantial evidence of the *corpus delicti,* independent of the confession and the two, together, are convincing beyond a reasonable'. . . .''

As to the second and very distinct proposition, *Wharton* is simply pointing out that one of the elements of the crime of arson, which needs be proved by the State, is the special *mens rea* that the burning be "willful and malicious." Again, however, the ill-advised language of *Wharton* has led latter-day readers astray, particularly where *Wharton* said:

> "Where nothing except the burning appears, the law presumes it to have been accidental, and not by criminal design; and the state must overcome this presumption of law, and prove a criminal design beyond a reasonable doubt."

This, of course, is not a presumption of law. It is simply a statement that the State has the burden of proving all elements of a crime. A legal presumption has the effect of shifting to an opponent the burden of going forward and producing evidence. 9 Wigmore on Evidence (3rd Ed. 1940) § 2490, "Presumptions." No burden, of course, shifted to the State in this instance, since the State had the burden from the very outset. This is one more instance of the word presumption having too many meanings. The presumption referred to here, like the so-called presumption of innocence, is neither a true presumption of law nor an inference of fact, but a statement as to the burden of proof. The State has the burden of proving the special *mens rea* in an arson case beyond a reasonable doubt, just as it has the burden of so proving every other element of this or any other offense. Indeed, later editions of Wharton have now refrained from this loose use of the word "presumption." 2 Wharton's Criminal Law and Procedure (Anderson Ed., 1957) now describes, in dealing with arson, the mental state at § 390.

> "In the absence of contrary evidence, it is assumed that every burning is accidental or is due to natural causes, and not criminal design. The burden is, accordingly, on the prosecution to establish that it was willful and malicious. Because of this, the *corpus delicti* is not established by proof of burning alone, since such fact does not prove the intent with which the act was done."

We note, moreover, that the words of the appellant at issue in this case were operative words of the then ongoing conspiracy spoken by the appellant to Credge and were not a confession within the contemplation of *Bollinger*. *Bollinger*, indeed, even considered the post-crime blurt, "I did it" made to a policeman as outside its notion of "confession," which it restricted to a response to a formal custodial interrogation.

Bollinger lends further support to our holding that there was in this case ample independent evidence to corroborate the testimony of an assumed accomplice. The defendants in *Bollinger* lived on a neighboring farm and regularly frequented a nearby store. Their presence, like the presence of the appellant here at the North

Howard Street store, was, therefore, unremarkable. Nonetheless, such presence near the scene of the crime was held to be independent evidence of guilt:

> "Appellants arrived in a car at the store two miles distant from the fire about twenty-five to thirty minutes after Mr. Keilholtz saw the small fire and from ten to twenty minutes after Mr. Valentine testified that the barn was ready to explode. In arson cases, evidence that the accused was seen in the vicinity of the fire, whether before or after it occurred, is always relevant. It is said in Underhill on Criminal Evidence, 3rd Ed., page 798, Section 564: 'It is always relevant, particularly in the case of the crime of arson, which is usually committed at night and with the greatest secrecy, to show that the accused was seen in the vicinity of the burned building about the time of the fire, whether before or after it occurred.' " *[Citation omitted.]*

Curiously opposite also in the *Bollinger* case was that the court there found relevant to the establishment of the *corpus delicti* the facts that the barn owners had been present in their barn a little more than an hour before the fire occurred, that neither of the owners smoked, and that the barn had recently been wired by an electrician and the wiring was apparently all right.

The appellant's third contention is that he was erroneously denied the following requested instruction:

> "The Court instructs the jury that to constitute the offense of arson it is necessary that the burning be willful and malicious. In the absence of contrary evidence it is assumed that every burning is accidental or due to natural causes and not criminal design. The burden is, accordingly, on the prosecution to establish that it was willful and malicious. Because of this the *corpus delicti* is not established by proof of the burning alone, since such fact does not prove the intent with which the act was done."

There was, to be sure, nothing improper about the requested instructions. *Hughes* v. *State*, 6 Md.App. 389, 251 A.2d 373. The adequacy of the instructions is determined, however, by viewing them as a whole. *State* v. *Foster*, 263 Md. 388, 283 A.2d 411; *Shotkosky* v. *State*, 8 Md.App. 492, 261 A.2d 171. The trial court's discretion as to wording, order and amount of detail in his instructions should not be disturbed on appeal absent a clear abuse of that discretion. *Mills* v. *State*, 12 Md.App. 449, 279 A.2d 473. The judge charged the jury that the burden of proof was upon the State to establish every element of the crime charged and that the appellant was presumed innocent until proved guilty beyond a reasonable doubt. The jury was instructed that the appellant was entitled to every inference that could reasonably be drawn in his favor from the evidence. The instructions went on:

"You are instructed that the crime charged by the State, the crime of arson, requires proof beyond a reasonable doubt of specific intent on the part of the defendant of specific intent before the defendant can be convicted. To establish specific intent, the State must prove that the defendant knowingly did an act which the law forbids purposely intending to violate the law; or that he wantonly and recklessly disregarded the law.

If a man does an unlawful act, the natural tendency of which is to set fire to and burn a building, and such consequences follow, the burning is to be regarded as intentional and malicious. In that regard, you are advised that to constitute the offense of arson, it is necessary that the burning be willful and malicious. The burden is upon the State to establish beyond a reasonable doubt that it was willful and malicious."

Setting down the nubs of the instruction requested and the instruction given side by side, it is clear that the difference is only stylistic.

Instruction Requested	**Instruction Given**
"[I]t is necessary that the burning be willful and malicious. In the absence of contrary evidence it is assumed that every burning is accidental or due to natural causes and not criminal design. The burden is, accordingly, on the prosecution to establish that it was willful and malicious."	"In that regard, you are advised that to constitute the offense of arson, it is necessary that the burning be willful and malicious. The burden is upon the State to establish beyond a reasonable doubt that it was willful and malicious."

Every essential point of law was properly covered. . . .

Judgments affirmed; costs to be paid by appellant.

LOWE, Judge (dissenting).

I must respectfully depart from the majority because it is obvious to me that the State failed to meet its burden.

Borza was convicted *solely* on the testimony of his accomplice, Credge, who, after a homosexual affair with Borza had been rejected by him in preference for a more natural relationship with one of the opposite sex. The testimony was gratuitously volunteered after two years of silence following the fire. Suffice to say, Credge was not a reluctant witness. More than that, his testimony at most established only that he and appellant had discussed burning the store the preceding April or May, that he, Credge, had twice traveled to Baltimore in September to burn the store, and that on the night before the fire, appellant had mentioned that he might set fire to the store. This factual posture presented at least four possibilities:

1. That Credge told the truth and appellant willfully and maliciously set fire to the store;
2. That Credge told the truth but that the fire was accidental and Credge wrongfully deduced appellant's guilt;
3. That Credge lied about the previous conversations;
4. That Credge himself set the fire knowing appellant would be in the store that day.

I recognize, of course, that fact-finding is left to the jury. The majority, however, inexplicably fails to recognize that it is for that reason the courts have engrafted on the fact-finding process certain fundamental legal protections for a criminal accused. To ensure that the second hypothesis is not the case, the State must prove the substantial fact that a crime has been committed, the *corpus delicti:*

> "The burden is on the state to show that the burning was with a criminal design, — this is the *corpus delicti*. The *corpus delicti* cannot be established by proof of the burning alone, or by the naked confession of the accused. Where nothing except the burning appears, the law presumes it to have been accidental, and *not by criminal design;* and the state must overcome this presumption of law, and prove a criminal design beyond a reasonable doubt."*[Emphasis added.]* Bollinger v. *State*, 208 Md. 298, 304 [117 A.2d 913, 916], quoting 2 Wharton's Criminal Law (12th Ed., Anderson), § 1063.

To minimize the possibility of the third and fourth hypotheses being the case, the State must introduce corroborative evidence when an accomplice testifies. The State totally failed in both regards.

Assuming Credge told the truth, his testimony did no more than establish that appellant was contemplating setting the fire, *not* that he set it. Had the State then proven that the fire was deliberately set — indicating a criminal act — the evidence may have been sufficient to convict. The State did just the opposite.

The only permissible evidence submitted by the State to meet its burden of overcoming the presumption of accident was that of Captain Richter of the Fire Investigation Bureau. From the majority's depiction of his testimony we are told:

> "Captain Richter could not fully ascertain the cause of the fire. . . .

He could not eliminate spontaneous combustion. . . .

He surmised that the fire probably resulted from careless smoking. . . .''
although the employee on the fifth floor had been a nonsmoker.

Even though the Captain had ''effectively eliminated electrical or heating fixtures as a cause of the fire'' certainly his testimony seemed more to establish that the fire arose from natural or accidental cause rather than to establish a deliberate burning.

The majority has transformed evidence of criminal agency, *i.e.*, if there was a crime, appellant probably committed it, into evidence of agency and *corpus delicti, i.e.,* the fire was deliberately set and appellant set it. That is a leap of faith wholly unwarranted by the evidence and diametrically opposed to the Court of Appeals's clear and simple expression of the State's burden. In *McDowell* v. *State,* 231 Md. 205, 208, 189 A.2d 611, 612* Chief Judge Brune said:

> ''As Wharton points out '[p]roof of the defendant's connection with the crime as the operative agent, although essential for conviction is not part of the *corpus delicti.*' '' [Reference here is to 1 Wharton's Criminal Evidence (12 Ed., Anderson 1955), § 17 at 48].**

The majority dismisses the requirement of independently proving the *actus reus* of arson simply because it is a crime requiring ''a special *mens rea,*'' *i.e.,* proof of ''willful and malicious'' intent. Because *mens rea* is also an element of criminal agency does not excuse the necessity of independently proving the act charged was a crime. ''Where nothing except the burning appears, the law presumes it to have been accidental, and not by criminal design; and the State must overcome this presumption of law, and prove a criminal design beyond a reasonable doubt.'' *Bollinger,* 208 Md. at 304, 117 A.2d at 916. However you read that sentence, whether in or out of context, its clarity dispels the notion that it is a ''freakish accident of murky exposition.'' It is not conceivable that the Court of Appeals meant that an accusation that a defendant discussed committing a crime, is sufficient to prove that a crime was committed. In laying their ''ghost to rest'' the majority may find themselves haunted by the apparition of a dormant *corpus delicti.*

Corroboration is the other fundamental legal protection that this case erodes.

---

*Recognizing that this was judicial dictum since that case focused on proof of criminal agency rather than the *corpus delicti,* it is not without precedent for this Court to adopt policy concepts in reliance upon Court of Appeals' dicta. *Cf., Transamerica Insurance Company* v. *Brohawn,* 23 Md.App. 186, 326 A.2d 758.

**The most recent edition of 1 Wharton's Criminal Evidence (13th Ed., Torcia 1972) § 17 at 27 reiterates the same concept:

> ''Proof that the defendant was the person who engaged in the unlawful conduct is of course necessary for a conviction, but it is not an element of the *corpus delicti.*''

The only corroborative evidence of Credge's testimony was the presence of the accused in his own store on the day of the fire.* Significantly, this fact in no way detracts from either of the third and fourth exculpatory hypotheses suggested above.

The test for corroboration of an accomplice has been many times stated. In *Luery* v. *State,* 116 Md. 284, 81 A. 681 the Court wrote:

"[T]he important matter is to have *[the accomplice]* supported in at least some of the material points involved. . . ."

In *Judy* v. *State,* 218 Md. 168, 176, 146 A.2d 29, 33, the Court held that corroborating testimony "need only support some of the material points of the accomplice's testimony." [Both cited with approval in *McDowell* v. *State,* 231 Md. 205, 211-212, 189 A.2d 611, 614].

Appellant's presence in his own store, though perhaps relevant, surely cannot be regarded as assuring Credge's verity by supporting "some of the material points" of his testimony. Presence of an accused may be adequate corroboration when it is "under unusual or suspicious circumstances." 3 Wharton, Criminal Evidence (Torcia Ed., 1973) § 649 at 368. But the presence of a furniture store owner at his own store can hardly be regarded as either unusual or suspicious. In that situation, mere presence is not sufficient corroboration. *Hatton* v. *Commonwealth,* 253 Ky. 103, 68 S.W.2d 780.

The majority alludes at length to language in *McDowell* suggesting that presence was there a corroborative circumstance. With that narrow statement I have no quarrel. Nonaccomplice testimony of the accused's presence took on special significance in *McDowell* because of the accused's statements which were "contradictory and untruthful . . . relating to his whereabouts at the time of the fire." The *McDowell* Court, relying on *Wright* v. *State,* 219 Md. 643, 150 A.2d 733, stressed the fact that the testimony of nonaccomplice witnesses was consistent with that of the accomplice and inconsistent with that of the defendant. That is a circumstance conspicuously absent here.

The potentially vengeful motivations of appellant's lone accuser and the unexplained delay of two and one half years before he made the accusation create a most tenuous setting in which to undermine the safeguards provided by the requisites of proof of *corpus delicti* and corroboration. Strain we have and strain we should to meet a technical mischance in a case where guilt and culpability of the accused is otherwise readily apparent. But what meaning can we give to the platitudes upon which our system of justice is said to rely if we need strain so hard

---

*Although prior to the fire Credge contacted an FBI agent and told him about his and appellant's fire discussions, it would be the ultimate in tautological thinking to treat such self-serving groundwork — whether or not termed "operative verbal acts" — as corroboration.

here, even to ascertain if a crime was committed. To permit a rejected homosexual paramour to supply that plus every other element necessary for conviction, is to allow the accuser to make a mockery of our time-honored safeguards. In trying to stabilize the tenuous structure upon which this case is built, the majority has exacted another chip from the foundation structure of the system.

MODEL ARSON PENAL LAW

In an attempt to lessen the rising incidence of arson, the Alliance of American Insurers, the American Insurance Association, and the National Association of Independent Insurers have proposed a *Model Arson Penal Law*. This model law has been adopted by statute either in whole or in part in many states. Following is the text of the *Model Arson Penal Law* — Offenses Against Property, Article 100, Arson, Criminal Mischief and Other Property Destruction, which describes and defines the degrees and classifications of the crime of arson.

§ 100.1 *Arson and Related Offenses*

(1) Aggravated Arson. A person is guilty of aggravated arson, a felony of the first degree, if he starts a fire or causes an explosion, or if he aids, counsels or procures a fire or explosion, with the purpose of:

(a) destroying an inhabited building or occupied structure of another; or
(b) causing, either directly or indirectly, death or bodily injury to any other person.

(2) Arson. A person is guilty of arson, a felony of the second degree, if he starts a fire or causes an explosion, or if he aids, counsels or procures the setting of a fire or causing of an explosion, with the purpose of:

(a) destroying or damaging a building or unoccupied structure of another; or
(b) destroying or damaging any real or any personal property having a value of $_____ or more, whether his own or another's, to collect insurance for such loss.

(3) Reckless Burning or Exploding. A person commits a felony of the third degree if he purposely starts a fire or causes an explosion, or if he aids, counsels or procures a fire or explosion, whether on his own property or another's, and thereby recklessly:

(a) places another person in danger of death or bodily injury; or
(b) places a building or structure of another, whether occupied or not, in danger of damage or destruction; or

(c) places any personal property of another having a value of $_____ or more in danger of damage or destruction.

(4) Failure to Control or Report Dangerous Fire. A person who knows that a fire is endangering life or property of another and fails to take reasonable measures to put out or control the fire, when he can do so without substantial risk to himself, or to give a prompt fire alarm, commits a misdemeanor if:

(a) he knows that he is under an official, contractual or other legal duty to control or combat the fire; or

(b) the fire was started, albeit lawfully, by him or with his assent, or on property in his custody or control.

(5) Definitions. "Occupied Structure" means any structure, vehicle or place adapted for overnight accommodation of persons, or for carrying on business therein, whether or not a person is actually present.

"Property of Another" means a building or other property, whether real or personal, in which a person other than the offender has an interest which the offender has no authority to defeat or impair, even though the offender may also have an interest in the building or property.

If a building or structure is divided into separately occupied units, any unit not occupied by the offender is an occupied structure of another.

§ 100.2   *Causing or Risking Catastrophe*

(1) Causing Catastrophe. A person who causes a catastrophe by explosion, fire, flood, avalanche, collapse of building, release of poison gas, radioactive material or other harmful or destructive force or substance, or by any other means of causing potentially widespread injury or damage, commits a felony of the second degree if he does so purposely or knowingly, or a felony of the third degree if he does so recklessly.

(2) Risking Catastrophe. A person is guilty of a misdemeanor if he recklessly creates a risk of catastrophe in the employment of fire, explosives or other dangerous means listed in Subsection (1).

(3) Failure to Prevent Catastrophe. A person who knowingly or recklessly fails to take reasonable measures to mitigate a catastrophe commits a misdemeanor if:

(a) he knows that he is under an official, contractual or other legal duty to take such measures; or

(b) he did or assented to the act causing or threatening the catastrophe.

§ 100.3   *Criminal Mischief*

(1) Offense Defined. A person is guilty of criminal mischief if he:

(a) damages or alters any tangible real or personal property of another purposely, recklessly, or by negligence in the employment of fire, explosives, or other dangerous means listed in Section 100.2(1); or

(b) purposely or recklessly tampers with tangible property of another so as to endanger person(s) or property; or

(c) purposely or recklessly causes another to suffer pecuniary loss by deception or threat.

(2) Grading. Criminal mischief is a felony of the third degree if the actor purposely causes pecuniary loss in excess of $_____, or a substantial interruption or impairment of public communication, transportation, supply of water, gas or power, or other public service. It is a misdemeanor if the actor purposely causes pecuniary loss in excess of $_____ or a petty misdemeanor if he purposely or recklessly causes pecuniary loss in excess of $_____.

§ 100.4   *Possession of Explosive or Incendiary Materials or Devices*

A person is guilty of a felony of the third degree when he shall possess, manufacture or transport any incendiary or expslosive device or material with the intent to use or to provide such device or material to commit any offense described in 100.1(1), (2), and (3).

§ 100.5   *Attempted Arson*

A person is guilty of attempted arson; a felony of the third degree, if he places or distributes any flammable or combustible material or any gas, radioactive material, or other harmful or destructive material or substance, in an arrangement or preparation with the intent to eventually start a fire or cause an explosion, or to procure the start of a fire or explosion, with the purpose of wilfully and maliciously:

(a) destroying or damaging any building or structure of another whether occupied or not; or

(b) destroying or damaging any personal property of another having a value of $_____ or more; or

(c) placing any person in danger of life or bodily harm.

§ 100.6   *False Reports*

A person is guilty of a misdemeanor if he knowingly conveys or causes to be conveyed to any person false information concerning the placement of any incendiary or explosive device or any other destructive substance in any place where persons or property could be endangered.

THE NFPA SUGGESTED MODEL ARSON LAW

The Suggested *Model Arson Law* of the Fire Marshal's Section of the NFPA was the forerunner of the *Model Arson Penal Law,* and is also a basis for much of the modern legislation which has expanded and clarified the boundaries of the crime of arson.

• Essentially, this model arson law defines four categories of arson and, in addition, proposes suggested criminal sentences for each category.

• Under this model arson law, arson in the first degree is committed when a dwelling is willfully and maliciously burned, whether it is one's own or that of another, and whether it is occupied or vacant.

• Arson in the second degree designates as arson the burning of all types of buildings and structures other than dwellings.

• Arson in the third degree is defined as the intentional burning of the personal property of another if the property is valued at $25 or more.

• Arson in the fourth degree addresses attempted arson.

• In addition, this model arson law suggests that the intentional burning of any property for the purpose of defrauding the insurer, although not labeled arson, be classified as a felony.

The following is an extract from a presentation made by William C. Braun, Chief Special Agent of the National Board of Fire Underwriters during the Eighth Annual Seminar in Arson Detection and Investigation at the Public Safety Institute of Purdue University, West Lafayette, Indiana, in 1952, regarding the *Model Arson Law:*[21]

LEGAL ASPECTS OF ARSON

Arson is a peculiar crime. It is committed covertly and in secrecy, and consequently, the admissibility and sufficiency of evidence are problems to be solved. A conviction, if possible, is generally the result of an accumulation of numerous bits of circumstantial evidence. The prosecuting attorney, of necessity, is impelled to offer evidence bordering on the imaginary line which separates the admissible from the inadmissible.

Now for a moment let us consider the old common law crime of arson which consisted of the willful and malicious burning of the house or outhouse of another man. It could not be committed by burning an unoccupied dwelling. While the house had to be occupied, it was not necessary that the occupant be in the building at the time of the fire. The crime also included setting fire to any outhouses which were used in connection with the dwelling and which were located close enough so that the flames would endanger the occupants of the dwelling. The crime was against the security of the home and not against property and, therefore, the occupant could not commit the crime by burning the house in

---

[21]William C. Braun, "Legal Aspects of Arson." *The Journal of Criminal Law, Criminology, and Police Science,* Vol. 43, Northwestern School of Law, Chicago, 1952-1953.

which he lived. It is easy to see that arson at common law was confined to very narrow limits. Other burnings were punishable, not as arson, but as high misdemeanors.

To remedy the inadequacies of the common law, the legislatures of the various states gradually began to extend and broaden the definition of the crime, until today it covers all kinds of buildings and structures, even including personal property. Now the crime of arson includes the burning of one's own property, whereas, at common law, it was limited to the burning of the dwelling house occupied by another person.

While the legislative enactments in some states were adequate, in others they fell short. The *Model Arson Law* was designed not only to bring about uniformity but to correct the deficiencies of the common law and the statutory laws. As will be seen later, the *Model Arson Law* covers practically any intentional burning that can occur and makes it much easier than before to convict the arsonist. At the present time, all states are operating under this law, with the exception of Maine, Minnesota, New York, Oklahoma, Tennessee, Texas and Washington.

The first section of the *Model Arson Law* covers the intentional burning of any dwelling houses or outhouses regardless of whether it is occupied, unoccupied or vacant. It is immaterial whether it is his own property or the property of another person.

The second section of the *Model Arson Law* covers the intentional burning of any building or structure, of any class or character, except those covered by the first section. Here again it is immaterial whether it is his own property or the property of another.

The third section covers the intentional burning of personal property, of any class or character, provided the value of the property is twenty-five dollars or more and is the property of another person. You will note that this section is limited to the burning of personal property of another person. For instance, if I wished to burn some old clothing or furniture, I could do so without committing arson, provided I did not do it with the intention of defrauding an insurance company. However, if I burned the clothing or furniture of some other person of the value of twenty-five dollars or more, I would be committing arson.

The fourth section covers an attempt to burn any of the buildings, structures or personal property referred to in the preceding section. It also defines what shall constitute attempted arson by declaring that the placing or distributing of any flammable explosive or combustible material or substance or any device in any building or property that I have mentioned, in an arrangement or preparation with the intent to willfully and maliciously burn same, or to procure the burning, shall constitute an attempt. At common law, and under the statutes of the various states, it was no crime to prepare a building for a fire. An overt act, such as striking a match, was necessary. Under the *Model Arson Law,* however, no overt act is necessary; mere preparation is sufficient. This makes it possible to prosecute in many instances where formerly a prosecution was impossible. For instance, suppose A, B, and C, enter the basement of a store building during the night, bringing with them several cans containing gasoline and varnish. After mixing them, they spray the mixture on the ceiling and walls of the basement and place a candle, but before they can light it police officers rush in and arrest them. They can be convicted, under the *Model Arson Law,* of attempted arson.

The final section covers the intentional burning of any property mentioned in the other sections, which is insured against loss or damage by fire, and with the intent to injure or defraud the insurer. The crime committed under this section is not termed arson but is a felony. In a prosecution under this section it is not only necessary to prove that the fire was intentionally set, but also that the specific intent to injure or defraud the insurer was present.

Under the *Model Arson Law,* any person who aids, counsels or procures is equally guilty with the person who actually starts the fire. Principals and accessories are included, and therefore, we need not concern ourselves with the definition of principals or accessories or their implication as such. . . .

In the case of the *Commonwealth* v. *Flagg* Chief Justice Morton, citing numerous authorities, said: "It is an indictable offense at common law for one to counsel and solicit another to commit a felony or other aggravated offense, although the solicitation is of no effect, and the crime counselled is not in fact committed.

In this case Flagg had endeavored to persuade an acquaintance to burn a barn belonging to a third person, urging him to do so on two or three occasions, promising him money for doing so and, in fact, advancing a small sum on account. The person solicited, however, apparently took no action looking to the burning of the barn, but the conviction of Flagg was sustained.

SUMMARY

Criminal law is, basically, statutory law. Because the field of criminal law is concerned with crimes against society as a whole, the field is a broad and complex one. Conduct considered so utterly unacceptable in a society that the government steps in on behalf of the society it governs and represents is, in general, the sphere that criminal law addresses. Because of this, the field of criminal law represents a society from moral, social, philosophical, and psychological points of view. Although states are free to define what conduct is considered acceptable for their own jurisdictions, each state is subject to the constitutional guarantees provided to each and every citizen by the federal Constitution. And though federal and state statutes may vary in wording and technicalities, they are more nearly similar than they are different. In every state, the elements that constitute a crime and the definitions of those who are criminally liable are virtually the same, although statutory wording may differ.

Obviously, the crime of arson is of vital consideration to the fire service. Without an understanding of the elements that constitute a crime and a basic knowledge of criminal law, members of the fire service will be less than successful in fulfilling their most necessary role — that of promoting the general welfare.

In a September 1976 report of the National Fire Prevention and Control Administration titled "Arson: America's Malignant Crime," arson's growth was estimated as follows:[22]

> Urban fire departments estimate that as much as half of all fire losses in America's cities are from fires that are set on purpose. Far from the cities, about a quarter of all forest fires are similarly set on purpose. Fires of incendiary origin are ripping off America at a rate so high that the metropolitan inner-city areas of some of our major population centers are beginning to resemble London after World War II.
>
> These losses are all around us, and we are beginning to feel their impacts on the American economy which each year may total as much as $10 billion* through higher insurance premiums, higher prices for what is not burned, lost jobs, and higher taxes. Each year, also, as many as a thousand of us may lose our lives to arson.**

ACTIVITIES

1. Discuss the reasons why members of the fire service should be familiar with criminal law. Give examples that might apply.
2. Explain why criminal laws are necessary in a society.
3. How do common law and statutory law affect criminal law in this country?
4. Describe the classifications of crime, and give examples of each degree.
5. Describe the elements necessary in order for a crime to have been legally committed. Include in your description reasons why a crime might not legally be considered a crime.
6. List and describe the various types of crimes.
7. Compare common-law arson with statutory arson.
8. List and describe the elements of arson.
9. Discuss with your classmates whether the *Model Arson Penal Law* addresses all kinds of arson and other illegal fires. Give examples from your own experience.
10. Describe an arson situation of which you know or in which you have been involved. Describe this situation from a legal point of view.

BIOGRAPHY

Goldstein, Abraham S. and Goldstein, Joseph, eds., *Crime, Law, and Society*, The Free Press, New York, 1971.

---

[22]"Arson: America's Malignant Crime," September 1976, National Fire Prevention and Control Administration, Washington, DC, p. 5.

*Based on insurance industry estimates. The Insurance Services Office, an insurance statistical, advisory, and rating organization, estimates that the total actual fire loss due to arson in 1976 could exceed $4 billion. According to the American Insurance Association, incendiary fires currently account for 21 percent of the number of fire insurance claims and 40-50 percent of dollars lost to fire.

**Estimate from International Association of Arson Investigators.

Haskins, Jim, *Your Rights Past and Present,* Hawthorn Books, Inc., New York, 1975.

Inbau, Fred E. and Aspen, Marvin E., *Criminal Law for the Layman,* Chilton Book Company, Philadelphia, 1970.

Inbau, Fred E. *et al., Cases and Comments on Criminal Law,* The Foundation Press, Inc., Mineola, NY, 1973.

_____, *Cases and Comments on Criminal Procedure,* The Foundation Press, Mineola, NY, 1974.

Levin, Molly Apple, *Violence In Society,* Houghton Mifflin Company, Boston, 1975.

Loeb, Robert H., Jr., *Your Legal Rights as a Minor,* Franklin Watts, Inc., New York, 1974.

Paulsen, Monrad G., *The Problems of Juvenile Courts and the Rights Of Children,* American Law Institute, Philadelphia, 1975.

Wright, Gene and Marlo, John A., *The Police Officer and Criminal Justice,* McGraw-Hill Book Company, New York, 1970.

For an in-depth study of the topics presented in this chapter, the following are recommended:

Besharov, Douglas J., *Juvenile Justice Advocacy,* Practising Law Institute, New York, 1974.

Curtis, Arthur F., *A Treatise on the Law of Arson,* Dennis & Company, Buffalo, 1936.

Fox, Sanford J., *Cases and Materials on Modern Juvenile Justice,* West Publishing Company, St. Paul, MN, 1972.

Nedrud, Duane R., *The Criminal; Law,* L E Publishers Inc., Chicago, 1976.

Perkins, Rollin M., *Criminal Law,* The Foundation Press, Mineola, NY, 1969.

Uviller, H. Richard, *The Processes of Criminal Justice: Investigation,* West Publishing Co., St. Paul, MN, 1974.

Varon, Joseph A., *Searches, Seizures and Immunitites,* 2nd Ed., The Bobbs-Merrill Company, Inc., Indianapolis, 1974.

VI

CRIMINAL PROCEDURE

OVERVIEW

The laws and rules of criminal procedure control the process through which the government seeks to convict and punish a person or persons for a criminal offense. This process is concerned not only with the criminal trial itself, but also with the events surrounding it. That aspect of criminal procedure — governmental procedure prior to arrest — is of major concern to members of the fire service. Criminal procedure laws and rules are drawn from state and federal constitutions, statutes, and court rules and interpretations involving criminal procedure. These rules and laws address themselves to four main goals, which are: (1) to ensure the justice of the guilt-assessing process; (2) to protect the rights of the accused; (3) to promote economy of time and money in adjudication; (4) to promote efficiency in law enforcement and in the judicial process.

THE AMERICAN CRIMINAL JUSTICE SYSTEM

The criminal justice system in any society represents the standards of conduct a society deems necessary to protect the individuals in that society and, ultimately, to protect the society itself. Generally, a society's criminal justice system has three objectives: (1) to remove people who are considered dangerous to the society, (2) to deter similar antisocial behavior in others, and (3) to give the society the opportunity to change lawbreakers into law-abiding citizens. Criminal justice systems may be characterized by the balance effected between the demands of the society as a whole and the rights of the individual. For example, in some societies there may be little or no attention paid to an individual's rights. *The Challenge of Crime in a Free Society*, a publication prepared by the President's Commission on Law Enforcement and Administration of Justice, states that in America the rights of the individual are carefully guarded to the extent that:[1]

> . . . our system of justice deliberately sacrifices much in efficiency and even in effectiveness in order to preserve local autonomy and to protect the individual.

The Challenge of Crime in a Free Society describes America's criminal justice system as follows:[2]

> The system of criminal justice America uses to deal with those crimes it cannot prevent and those criminals it cannot deter is not a monolithic, or even a consistent, system. It was not designed or built in one piece at a time. Its philosophic core is that a person may be punished by the government if, and only if, it has been proved by an impartial and deliberate process that he has violated a specific law. Around that core layer upon layer of institutions and procedures, some carefully constructed and some improvised, some inspired by principle and some by expediency, have accumulated. Parts of the system — magistrate's courts, trial by jury, bail — are of great antiquity. Other parts — juvenile courts, probation and parole, professional policemen — are relatively new. The entire system represents an adaptation of the English common law to America's peculiar structure of government, which allows each local community to construct institutions that fill its special needs. Every village, town, county, city, and state has its own criminal justice system, and there is a federal one as well. All of them operate somewhat alike. No two of them operate precisely alike.

Figure 6.1 is a simplified version of a chart prepared by the President's Commission on Law Enforcement and Administration of Justice. This chart outlines the American criminal justice system and the many decision points in this system. On the chart, felonies, misdemeanors, petty offenses, and juvenile cases are shown as separate subsystems because they tend to follow different paths in the criminal justice system.

GOVERNMENTAL PROCEDURE PRIOR TO ARREST

As stated in the "Overview," the stage of criminal procedure of particular concern to members of the fire service is governmental procedure prior to arrest. This aspect of criminal procedure is concerned with ascertaining that a crime has been committed, and with gathering evidence for presentation *in court* to support the contention that a crime has been committed and that a certain party or parties is

---

[1] The President's Commission on Law Enforcement and Administration of Justice, *The Challenge of Crime in a Free Society*, Government Printing Office, Washington, DC, 1967, p. 7.
[2] *Ibid.*

responsible. Proof of the fact that a crime has been committed, the resulting conviction, and punishment of the party or parties responsible for the crime are dependent upon employing proper evidence-gathering procedures and safeguarding the constitutional rights of all involved.

Determining That a Crime Has Been Committed: A crime is a violation of the public peace and welfare that has been defined and made punishable by law. Statutes must clearly and carefully define the crime. A poorly drawn statute (and any convictions based on it) can be struck down in court for the vagueness in its wording.

In addition to definition by statute, a crime must involve an *act* as well as an *intent* to commit a crime, as described in the following excerpt from Rollin M. Perkins's *Criminal Law and Procedure*:[3]

> A crime is any social harm defined and made punishable by law. . . . this definition, or any other acceptable one, makes clear the necessity of an act for the existence of any crime. The intent with which a harmful act was done is a matter of special interest to the criminal law, but a wrongful intent which has no consequences in the external world — which exists only in the secret recesses of the mind — is not. Ordinarily such an intent would be known only to him who entertained it, but if he freely admits that such a thought was once in his mind, no crime has been established. There has been no "social harm." Few [persons] are so upright as to be able to exclude any criminal thought from entering the mind under any and all circumstances. The average law-abiding citizen is not one who never has a criminal intent, but one who never permits such a thought to rule his conduct.

The various degrees of act and intent (and also legal responsibility for one's acts) are virtually separate studies in themselves. The following excerpt regarding the appeal of a defendant in an arson case illustrates the ways that statutory wording may be argued in court. (The remainder of this decision will be presented later in this chapter.)

STATE v. ISENSEE

Supreme Court of North Dakota. Aug. 24, 1933.
249 N.W. 898

Syllabus by the Court.

1. Chapter 115, Laws of North Dakota 1929, entitled "An Act defining Arson, prescribing punishment for burning or attempting to burn buildings or other property, and burning of buildings or other property to defraud insurer, and

---

[3]Rollin M. Perkins, *Criminal Law and Procedure,* The Foundation Press, Inc., Mineola, NY, 1972, p. 546.

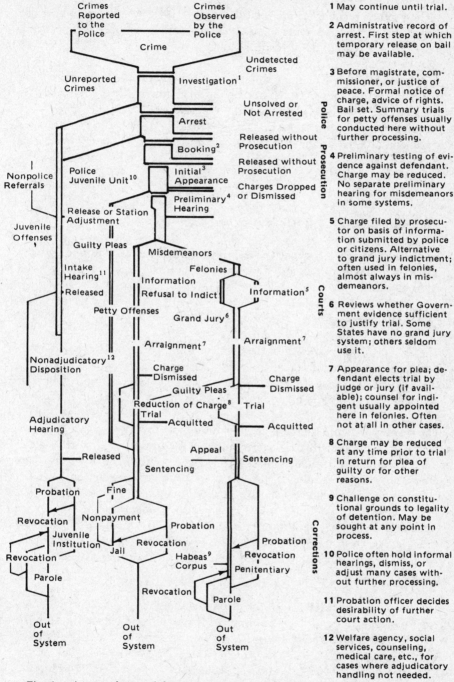

Fig. 6.1. A general view of the criminal justice system. (Adapted from the *Report by the President's Commission on Law Enforcement and Administration of Justice: The Challenge of Crime in a Free Society*, 1967, pp. 8-9)

repealing certain sections of the Compiled Laws of 1913 (relating to similar acts and offenses),'' does not embrace more than one subject in contravention of section 61 of the Constitution of North Dakota.

2. For reasons stated in the opinion, no error prejudicial to the substantial rights of the defendant was committed by the amendment of the information by striking out the word "or," and inserting in place thereof the word "and," and proceeding to trial without having the information again verified after such amendment, and having the defendant again arraigned upon the information as so amended. . . .

Appeal from District Court, Cass County; Kneeshaw, Judge.

Wesley L. Isensee was convicted of willfully setting fire to and burning personal property with intent to defraud the insurer (Chapter 115, Laws 1929), and he appeals. . . .

CHRISTIANSON, Judge.

The defendant was convicted of the crime of willfully setting fire to and burning personal property with intent to defraud the insurer in contravention of section 4, chapter 115, Laws of North Dakota 1929, and appeals from the judgment of conviction and from the order denying his motion for a new trial.

The first assignment of error is predicated upon the overruling of defendant's demurrer to the information. The real ground of the demurrer is that the statute under which the prosecution was had is violative of section 61 of the State Constitution, and that, consequently, the information did not state facts sufficient to constitute a public offense.

Section 61 of the State Constitution reads: "No bill shall embrace more than one subject, which shall be expressed in its title, but a bill which violates this provision shall be invalidated thereby only as to so much thereof as shall not be so expressed."

The title of the act under which the prosecution was had reads as follows: "An Act defining Arson, prescribing punishment for burning or attempting to burn buildings or other property, and burning of buildings or other property to defraud insurer, and repealing Sections 9849, 9850, 9851, 9852, 9853, 9854, 9855, 9856, 9857, 9858, 9859, 9860, 9861, 9862, 9863, 9864, 9865, 9866, 9867 of the Compiled Laws of 1913."

Section 1 of the act defines arson and prescribes the punishment therefor. Section 2 makes it an offense to willfully and maliciously set fire to, or burn, buildings other than a dwelling whether the property is that of the perpetrator or another. Section 3 makes it an offense for any person to willfully and maliciously set fire to, or cause to be burned, any barrack, crib, rick or stack of hay, corn, etc. Section 4 of the act (under which the defendant is charged) reads as follows: "Any person who willfully and with intent to injure or defraud the insurer sets fire to, or burns, or causes to be burned, or who aids, counsels, or procures the burning of any goods, wares, merchandise, or other chattels or personal property of any kind, whether the property of himself or of another, which shall at the time be insured by any person or corporation against loss or damage by fire, shall upon conviction thereof, be sentenced to the county jail not to exceed one year, or to the penitentiary not to exceed five years." (Section 4, c. 115, Laws of North Dakota, 1929.)

Section 5 makes it an offense for any person to willfully and maliciously attempt to set fire to, or burn, or aid, counsel, or procure the burning of, any of the buildings or property mentioned in the preceding sections. Section 6 makes it a crime to place or distribute any inflammable explosive material or substance or any device in any building or property mentioned in the foregoing sections in an arrangement or preparation with intent to eventually willfully and maliciously set fire to or burn the same.

Appellant contends that the statute embraces more than one subject, and, consequently, is violative of section 61 of the Constitution. It is argued that section 4 is a new statute with a different end and purpose than that of the other provisions in the act. It is said that the object and purpose of this statute was the protection of insurers and not the protection of property. It is, also, argued that the statute is violative of section 61 of the Constitution in this: that section 1 of the act alters the definition of arson and makes it an offense to set fire to or burn certain kinds of houses "regardless of habitation or possession"; that section 2 of the act creates a new and distinct offense; and the same argument is advanced as regards section 3. In our opinion, the contentions thus advanced cannot be sustained. We are aware of no reason why the Legislative Assembly may not in one statute enumerate the several acts which it seeks to inhibit so far as wrongful destruction of property by fire is concerned, and define the several acts inhibited as different offenses. It seems rather that it is desirable to place all inhibited acts of this nature in one legislative enactment. An act which seeks to deal with and prescribe appropriate penalties for the destruction of property by fire cannot, we think, be said to embrace more than one subject within the constitutional rule. Such statute has one main object and purpose in view: namely, to deal with and prescribe appropriate penalties for those who, by means of fire, commit, or attempt to commit, wrongful acts.

Constitutional provisions similar to section 61 are in force in many of the states in the Union. The cases that have arisen cover a multitude of subjects. *Corpus*

Juris states the general rule to be deduced from the several decisions thus: "All matters which are germane to and connected with the general subject of a statute may be included in its provisions without rendering it violative of a constitutional provision prohibiting a statute from embracing more than one subject, and a statute, no matter how comprehensive it may be or how numerous its provisions, complies with the constitutional requirement if a single main purpose is held in view and nothing is embraced in the act except what is naturally connected with and incidental to that purpose. Thus, if desired, the entire statutory law upon a subject may be incorporated in one statute; and civil and criminal provisions may be incorporated in the same act." 59 C. J. pp. 800, 802.

This court has, also, considered the purpose and effect of section 61 of the Constitution in many cases. *[Citations omitted.]*

In *Chaffee* v. *Farmers' Co-operative Elevator Co., supra,* this court said:

"The requirement that the subject shall be expressed in the title of the act relates to substance and not to form. The requirement is addressed to the subject, and not to the details of the act. None of the provisions of a statute will be held unconstitutional when they are related, directly or indirectly, to the same subject having natural connection, and are not foreign to the subject expressed in the title. As very frequently expressed by the courts, any provisions that are germane to the subject expressed in the title may properly be included in the act. *[Citation omitted.]* The Constitution does not contemplate that the title shall employ anything more than general terms, leading to an inquiry into the body of the act. It does not contemplate that the title shall be an index, or furnish an abstract of the contents of the act. Generality or comprehensiveness of the title is not objectionable, provided the title is not misleading and is sufficient to give notice of the general subject of the proposed legislation and the interests likely to be affected. The choice of language is a matter within the legislative discretion. And if the title chosen fairly indicates the general subject of the act and is comprehensive enough in its scope reasonably to cover all the provisions thereof and is not calculated to mislead either the Legislature or the public, it is a sufficient compliance with the constitutional requirement, even though it be not the most appropriate that could have been selected. *[Citations omitted.]*

"The title must state the subject of the act for the purpose of information to members of the Legislature and the public, while the bill is going through the forms of enactment. It is not required that the title should be exact or couched in the most precise language. 'It is sufficient if the language used in the title, on a fair construction, indicates the purpose of the Legislature to legislate according to the constitutional provision; so that making every reasonable intendment in favor of the act it may be said that the subject or object of the law is expressed in the title.' Sutherland, St. Constr. (2nd Ed.) § 121. As was said by this court in *State ex rel. Erickson* v. *Burr [citations omitted]:* 'The title should be liberally, and not technically, construed. The construction should be reasonable. Conflict with the constitutional provision must appear clear and palpable, and, in case of doubt as to whether the subject is expressed in the title, the law will be upheld. Titles should be construed in connection with the law with the view of remedying the evil intended to be obviated by the Constitution makers — that of preventing fraud or surprise upon the members of the Legislature and the people by introducing provisions into the law independent of or foreign to the subject expressed in the title. If the subjects in the law are germane or reasonably connected with the subject expressed in the title, the constitutional requirement is sufficiently met.' '' *[Citations omitted.]*

In *Great Northern Ry. Co.* v. *Duncan, supra,* this court held:

"The requirements of section 61 of the Constitution to the effect that no bill shall embrace more than one subject which shall be expressed in its title, are not violated where the title fairly indicates the general scope of a bill designed to accomplish a single object." Syllabus, par. 7.

"Where a legislative bill, considered in the light of facts generally known, is designed to accomplish one general object, and this is fairly indicated in the title, the title is not multifarious within the inhibition of section 61, although it indicates that several subjects related to the general object are embodied in the bill." Syllabus, par. 8. *[Citations omitted.]*

By applying these principles in the case at bar we inevitably reach the conclusion that the statute in question here does not contravene the provisions of section 61 of the Constitution. That statute deals with one general subject — the wrongful destruction of property by fire, and the several provisions thereof are germane to, and reasonably connected with, the general subject, which is, we think, fairly expressed in the title of the act.

The error next assigned is predicated upon the allowance of an amendment of the information. The facts on which this assignment of error is predicated are as follows: The state was permitted to and did file an amended information. The charging part of this information read in part as follows: "That at said time and place the said defendant did willfully and feloniously and with intent to injure, damage, or defraud the insurers thereof, set fire to and burn, *or* cause to be burned," etc. The defendant at once filed a demurrer to the amended information. Defendant's attorney stated that the demurrer then filed was practically the same as the one that had been filed to the original information. Immediately thereafter the trial judge stated that he would allow the state to amend the new information by inserting the word "and" in place of the word "or," where the latter word is italicized in the foregoing quotation from the information. The Attorney General thereupon moved that the information be so amended. The motion was granted. No objection was made by or on behalf of the defendant whatsoever to the proposed amendment or the allowance thereof. The court thereupon inquired whether defendant was ready to plead. The defendant at once entered a plea of not guilty, whereupon the court inquired whether the parties were ready for trial. Defendant's counsel answered in the affirmative, whereupon a jury was impaneled and the cause proceeded to trial. Appellant asserts that the procedure thus adopted was erroneous; that the information as amended should have been again verified and the defendant arraigned thereon.

We are agreed that the record discloses no prejudicial error. *[Citations omitted.]* The information in question here was verified by the state's attorney of Cass County. He was present and participated actively in the trial, and, so far as the record discloses, he was the person who made the interlineation. The amendment here did not substantially change the charge made against the defendant. The crime charged remained in its essentials the same after as before the amendment. There was no objection by the defendant or his counsel to what took place, or suggestion that the defendant desired to have a copy of the information as amended delivered to him and to be again arraigned. The interlineation was made before he entered a plea. He pleaded not guilty to the information as amended and put in issue the material averments thereof. That issue was tried without even a suggestion that the amendment was improperly made. There is not the slightest reason to believe that the defendant was prejudiced in the slightest degree by the procedure adopted or that he or his counsel at the time desired any other procedure to be had.

Our laws provide that an information may be amended in matter of form without a new verification at any time before the defendant pleads, without leave of the court; and that it may be amended at any time thereafter or during the trial as to all matters of form at the discretion of the court when the same can be done without prejudice to the rights of the defendant. C. L. 1913, § 10633.

Identifying the Crime of Arson: Every fire must be presumed to be accidental until or unless evidence can be obtained to indicate that the fire was caused through criminal intent. This evidence can then be used in a court as the basis for proving arson. John Kennedy's book, *Fire and Arson Investigation*, details three requirements for a conviction in an arson prosecution:[4]

> 1. Proof that a burning occurred. Smoke discoloration is not sufficient. There must be a burning.
> 2. Proof that this burning resulted from a criminal action. This involves the *corpus delicti*, or proof of the incendiary origin of the fire or that the fire was willful and malicious.
> 3. Proof that the person charged caused the criminal burning and . . . is legally responsible.

---

[4]John Kennedy, *Fire and Arson Investigation*, Investigations Institute, Chicago, 1962, pp. 68-70.

In an article titled "How to Identify Fire Causes," Chief Investigator C. W. Stickney of the Portland, Oregon, Fire Department comments as follows:[5]

> "*Corpus Delicti*" is the body of the crime; and in arson this means establishing that a building or property did actually burn and that the burning was the result of the willful and criminal act of some person. We must show *intent,* or that the action was intended to result in the willful and wanton or malicious destruction of property by fire. This is the foundation of all arson prosecutions; and is the first step taken in a criminal proceeding since if it cannot be shown that the action was *intended* to destroy property maliciously, then a crime does not exist and the case would stop right there. Building this foundation for arson prosecution is usually the sole responsibility of the fire service, and dischaging this responsibility can only be accomplished by a thorough investigation and understanding of the causes of all fires.

Gathering Evidence: Evidence gathered and produced in court must prove that:
- A crime has been committed.
- The accused has committed the crime in question.

To be properly admitted in court, evidence must conform to a large body of procedural and evidential rules drawn to ensure a fair and proper trial for the accused. According to criminal procedure, the burden is on the government to prove the matters placed before the court.

In *Coffin* v. *United States,* Justice White points out the safeguards in the criminal judicial procedure and the ultimate goal of evidence presented at trial.

COFFIN v. UNITED STATES

Supreme Court of the United States, 1895.
156 U.S. 432, 15 S.Ct. 394.

[One of the instructions to the jury that was requested by the defendants, but refused by the court, read as follows:]

> "The law presumes that persons charged with crime are innocent until they are proven by competent evidence to be guilty. To the benefit of this presumption the defendants are all entitled, and this presumption stands as their sufficient protection unless it has been removed by evidence proving their guilt beyond a reasonable doubt."

Although the court refused to give this charge, it yet instructed the jury as follows: "Before you can find any one of the defendants guilty you must be satisfied of his guilt as charged in some of the counts of the indictment beyond a

---

[5]C. W. Stickney, "How to Identify Fire Causes," *Arson: Some Problems and Solutions,* NFPA, Boston, 1976, p. 14.

reasonable doubt." And, again: "You may find the defendants guilty on all the counts of the indictment if you are satisfied that beyond a reasonable doubt the evidence justifies it." And, finally, stating the matter more fully, it said: "To justify you in returning a verdict of guilty, the evidence must be of such a character as to satisfy your judgment to the exclusion of every reasonable doubt. If, therefore, you can reconcile the evidence with any reasonable hypothesis consistent with the defendants' innocence, it is your duty to do so, and in that case find the defendants not guilty. And if, after weighing all the proofs and looking only to the proofs, you impartially and honestly entertain the belief that the defendants may be innocent of the offenses charged against them, they are entitled to the benefit of that doubt and you should acquit them. It is not meant by this that the proof should establish their guilt to an absolute certainty, but merely that you should not convict unless, from all the evidence, you believe the defendants are guilty beyond a reasonable doubt. Speculative notions or possibilities resting upon mere conjecture, not arising or deducible from the proof, or the want of it, should not be confounded with a reasonable doubt. A doubt suggested by the ingenuity of counsel, or by your own ingenuity, not legitimately warranted by the evidence or the want of it, or one born of a merciful inclination to permit the defendants to escape the penalty of the law, or one prompted by sympathy for them or those connected with them, is not what is meant by a reasonable doubt. A reasonable doubt, as that term is employed in the administration of the criminal law, is an honest, substantial misgiving, generated by the proof or the want of it. It is such a state of the proof as fails to convince your judgment and conscience, and satisfy your reason of the guilt of the accused. If the whole evidence, when carefully examined, weighed, compared, and considered, produces in your minds a settled conviction or belief of the defendants' guilt — such an abiding conviction as you would be willing to act upon in the most weighty and important affairs of your own life — you may be said to be free from any reasonable doubt, and should find a verdict in accordance with that conviction or belief."

The fact, then, is that whilst the court refused to instruct as to the presumption of innocence, it instructed fully on the subject of reasonable doubt.

The principle that there is a presumption of innocence in favor of the accused is the undoubted law, axiomatic and elementary, and its enforcement lies at the foundation of the administration of our criminal law. . . .

. . . Greenleaf traces this presumption to Deuteronomy, and quotes Mascardus De Probationibus to show that it was substantially embodied in the laws of Sparta and Athens. Whether Greenleaf is correct or not in this view, there can be no question that the Roman law was pervaded with the results of this maxim of criminal administration, as the following extracts show:

"Let all accusers understand that they are not to prefer charges unless they can be proven by proper witnesses or by conclusive documents, or by circumstantial evidence which amounts to indubitable proof and is clearer than day." Code, L. IV, T. XX, 1, 1.25.

"The noble *(divus)* Trajan wrote to Julius Frontonus that no man should be condemned on a criminal charge in his absence, because it was better to let the crime of a guilty person go unpunished than to condemn the innocent." Dig. L. XLVIII, Tit. 19, 1.5.

"In all cases of doubt, the most merciful construction of facts should be preferred." Dig. L. L. Tit. XVII, 1.56.

"In criminal cases the milder construction shall always be preserved." Dig. L. L. Tit. XVII, 1.55, s. 2.

"In cases of doubt it is no less just than it is safe to adopt the milder construction." Dig. L. L. Tit. XVII, 1.92, s. 1.

. . . The rule thus found in the Roman law was, along with many other fundamental and humane maxims of that system, preserved for mankind by the canon law. Exactly when this presumption was in precise words stated to be a part of the common law is involved in doubt. The writer of an able article in the *North American Review*, January, 1851, tracing the genesis of the principle, says that no express mention of the presumption of innocence can be found in the books of the common law earlier than the date of McNally's Evidence (1802). Whether this statement is correct is a matter of no moment, for there can be no doubt that, if the principle had not found formal expression in the common law writers at an earlier date, yet the practice which flowed from it has existed in the common law from the earliest time. . . .

. . . It is well settled that there is no error in refusing to give a correct charge precisely as requested, provided the instruction actually given fairly covers and includes the instruction asked. The contention here is that, inasmuch as the charge given by the court on the subject of reasonable doubt substantially embodied the statement of the presumption of innocence, therefore the court was justified in refusing in terms to mention the latter. This presents the question whether the charge that there cannot be a conviction unless the proof shows guilt beyond a reasonable doubt, so entirely embodies the statement of presumption of innocence as to justify the court in refusing, when requested, to inform the jury concerning the latter. The authorities upon this question are few and unsatisfactory. . . .

. . . This confusion makes it necessary to consider the distinction between the presumption of innocence and reasonable doubt as if it were an original question. In order to determine whether the two are the equivalent of each other, we must first ascertain, with accuracy, in what each consists. Now the presumption of innocence is a conclusion drawn by the law in favor of the citizen, by virtue whereof, when brought to trial upon a criminal charge, he must be acquitted, unless he is proven to be guilty. In other words, this presumption is an instrument of proof created by the law in favor of one accused, whereby his innocence is established until sufficient evidence is introduced to overcome the proof which the law has created. This presumption on the one hand, supplemented by any other evidence he may adduce, and the evidence against him on the other, constitute the elements from which the legal conclusion of his guilt or innocence is to be drawn. Greenleaf thus states the doctrine: "As men do not generally violate the penal code, the law presumes every man innocent; but some men do transgress it, and therefore evidence is received to repel this presumption. This legal presumption of innocence is to be regarded by the jury, in every case, *as matter of evidence, to the benefit of which the party is entitled.*" 1 Greenl.Ev. § 34.

Wills on Circumstantial Evidence says: "In the investigation and estimate of criminatory evidence there is an antecedent *prima facie* presumption in favor of the innocence of the party accused, grounded in reason and justice, not less than in humanity, and recognized in the judicial practice of all civilized nations; which presumption must prevail until it be destroyed by such an overpowering amount of legal evidence of guilt as is calculated to produce the opposite belief." Best on Presumptions declares the presumption of innocence to be a *"presumptio juris."* The same view is taken in the *Criminal Law Magazine* for January, 1889, to which we have already referred. It says: "This presumption is in the nature of evidence in his favor *[i.e., in favor of the accused]*, and a knowledge of it should be communicated to the jury. Accordingly, it is the duty of the judge in all jurisdictions, when requested, and in some when not requested, to explain it to the jury in his charge. The usual formula in which this doctrine is expressed, is that every man is presumed to be innocent until his guilt is proved beyond a reasonable doubt. The accused is entitled, if he so requests it . . . to have this rule of law expounded to the jury in this or in some equivalent form of expression." . . .

. . . The fact that the presumption of innocence is recognized as a presumption of law and is characterized by the civilians as a *presumptio juris* demonstrates that it is evidence in favor of the accused. For in all systems of law legal presumptions are treated as evidence giving rise to resulting proof to the full extent of their legal efficacy.

Concluding, then, that the presumption of innocence is evidence in favor of the accused introduced by the law in his behalf, let us consider what is "reasonable doubt." It is of necessity the condition of mind produced by the proof resulting from the evidence in the cause. It is the result of the proof, not the proof itself; whereas the presumption of innocence is one of the instruments of proof, going to bring about the proof, from which reasonable doubt arises; thus one is a cause, the other an effect. To say that the one is the equivalent of the other is therefore to say that legal evidence can be excluded from the jury, and that such exclusion may be cured by instructing them correctly in regard to the method by which they are required to reach their conclusion upon the proof actually before them. In other words, that the exclusion of an important element of proof can be justified by correctly instructing as to the proof admitted. The evolution of the principle of the presumption of innocence and its resultant, the doctrine of reasonable doubt, makes more apparent the correctness of these views, and indicates the necessity of enforcing the one, in order that the other may continue to exist. Whilst the Romans and the Medievalists taught that wherever doubt existed in a criminal case, acquittal must follow, the expounders of the common law, in their devotion to human liberty and individual rights, traced this doctrine of doubt to its true origin, the presumption of innocence, and rested it upon this enduring basis. The inevitable tendency to obscure the results of a truth, when the truth itself is forgotten or ignored, admonishes that the protection of so vital and fundamental a principle as the presumption of innocence be not denied, when requested, to any one accused of crime. The importance of the distinction between the two is peculiarly emphasized here, for, after having declined to instruct the jury as to the presumption of innocence, the court said: "If after weighing all the proofs and looking only to the proofs, you impartially and honestly entertain the belief," *etc.* Whether thus confining them to "the proofs" and only to the proofs would have been error if the jury had been instructed that the presumption of innocence was a part of the legal proof need not be considered, since it is clear that the failure to instruct them in regard to it excluded from their minds a portion of the proof created by law, and which they were bound to consider. "The proofs and the proofs only" confined them to those matters which were admitted to their consideration by the court, and among these elements of proof the court expressly refused to include the presumption of innocence, to which the accused was entitled, and the benefit whereof both the court and the jury were bound to extend him. . . .

. . . Judgment reversed and case remanded with directions to grant a new trial.

Gathering the type of evidence that will prove that arson has been committed and secure a conviction is no easy task. In "How to Identify Fire Causes," Chief Investigator C. W. Stickney warns:[6]

> Good investigation requires many more man-hours than is generally realized; and even though an extensive fire may be put out in a few hours, the investigation could take several hundred man-hours for such items as examining the premises for point of origin; reconstructing and analyzing materials in this area (glass, wood, rugs, plastics, chemicals, etc.); examining electrical and mechanical equipment, such as wiring, appliances, clocks, and thermal controls; searching out and interviewing the person who turned in the alarm, occupants, owners, adjusters, and many other possible witnesses — not to mention the photographs, sketches, preserving of evidence, report writing, and court appearances.

---

[6]*Ibid.*, p. 13.

JUDICIAL PROCEDURE FROM ARREST THROUGH TRIAL

Procedures used in criminal cases are similar in all states, as well as in the federal court system.

Procedure Between Arrest and Trial: A person who is arrested must be taken without unnecessary delay before a judge or magistrate. If the charge is a misdemeanor, the judge or magistrate may conduct the trial at this point unless the accused demands a jury trial or requests a delay for a valid reason. If the crime is a felony, the judge conducts a preliminary hearing at this point to determine if there are reasonable grounds for believing that the person who is accused has committed the crime.

If the judge or magistrate determines that there is no "probable cause" or reasonable belief that the accused has committed the crime, the case is dismissed and the defendant is released. If, on the other hand, the judge or magistrate determines that probable cause does exist, the accused is "held over" or "bound over" for the grand jury. If the offense is a bailable offense, the accused may post bond and be released until the grand jury has acted. If the offense is not a bailable offense, the accused is held in jail.

If the accused is not charged with a specific crime or brought before a judge or magistrate without unnecessary delay, the judge may be petitioned through a writ of *habeus corpus,* which requests a release on bail or that a specific charge be made, or both.

Many states mandate that a grand jury consider the evidence against any person accused of a felony. The Constitution requires grand jury consideration in all federal cases. Grand jury hearings do not constitute a trial, for only the state's evidence is presented. In states that require grand jury hearings, no felony case can be brought to trial unless the grand jury has returned an indictment. This is true even though the judge or magistrate has determined at the preliminary hearing that probable cause does exist.

Following the indictment, the accused is arraigned before a judge who is empowered to try felony cases. At the arraignment, the accused is advised of the criminal charge that is being brought. The accused then pleads "guilty" or "not guilty." If the accused pleads "guilty," the judge can pass sentence immediately. If the plea is "not guilty," a date is then set for the trial. The accused may also plead *nolo contendere,* which has the same effect as pleading "guilty" except that the admission cannot be used as evidence in any subsequent actions.[7]

The Criminal Trial: Potential jurors are usually selected by lottery from a list of qualified residents. Jurors are chosen from a group of about 20 to 30 prospective jurors who have been called. Each potential juror is questioned first by the plaintiff's attorney, then by the defendant's attorney. The purpose for this procedure is to give each side an opportunity to select jurors whom the attorneys feel will be

---

[7]Fred E. Inbau and Marvin E. Aspen, *Criminal Law for the Layman,* Chilton Book Company, Philadelphia, 1970.

fair and unprejudiced when hearing the evidence to be presented at trial. Both attorneys are permitted to challenge the selection of a prospective juror, either "for cause" or without giving a reason, although this latter type of challenge can only be used a certain number of times.

After the jury has been selected, there are several steps that must be taken. In his book titled *Fire Service and the Law,* Charles W. Bahme outlines these steps as follows:[8]

(1) The trial proceeds as follows:
   (a) Prosecutor opens his presentation and produces witnesses and evidence to prove his case.
   (b) Defendant's attorney cross-examines the complaining witnesses. The prosecutor may then reexamine them if he desires.
   (c) Attorney for the accused then puts on the case in defense.
   (d) Prosecutor cross-examines defendant's witnesses, after which the defendant's counsel may reexamine them.
   (e) Defendant's attorney and the prosecutor give their closing arguments.
(2) Judge charges the jury and instructs them as to the law which should be applied in the case; he may also comment on the evidence and the witnesses.
(3) After considering all the evidence and applying the law contained in the instructions given them, the jury announces its verdict. All of the jurors must agree or else there can be no conviction; if even only one juror dissents, there is a "hung jury," and the accused must be released or be tried again. If a verdict of "guilty" is found, the judge gives the sentence.
(4) The accused may appeal the verdict or ask for a new trial on various grounds which tend to show that he did not have a fair trial. The state, however, cannot appeal if the accused is found "not guilty."

EVIDENCE

Because the crime of arson is usually committed in secrecy, witnesses to the actual commission of the crime are most often unavailable. Proof of the *corpus delicti* and intent as well as identification of the perpetrator or perpetrators of the crime must be carefully constructed through available evidence. In addition, the evidence found must comply with certain rules which, in judicial procedure, are extensive. Evidence in criminal cases must be relevant, that is, directly tending to prove or disprove a contention in court. Irrelevant evidence or evidence that is immaterial to the matter at hand will not be admitted by the court. The evidence presented by the state in an arson case must also establish beyond a reasonable doubt all three elements required to secure an arson conviction:

1. That burning occurred.
2. That the burning was the result of a criminal act.
3. That the person or persons accused committed the crime.

---

[8]Charles W. Bahme, *Fire Service and the Law,* NFPA, Boston, 1976, p. 22.

The following case study concerns the burning of an automobile with intent to injure the insurer of the vehicle. It demonstrates the need for all of the preceding elements to be present in order to secure a conviction for arson.

Joyce Annette JENKINS
v.
COMMONWEALTH of Virginia

Supreme Court of Virginia.
April 23, 1976.
223 S.E.2d 880

Defendant was convicted before the Circuit Court, Gloucester County, John E. DeHardit [Judge], of burning her automobile with intent to injure insurer of such vehicle, and writ of error was granted. The Supreme Court, Harrison, J., held that evidence that analysis, made weeks after fire, of floor matting found in rubbish of burned vehicle disclosed substance consistent with residue from fuel oils, was insufficient to sustain conviction.

Reversed and remanded.

HARRISON, Justice.

We consider the sufficiency of the evidence to sustain the conviction by the court below, sitting without a jury, of Joyce Annette Jenkins of burning her automobile with intent to injure the insurer of the vehicle, in violation of Code §18.1-85, then in effect and applicable.

The vehicle involved, a 1973 Chevrolet Monte Carlo, was purchased by the defendant in February, 1973, for $5,200 and had been operated approximately 13,000 miles. On the afternoon of April 2, 1974, Mrs. Jenkins had gone to the Riverside Hospital in Newport News for therapy. Thereafter, she met a friend for dinner and shopping. En route to take her friend home, she thought she smelled "rubber burning." Defendant said: "The motor was racing so fast that . . . you couldn't hardly stop at the stop lights. . . ." A service station attendant adjusted the carburetor and "idled it down some." Defendant testified that on the way to her home in Gloucester the car "started doing the same thing again, the motor started racing and I could smell rubber. . . ." She said that about a mile before reaching her home in a relatively uninhabited section where highway number 1104 is bordered on both sides by marsh lands, "the car all of a sudden, it went up, black smoke underneath the hood and it was coming all inside." Mrs. Jenkins said that she had gotten out of the car and had been standing in the road for four or five minutes when William James Green came along in his truck. He picked her up and the two went to Green's home, called the fire department, and then, together with her husband and daughter, returned to the scene of the fire.

Green testified that when he came upon Mrs. Jenkins in the road and asked the trouble, she said, "my motor is on fire or something, smoke is coming out from under the hood." Green noticed that the fire was coming from the motor and from underneath the hood. He was very positive in his testimony about the absence of fire inside the car when they left the scene to call the fire department. He also observed that the horn had started a continuous blowing. Green said that by the time they got back the car was "nearly burnt up."

Robert F. Berry, Jr., assistant fire chief of the Abington Volunteer Fire Department in Bena, Virginia, testified that the department responded to the fire about 10:45 p.m. He said that when he arrived the car was completely engulfed in flames. He described the fire as "a very intense fire" that completely "gutted out" the car. He stated that the fire was most intense in the engine compartment and the front of the vehicle, and that the rear was the least affected. He observed a number of bystanders and a state trooper at the scene.

The loss was reported to the Pennsylvania National Mutual Casualty Insurance Company, which had insured the vehicle against loss or damage by fire. The claim was investigated by William E. Kahn, Sr., a company claims investigator. The company settled with Mrs. Jenkins for $3,425. She was dissatisfied with the settlement, testifying that she could have sold the car for $1,000 more than the amount she received. Kahn was asked if he found anything "to show there was any arson involved . . . in your investigation," and he replied, "No, we found nothing."

James H. Jessup, district arson supervisor for the state, was requested by the local authorities to investigate the fire. He went to Gloucester County to make the investigation on April 16, 1974. At that time the car was located in a lot behind a garage operated by James E. Harris, to which point it had been towed. It appears from the evidence that the car was then in the same condition it was in following the fire.

Jessup testified that he did a salvage examination of the Jenkins automobile to determine what had happened and where the fire had originated. He said all of the electrical system was "just about completely consumed"; that the power brake unit, which was on the left side directly in front of the driver, had been completely consumed by the fire; that the only thing he found left on the engine was the block. He testified that there was gas in the tank and that the line leading from the fuel filter was intact, but that the carburetor had been completely consumed and the nut which held the fuel line onto the carburetor had been melted.

In the course of making his investigation, Jessup said that he dug out padding from under all four floor mats. These samples were sent to the Forensic Laboratory of the Commonwealth of Virginia where they were examined by Mrs. Mary Jane Burton, a forensic scientist. The material which Jessup had retrieved from the vehicle was sent to the laboratory in four separate containers, thereby keeping separate the material taken from under each mat. However, at the laboratory the matting in each of the containers was placed inside a big pot of water and boiled together.

Mrs. Burton testified that the boiling process removes any materials that are lighter than water, like petroleum products, and that they collect on top; that various tests and the analysis indicated that the "material was consistent with residue from fuel oils"; and that "it is consistent with the majority of materials that were in this boiling range that we call Number 2 fuel oils and this would include the diesel fuel, diesel oils and some kerosines." She said the material "was not consistent with whole fuel oil, fuel oils as they would be bought or purchased."

Mr. Jessup was called as a witness and thereafter, over objection, the Commonwealth's Attorney was permitted to recall him three times. It is clear from Jessup's testimony that he believed the fire started inside the car and was of incendiary origin. It was his opinion that the defendant used a petroleum-based product and "struck a match to it."

The testimony here discloses none of the indicia commonly associated with arson cases. Mrs. Jenkins was employed, was not financially embarrassed and was not in dire need of funds. She testified without contradiction that she owned a $60,000 home and a new trailer. Repossession of the car was not threatened or involved. There was no lien on the vehicle. There is no evidence that the defendant was dissatisfied with her automobile; that prior to the night of the fire it had given her any trouble; or that there was any reason why she would want to rid herself of it.

Mrs. Jenkins fully accounted for her actions on the afternoon and evening of April 2, 1974. She gave the names of two individuals who had observed some malfunction of the car on the night of April 2, 1974, one being a service station attendant. The Commonwealth did not refute any of this testimony.

No witness testified to smelling any petroleum fumes on or about the defendant, or at the site of the fire. The state trooper who was present at the scene was not called as a witness. Both the defendant and Green testified that the smoke and fire were coming from the engine and under the hood of the vehicle, not from the inside, when they left the scene to call the fire department.

In essence, the Commonwealth relies upon the fact that an analysis made weeks after the fire of the floor matting found in the rubbish of the burned vehicle disclosed a substance "consistent with residue from fuel oils." It is a matter of common knowledge that an automobile does not function without the presence of petroleum-type products — gasoline, motor oil, grease, transmission fluid, etc. The intensity of this fire, whether accidental or of incendiary origin, was necessarily augmented by the petroleum products that were incident to the operation of the car. There is no evidence that Mrs. Jenkins ever purchased any petroleum products of any kind. Assuming that the matting under one or all of the floor mats did contain a residue from fuel oil, this fact alone does not establish the incendiary nature of or connect the defendant with the fire. There remains the possibility that the substance found in the matting had its origin from products used incident to the operation of the car, or, in any event, from a source other than Mrs. Jenkins.

The Commonwealth also attaches some significance to testimony that the defendant appeared to be "very calm" during the fire and that one time she was observed seated in a truck with a gentleman to whom she turned "and smiled. . . ."

While there may be a suspicion of guilt in this case, the evidence does not exclude all reasonable conclusions inconsistent with that of defendant's guilt. It does not overcome the presumption of innocence to which she is entitled; rebut the presumption that the fire was of accidental origin; or prove her guilt beyond a reasonable doubt. The evidence is not sufficient to establish that the defendant feloniously set fire to her automobile with intent to injure the insurer of the vehicle.

The judgment of the lower court is reversed, and the case is remanded for a new trial if the Commonwealth be so advised.

Reversed and remanded.

***

DIRECT AND REAL EVIDENCE

Direct evidence proves a fact directly without the necessity of any inference or presumption based on the evidence presented. For example, a witness who sees a person set a fire can provide direct evidence.

Real evidence, on the other hand, represents tangible evidence such as a piece of burned building to prove that burning did indeed occur, articles used to start a fire, or proof that a fire was started by unnatural means through the presentation of evidence such as kerosine-soaked earth taken from beneath a burned building or, as described in the preceding case, the charred remains of padding that had been placed under the car floor mats. In a courtroom, all of these factors would be considered real evidence and would be vital in proving the matter before the court if, in fact, relevant to the issue at hand.

CIRCUMSTANTIAL EVIDENCE

It is well recognized by the courts that, by their very nature, crimes of arson can seldom be successfully prosecuted on direct evidence alone and that reliance must often be placed on circumstantial evidence. In the following excerpt concerning

Fig. 6.2. A chemist at the state police laboratory in Little Falls, NJ, injects an air sampling into a gas-liquid chromatograph. The sampling was taken from fire remains brought to the laboratory in the paint can shown. Beside the chromatograph is a portable sniffer used in arson investigations.

circumstancial evidence from "Legal Aspects of Arson," author William C. Braun states:[9]

> Circumstantial evidence . . . takes a more round-about route to prove the fact in issue by uncovering facts here and there; and all these facts and circumstances, when considered together, must point to only one conclusion — the guilt of the accused beyond a reasonable doubt. The courts have held that a strong set of circumstances are just as convincing as direct evidence, and, in some cases, even more so.

A good rule of thumb is: If the circumstantial evidence presented by the state may point to another person equally as well as to the accused, then the circumstantial evidence is not enough to convict the accused.

In court, arson may be proved by direct or circumstantial evidence. Most often, however, proof of the crime of arson and the person or persons who committed it must be demonstrated through circumstantial evidence. This evidence must be able to withstand judicial procedures in court. The decision regarding the appeal

---

[9]William C. Braun, "Legal Aspects of Arson," *The Journal of Criminal Law, Criminology, and Police Science,* Vol. 43.

of Wesley L. Isensee, which was presented earlier in this chapter, was reached primarily because of circumstantial evidence presented in court and the defendant's right to refute it. As you read the remainder of this decision, be prepared to discuss the difficulties in presenting a case based on circumstantial evidence.

STATE v. ISENSEE

Supreme Court of North Dakota
Aug. 24, 1933.
249 N.W. 898

Syllabus by the Court

3. Ordinarily it is relevant to put in evidence any circumstance which tends to make the proposition at issue either more or less probable.

4. In a criminal action it is competent for the accused to prove any fact which may tend to explain or answer any incriminating evidence against him.

5. Where one party has adduced evidence tending to establish a fact adverse to the other, and such other party offers evidence in reply, the test of relevancy is whether the evidence offered tends to cut down, eliminate, explain, or obviate the force of the evidence that the former party adduced, or the fact which such evidence tended to establish.

6. For reasons stated in the opinion, it is *held* that it was error to exclude certain evidence offered by the defendant, which had a tendency to explain or answer incriminating evidence against him.

7. In the instant case the defendant is accused of willfully setting fire to and burning personal property with intent to defraud the insurer. On cross-examination of the accused the state was permitted to show that property belonging to a corporation in which defendant was a stockholder had sustained a loss by fire more than four years before the alleged offense in controversy was committed; and that another corporation of which defendant was president had sustained a loss by fire, partly covered by insurance, more than eight years before the alleged offense is claimed to have been committed. It is *held,* for reasons stated in the opinion, that the cross-examination was improper, and that the testimony elicited was not relevant to prove the guilt of the accused, nor was the cross-examination permissive for the purposes of impeachment, or as affecting the credibility of the defendant as a witness.

The next assignments of error are predicated upon rulings in the admission and exclusion of evidence. A proper understanding of these assignments requires a brief statement of facts.

The defendant was engaged in the automobile business in the city of Fargo, known as the Isensee Motor Company. Both new and secondhand cars were handled. The used cars were principally those which had been taken in on trade. The business was conducted in a two-story building with a full basement beneath. Defendant's business occupied the ground floor and the basement, and the second story was occupied as a rooming house. The defendant also had another place of business in the city of Jamestown.

The principal witness for the prosecution was one Hart who, at the time of the fire, and for some time prior thereto, had been in the employment of the defendant as manager of the service and parts department.

The fire took place July 23rd shortly after 11 o'clock p.m. Hart testified to several conversations with the defendant before the fire wherein defendant discussed with him plans which he then had or was formulating looking toward the procuring of additional insurance and the proposed destruction of the various cars and other property by fire so as to enable him to collect the insurance. One conversation, according to Hart, occurred some three or four weeks before the fire on a trip which the defendant and Hart made from Fargo to a village some miles distant therefrom. According to Hart's testimony, on their return from this trip they went into the basement and the defendant then pointed out the combustible condition of the building, stating in effect that it would be practically impossible for a fire department to prevent complete destruction if a fire got started. According to Hart's testimony, the defendant at that time even discussed with him a proposed new business to be established after the fire and with the proceeds received from the insurance policies. Hart further testified that shortly thereafter on that same day the defendant proposed that he (the defendant) write up a contract for the sale to Hart of a certain Oldsmobile coupé in order that the defendant might obtain some immediate cash from the finance company. It was further intimated that the finance company would be paid off from the proceeds of the anticipated fire. It is undisputed that such contract was made and that Hart gave a note as evidence of a part of the purchase price, and that the contract was negotiated with the finance company. The defendant Isensee, however, testified that the deal was exactly what it purported to be, and that it was not in any sense a subterfuge; that Hart had sought repeatedly to purchase a car, but that it had not been deemed good business to sell him one. Hart further testified that Isensee at other times thereafter approached him and discussed the anticipated fire with him and the plan he had to defraud the insurers. Hart also testified that in one of these conversations the defendant offered him a commission if he (Hart) would set the fire; that upon Hart's refusing to enter into any such agreement the defendant said to him: "Bill, I had a fire in Pine River, and if another fire happened to me while I was in town, it would make it appear to the insurance companies that I had something to do with it and it would make it hard to collect insurance." Hart further testified that the defendant told him he was going to Minneapolis and directed Hart to have certain used cars then in storage in the city of Hillsboro brought down and put in the basement, all in anticipation of the contemplated fire. The evidence discloses that about noon on July 20th the defendant left Fargo and went first to Duluth and then to Minneapolis. He took with him some men, including one Skiem, who drove new cars that were received at Duluth from Duluth to Fargo. The defendant, Isensee, thereafter, on July 21st, went to Minneapolis. While there he went to see one Supornick, a private adjuster. The evidence discloses that Supornick later represented the defendant in the adjustment of the fire loss. The defendant returned to Fargo in the afternoon of July 22nd. While the defendant had been away the witness Hart had brought some four or five cars down from Hillsboro, also one or two cars from another garage in the city of Fargo. Hart testified that before leaving for Minneapolis on July 20th the defendant had instructed him to order fifty gallons of gasoline and put it in a tank in the basement in which kerosine was usually kept; that before leaving for Minneapolis defendant said to Hart: "I will get the stuff to do this job with while I am gone, it has got to be done right away; it is getting to the point where I can't even pay my help the first of the month." Hart further testified that on the morning of July 23rd the defendant also told him (Hart) that "he was going to pull the job right away as he had to have some money and that was the only way he could get it." Hart testified that at noon on July 23rd the defendant instructed Hart to have another employee wash the motor of a certain car with gas and air in the basement so as to create an odor of gasoline there.

The evidence discloses that on the evening of July 23rd the defendant, together with one Murphy, a representative of General Motors Acceptance Corporation, went into the country to repossess a car and returned to the defendant's place of business about 9:30 in the evening. Hart testified that defendant directed him to stay at the garage, or in the vicinity, as he wanted to see him later in the evening. The testimony discloses that Murphy and the defendant stayed in defendant's office and discussed various business matters until about 11 o'clock. About this time Mr. Skiem, who had driven one of the cars from Duluth, came to the garage and called for the defendant.

Skiem testified that he had just come in from Duluth; that when he came into the garage the defendant and Murphy were still engaged in the defendant's private office; that immediately upon Murphy's leaving he (Skiem) went in and reported to the defendant that he had brought the car from Duluth and wanted him to check up the car and see if it was O. K. and also take him (Skiem) home at the same time; that thereupon they went out of the building and entered the car and started to drive it, with Isensee driving. The evidence discloses that they drove around the block, returned, and stopped again in front of the garage; that Isensee went out, entered the building, stayed there for a short time, came out and drove away with the car.

Hart testified that while the defendant and Skiem were out in front of the garage he went into the basement, turned on the light and stepped in a puddle of gasoline at the foot of the stairway; that the doors of the cars in the basement were open and the upholstery of one of the cars saturated with gasoline; that a pile of tires and inner tubes near the foot of the stairway were wet with gasoline. Hart further testified that he thereafter came back to the first floor and sat down at a desk in the parts room; that the defendant Isensee came in, went through the parts room and down in the basement; that in a few minutes he came up and went out through the parts room; that on the way out he had the following conversation with Hart: "*Q.* What did you say? *A.* 'Looks like you got everything fixed around here' I said. He grinned or smiled and said 'Yes, I have; I want to see you in a little bit, you wait here until I get back.'" Hart further testified that in a few minutes thereafter an explosion occurred, followed by the fire; that he was partly stunned, but succeeded in escaping from the building; that after the fire had commenced the defendant returned to the scene; that he then said to Hart: "If anybody asks where you were at, tell them you were blown out of the building"; — and that he gave him the same instruction the following morning; that he further suggested that Hart go back to his former home in North Carolina, and furnished him with a car and money to enable him to make the trip.

The evidence discloses without dispute that Hart did go back to North Carolina, but the defendant denies that he suggested to Hart that he do so or that he furnished him with a car. He claims, on the other hand, that Hart obtained the car surreptitiously and by deceit from the manager of his business at Jamestown.

Hart also testified that after he had been in North Carolina for only a few days the defendant telegraphed him to return and also telegraphed him money for the expenses back to Fargo; that after he got back the defendant told him that he had sent for him because the adjuster "was raising Cain" because Hart had left town, and that the defendant wanted him back so that a quick settlement could be effected with the insurance companies. The defendant Isensee denied much of the testimony given by Hart as regards the incidents connected with the trip to North Carolina. Defendant also offered certain testimony concerning this which was excluded. The particular facts as regards the North Carolina trip will be dealt with in more detail in connection with the assignment of error based upon the exclusion of such testimony.

One of the assignments of error is predicated upon the court's ruling in refusing to permit the defendant to show why he called on the adjuster Supornick in Minneapolis on July 21st. The defendant sought to show, both by his own testimony and the testimony of Supornick, the reason for and the nature of the business transacted with Supornick on that occasion. He was denied the opportunity to introduce this testimony, apparently on the ground that this testimony would be in the nature of a self-serving declaration. It is clear from the record that the prosecution laid considerable stress upon this circumstance; namely, that the defendant made the trip to Minneapolis and had a conference or conversation with Supornick, the private adjuster who afterwards represented him in adjusting the fire loss. We are of the opinion that the defendant was entitled to introduce this evidence.

Ordinarily, "it is relevant to put in evidence any circumstance which tends to make the proposition at issue either more or less improbable. . . . Whatever is a condition, either of the existence or nonexistence of a relevant hypothesis, may be thus shown. But no circumstance is relevant which does not make more or less probable the proposition at issue." *[Citations omitted.]*

In a case of this kind, where one party has adduced evidence tending to establish a fact adverse to the other, and such other party seeks to introduce evidence in reply, the test of relevancy is whether the evidence offered tends to cut down, eliminate, explain, or obviate the force of the evidence that the other party adduced or the fact which such evidence tended to establish. *[Citations omitted.]*

In the brief on this appeal counsel for the prosecution calls attention to the defendant's trip to Minneapolis and his meeting there with Supornick as evidence of defendant's scheme to defraud the insurers, and as a circumstance tending to establish his guilt of the offense charged. Evidence of the trip to Minneapolis and of the meeting between the defendant and Supornick having been introduced by the prosecution, it was competent for the accused to prove any fact which reasonably tended to explain and answer the incriminating evidence that the prosecution had brought forth against him. 5 C. J. 578: 1 Wigmore on Evidence (2d Ed.) § 34.

Reference has been made to Hart's purchase of an automobile (which transaction, according to Hart, was merely a device to enable the defendant to obtain money from the finance company), and to Hart's trip to his former home in North Carolina. The defendant sought to introduce evidence to the effect that the transaction relating to the purchase of the automobile by Hart, among other matters, included a proposed sale by Hart to the defendant of certain furniture (the purchase price of the furniture to be applied on Hart's indebtedness for the car); that during their negotiations Hart had pointed out to the defendant in a certain catalogue the kind and class of his furniture; that check marks identifying the furniture were placed in the catalogue at the time, and which catalogue was produced upon the trial. The defendant further sought to show that the reason he wired money to Hart was that he had learned that Hart had made the trip to North Carolina in a car belonging to the defendant, which Hart had obtained from the manager of the defendant's establishment at Jamestown; that Hart had informed the defendant, from North Carolina, that the car was broken down, that he needed money to procure necessary repairs, and that the money that defendant wired was to enable Hart to procure the necessary repairs and enable him to bring the car back to North Dakota.

We are of the opinion that the defendant was entitled to introduce such evidence. The evidence relating to the furniture was a circumstance in a transaction that, according to the theory of the prosecution, was to some extent entwined with defendant's alleged fraudulent scheme. The evidence relating to defendant's reasons for wiring the money had a tendency to overcome the impression created by Hart's testimony.

The undisputed evidence is to the effect that the defendant on the night of the fire left his place of business with one Skiem who had brought one of the new cars from Duluth. Witnesses both for the state and for the defense testified to this effect. It is, also, undisputed that after the defendant and Skiem had driven around the block they returned and stopped in front of defendant's place of business; that the defendant then went into the building, where he remained for a short time; that on coming out he and Skiem left in the car with defendant driving, and that later, when notified of the fire, defendant was at his home.

The street in front of defendant's place of business runs north and south. The defendant sought to show that, after he and Skiem drove away the first time, and when the defendant started to drive in a northerly direction, Skiem informed him that he no longer lived where he formerly had lived, but that he had moved, and stated as his then place of residence a location south from the defendant's place of business and in the same general locality and direction from defendant's place of business that defendant himself resided.

Much of this proffered evidence was excluded, and the exclusion is assigned as error. What has been said as regards the admissibility of the evidence offered by the defendant concerning the meeting between the defendant and Supornick is equally applicable here. The incident of defendant's return to the garage before he took Mr. Skiem to his home, the drive around the block, the return to and defendant's entry into his place of business, are matters upon which the prosecution apparently laid considerable stress. The theory of the prosecution is that the defendant set the fire when he went into the building after he returned from the trip around the block. Unexplained, the short trip, apparently with no objective, was, to say the least, somewhat suspicious. It might readily be inferred that this aimless trip was somewhat of a ruse. In our opinion the defendant was entitled to introduce evidence tending to explain the reason for the apparently aimless trip and why he returned to the building as he did. In short, he is entitled to introduce any evidence that will tend to overcome the adverse probative effect of the incident if it were left unexplained. The defendant will, of course, not be permitted to introduce proof as to anything that was said or done during the course of the trip, except such as is directly connected with and has some reasonable tendency to explain the occurrence of the act in question.

After defendant had introduced his evidence and defendant's counsel had announced that the defendant rested his case, counsel for the prosecution asked that the defendant be recalled for further cross-examination. The request was granted, and the defendant was subjected to additional cross-examination. Error is assigned upon the rulings made during such cross-examination. The cross-examination related to matters not alluded to in the direct examination or in the former cross-examination. In response to questions propounded by counsel for the prosecution the defendant stated that at some time in the past he had had some connection with a corporation engaged in business at Pine River, MN; also that his brother had had some connection with this corporation, but that the defendant had severed all connection with it some four years before the fire in question here occurred. The defendant further stated, in answer to questions propounded by counsel for the prosecution, that he at one time was president of a corporation known as Isensee Auto Company, Inc., which operated a garage business at Hawley, MN. The defendant was thereupon required to testify further on such cross-examination, over appropriate objection, that the corporation at Pine River had sustained a loss by fire, also that the corporation at Hawley, MN, had sustained a loss by fire; that there was a large fire loss and some insurance coverage. The fire at Hawley, MN, occurred some eight years before the fire in question here.

In our opinion it was error prejudicial to the substantial rights of the defendant to compel him to give, or to admit, this testimony.

Obviously evidence of the fires at Pine River and Hawley was not admissible as tending to establish defendant's guilt of the offense with which he is charged in this case. The fires at Pine River and Hawley were wholly disconnected from, and had no relation whatever to, the fire in question here. Evidence regarding these fires could in no sense tend to establish a motive, or identify the perpetrator, or tend to show any general scheme or design. There is not even the slightest proof that the fires were incendiary in origin, or that the defendant had any connection with them. The authorities are all agreed that, in the circumstances here, evidence of the other fires was not admissible as tending to establish plaintiff's guilt. *[Citations omitted.]*

Neither was the cross-examination permissive on the theory that the testimony sought to be elicited, and which was elicited, had a tendency to impeach, or affect the credibility of, the accused. It is true the defendant, by becoming a witness, waived his constitutional privilege and rendered himself subject to the same rules of cross-examination that apply to other witnesses *[citations omitted]*; but the questions propounded to the defendant here would have been improper for purposes of impeachment, or as affecting the credibility, if they had been propounded to a witness other than the accused. They were equally improper when propounded to him. They did not even relate to any acts of the defendant. They did not even go to the extent of stating that defendant had been accused of any wrongdoing in connection with the fires *[citations omitted]*, yet the questions and the answers returned thereto left room for the play of imagination and suspicion in the deliberations of the jury. The cross-examination was improper. The questions propounded, and the testimony which they called for and produced, were not allowable for the purpose of impeachment, or as affecting credibility. *[Citations omitted.]*

The appellant, also, contends that the evidence is insufficient to sustain the verdict. This contention is, we think, without merit. If the testimony of the witness Hart is true — and his testimony is not without substantial corroboration — there can be little doubt but that the defendant deliberately planned the destruction of the property in question here by fire for the purpose of defrauding the insurer. There is also ample evidence from which the jury could find that defendant had the opportunity, as well as the will and the motive, to set the fire, and that he did set it.

The remaining assignments of error fall into three classes: some are without merit; as regards others the record is such as to preclude a review on this appeal; the others are not likely to arise upon another trial.

As indicated, we are of the opinion that the errors committed in the exclusion and admission of evidence were prejudicial to the substantial rights of the defendant and prevented him from having a new trial. This necessitates a reversal of the judgment and a new trial. Accordingly it is ordered that the judgment appealed from be reversed, and the cause remanded for a new trial.

NUESSLE, C. J., and BURR, BIRDZELL, and BURKE, JJ., concur.

EVIDENCE OF ARSON

Members of the fire service are vital evidence-gathering agents. The following recommendations by William Braun apply to all those connected with fire investigations:[10]

> It is the duty of the investigator to gather the evidence and present it to the prosecuting attorney, whose duty it then becomes to institute whatever criminal prosecution the facts may warrant. Every investigator should become familiar with state laws pertaining to arson and related crimes. Investigators should be able to distinguish between the various offenses and the ingredients of each in order to know which law has been violated and what evidence to look for.
> In some states there is a fire marshal's law and, when operating under that law, the investigator has more power and authority than any other peace officer. In order to discharge duties effectively, investigators should not only be familiar with the law itself, but also with any court decisions that may have been rendered.

---

[10]*Ibid.*

The success or failure of an arson investigation depends not only on how competent the arson investigator is, but also on the degree of cooperation that exists between the investigator and the police and fire services, prosecuting attorneys, and others with whom the investigator is obliged to work. Unfortunately, in many instances, this cooperation is lacking, possibly due to envy, jealousy or other reasons; such a situation might result in losing a case. The investigation of an arson case is not an easy one by any means and the doing it made more difficult when there is not the necessary cooperation. It is only when the spirit of friendliness and hearty cooperation prevails that we can make an honest, all-out effort to curb arson.

Determining Fire by Arson: Intent is the determining factor of arson fire. Without intent, a fire is classified as accidental. In some fires, however, intent can be established without the fire being determined to be arson. For example, a fire that is set in a homeowner's fireplace which, through human negligence, burns down the house, is not an arson fire. The deliberate *setting* of a fire involves no act of arson if the intent to make the fire rage beyond control is not present. However, if there is no clear evidence that the fire was accidental, the fire investigator should view the fire as the result of arson.

Responsibility for the Investigation: In many states and in all Canadian provinces, the chief responsibility for the investigation of arson fires rests with the state or provincial fire marshal. In addition to investigation of the fire, the state fire marshal also has the authority to summon persons to an inquest, start proceedings to compel testimony under oath, and punish persons in contempt.

In the absence of a state fire marshal, police and fire departments investigate the fire and collect evidence to be presented to a prosecuting attorney. Specialists in arson investigation (such as arson squads and laboratory technicians) operate in cooperation with police and fire departments. In some states, the head of the fire department or a municipal fire marshal may be granted powers equal to those of the state fire marshal.

Details of the Investigation: The NFPA's *Fire Protection Handbook* recommends the following points as essential to the investigation of incendiary fires, the preparation of evidence from those fires, and the securing of convictions:[11]

It must be established that the fire in question actually occurred. This may be done through witnesses, fire records, photographs, sketches, and similar evidence.

A description of the building in which the fire occurred must be given, together with a statement that gives an accurate picture of the fire.

It must be indicated that the fire was caused by criminal design. This may be done by positive or circumstantial evidence. Confessions alone are not enough to establish criminal origin; they must be supported by corroborative facts such as the presence of gasoline, kerosine, or other flammables, or explosives, by the presence of more than one fire, by obstructions deliberately placed to impede fire

[11]*Fire Protection Handbook,* 13th Ed., NFPA, Boston, 1969, p. 1-34.

fighting operations, by the removal of valuables just prior to the fire, etc. . . .
. . . It must be proved that the fire did not occur from accidental causes, except
where incendiary origin is so clearly established that further proof is unnecessary.
This may be done by eliminating the possible sources of accidental origin.

Establishment of a motive greatly strengthens other evidence. For example,
facts may be presented to show the financial straits of the defendant, or motive
may be suggested by proving the existence of a desire to move or to break a lease
on the part of the person who had the fire.

Responsibility for the fire must be directly connected with an individual. This
may be done by a confession that is suitably corroborated. Evidence of the circumstances of the fire may also be used to establish the guilt of an individual.
Possession of the only available keys to a locked building, for example, constitutes
evidence of the sort which may be used to connect a given individual with the fire.

Any fire of a suspicious nature should be carefully outlined in a report on the
fire by the fire department. The local police or the state fire marshal's office
should then be called in to assist in a further investigation. To ensure a successful
arson prosecution, the district attorney should also be notified for guidance.

The fire investigator may be able to determine the *modus operandi* of the arsonist from the investigation, which may prove valuable to tracing the arsonist and
producing information for a trial. For example, an absence of breaking and entering can place suspicion on persons who have access to and familiarity with a
building; wanton destruction that accompanies a fire might point to an individual
seeking revenge on the property owner. This information, when obtained, is of
great help to the entire fire investigation and to the prosecution of an arsonist.

CRIMINAL SEARCH AND SEIZURE

The Fourth Amendment of the Constitution of the United States provides
that:[12]

The right of the people to be secure in their persons, houses, papers, and effects, against unreasonable searches and seizures shall not be violated, and no
Warrants shall issue, but upon probable cause, supported by oath or affirmation,
and particularly describing the place to be searched, and the person or things to
be seized.

Originally this amendment, as well as the other amendments in the Bill of
Rights, was interpreted by the courts to apply to the federal government alone. In
recent years, however, the Supreme Court has interpreted this provision of the
Fourth Amendment to apply to criminal procedures in states as well, based on the
provision in the Fourteenth Amendment that states:[13]

---

[12]U. S., *Constitution,* Amend. 4.
[13]*Ibid.,* Amend. 14.

... nor shall any State deprive any person of life, liberty, or property, without due process of law. ...

Courts have repeatedly refused to allow the admission of evidence — no matter how powerful or convincing — obtained in violation of the defendant's constitutional rights as a result of an illegal search or seizure. The most obvious violation of the rules regarding criminal search and seizure is action taken without a search warrant when one is required. Improper warrants also invalidate evidence or the legality of an arrest.

Necessary procedures involving search and seizure go beyond the necessity of a proper warrant, as the decision in the following case demonstrates. As you read the case, consider what elements are necessary to effect proper search and seizure of a premise:

STATE v. Steven P. SLEZAK

Supreme Court of Rhode Island.
Jan. 30, 1976.
350 A.2d 605

Defendant was convicted in the Superior Court, Washington County, Carrellas, [Judge], for the possession of cannabis, and defendant appealed. The Supreme Court, Doris, J., held that even though police officers were in possession of a valid search warrant when they entered defendant's premises, where the police officers entered without knocking on the door and announcing themselves and the purpose of their presence and where there was no evidence that persons within the house attempted to dispose of the marijuana, the entry of the police was unreasonable and illegal, and thus the subsequent search was unlawful.

Reversed and remitted.

OPINION

DORIS, Justice.

Steven P. Slezak was indicted on May 18, 1972, for the possession of cannabis in violation of G.L.1956 (1968 Re-enactment) §§ 21-28-3 and 21-28-31. The defendant filed a pretrial motion to suppress evidence which was based on a so-called "no-knock" entry pursuant to a valid search warrant. After a denial of the motion to suppress, trial was held before a Superior Court justice sitting without a jury. The defendant was found guilty and placed on probation for two years. The defendant has appealed to this court.

The record indicates that on May 18, 1972, State Police Detectives Leon Blanchette, Richard Sullivan, and other police officers participated in a narcotics raid at premises located on Post Road in North Kingstown. The premises were occupied by defendant and several other persons to whom Slezak had rented rooms. The police had a valid warrant to search the North Kingstown premises.

Detective Leon Blanchette testified that he and Detective Richard Sullivan entered the back porch on the premises and observed a screen door which was closed but unlocked. He opened the screen door and discovered that the inner door, the entrance to the kitchen, was open. He then saw Slezak who he recognized as a person who formerly had lived in his neighborhood. Blanchette stated that Slezak knew that he was a State Police officer involved in drug investigations. He then stated that as he was crossing the threshold into the kitchen he identified himself and gave Slezak a copy of the search warrant and further identified himself and Detective Sullivan. Blanchette stated that after entering onto the porch, opening the screen door, and finding the inner door open and because he was in plain view of defendant, it would have been senseless to knock. He further stated that the information in possession of the police was that the drugs would be located in the bathroom in position to be discarded quickly, and also behind bricks in the fireplace as well as in individual rooms.

The defendant, Slezak, testified that he recognized Blanchette and Sullivan when they entered the kitchen, but that as they were coming in the door, he didn't know what was happening. The record indicates that the police found marijuana in a teapot in the kitchen, a few feet from where defendant was situated. The defendant told police that the teapot contained marijuana and that they would find a "little stuff" in the house. It is not disputed that a valid search warrant was issued to search the premises rented by defendant.

The defendant argues that since the police officers, although in possession of a valid search warrant, entered the premises without actually knocking on the door and announcing themselves and the purpose of their presence, such an entry was illegal and thereby invalidated the subsequent search which resulted in the discovery of marijuana.

It is well settled that in Rhode Island we follow the common law rule that an officer must first knock and announce his identity and purpose and wait a reasonable period of time before he may break and enter into the premises to be searched. *[Citations omitted.]*

It is equally well settled that, under exigent circumstances, the rule is subject to certain qualifications and exceptions. An officer need not knock and announce when doing so will lead to destruction of the evidence or increase the peril to the officer's safety, where persons on the premises will escape, and finally, where the facts make it evident that the officer's purpose be known to those against whom the search is directed. *[Citation omitted.]*

The question to be decided is whether or not the circumstances are sufficient to bring the case within the exception set forth in *Carufel* and *Johnson*. *[Citations omitted.]*

Detective Blanchette testified that he entered without knocking, stating that it would have been senseless to knock after opening the screen door and observing that the inside door was open, and also because the police were in possession of intelligence information that the drugs would probably be located in bathrooms in positions to be discarded quickly.

The defendant argues that the mere fact that a suspect may flush drugs down a toilet is insufficient to authorize a so-called "no-knock" entry. He also points out that the court confirmed this theory in *State* v. *Carufel*. *[Citation omitted.]* Compliance with the announcement requirements of the common law is excused not because of the particular crime involved, but because of the particular circumstances of the case giving rise to a reasonable belief that immediate action is necessary to prevent the destruction of physical evidence. *People* v. *De Santiago*. *[Citations omitted.]* Thus, a no-knock search may be sustained where the police, on the basis of previously obtained information, supported by facts occurring at the scene, are aware that at the time they approach a particular premises to effectuate entry, they are confronted with an emergency situation. *People* v. *Dumas*. *[Citations omitted.]* Here, apart from the information given Blanchette, the record is devoid of any evidence that persons within the house attempted to dispose of the marijuana. The case at bar differs from *State* v. *Johnson [citations omitted]*, in that in that case there was movement within Johnson's residence which justified the officers' belief that heroin was being destroyed.

Detective Blanchette testified that as he came across the kitchen threshold he saw defendant and said, "This is Detective Sullivan. I think you know what we're here for, Steve. We're going to advise you of your constitutional rights. Why don't you have a seat in the kitchen?" The state, arguing that there was no need to knock and announce, argues, however, that by this statement Detective Blanchette did go through the formality of identifying himself and the purpose of his presence on the premises. We do not agree since the testimony of Detective Blanchette does not show any clear announcement of the purpose for which he was on the premises. We find nothing in the record to justify a conclusion that the officers announced their purpose, nor do the facts clearly show that the purpose of the officers' presence on the premises was known to defendant. There is nothing in the record to justify the police officers' failure to comply with the common law requirement of knocking before entry as announced by this court in *State* v. *Carufel* and *State* v. *Johnson*. *[Citations omitted.]*

When balancing an individual's right to privacy and the public's interest in the prosecution of crime and the apprehension of criminals, we are of the opinion that the action of the police in the circumstances of this case was unreasonable. The entry by the police was illegal thereby invalidating the subsequent search and the evidence obtained should have been suppressed. The trial justice erred in denying defendant's motion to suppress since the evidence was the result of an illegal search.

The defendant's appeal is sustained, the judgment appealed from is reversed and the case is remitted to the Superior Court.

THE "EXCLUSIONARY RULE"

The court-developed rule that requires the rejection of evidence obtained by an *unreasonable* search and seizure is called the "Exclusionary Rule." If an illegal search or seizure is considered an unreasonable one, it falls under the court's exclusionary rule. The effect this can have on members of the fire service is demonstrated in the following decision:

STATE of Indiana, Appellant,
v.
Loran O. BUXTON, Appellee

Supreme Court of Indiana.
March 20, 1958.
148 N.E.2d 547

Prosecution for arson against owner of restaurant which was damaged by fire. The Circuit Court, Scott County, S. Morris Wilson, Special Judge, sustained objection to admission of evidence by state, owner was acquitted, and state appealed on reserved question of law only. The Supreme Court, Achor, [Judge], held that, where investigation of cause of fire in restaurant by deputy state fire marshal, member of National Board of Fire Underwriters, and state trooper was not made at time of fire and was made without a search warrant, hot plate, pile of torn newspapers and gunny sack soaked in fuel oil, which were discovered at place where fire had originated and which were seized as evidence, were inadmissible.

Order accordingly.

ACHOR, Judge.

In September, 1955, a restaurant building owned by appellee, located near the town of Blocher, Indiana, caught fire causing the interior to be badly burned. On September 7, 1955, Howard Boegaholtz, a deputy state fire marshal, and Robert Campbell, a member of the National Board of Fire Underwriters, investigated the fire. The two men arrived at approximately 10 o'clock a.m. and entered through an unlocked door. During the investigation appellee Buxton arrived at the scene.

He talked with the men for a few minutes. He neither consented nor objected to the investigation which continued after his departure.

The investigation indicated that the fire had originated in the utility room at a hole in the floor 10 to 12 inches in diameter. Boegaholtz and Campbell cleared the debris from the hole and found no defective wiring or other evidence that the fire was the result of accident; neither did they find evidence which pointed to an incendiary origin. At this point Boegaholtz went to Scottsburg, a distance of approximately eight miles, to call in a state police photographer at Seymour. No attempt was made to obtain a search warrant and none was had any time during the investigation.

Later Boegaholtz, Campbell and state trooper Loy, the photographer, met at the scene. While Loy was taking a picture of the hole he discovered a hot plate at the bottom. Further examination revealed a cord for the hot plate, a pile of torn newspapers and a gunny sack soaked in fuel oil. These materials were seized as evidence.

Upon the strength of this evidence Buxton was arrested and placed on trial for arson. At the trial the state attempted to introduce its Exhibit No. 5, which included the hot plate and its cord, the oil-soaked gunny sack and the torn newspapers. Appellee had previously made a motion to suppress the evidence and objected on the ground that it was obtained in violation of Art. 1, § 11 of the Indiana Constitution, which prohibits unreasonable search and seizure. The court sustained the objection and Buxton was acquitted. The state appeals to this court on a reserved question of law only, claiming that under the Fire Marshal Act no search warrant was necessary and therefore that the court erred in sustaining the objection of appellee to the admission of the above evidence.

The general question then which this court must determine is whether the evidence so obtained was taken unlawfully in violation of appellee's constitutional rights under Art. 1, § 11, *supra,* which provides as follows:

> "The right of the people to be secure in their persons, houses, papers, and effects, against unreasonable search, or seizure, shall not be violated; and no warrant shall issue, but upon probable cause, supported by oath or affirmation, and particularly describing the place to be searched, and the person or thing to be seized."

Incidental to the above major question the following specific questions must be determined. (1) Whereas the Fire Marshal Act provides that the fire marshal or his deputies ". . . may, at all reasonable hours enter any building, property or premises . . . for the purpose of making an inspection or investigation which, under the provisions of this act . . . he or they may deem necessary to be made." § 20-808, Burns' 1950 Repl. (Acts 1927, ch. 115, § 5, p. 298). Does said provision purport to authorize searches and seizures *without a search warrant* and the subsequent use of the evidence so obtained in a criminal prosecution for arson against the owner of the property? (2) If so, do this and similar provisions of the Fire Marshal Act make the Act unconstitutional?

The right against unreasonable search and seizure is a right guaranteed by governments of free peoples as distinguished from the power formerly exercised by English kings and by the dictators of police states today. Since the advent of the writ of assistance issued in England under the statute of 12 Chas. II (1672), allowing the king's messengers to enter any and all places and search and seize papers and evidence of any kind, such infringement upon liberty was and has been continually fought. The early decision of Lord Camden in *Entick* v. *Carrington* (1765), 19 Howell's State Trials 1029, held these general writs to be invalid and laid the basis for the Fourth Amendment. Thus the doctrine "Every man's home is his castle" was founded. *[Footnote omitted.]*

Our courts have consistently maintained the position that evidence obtained through unreasonable search and seizure may not be used in evidence in a criminal case. . . . *[Citations omitted.]*

> . . . Evidence obtained as a result of an unreasonable search and seizure in violation of the constitutional prohibition, whether it be the instruments used to commit the crime, or oral evidence of what was found or seen in such unlawful search — is incompetent against the accused, and a conviction based thereon ought to be reversed. *[Citations omitted.]*

The problem which we must resolve is that of harmonizing the constitutional right against unreasonable search and seizure with the rights of people collectively to life, liberty, *safety* and the pursuit of happiness as guaranteed by our state and federal constitutions. Art. 1, § 1 of the Constitution of the State of Indiana, among other things, provides as follows: ". . . that all power is inherent in the People; and that all free governments are, and of right ought to be, founded on their authority, and instituted for their peace, *safety,* and *well-being.* . . . *(Our italics.)*

What kind of legislation may be passed for the safety of its people without violating other sections of the Constitution? Admittedly the safety of the people is the first law of the land and will prevail as against private rights provided by the Constitution. *[Citations omitted.]* But, to be justified, such encroachment upon private rights must be reasonable and necessary.

Admittedly the authority which the fire marshal or his deputies had, if any, to enter said premises and to make inspections and investigations is by virtue of and pursuant to the provisions of what is known as the Fire Marshal Law of the State of Indiana. Relevant parts of such legislative enactments deemed to have important bearing on this case are as follows:

§ 20-802, Burns' 1950 Repl.:

"It shall be the duty of the state fire marshal to enforce all laws of the state and the ordinances of the several cities and towns in Indiana, providing for any of the following: . . .

"4. The investigation, prosecution and suppression of the crime of arson and other crimes connected with the destruction or attempted destruction of property by fire or explosion, and the crime of swindling or defrauding an underwriter or attempting to do so; and for the investigation of the cause, origin and circumstances of fires.

"It shall be the further duty of the state fire marshal to make, . . . inspections of property and place orders thereon, when needed, for the prevention of fires and fire losses, and to enforce and carry out such orders, . . ."

§ 20-807:

". . . The state fire marshal, his deputies or assistants, upon the complaint of any person or whenever he or they shall deem it necessary, shall inspect all buildings, premises, property, conditions, and things comprehended in this Act within their jurisdiction."

§ 20-808:

"The state fire marshal or his deputies may, in addition to the investigation made by any of his assistants, at any time investigate as to the origin or circumstances of any fire occurring in this state. . . .

"The state fire marshal or his deputies or any of his assistants may at all reasonable hours enter any building, property or premises within his jurisdiction for the purpose of making an inspection or investigation which, under the provisions of this Act or any law which may have been or may be from time to time enacted requiring the fire marshal to enforce or carry out, he or they may deem necessary to be made."

It is urged by the state that investigations as to the cause of fires, authorized by the law, are necessary to the public safety as a means of fire prevention. Furthermore, the state argues that the circumstances surrounding such a situation make it necessary that such investigations be initiated without a search warrant because, the cause of fire being unknown, there is no probable cause upon which a search warrant could issue. *[Footnote omitted.]*

Finally the state contends that, it being necessary to initiate such investigations without a search warrant, evidence so obtained as to the cause of such fire, whether it was of accidental or incendiary origin, should be admissible in evidence by the state in any action wherein the cause of such fire is an issue.

The state contends that to hold otherwise would, for all practical purposes, be to deny the state the right to prosecute persons for arson in every instance where, at the outset, the cause of a fire is unknown. Upon this issue the state's contention is not well taken. Under the facts before us, it cannot well be maintained that the search made was a mere civil inspection and, even if it began as such as the state contends it may have, no good reason is shown why a search warrant was not obtained at the point in time when evidence believed to be incriminating was discovered. From the point in time when the investigation was conducted for the purpose of obtaining incriminating evidence against the owner of the property, a search warrant was necessary. In support of its position to the contrary the state has cited numerous cases in which the right of inspection has been upheld without the necessity of a search warrant.* However, it must be noted that in each instance referred to the inspection or investigation was civil and not criminal in nature. It was permitted as relating to the regulation of the use or occupancy of the property itself. In no instance was the information obtained or the evidence seized used in a criminal prosecution against the proprietor of the property.

Having held that a search warrant is required by the fire marshal and his deputies in criminal investigations, regarding the cause of fires, not made at the time of the fire we are now required to consider whether or not the Fire Marshal Law is unconstitutional because of the duty and authority it vests in the fire marshal and his deputies without reference to a search warrant. If the law is construed to provide that criminal investigations may be made without a search warrant, the law is unconstitutional, for the reasons above stated. If, however, the law can be construed to have adopted and incorporated by implication the constitutional requirement that the fire marshal and his deputies obtain a search warrant before making an investigation intended to lead to a criminal prosecution, it is constitutional. In deciding this question we are committed to holding the statute constitutional, if we can. . . .

In construing this law, there being no clear expressed and intendment to the contrary, we must assume that the legislature when they passed this statute intended that all the constitutional safeguards were to be incorporated in it. In *Klink* v. *State ex rel. Budd, [citations omitted]*, this court stated: " . . . It is our duty, if possible, to construe statutory enactments as conforming to the limitations imposed by the Constitution. . . ." It is the duty of the court to adopt a con-

---

*The right of inspection has been sustained in the following instances: (1) inspections relating to the supply of milk for human consumption, *Albert* v. *Milk Control Board of Indiana*, 1936, 210 Ind. 283, 200 N.E. 688; (2) inspections relating to fitness of houses for human habitation, *Richards* v. *City of Columbia*, 1956, 227 S.C. 538, 88 S.E.2d 683; (3) inspections of cattle for tuberculosis without a warrant, *Dederick* v. *Smith*, 1936, 88 N.H. 63, 184 A. 595, appeal dismissed 299 U.S. 506, 57 S.Ct. 28, 81 L.Ed. 375.

Courts have also upheld the right to enter premises for the purpose of inspecting places of business, such as small loan companies, junk dealers, optometrists, and public lodging houses. *[Citations omitted.]*

struction which will uphold a statute and bring it in harmony with the constitution if its language makes it possible. *[Citations omitted.]*

Upon examination of the Fire Marshal Law we note that it makes no reference to the use or nonuse of search warrants in connection with the inspections and investigations authorized by the fire marshal and his deputies. Therefore, we cannot say that it clearly purports to grant authority to make criminal investigations without resort to this constitutional limitation upon the police power of the state. On the other hand, the law does expressly describe the police power of the fire marshal and his deputies ". . . in all cases . . . of arson . . ." as that ". . . possessed by a constable, sheriff or police officer of the state." § 20-802, *supra*. The above named officers have no authority of search and seizure, preliminary to a criminal prosecution, without a search warrant.

We therefore hold that under the Fire Marshal Law the fire marshal and his deputies and agents must obtain a search warrant prior to searching a person's dwelling, effects or possessions for the purpose of obtaining evidence, to be used in a criminal prosecution and that this basic constitutional requirement can be and is by implication incorporated as a necessary part of the responsibility and activity authorized by the law.

We conclude that the state's Exhibit No. 5 was properly excluded from the evidence.

EMMERT, C. J., concurs in the result.

ARTERBURN, BOBBITT and LANDIS, JJ., concur in the opinion.

INSPECTION, INVESTIGATION, AND SEARCH

In the preceding case, judicial reference in the footnote was made to the permissibility of inspections. The topic of fire investigation inspections will be fully presented later in this book in Chapter VII, "Administrative Law," in the section titled "Administrative Search."

The distinction between an investigation and a search has particular importance to members of the fire service when the possibility of arson is involved. In the following case, a distinction was drawn between the statutory authority to investigate and the constitutional prohibition against warrantless searches when a crime is involved.

<hr />

STATE of Iowa, Petitioner,
v.
Warren J. REES, Judge, Respondent

Supreme Court of Iowa.
Jan. 11, 1966.
Rehearing Denied March 8, 1966.
139 N.W.2d 406

Original proceeding in *certiorari* to determine legality of action of Linn District Court, Warren J. Rees, [Judge], in sustaining a motion to suppress evidence. The Supreme Court, Snell, J., held that defendant indicted for arson did not meet burden of demonstrating that evidence against him had been illegally procured where officers, at time they entered burned premises where evidence was obtained were acting under explicit statutory authority, there was no claim of any unnecessary or arbitrary force, subterfuge, coercion or objection, and nothing unreasonable or violative of any constitutional rights was present in statutory procedure allowing such entry.

Writ sustained order suppressing evidence reversed, case remanded.

RAWLINGS, LARSON, THORNTON and BECKER, JJ., dissented.

SNELL, Justice.

This cause comes before us on *certiorari* to review an order of the trial court sustaining a motion to suppress evidence claimed to have been secured by an unreasonable search and seizure.

The case arose from criminal cause number 20814 entitled *State of Iowa* v. *Joseph W. Grant, Jr.* in Linn District Court.

On August 15, 1964, a fire was reported at 324 7th St., SE, in Cedar Rapids. The fire department responded and extinguished the flames at 5:50 a.m. The building involved was being used as a printing plant operated by Citizen's Publications, Inc., lessee. One Joseph W. Grant, Jr., the defendant in the criminal action, had an interest in the business and occupied an apartment on the premises.

Immediately after the fire had been extinguished the fire chief and his employees, a city electrical inspector, a foreman of the light and power company, a deputy state fire marshal, an agent of the National Board of Fire Underwriters, and others entered the premises for purpose of investigating the cause and origin of the fire. The investigation was prolonged and the building reentered several times as a part of a continuing investigation. The defendant said no search warrant was ever requested or secured. However, there is no claim of harassment, abuse, subterfuge, force or even objection.

Based in part at least upon the evidence obtained by this extended investigation, the Linn County grand jury returned an indictment charging accused with the crime of arson. He was arraigned and entered a "not guilty" plea. Prior to trial he filed the subject motion to suppress testimony of grand jury witnesses, Jesse G. Hunter *(chief of Cedar Rapids Fire Department)*, Kenneth A. Anderson *(a fire department captain)*, James R. Kuta *(city electrical inspector)*, Harry Billings *(an employee of Iowa Light and Power Company)*, and M. D. Huffman *(an agent in the Arson Department of the National Board of Fire Underwriters)*. By his motion accused requested the court to suppress all evidence gained by these persons as a result of their investigation in connection with the fire.

Pursuant to court order there was a hearing on the motion, at which time accused appeared and testified, the State offering no evidence. The indictment, with minutes attached, is made a part of the record for review. The trial court promptly entered order sustaining the motion, the material portion of which provided as follows: ". . . All evidence obtained, and all testimony of witnesses having to do with any search of the premises conducted subsequent to 5:50 a.m. of August 15, 1964, would be, therefore, found to be inadmissible, and such is the Order, Judgment and Ruling of the Court upon the motion to suppress."

The record now before us discloses accused not only had an interest in the business operated upon the subject property, but also had and occupied an apartment somewhere on the premises. He was clearly a person aggrieved by the search. Furthermore, his indictment subsequent to the search, based in part at least upon the evidence so obtained, gave meaning to his status as a person aggrieved.

The statutory authority for investigation into the origin and cause of fires is found in chapter 100, Code of Iowa 1962.

Section 100.1, subparagraph 2, requires an investigation by the state fire marshal.

Section 100.2 provides: "The chief of the fire department of every city or town in which a fire department is established . . . shall investigate into the cause, origin and circumstances of every fire occurring in such city . . . and determine whether such fire was the result of natural causes, negligence or design. The state fire marshal may assist . . . superintend and direct . . . ."

Section 100.3 requires a report to the state fire marshal.

Section 100.9 provides that when the fire marshal is of the opinion that there is evidence sufficient to charge any person with arson or related offenses he shall cause arrest and prosecution and shall furnish to the county attorney all evidence.

Section 100.10 authorizes the fire marshal and his subordinates to enter any building and examine the same and the contents.

Section 100.12 authorizes entry and examination by the chief of the fire department.

Under these statutes the entry and examination by the officers was legal and mandatory.

What was done here is exactly what is required by the statutes, *i.e.,* investigation, determination of opinion as to cause of the fire, prosecution and furnishing of evidence. The trial court held that any evidence obtained after the date and time of the extinguishment of the fire would be inadmissible as having been the fruits of an unlawful search and seizure.

Here there was no unlawful search and seizure. What was done was pursuant to statute. We use the word unlawful as meaning without statutory support or in violation of statute. The question is was it unreasonable and violative of constitutional limitations.

Statutes and ordinances authorizing civil inspections have long been acknowledged and sanctioned as incident to the police power of a state or municipality but they must be within constitutional limits. *[Citations omitted.]*

In *Frank* v. *State of Maryland,* 359 U.S. 360, 79 S.Ct. 804, 3 L.Ed.2d 877, it was held that a health officer under the authority of a city ordinance could go on property at reasonable times without the aid of a search warrant for the limited purpose of an inspection to ascertain whether conditions are present which do not meet minimum standards and might be inimical to the health, welfare and safety of the public.

The case recognizes two basic constitutional protections.

''(1) the right to be secure from intrusion into personal privacy, the right to shut the door on officials of the state unless their entry is under proper authority of law. The second, and intimately related protection is self-protection: the right to resist unauthorized entry which has as its design the securing of information to fortify the coercive power of the state against the individual, information which may be used to effect a further deprivation of life or liberty or property. Thus, evidence of criminal action may not, save in very limited and closely confined situations, be seized without a judicially issued search warrant.''

Mr. Justice Frankfurter traced the history of and necessity for certain inspections and held there was no violation of due process.

A well-reasoned opinion clearly in point appears in *Dederick* v. *Smith [citations omitted]*, a New Hampshire case decided in 1936. Although prior to *Mapp* v. *Ohio*, discussed *infra*, the reasoning is not in conflict therewith.

Plaintiff sought an injunction restraining the state veterinarian "his aiders, abettors, agents, and employees" from trespassing upon the property of plaintiff or from entering or breaking into her premises to test plaintiff's cattle for bovine tuberculosis. The defendant had requested plaintiff's permission to make the tests. Plaintiff had refused. Defendant with assistants forced entrance. Because of plaintiff's objections extra trips for injection of testing materials were necessary. Defendant's testing and actions were pursuant to statute.

Plaintiff claimed (1) defendant was without authority to break and enter plaintiff's barn over her protest or commit any trespass without judicial authorization, (2) plaintiff was entitled to a prior hearing, (3) defendant's entry was without due process of law, and (4) the statute authorizing defendant's procedure was unconstitutional. The court said:

"The contention of the plaintiff that the statute does not authorize an entry without the supplementary aid of judicial process in the nature of a search warrant is without merit. On the contrary, the act was apparently designed to dispense with the necessity of judicial process in its enforcement. The statute, in effect, gives to the named officers a blanket search warrant which permits them to investigate all premises where cattle are kept. To the validity of this warrant the consent of the owner is not necessary, and authority to use such force as may be necessary to effect an entrance to buildings which have been locked for the purpose of preventing an investigation is necessarily implied.

"The specific claim of the plaintiff that before such entry she 'was entitled to a hearing as to whether or not the keeping by her of cattle that had not been tested for tuberculosis came within the provisions of P.L. c. 187, § 46, as amended, and was within the police power of the state,' finds no support in the provisions of the statute or the established procedure in regard to the issuance of such warrants. . . ." *(loc. cit.* 597 and 598.)

The statute was approved as a valid exercise of police power and its constitutionality approved. The statute was held reasonable and unobjectionable.

In discussing the 14th Amendment the court cited numerous authorities and said:

" 'Neither the amendment — broad and comprehensive as it is — nor any other amendment, was designed to interfere with the power of the state, sometimes termed its police power, to prescribe regulations to promote the health, peace, morals, education, and good order of the people, and to legislate so as to increase the industries of the state, develop its resources, and add to its wealth and prosperity.' *[Citations omitted.]*

" 'It cannot be supposed that the states intended, by adopting that amendment, to impose restraints upon the exercise of their powers for the protection of the safety, health, or morals of the community.' *[Citations omitted.]*

"The contention of the plaintiff that she had a constitutional right to notice and an opportunity to be heard before a judicial tribunal before her cattle could lawfully be tested without her consent is, therefore, without foundation . . ." *(loc. cit.* 600.)

"The statute explicitly gave to the defendant, as the lawful agent of the commissioner of agriculture, specific authority to enter the plaintiff's 'premises' for the purpose of investigating the existence of infectious disease amongst her cattle. As above stated, we think that this necessarily included authority to break into any barn from which the plaintiff undertook to exclude him or his agent. This is in accordance with the general rule in such cases which has been stated, as follows: 'One who is privileged to enter land . . . pursuant to legislative duty or authority . . . is further privileged to break and enter a fence or other enclosure or a dwelling or other building if to do so is reasonably necessary or is reasonably believed by the actor to be necessary to accomplish the purpose of the privilege.' " *(loc. cit.* 601.)

Plaintiff's case was dismissed.

In *State* v. *Mehlhaff [citations omitted],* the Supreme Court of South Dakota considered the admissibility of evidence unrelated to a search warrant. The defendant was convicted of larceny. On appeal one of his assignments of error related to search and seizure. The court took notice of both the federal and state constitutions and held that to constitute an unreasonable search and seizure two facts are essential, namely: (1) entry by the witnesses upon defendant's property, and their presence there must have been unlawful, and (2) discovery of the facts to which the witnesses testified must have been unlawful.

Defendant's conviction was affirmed.

The reasoning of the court was sound and appropriate to the case before us.

In the case at bar except for the fact that the evidence was obtained pursuant to a statutory mandate instead of a search warrant there is not a word in the record to support the order of suppression.

It is argued that when the investigation uncovered evidence of crime its investigatory status ended; that the statutory authority came to an end; that it became an accusatory search and could be supported only by a search warrant. We do not agree. The statutes clearly contemplate the discovery of evidence of crime, the arrest and prosecution of the person to be charged, and the delivery of all evidence, names of witnesses and all information to the county attorney. The statutory authority of the investigating officials did not terminate when evidence of arson was found.

We recently considered the changing concept during a search. See *State* v. *Hagen. [Citation omitted.]* In that case the original entry was a trespass and was illegal. We said: "A search is good or bad when it starts and does not change character from its success. . . . An unlawful search taints all evidence obtained at the search or through leads uncovered by the search." We held the initial entry unreasonable and the search that followed unlawful.

In the case at bar the contrary appears. The original entry was lawful and mandatory under the statute. No one contends that the original entry was unreasonable. As we said in *Hagen, supra,* it did "not change character from its success." The entry did not become unlawful or the search become unreasonable just because evidence of arson was found. There is nothing in the record before us, except the finding of evidence of arson, to support a holding that there was a change from a lawful to an unlawful procedure.

The fact that an investigation becomes accusatory does not make it unconstitutional. It does go beyond proper limits when it extends into fields unrelated to the authorized investigation or is unduly prolonged over the objection of the accused. Not even a search warrant is good indefinitely. See section 751.12, Code of Iowa. The question is not before us but it may be that the defendant could have, after a reasonable time for completion, terminated the investigation by telling the investigators to stay out. In that event further search and seizure would have required a search warrant. However, the record is silent as to any search beyond the limits of the statutory mandate or over any objection, timely or otherwise, by the defendant.

We find nothing in the Constitution of the United States, the pronouncements of the United States Supreme Court, the Constitution of the State of Iowa or our pronouncements requiring the suppression of the evidence obtained during the search involved.

Amendment 4 to the Constitution of the United States provides:

"The right of the people to be secure in their persons, houses, papers, and effects, against unreasonable searches and seizures, shall not be violated, and no Warrants shall issue, but upon probable cause, supported by Oath or affirmation, and particularly describing the place to be searched, and the persons or things to be seized.''

Amendment 5 protects, among other things, against coerced self-incrimination.

Amendment 6 to the Constitution of the United States provides for procedural guarantees and, among other things, the right to counsel.

Amendment 14 provides for due process.

These are the basic constitutional provisions relating to criminal prosecutions.

Amendment 6 is not involved in the case before us.

Section 8, Article I of the Constitution of the State of Iowa is identical in its provisions and almost identical in its wording to Amendment 4 to the Constitution of the United States. Other provisions in our State Constitution provide protection in a manner comparable to our Federal Constitution.

The question before us is the reasonableness of the search and seizure of evidence.

The trial court apparently proceeded from the premise that a reasonable search and seizure can only proceed from the foundation of a search warrant issued on an affidavit showing probable cause. We do not think that is what the Constitution says nor what the cases hold.

Mapp v. *Ohio*, 367 U.S. 643, 81 S.Ct. 1684, 6 L.Ed.2d 1081 holds that evidence obtained by unconstitutional search is inadmissible. The rights guaranteed by Amendment 4 are enforceable against the states by the Due Process Clause of Amendment 14. With these basic rights of the people no one has any legitimate quarrel. However, neither the Constitution nor the cases decided thereunder say that there can never be a legal or reasonable search without a warrant. Neither do they say that evidence obtained during a legal investigation is inadmissible. The constitutional guarantees do protect against indiscriminate and unreasonable searches and against warrants issued without probable cause.

It is the indiscriminate search and seizure without benefit of a warrant issued on affidavit of probable cause that is proscribed. Nowhere do we find any proscription against the use of evidence obtained during a reasonable and legally authorized investigation. It is unreasonable and illogical to say that when officers are carrying on a legal or as in this case a mandatory investigation they must stop and get a search warrant before they can seize and later use the evidence for which they were making their original investigation.

The present case is a far cry factually from *Mapp*.

In *Mapp* a reenforced group of seven or more officers, in search of a person wanted for questioning on an unrelated matter and for possible evidence of an unrelated offense, forcibly entered a house over the objection of defendant, pretended to have a warrant, manhandled the defendant and refused to permit defendant's lawyer to enter or see defendant. Proceeding "in this highhanded manner" the officers searched thoroughly and incident thereto found some obscene materials for possession of which defendant was ultimately convicted.

If there was anything to justify the "highhanded" procedure of the police it does not appear from the opinion. The Supreme Court held the evidence so obtained inadmissible. With that we agree but it does not follow that there is anything sacrosanct about the words "search warrant," the need therefor or the use thereof. The constitutional test is the reasonableness of the search and the propriety of the authority under which the search is made. The tests are strictly construed. If made under authority of a search warrant the warrant must be based on an affidavit of probable cause. In other words, there must be some legal reason why the officers are there.

In the case at bar there was a legal reason why the officers were there. There had been a fire. An investigation was required by the statute. It was made. There is no claim that there was any violation of the statute.

The 4th Amendment to the Constitution of the United States protects against unreasonable search and seizure. It protects the right of privacy against unwarranted invasion. It prohibits the issuance of warrants except upon oath or affirmation showing probable cause. It creates no magic in the use of the words "search warrant." A search warrant without the constitutional prerequisites is invalid but there is nothing in the constitution that says there can be no valid search under authority other than a search warrant.

In *State* v. *Shephard*, 255 Iowa 1218, 1226, 124 N.W.2d 712, 717, we said:

"The restrictions upon searches and seizures were obviously designed for protection against official invasion of privacy and the security of property. They are not exclusionary provisions against the admission of kinds of evidence deemed inherently unreliable or prejudicial. . . .

. . . "The 4th and 14th Amendments are not designed to help a guilty party escape the consequences of his act. Their purpose is to protect a person and his property from arbitrary and unreasonable searches and seizures. It is the 'right of privacy' that is protected."

What is proscribed by the Amendment is not the seizure of incriminating evidence but the unreasonable invasion of the right of privacy. If the search is unreasonable and without authority then in that event seizure of the evidence is without authority and the evidence is inadmissible.

In the case at bar the officers had a right to investigate. There is nothing unreasonable about searching for the cause of a fire. The evidence found was the work product of a reasonable and statutory investigation. When evidence is so found it is not necessary for officers to desist from further action and wait for a search warrant before they may pick up what they have found, or testify as to what they have seen, or complete their investigation.

The test is the reasonableness of the search under proper authority and not the source of the authority under which the search is made. A reasonable search mandatory under a legislative enactment is clothed with as much dignity and is entitled to as much consideration as a search under a warrant issued by a Justice of the Peace. It would be illogical to say that the evidence would have been admissible if seized under a search warrant but inadmissible if seized under the statute. This assumes, of course, that the entry was reasonable and legal and no one contends otherwise. Within the constitutional limitations as to reasonableness the legislature may and has authorized such an investigation as was made here.

A confession obtained by police interrogation when the accused is denied the right to counsel is inadmissible. *Escobedo* v. *State of Illinois,* 378 U.S. 478, 84 S.Ct. 1758, 12 L.Ed.2d 977. The case involved the right to counsel and the protection against self-incrimination under the 5th Amendment. The interrogation was accusatory and the accused's rights under the 5th Amendment were violated. The distinction between a routine investigation and an accusation under the 5th Amendment is not applicable to the right to privacy under the 4th Amendment. Under the 4th Amendment if the search is reasonable and lawful the fact that it becomes accusatory by the finding of incriminating evidence does not make it invalid.

Escobedo, supra, is not relevant to our problem, except to note the distinction between the rights involved and what was there said. "Nothing we have said today affects the powers of the police to investigate 'an unsolved crime,' *[citation omitted]* by gathering information from witnesses and by other 'proper investigative efforts.' *[Citation omitted.]*"

The burden of demonstrating that evidence has been illegally procured normally devolves upon the accused in a motion to suppress such evidence. *[Citations omitted.]* . . . What we have here is a case where evidence has been suppressed because obtained as a part of a statutory investigation rather than under a search warrant. The basic premise or foundation stone supporting a search under a warrant is an affidavit of probable cause. The constitution so provides. The whim, caprice or curiosity of officers will not suffice. In other words, the reason must appear and the search must be reasonable. That does not make procedure under a document called a search warrant exclusive.

The foundation for procedure after a fire is the statute. The legislature has said that when there has been a fire there shall be an investigation. There is nothing in the constitution that says that the legislature may not, within the limits of reasonableness and by positive mandate, direct a search upon the happening of an event of such public interest as a fire. The fire is the reason and the statute is the authority for the procedure.

It is unrealistic and unsound to say that when a constitutionally reasonable investigation or search has been directed by the legislature upon the happening of a certain event, *i.e.,* a fire, there must also be a search warrant that can only be issued upon affidavit of probable cause.

It is not logical to say that a mandatory entry and examination of the cause of a fire is unreasonable. The fact that the statutory investigation may uncover evidence of a crime does not mean that further search becomes constitutionally unreasonable or that evidence obtained thereby is inadmissible.

It is unlikely that before beginning the statutory investigation there would be information as to "probable cause" sufficient to support a search warrant. The constitution does not say that there can be no investigation without "probable cause" as to a specific crime when "probable cause" can only be determined by investigation or that when probable cause develops the right to search ends. That would draw too fine a line between investigation and search.

In the case at bar the officers were acting under explicit statutory authority. We find nothing unreasonable or violative of constitutional rights in the statutory procedure. There is no claim of any unnecessary or arbitrary force, subterfuge, coercion or objection.

The accused did not meet the burden of demonstrating that the evidence had been illegally procured.

There is no showing that the search was violative of any constitutional rights or that the evidence is inadmissible on the grounds urged.

The writ of *certiorari* is sustained, the order of the trial court suppressing the evidence is reversed and the case is remanded to the trial court for entry of an order in harmony herewith.

Writ sustained and case remanded.

GARFIELD, C. J., and MOORE, STUART and MASON, JJ., concur.

RAWLINGS, LARSON, THORNTON, and BECKER, JJ., dissent.

RAWLINGS, Justice (dissenting).

I concur in Divisions I and II of the majority opinion, but respectfully disagree with the factual analysis, reasoning, and conclusions reached in subsequent divisions.

We are here dealing with fundamental constitutional rights of the people and should strive for ways and means by which to not only comply with, but to promote those all important safeguards.

We said in *Lewis Con. Sch. Dist.* v. *Johnston, Iowa [citation omitted]:*

"Perhaps the most efficient form of government is an intelligent and benevolent dictatorship. But, passing the point that such dictatorships rapidly lose their intelligence and benevolence, we must observe that it is not the kind of government provided for by our constitution. Some check must be put upon administrative bodies; they must be required to follow some sort of pattern designed by the legislature. The law-making body may not entirely abrogate its functions, and surrender them to administrative officials."

By the same token some check must be maintained upon all departments of government. They, as well as administrative officers and agencies, must be required to follow some sort of constitutional pattern. Neither the Constitution of the United States nor of the State of Iowa can be deprived of vitality or meaning by any court, executive or legislative body. No law-making body can by legislative process strip from the people their constitutional rights, privileges and immunities, nor arbitrarily delegate them to some administrative or law enforcement officer. To me this basic precept is at the heart of the case now before us.

The majority opinion takes the position that "the present case is a far cry factually from *Mapp.*" This may be so, but I respectfully submit the interpretations of the Constitution of the United States of America and the legal principles so clearly declared in *Mapp* are not only in point but are here most persuasive if not controlling.

For good or for bad *Mapp* is undoubtedly here to stay and we should accord it full and fair consideration. . . .

. . . The factual situation set forth in the majority opinion is substantially correct, but omits certain important factors. Therefore some repetition of the facts is here unavoidable.

The record in this case is meager, but at the same time, for reasons hereafter discussed, is probably as complete as defendant could make it.

On August 15, 1964, a fire was reported at 324 7th St., SE, in Cedar Rapids. The fire department responded to a call and extinguished the flames at 5:50 a.m.

The building involved was being used as a printing plant operated by Citizen's Publications, Inc., lessee. However, one Joseph W. Grant, Jr., hereafter referred to as accused, had an interest in the business and occupied an apartment on the premises, this being his sole place of residence.

Immediately after the fire had been put out, the fire chief and his associates, a city electrical inspector, a foreman of the light and power company, a deputy state fire marshal, an agent of the National Board of Fire Underwriters, and others entered the premises for purpose of "investigation," and over a period of weeks reentered the building several times in pursuit of their inquiry. The defendant said no search warrant was ever requested or secured.

Based in part at least upon the evidence obtained by this extended investigation, the Linn County grand jury returned an indictment charging accused with the crime of arson. He was arraigned, entered a "not guilty" plea, and in due time, prior to trial, filed the subject motion to suppress testimony of grand jury witnesses, Jessee G. Hunter *(Chief of Cedar Rapids Fire Department)*, Kenneth A. Anderson *(a fire department captain)*, James R. Kuta *(city electrical inspector)*, Harry Billings *(an employee of Iowa Light and Power Company)*, and M. D. Huffman *(an agent in the Arson Department of the National Board of Fire Underwriters)*. By his motion accused requested the court to suppress all evidence gained by these persons as a result of their "investigation" in connection with the fire, *because it was obtained by an unconstitutional search*. This was not denied by the State.

Pursuant to court order there was a hearing on the motion, at which time accused appeared and testified. *Defendant stated his place of residence was on the subject premises; he had an interest in the business there operated; and no search warrant had ever been served on him. The State offered no evidence.* The indictment, with minutes attached, and what is designated a bill of particulars, is made a part of the record for review. The trial court promptly entered order sustaining the motion, the material portion of which provided as follows: ". . . All evidence obtained, and all testimony of witnesses having to do with any search of the premises conducted subsequent to 5:50 a.m. of August 15, 1964, would be, therefore, found to be inadmissible, and such is the Order, Judgment and Ruling of the court upon the motion to suppress."

The foregoing constitutes substantially the whole record before us. I shall return to this later.

Mapp v. *Ohio*, 367 U.S. 643, 81 S.Ct. 1684, 6 L.Ed.2d 1081, clearly served to impose a strict constitutional ban upon unreasonable searches and seizures as a protection against invasion of privacy, and for security of property, then implicitly placed an equally strict quarantine upon the use in all courts, state and federal, of any evidence secured through any constitutionally prohibited intrusions.

At this point we encounter *Frank* v. *State of Maryland,* 359 U.S. 360, 79 S.Ct. 804, 3 L.Ed.2d 877. In the cited case, 359 U.S. at 365-367, 79 S.Ct. at 808-809, the court said:

". . . two protections emerge from the broad constitutional proscription of official invasion. The first of these is the right to be secure from intrusion into personal privacy, the right to shut the door on officials of the state unless their entry is under proper authority of law. The second, and intimately related protection, is self-protection: the right to resist unauthorized entry which has as its design the securing of information to fortify the coercive power of the state against the individual, information which may be used to effect a further deprivation of life or liberty or property. Thus, *evidence of criminal action may not, save in very limited and closely confined situations, be seized without a judicially issued search warrant."* (Emphasis supplied.)

"But giving the fullest scope to this constitutional right to privacy, its protection cannot be here invoked. *The attempted inspection of appellant's home is merely to determine whether conditions exist which the Baltimore Health Code proscribes.* If they do appellant is notified to remedy the infringing conditions. *No evidence for criminal prosecution is sought to be seized.* Appellant is simply directed to do what he could have been ordered to do without any inspection, and what he cannot properly resist, namely, act in a manner consistent with the maintenance of minimum community standards of health and well-being, including his own. Appellant's resistance can only be based, not on admissible self-protection, but on a rarely voiced denial of any official justification for seeking to enter his home. The constitutional 'liberty' that is asserted is the absolute right to refuse consent for an inspection designed and pursued solely for the protection of the community's health, even when the inspection is conducted with due regard for every convenience of time and place."

". . . Moreover, the inspector has no power to force entry and did not attempt it. A fine is imposed for resistance, but officials are not authorized to break past the unwilling occupant.

"Thus, *not only does the inspection touch at most upon the periphery of the important interests safeguarded by the Fourteenth Amendment's protection against official instrusion, but it is hedged about with safeguards designed to make the least possible demand on the individual occupant, and to cause only the slightest restriction on his claims of privacy."* (Emphasis supplied.)

The court did not say an officer may enter the premises of another at any time, on any pretense, or over the refusal of the occupant to permit peaceable entry, without benefit of a warrant to arrest or to search.

The court simply said a health officer, absent objection by an occupant, may enter a building or go upon property at reasonable times, on reasonable cause, without aid of a warrant for the limited purpose of there making an inspection to ascertain whether conditions are present which do not meet minimum standards and may be dangerous to the health, welfare and safety of the public.

Admittedly statutes and ordinances authorizing such civil inspections have long been acknowledged as incident to the police power of a state or municipality, and thus sanctioned.

So, on the one hand we are confronted with a bar to unreasonable searches, and on the other, a qualified right to enter for health, safety and welfare inspections.

Without question our general assembly may enact any legislation deemed proper, *provided it is not violative of some provision of our state or federal constitutions. [Citation omitted.]* And we must, if possible, so construe the state fire marshal law as to avoid finding it, or any portion of it, unconstitutional. *[Citation omitted.]* Also we are required to resolve any doubts in favor of constitutionality of all statutes. *[Citation omitted.]*

As I see it, our task is to harmonize, if reasonably possible, the provisions of chapter 100 and the privacy rights of the people guaranteed by article I, section 8, Constitution of Iowa, and Amendments 4, 5, and 14, Constitution of the United States.

An examination of our state fire marshal law discloses no provision which automatically directs or authorizes searches and seizures. And, in event this law were to be so construed as to arbitrarily order, direct or permit a criminal investigation in any case, under any circumstances, or to allow such under the guise of civil inspection, then the law would, to that extent, be unconstitutional. *No law can so relax constitutional prohibitions as to permit searches and seizures at the whim and fancy of any individual. [Citations omitted.]* To hold otherwise would, in effect, serve to so construe chapter 100 as to authorize a *carte blanche* general warrant which is clearly prohibited. *[Citation omitted.]*

In *Johnson* v. *United States,* 333 U.S. 10, 13, 68 S.Ct. 367, 369, 92 L.Ed. 436, a case involving search of premises and arrest without a warrant, that court said: "The point of the Fourth Amendment, which often is not grasped by zealous officers, is not that it denies law enforcement the support of the usual inferences which reasonable men draw from evidence. Its protection consists in requiring that those inferences be drawn by a neutral and detached magistrate instead of being judged by the officer engaged in the often competitive enterprise of ferreting out crime. *Any assumption that evidence sufficient to support a magistrate's disinterested determination to issue a search warrant will justify the officers in making a search without a warrant would reduce the Amendment to a nullity and leave the people's homes secure only in the discretion of police officers.* Crime, even in the privacy of one's own quarters, is, of course, of grave concern to society, and the law allows such crime to be reached on proper showing. The right of officers to thrust themselves into a home is also a grave concern, not only to the individual but to a society which chooses to dwell in reasonable security and freedom from surveillance. *When the right of privacy must reasonably yield to the right of search is, as a rule, to be decided by a judicial officer, not by a policeman or Government enforcement agent."* (Emphasis supplied.)

As stated in *Frank* v. *State of Maryland,* 359 U.S. at 372-373, 79 S.Ct. at 812: "If a search warrant be constitutionally required, the requirement cannot be flexibly interpreted to dispense with the rigorous constitutional restrictions for its issue."

However if we, in effect, read relevant constitutional provisions into our state fire marshal law, then to that extent the law would be constitutional. *[Citations omitted.]*

The matter of civil inspections and criminal investigations should next be considered.

In its legal sense, a search and seizure is made and effected by entry of an officer into or upon the premises or property of another person and who there, by some force, actual or constructive, seeks out, reaches for, pries into or explores for, and thus physically or mentally obtains, takes, retains and holds the thing or condition so discovered, all with the intent and for the purpose of using such thing or condition in the criminal prosecution of some person or persons. It is a seeking out, a quest for incriminatory evidence. It may well be classified as accusatory in nature. It is an invasion of privacy. If effected with legal right, it is reasonable and lawful, but if done without such right, it is unreasonable and unlawful. *[Citations omitted.]*

On the other hand, a civil inspection as the term is here employed, is made and effected by entry of a qualified public official into or upon the premises or property of another person for the limited purpose of there making or effecting an observation only of some thing there present or some condition which may there exist, which may be inimical to the health, safety and welfare of the community, to the end that some appropriate action or sanction may be had or exercised in order to effect a cure, correction or alleviation of any such condition there found to exist. *Unlike searches, it is investigatory in nature. [Citations omitted.]*

Clearly and unmistakably, *Mapp* said there is a constitutional ban upon *unreasonable searches and seizures.* It would therefore appear that since a good-faith lawful civil inspection, so limited, does not stand in the shoes of an unreasonable search and seizure, an officer lawfully upon the premises or property of another, for such purpose, who there sees, observes or becomes aware of self-evident things or conditions, is not prohibited from thereafter giving testimony in a criminal prosecution as to such things or conditions. *[Citations omitted.]*

By the same token, an officer lawfully engaged in fighting a fire, or a duly authorized public official legitimately conducting a good-faith post-fire civil inspection upon the premises of another person, should be permitted to employ evidence of self-evident things or conditions in support of an application for a search warrant, and testify as to same in either civil or criminal proceedings.

Briefly stated, it is not a search to merely see, smell, hear, touch, or taste that which is plain, obvious, visible, patent or apparent to the senses. *[Citations omitted.]*

The crucial question presented is whether a person authorized by law to go upon the premises of another to conduct a civil inspection or investigation may at any time, or on any occasion there undertake a search and seizure.

In effect the majority opinion holds that when peace officers, administrative agents or others are in a place where they have a lawful right to be for conduct of a civil investigation they are, by the same token, in a place where they have a lawful right to be for a search and seizure. This cannot be.

Despite the contrary view expressed in the majority opinion, I, for one, find it impossible to believe an official acting under authority of chapter 100, having gained entrance to the premises or property of another for the purpose of then and there fighting a fire, or effecting a good-faith post-fire civil or administrative investigation, can *ipso facto* convert such fire fighting efforts, or any such investigation into a constitutionally proper search and seizure. Surely an officer so entering or being upon the premises of another person must ordinarily obtain proper consent or valid warrant prior to any search of the premises or possessions of another for the purpose of obtaining evidence incriminatory or accusatory in nature, and particularly so where a dwelling place is involved. *[Citations omitted.]*

By this I do not mean consent or a warrant is *always* a prerequisite to a lawful search. The test is reasonableness as to place, time, and area, which must be resolved according to the facts and circumstances of each case. *[Citations omitted.]* And, fine lines of distinction must usually be drawn requiring a full exploration of the factual situation in each case.

Assuming then that an officer conducting a lawfully proper investigation under chapter 100 is in a place where he has a legal right to be, it would logically follow that if while there one or more of his five senses discloses to him *reasonably probable cause to believe a crime has been or is being there committed,* he may conceivably proceed with a reasonably limited lawful search and seizure without first securing consent or warrant to search. *[Citations omitted.]*

With reference to probable cause, *Brinegar* v. *United States,* 338 U.S. 160, 175, 69 S.Ct. 1302, 1310, 93 L.Ed. 1879 says:

"In dealing with probable cause, however, as the very name implies, we deal with probabilities. These are not technical; they are the factual and practical considerations of everyday life on which reasonable and prudent men, not legal technicians, act. The standard of proof is accordingly correlative to what must be proved.

" 'The substance of all the definitions of probable cause is a reasonable ground for belief of guilt.' . . . Probable cause exists where 'the facts and circumstances within their *[the officers']* knowledge and of which they had reasonable trustworthy information *[are]* sufficient in themselves to warrant a man of reasonable caution in the belief that' an offense has been or is being committed."

However, exploratory, repetitive or delayed searches, any of which are with or without consent or warrant, have generally been held to be unreasonable. *[Citations omitted.]*

It seems to me to be now well settled that where a motion to suppress evidence is filed, the burden in the first instance is upon an accused to establish his standing as a person aggrieved by a challenged search, insufficiency of the affidavit for the warrant, or that a search was effected without lawful authority. *[Citations omitted.]* This done, the burden is clearly upon the state to show what, if any, civil inspection had been made, the basis or foundation for it, the extent of such investigation, and if a search was in fact made to show proper authority, cause or *reasonableness. [Citations omitted.]*

And reasonableness of any search is always for the court to determine. *[Citations omitted.]*

But in the case now before us the showing made may not have been, *arguendo,* sufficient to permit a fair appraisal and determination of the matter of reasonableness.

Admittedly nothing is shown as to extent of the fire, damage to the structure, its condition after the fire, or whether defendant was or was not present at any material time or times. There is no showing made as to whether the chief of the fire department, and others concerned, conducted a civil investigation and if so their cause or reason, time devoted, and area inspected; and if such investigation was undertaken whether it developed into a search and seizure, and if so why and when, the area searched, and time devoted to it; and whether defendant's living quarters were inspected or searched. *Also, no showing is made as to the authority of the five challenged witnesses, or any of them, to investigate or search.* These elements may or may not be essential to a fair determination of the defendant's motion, but they are elements which should be disclosed by the State, if disclosed at all.

The basic thrust of the majority opinion is that *defendant* failed to sustain his burden of proof; that the meagerness of the record is *defendant's* fault; that *defendant* failed to produce all evidence necessary to support his claim or to make his case; and that this is fatal to his motion to suppress.

Actually it would appear defendant produced all evidence within his limited command. This being a criminal case pretrial discovery was of questionable value to him.

This then means defendant's only real source of information with which to act in support of his motion to suppress was a motion for a bill of particulars. But he [defendant] used this method of approach and was simply told that an "investigation" had been repeatedly conducted on the premises over a period of about five or six weeks.

The result is, defendant found himself on the horns of a dilemma, but the majority says he did not prove enough.

I cannot agree. Defendant proved he was a person aggrieved; that in truth and in fact four or five people, some with no apparent right or authority, had participated in what can only be classified as repetitive on-the-spot searches, all done without evident authority. This the *State* neither refuted nor explained.

I am satisfied defendant made his case, that the burden was then upon the *State,* and the *State* failed to meet its burden. *[Citation omitted.]* Upon this premise the writ of *certiorari* should be annulled.

However, it is my belief that both the State and defendant should, as a matter of fairness, be permitted to present such additional testimony and make such further record as is possible and proper.

I believe manifest justice dictates this case be remanded.

LARSON, THORNTON and BECKER, JJ., join in this dissent.

The decision in *State* v. *Rees* was 5 to 4.

1. Why was the decision so close?
2. What point of disagreement did the dissenting judges present?
3. What conclusion can be drawn as a result of this close decision?

SUMMARY

Criminal law and the rules, regulations, and laws of criminal procedure go hand-in-hand. Criminal law *defines* a crime, its various degrees, and its punishment. Criminal procedure is the legal *method* by which a criminal may be investigated, arrested, and prosecuted. Those who commit acts in violation of criminal law are concerned primarily with their own individual and personal welfare, and not with promoting the general peace and welfare of society as a whole. The rules of criminal procedure are concerned both with society as a whole and with the individual, and therefore the rules and regulations of criminal procedure are often stringent in order to promote and maintain general peace and harmony and, at the same time, protect the rights of the individual who has wronged society.

Constitutional law and judicial decisions regarding the rights of individuals under the Constitution have had a profound effect on criminal law and criminal procedure in recent years. As Attorney Peter J. Connelly, Jr., stated when referring to the impact recent judicial decisions have had on criminal law and its companion, criminal procedure:[14]

> Criminal law is much more difficult now in almost every respect. . . . The rights of the individual are now more important than the rights of the general public.

Following are some of the constitutional protections afforded individuals in proceedings involving criminal law and criminal procedure. These protections are afforded by the Constitution of the United States. *(See Appendix A.)* In his book titled *Fire Service and the Law,* Charles W. Bahme describes some of these important principles as follows:[15]

FIRST   Every person is presumed to be innocent until proved guilty, and that person is entitled to any benefit of the doubt where any reasonable doubt exists concerning that person's guilt or innocence.

SECOND   No person can be brought to trial until there has been a proper indictment by a grand jury (though in some states, as in California, the alternative procedure of filing of an information or a complaint is now permissible).

THIRD   The accused is entitled to a trial by a jury of peers (equals in rank), selected impartially from among the people, and whose decision on questions of fact is final.

FOURTH   In contrast to the European system, the accused must be tried on the basis of the facts presented and not on character and prior conduct.

FIFTH   The prisoner [accused] cannot be required to give testimony which will *incriminate* him (subject him to possible punishment for the commission of a crime). This rule does not mean, however, that an ac-

---

[14]Quoted from *The Times Union,* Rochester, NY, Jan. 15, 1977.
[15]Bahme, *Fire Service and the Law,* pp. 10-11.

cused who has testified in his own behalf cannot be made to answer questions on cross examination.

SIXTH   He cannot be tried for the commission of an act which was not a crime at the time he committed it, nor can the punishment imposed be increased over what it was at that time. (A law that violates these rights is an *ex post facto* law.)

SEVENTH The accused cannot be put in jeopardy twice for the same offense. (The Fifth Amendment to the Constitution of the United States [*see Appendix A*] forbidding double jeopardy only prohibits double punishment for the same offense in the federal courts, but does not forbid the states from also punishing for the act when the same act may be considered a separate and distinct offense against the state; nor does it prohibit either the state or federal government from passing a law authorizing the separate punishment of each step leading to the consummation of the crime and also punishment of the completed crime.)

ACTIVITIES

1. Discuss the reasons why members of the fire service should be familiar with the rules of criminal procedure. Give examples that might apply.
2. Why are laws and rules of criminal procedure necessary?
3. Why is a knowledge of proper evidence-gathering techniques important to members of the fire service? Give examples from your own experience or from situations of which you are aware.
4. What two major elements must the government prove at a criminal trial?
5. How can a poorly drawn statute affect a criminal trial?
6. Discuss "presumption of innocence" and "reasonable doubt" based on the decision in *Coffin* v. *United States*. Explain the role of evidence in a courtroom proceeding based on your discussion.
7. Describe the elements of criminal procedure from arrest through trial.
8. What kinds of evidence can be presented in court to secure a conviction for arson? Give examples of the kinds of evidence that would not be admissible in court.
9. Explain the search and seizure rule with regard to criminal prosecutions.
10. Compare the principles of the American system of criminal procedure listed in the "Summary" section of this chapter with the constitutional safeguards described in *State* v. *Rees*. What principles would you add to the Summary?

BIBLIOGRAPHY

Chadman, Charles E., ed., *A Treatise on Criminal Law and Criminal Procedure*, De Bower-Elliot Company, Chicago, © 1976 by AMS Press, Inc., New York.

Goldstein, Abraham S. and Goldstein, Joseph, eds., *Crime, Law, and Society*, The Free Press, New York, 1971.

Haskins, Jim, *Your Rights Past and Present,* Hawthorn Books, Inc., New York, 1975.

Inbau, Fred E. and Aspen, Marvin E., *Criminal Law for the Layman,* Chilton Book Company, Philadelphia, 1970.

Inbau, Fred E., *et al., Cases and Comments on Criminal Law,* The Foundation Press, Inc., Mineola, NY, 1973.

_____, *Cases and Comments on Criminal Procedure,* The Foundation Press, Inc., Mineola, NY, 1974.

Levin, Molly Apple, *Violence In Society,* Houghton Mifflin Company, Boston, 1975.

Loeb, Robert H., Jr., *Your Legal Rights as a Minor,* Franklin Watts, Inc., New York, 1974.

Paulsen, Monrad G., *The Problems of Juvenile Courts and the Rights of Children,* American Law Institute, Philadelphia, 1975.

Vorenberg, James, *Criminal Law and Procedure,* West Publishing Co., St. Paul, MN, 1975.

Wright, Gene and Marlo, John A., *The Police Officer and Criminal Justice,* McGraw-Hill Book Company, New York, 1970.

For an in-depth study of the topics presented in this chapter, the following are recommended:

Besharov, Douglas J., *Juvenile Justice Advocacy,* Practising Law Institute, New York, 1974.

Curtis, Arthur F., *A Treatise on the Law of Arson,* Dennis & Company, Buffalo, 1936.

Fox, Sanford J., *Cases and Materials on Modern Juvenile Justice,* West Publishing Company, St. Paul, MN, 1972.

Goldstein, Abraham S. and Orland, Leonard, *Criminal Procedure,* Little, Brown and Co., Boston, 1974.

Perkins, Rollin M., *Criminal Law,* The Foundation Press, Inc., Mineola, NY, 1969.

Uviller, H. Richard, *The Processes of Criminal Justice: Investigation,* West Publishing Co., St. Paul, MN, 1974.

Varon, Joseph A., *Searches, Seizures and Immunities,* 2nd Ed., The Bobbs-Merrill Company, Inc., Indianapolis, 1974.

VII

ADMINISTRATIVE LAW

OVERVIEW

The branch of American law called administrative law has become an integral part of local, state, and national government. Administrative law is the law that controls governmental machinery for implementing governmental programs. The presence of administrative agencies and commissions is an ever-growing fact of life at all levels of government. Although administrative law touches virtually every aspect of citizens' lives, the topic of administrative search is of particular concern to members of the fire service, and will be emphasized in this chapter.

WHAT IS ADMINISTRATIVE LAW?

Administrative law is the body of law that concerns the powers and procedures of administrative agencies — those agencies that carry out government programs. Administrative agencies are described by Kenneth Culp Davis in his book titled *Administrative Law,* as follows:[1]

An administrative agency is a governmental authority, other than a court and other than a legislative body, which affects the rights of private parties through either adjudication, rule-making, investigating, prosecuting, negotiation, settling, or informally acting. An administrative agency may be called a commission, board, authority, bureau, office, officer, administrator, department, corporation, administration, division, or agency. Nothing of substance hinges on the choice of

---

[1]Kenneth Culp Davis, *Administrative Law,* West Publishing Co., St. Paul, MN, 1973, p. 1.

name, and usually the choices have been entirely haphazard. When the President, or a governor, or a municipal governing body exercises powers of adjudication or rule-making, he or it is to that extent an administrative agency.

The main components of administrative law are constitutional law, statutory law, common law, and agency-made law. Agency-made law or rule-making will be considered in depth in Chapter VIII, "Administrative Procedure."

There are two basic aspects of statutory law that are involved in administrative law. Administrative agencies are created only by statute, by executive orders authorized by statute, or by state constitutional provisions that can be said to have the effect of a statute. The chief powers and functions of the agency or agencies created by statute are defined within the statute.

In addition to the fact that administrative agencies can only be created by statute, administrative agencies are generally empowered to interpret statutes that are related to or which affect the duties of these agencies. These interpretations are the basis of the "discretionary power" that many agencies possess. Decisions made by agencies using this discretionary power can have the effect of law. Administrative discretion is described as follows by Professor Ernst Freund in Milton M. Carrow's book *The Background of Administrative Law*:[2]

> When we speak of administrative discretion, we mean that a determination may be reached, in part at least, upon the basis of considerations not entirely susceptible of proof or disproof. A statute confers discretion when it refers an official for the use of his power to beliefs, expectations, or tendencies instead of facts, or to such terms as "adequate," "advisable," "appropriate," "beneficial," "competent," "convenient," "detrimental," "expedient," "equitable," "fair," "fit," "necessary," "practicable," "proper," "reasonable," "reputable," "safe," "sufficient," "wholesome," or their opposites.

The various methods by which administrative agencies carry out their duties is called the administrative process, which has as its primary purpose solving immediate problems in the fastest and most effective way possible:[3]

> The fundamental reason for resort to the administrative process has been the undertaking by government of tasks which from a strictly practical standpoint can best be performed through that process. In each instance, practical men have been seeking practical answers to immediate problems, and their concern has been with how to get the substantive job done, not with inventing some new theory to take the place of the theories that have seemed to stand in the way of what they wanted to accomplish.
>
> For instance, in 1789, the first Congress wanted to provide benefits to the veterans of the Revolutionary War, and someone had to decide which individuals would be paid; instead of using high-paid judges and all the trappings of the

---

[2]Milton M. Carrow, *The Background of Administrative Law*, Associated Lawyers Publishing Co., Newark, NJ, 1948, p. 19.

[3]Davis, *Administrative Law*, p. 13.

courtroom, the Congress assigned the task to an agency that could use a staff of low-paid clerks, because that seemed to be the best way to get the job done. . . .

More and more, legislatures are creating statutes that describe the main outlines of programs and leave to administrative agencies the task of developing the rules and policies for a program through administrative processes and procedures. Very often, in order to implement a program of continuing concern, an agency must be created by statute to administer a specific statute. The following is a brief description of the creation of an agency to fulfill a certain need, as summarized by Kenneth Culp Davis in *Administrative Law*:[4]

In 1968, President Johnson appointed a National Commission on Product Safety to study the problem. The seven-man Commission was headed by Arnold B. Elkind, a New York lawyer. It was assisted by a staff of 58. In 1970 it brought in a report of 167 large double-column pages, plus a proposed bill in 32 pages.

The findings were that 20 million Americans are injured each year in the home as a result of incidents connected with consumer products, 110,000 are permanently disabled, and 30,000 are killed. The annual cost to the nation of product-related injuries may exceed $5.5 billion.

Products involving "unreasonable hazards" include architectural glass, color television sets, fireworks, floor furnaces, glass bottles, high-rise bicycles, hot-water vaporizers, household chemicals, infant furniture, ladders, power tools, protective headgear, rotary lawnmowers, toys, unvented gas heaters, and wringer washers. Self-regulation by producers and their trade associations was found to be "patently inadequate."

A former Trade Commissioner has responded to the question whether competition can solve the problem: "The answer, very simply, is that competition and voluntary actions of businessmen do not always suffice to safeguard the public interest. Competition does not inevitably take the form of a rivalry to produce the safest product. Indeed, the competitive struggle may sometimes lead to a 'shaving' of the costs of manufacture involving some sacrifice of safety. Nor does competition always reward, in the form of greater volume and higher profits, the manufacturer who tries to sell 'safety' as a feature of his product."

The National Commission dealt easily with other problems:

"Without central leadership, States and municipalities are unable to chart broad spectrum product safety programs. Balkanized jurisdiction plagues some manufacturers with diverse manufacturing specifications that interfere with distribution of their products. . . .

"Despite its humanitarian adaptations to meet the challenge of product-caused injuries, the common law puts no reliable restraint upon product hazards.

---

[4]*Ibid.*, pp. 10-12.

"Because of the inadequacy of existing controls on product hazards, we find a need for a major Federal role in the development and execution of methods to protect the American consumer."

Congress enacted the Consumer Product Safety Act in 1972. 15 U.S.C. § 2051. The bill went through Congress without controversy about the question whether federal legislation should establish a system of regulation; the principal controversies within Congress had to do with choices among various kinds of administrative machinery. For instance, in the House Committee, a minority report was filed, but its first sentence said: "The concept that appropriate Federal legislation can result in reducing the risks resulting from the use of some consumer products is generally agreed upon." H. R. Rep. No. 92–1153, 92d Cong., 2d Sess. 69 (1972). The main dissent went to the question whether an independent agency* should be created, as the majority of the Committee recommended, or whether the new regulatory authority should be conferred upon the Food and Drug Administration. In the Senate Committee, only two Senators dissented; one wanted regulation by a Consumer Protection Agency and not by a Consumer Safety Agency, and the other wanted regulation by the Food and Drug Administration instead of by an independent agency. Sen. Rep. No. 92–749, 92d Cong., 2d Sess. 137, 153 (1972). No one in Congress was asserting rugged individualism. No one wanted *laissez faire*. No one asserted: "Let's not interfere with businessmen." No one argued for cutting back government regulation instead of increasing it.

In the statute, Congress states findings of unreasonable risks of injury from consumer products, that the public should be protected, that control by state and local governments is inadequate, and that regulation is necessary. The Act provides for product safety information and research and for public disclosure of information. It creates the CPSC and gives it ample powers; it also creates a Product Safety Advisory Council. It establishes elaborate procedural protections for those affected by the regulation. It provides civil and criminal penalties, and it authorizes seizure of products.

The two major powers of the Commission are to formulate safety standards for particular products and to ban the sale of particular products. The Act provides, 15 U.S.C. § 2056: "The Commission may by rule . . . promulgate consumer product safety standards. A consumer product safety standard shall consist of one or more of any of the following types of requirements: (1) requirements as to performance, composition, contents, design, construction, finish, or packaging of a consumer product; (2) requirements that a consumer product be marked with or accompanied by clear and adequate warnings or instructions. . . ." And § 2058 provides that when the Commission finds that a product "presents an unreasonable risk of injury" and that no feasible standard would adequately protect the public, the Commission may "promulgate a rule declaring such product a banned hazardous product."

What turned out to be the main issue as the bill went through Congress was the

---

*An independent agency is any agency outside the executive department. It is independent in that the President has no formal power to give orders to its members.

choice between an independent agency, as recommended by the National Commission, and locating the new power in the Department of Health, Education and Welfare, principally in the Food and Drug Administration, as recommended by the Nixon administration. Congress chose the independent agency. The House Committee explained: "This decision reflects the committee's belief that an independent agency can better carry out the legislative and judicial functions contained in this bill with the cold neutrality that the public has a right to expect of regulatory agencies formed for its protection. Independent status, and bipartisan commissioners with staggered and fixed terms, will tend to provide greater insulation from political and economic pressures than is possible or likely in a cabinet-level department. The Commission's decisions under this legislation will necessarily involve a careful meld of safety and economic considerations. This delicate balance, the committee believes, should be struck in a setting as far removed as possible from partisan influence." House Rep. at 24-25.

HISTORICAL BACKGROUND

Administrative law is generally considered to be the outstanding legal development of the 20th century. (The federal agencies involved in fire protection and prevention are listed in Appendix D at the end of this book.) Beginning with the creation of the Interstate Commerce Commission in 1887, administrative agencies have grown in number and scope until now they affect the life of virtually every citizen. As explained by Milton M. Carrow in *The Background of Administrative Law*:[5]

It is taken for granted, for instance, that a pharmacist, desiring to engage in the profession for which he studied, like most other professional people, must obtain a license from an administrative board, which may refuse to issue it, or, after issuance, may suspend or revoke it.

In the following excerpt from an article by Cuthbert W. Pound, Mr. Pound describes administrative law in its broad interpretation and also indicates the scope and continuing growth of modern administrative law.[6]

In its widest sense it includes the entire system of laws under which the machinery of the state works and by which the state performs all governmental acts, such as the administration of justice, the collection of taxes, duties, imports and excises, the regulation of trade and commerce, the raising and supporting of armies and navies, the government of territories and foreign possessions, and the promotion of the general welfare by regulative measures of all sorts. In a narrower sense, and as commonly used today, administrative law implies that branch of

---

[5]Carrow, *The Background of Administrative Law*, p. 1.

[6]Cuthbert W. Pound, "Constitutional Aspects of Administrative Law," in *The Growth of American Administrative Law*, by Ernst Freund, *et. al.*, Thomas Law Book Company, St. Louis, 1923, pp. 110-112.

modern law under which the executive department of government, acting in a quasi-legislative or quasi-judicial capacity, interferes with the conduct of the individual for the purpose of promoting the well-being of the community, as under laws regulating public utility corporations, business affected with a public interest, professions, trades and callings, rates and prices, laws for the protection of the public convenience and advantage. In the states we have examining boards and other bodies to pass on the competency, responsibility or other qualifications of private schools, chauffeurs, engineers, surveyors, private detectives, real estate brokers, stockbrokers, teachers, chiropodists, nurses, public accountants, shorthand reporters, physicians and surgeons, midwives, peddlers, lawyers, dentists, pharmacists, plumbers, undertakers, embalmers, veterinarians, optometrists, architects, employees of the state and its civil divisions, and other professions, trades and callings; also boards and commissioners of education, public service commissions, probation commissions, parole boards, athletic commissions to regulate boxing and wrestling contests, racing commissions, bank examiners, insurance departments, transit commissions, health boards with divisions for the safeguarding of motherhood, saving of infant life and instruction in child hygiene, child welfare boards, tax commissions, tenement house commissions, building commissions, water power commissions, water control commissions, commissions for the blind and for mental defectives, recreation commissions, boards of charities, agricultural commissions with power to grant indemnities for diseased animals destroyed and for animals killed by dogs, conservation commissions, industrial courts ("Miscalled a court," Taft, Ch. J., *Howat* v. *State of Kansas,* 42 Sup. Ct., 277, 278), workmen's compensation and industrial commissions, boards of child welfare for the granting of mothers' allowances, and motion picture commissions. In the United States the Interstate Commerce Commission, the Federal Trade Commission, the Railroad Labor Board and other similar bodies exercise vast powers and carry grave responsibilities.

THE ADMINISTRATIVE PROCEDURE ACT OF 1946

Administrative law incorporates many aspects of the powers of the three major branches of American government: legislative, judicial, and executive. As can readily be imagined, the scope and range of authority and power vested in administrative law and administrative agencies, if unchecked, could cause chaos in our governmental system. The Constitution of the United States created a system of government consisting of three branches — executive, legislative, and judicial — and a system of "checks and balances" so that none of the branches of government as defined could delegate its authority to another branch or another agency. This system was designed to prevent the possibility of tyranny or dictatorship from arising because of unbalanced power in the government. However, Kenneth Culp Davis considers a different view of this system in *Administrative Law*:[7]

> In our theoretical discussions we should frankly recognize that we have abandoned the basic idea that executive, legislative and judicial power should be separated from each other in order to protect against tyranny. We purposefully

[7]Davis, *Administrative Law,* pp. 25-26.

combine the three kinds of powers in particular agencies. The protection against tyranny comes, not from separating the powers, but from our system of legislative supervision of administration and from our system of judicial review of administrative action. If we were framing a new Constitution, we probably would have three branches, plus administrative agencies that combine the three kinds of powers. We have learned that danger of tyranny or injustice lurks in unchecked power, not in blended power.

Administrative law and administrative agencies can function using some of the powers of each or every branch of government. Although the concept of administrative law is now accepted in this country, it has in the past caused great alarm because of its rapid, almost unchecked growth in this century. In 1946, the Administrative Procedure Act was made law by the Congress. The purpose of this Act is described by its full title: "An Act to Improve the Administration of Justice by Prescribing Fair Administrative Procedure." The general motive of the Act was to curb the administrative branch of government by providing guidelines so that "the governors shall be governed and the regulators shall be regulated."[8] The Administrative Procedure Act (which is reproduced in Appendix C) does not try to regulate every administrative procedure used by administrative agencies; it is primarily directed to the rule-making and adjudication (decision-making) procedures used in the administrative process.

ADMINISTRATIVE AGENCIES

Administrative agencies, whether at the federal, state, or local level, have been referred to as "the fourth branch of American government." Cuthbert Pound's description of the scope and resultant power of administrative agencies in the previous section suggests one of the reasons for this title. Although the Constitution delegates definite powers and authority to the legislative, judicial, and executive branches of government, many people feel that the administrative area of government at all levels rivals the original three branches of government because of the administrative branch's extensive growth in both scope and power. And, although all three branches of government exert influence over administrative agencies, the rule-making authority granted to administrative agencies allows these agencies to — in effect — make laws. In addition, the decision-making powers granted to administrative agencies have diminished the role of the courts in the field of administrative law. In administrative procedure, access is available to the courts but usually only if a decision of an administrative agency is appealed. Even though a decision of an administrative agency may be appealed in court, judges are often reluctant to overturn administrative decisions. The reason given for the reluctance of the judiciary to interfere in the decisions of administrative agencies is generally the same reason that administrative agencies were formed in

[8]Sen. Doc. No. 248 at 244.

the first place: neither the legislature nor the judiciary has the expertise in specialized areas to deal adequately with certain subject areas, and must therefore rely on experts in the field for rules and decisions. For example, decisions, codes, and rules regarding firesafety codes require specialized expertise that is not usually available from the three branches of government.

The major federal agencies are well-known to most people. Among the major federal agencies are the Interstate Commerce Commission (ICC), which regulates rail, motor, water carriers, and pipelines; the Civil Aeronautics Board (CAB), which regulates air carriers; the Federal Power Commission (FPC), which regulates electric and gas utilities; and the Federal Communications Commission (FCC), which regulates radio, telephone, television, and telegraph. In addition to these major federal agencies, there are many, many more agencies that bring most aspects of a citizen's life under federal regulation in one way or another. In addition, the average state has about one hundred agencies with rule-making and/or adjudication powers.

FIRE PROTECTION AGENCIES

Administrative agencies play a major role in the fire protection field on the federal, state, and local level. Regulations relating to firesafety are determined and enforced by different levels of government. While some functions overlap at each level, federal and state laws generally govern those areas that cannot be regulated at the local level. The following three sections of this chapter discuss federal, state, and local-level administrative agencies and their powers with regard to fire protection.[9]

Federal-level Authority: There is a substantial amount of federal regulation with respect to firesafety. (Refer to Appendix D for a listing of federal agencies involved in fire protection.) Under the Constitution, the Congress has the power to regulate interstate commerce; this power has been interpreted to permit Congress to pass laws authorizing various federal departments and agencies to adopt and enforce firesafety regulations.

Any federal department or agency can promulgate firesafety regulations only if authority to do so is granted by a specific act of Congress. Once promulgated, these regulations carry the force of law, with violations resulting in serious penalties. Generally, such federal law can be enacted to provide: (1) that all state laws on the same subject are superseded by the federal law, (2) that state laws not conflicting with the federal law remain valid, or (3) that any state law will control if it is more stringent than the federal law.

All federal fire regulations affecting the general public must be published in the *Federal Register* before they become law so that interested citizens can make favorable or unfavorable comments to the proper governmental department or agency prior to the adoption of such regulations into the law. After their adop-

---

[9]Percy Bugbee, *Principles of Fire Protection*, NFPA, Boston, 1978, pp. 276-278.

tion, all fire regulations are published in the *Code of Federal Regulations*, which is revised and updated annually.

Probably the most extensive set of fire regulations ever adopted by a federal administrative agency are the regulations promulgated by the Occupational Safety and Health Administration (OSHA) of the Department of Labor. These regulations govern the health and safety of employees in industry and commerce.

The Consumer Product Safety Commission (CPSC) an independent agency, has broad regulatory authority over the firesafety aspects of products sold to consumers. It can prescribe and enforce mandatory product standards.

The Public Health Service (PHS) of the Department of Health, Education, and Welfare (HEW) issues firesafety standards for hospitals, nursing homes, and other health care facilities. These standards must be met as a condition for grants and loans for building and improving public and private health care facilities. The Social Security Administration (SSA), also a division of the Department of Health, Education, and Welfare, requires that hospitals and nursing homes participating in Medicare and Medicaid programs comply with the National Fire Protection Association's (NFPA) *Life Safety Code*.[10]

State-level Authority: The Tenth Amendment to the U.S. Constitution reserves certain police powers affecting public health and safety to the state level. State and local laws concerned with public firesafety are based on exercise of these powers.

Although all of the states have adopted some firesafety legislation, the amount and effectiveness of such legislation varies widely. For example, one state has adopted the latest edition of all the NFPA *National Fire Codes*, while another state leaves firesafety legislation largely to local fire officials in its cities and towns. Most of the states have laws covering such areas as the storage, use, and role of combustibles and explosives; installation and maintenance of automatic and other fire alarm systems and fire extinguishing equipment; construction, maintenance, and regulation of fire escapes; means and adequacy of fire exits in factories, asylums, hospitals, churches, schools, halls, theaters, nursing homes, and all other places in which numbers of persons live, work, or congregate from time to time for any purpose; suppression of arson; and the investigation of the cause, origin, and circumstances of fires.

In all but four of this country's fifty states (Colorado, Idaho, New Jersey, and New York), the principal fire official is the state fire marshal. In eight states, the office of the state fire marshal is an independent agency of the state government; in fifteen states, the office is part of the Department of Public Safety or of the State Police. In eleven states, the office of the state fire marshal is in the office of the State Insurance Commissioner. In the remaining states, the office is under various state departments such as the Attorney General's Office or the State Department of Commerce.

In most states, the fire marshal has the legal power to draft rules and regulations covering various fire hazards. Such regulations have the effect of the law. Usually the office of the state fire marshal serves as a central agency for sponsoring and

[10]NFPA 101, *Code for Safety to Life from Fire in Buildings and Structures*, NFPA, Boston, 1976.

promoting all kinds of fire prevention activities. Some state fire marshals are also responsible for reviewing construction or remodeling plans of state buildings, schools, nursing homes, and hospitals. Generally, however, the responsibilities of the office of state fire marshal are in the following general areas:

- Prevention of fires.
- Storage, sale, and use of combustibles and explosives.
- Installation and maintenance of automatic alarms and sprinkler systems.
- Construction, maintenance, and regulation of fire escapes.
- Means and adequacy of exits in case of fire in public places or buildings in which a number of persons live, work, or congregate (such as schools, hospitals, and large industrial complexes).
- Suppression of arson, and the investigation of the cause, origin, and circumstances of fire.

While the state fire marshal's office has legal authority for fire prevention, much of this power is delegated to the local fire departments and to local government. Fire departments may carry out inspection of private properties to determine if there are fire hazards or code violations on the premises, and local authorities are given the power — through "enabling acts" — to adopt their own regulations relating to fire prevention.

Local-level Authority: Local fire legislation is for the most part embodied in a fire prevention code administered by the fire department and a building code administered by the building department. Most localities use nationally recognized standards and codes as the basis for their codes, sometimes modifying them to fit local circumstances and needs.

Special "fire laws" may be enacted locally to handle particular life and firesafety problems in a community, such as piers and wharves, oil refineries, grain elevators, chemical plants, *etc.* The local government has the power to enforce state regulations that support fire codes where they exist, and to enact its own ordinances. Laws for local firesafety generally fall into two categories: (1) those relating to buildings, and (2) those relating to hazardous materials, processes, and machinery that may be used in buildings.

CONSTITUTIONAL QUESTIONS

Because of the particular and specialized nature of administrative agencies, many constitutional questions are involved in the field of administative law. Basically the constitutional questions involve two aspects: (1) the laws governing the powers and processes of administrative agencies, and (2) an individual's rights, under the Constitution, to "due process of law," the latter discussed by Peter Woll in *Administrative Law:*[11]

> Generally speaking, due process of law in an administration requires that fundamental procedures such as notice, hearing, adequate record, and appeal be

---

[11]Peter Woll, *Administrative Law*, University of California Press, Berkeley, CA, 1963, p. 22.

maintained. These procedures are formal and are characteristic of the formal stage of administrative adjudication in theory. As presently defined, in order for due process to be operative in administrative proceedings, private parties must have opportunity for recourse to the formal hearing stage within the agency, for only at that stage are procedures sufficiently formalized and opportunity present for further appeal to the judicial branch.

Another constitutional consideration is one involving the constitutional requirement that the Congress cannot delegate its law-making powers. State constitutions also delegate law-making powers only to the legislature. Chapter Eight, "Administrative Procedure," will detail the source and scope of the decision-making and rule-making powers of administrative agencies. This aspect of constitutional consideration has little current validity, but the constitutional question of whether an administrative agency has acted *beyond* the scope of its delegated power is increasingly being tested in the courts.

It is this aspect of delegation of power to administrative agencies which is, increasingly, affecting the fire service and related organizations. These considerations are most closely felt in the area of administrative search.

ADMINISTRATIVE SEARCH

The cases of *See* v. *City of Seattle* and *Camara* v. *Municipal Court of the City and County of San Francisco* hinge on the right of private citizens to refuse to permit entry of governmental inspectors and to require such inspectors to obtain a search warrant, thus subjecting the purposes and inspectional procedures of the governmental agency to judicial review. As explained by Robert W. Grant in *Public Fire Safety Inspections*:[12]

> In both cases, the Appellants (See and Camara) challenged the right of inspectors to enter their occupancies to make routine inspections without a search warrant. In both cases, the Appellants claimed that the city ordinances in question were unconstitutional in permitting entry without a search warrant or proper authorization by the owner or occupant.

In his conclusion, Mr. Grant states:[13]

> The recent U.S. Supreme Court decisions make it highly desirable that public fire prevention authorities review their inspection operations in order to determine that all inspection procedures fall within the U.S. Supreme Court's guidelines of reasonableness. A main point made by the Court is that judicial review of the reasonableness of an inspection should be made where entry is refused an inspector.

---

[12]Robert W. Grant, *Public Fire Safety Inspections,* NFPA, Boston, 1967, p. iii.

[13]*Ibid.,* p. 23.

OCCUPANCIES

As you read the following two cases of *Camara* v. *Municipal Court City and County of San Francisco* and *See* v. *City of Seattle*; consider their effects upon a fire department's fire prevention programs and code enforcement efforts. Does the enforcement of codes satisfy "probable cause" to inspect properties? Consider also an individual's constitutional rights *vs.* the protection of society as a whole from fire. How do these cases affect building owners or occupants who are adjacent to a property owned by an individual who refuses entry to fire service inspectors?

CAMARA v. MUNICIPAL COURT OF THE CITY AND COUNTY OF SAN FRANCISCO

Appeal from the District Court of Appeal of
California, First Appellate District.
387 U.S. 523
No. 92. Argued February 14, 1967. — Decided June 5, 1967.

Opinion of the Court.

MR. JUSTICE WHITE delivered the opinion of the Court.

In *Frank* v. *Maryland,* 359 U. S. 360, this Court upheld, by a five-to-four vote, a state court conviction of a homeowner who refused to permit a municipal health inspector to enter and inspect his premises without a search warrant. In *Eaton* v. *Price,* 364 U. S. 263, a similar conviction was affirmed by an equally divided Court. Since those closely divided decisions, more intensive efforts at all levels of government to contain and eliminate urban blight have led to increasing use of such inspection techniques, while numerous decisions of this Court have more fully defined the Fourth Amendment's effect on state and municipal action; *e.g.,* *Mapp* v. *Ohio,* 367 U. S. 643; *Ker* v. *California,* 374 U. S. 23. In view of the growing nationwide importance of the problem, we noted probable jurisdiction in this case and in *See* v. *City of Seattle, post,* p. 541, to reexamine whether administrative inspection programs, as presently authorized and conducted, violate Fourth Amendment rights as those rights are enforced against the States through the Fourteenth Amendment. 385 U. S. 808.

Appellant brought this action in a California Superior Court alleging that he was awaiting trial on a criminal charge of violating the San Francisco Housing Code by refusing to permit a warrantless inspection of his residence, and that a writ of prohibition should issue to the criminal court because the ordinance authorizing such inspections is unconstitutional on its face. The Superior Court denied the writ, the District Court of Appeal affirmed, and the Supreme Court of

California denied a petition for hearing. Appellant properly raised and had considered by the California courts the federal constitutional questions he now presents to this Court.

Though there were no judicial findings of fact in this prohibition proceeding, we shall set forth the parties' factual allegations. On November 6, 1963, an inspector of the Division of Housing Inspection of the San Francisco Department of Public Health entered an apartment bulding to make a routine annual inspection for possible violations of the city's Housing Code. * The building's manager informed the inspector that appellant, lessee of the ground floor, was using the rear of his leasehold as a personal residence. Claiming that the building's occupancy permit did not allow residential use of the ground floor, the inspector confronted appellant and demanded that he permit an inspection of the premises. Appellant refused to allow the inspection because the inspector lacked a search warrant.

The inspector returned on November 8, again without a warrant, and appellant again refused to allow an inspection. A citation was then mailed ordering appellant to appear at the district attorney's office. When appellant failed to appear, two inspectors returned to his apartment on November 22. They informed appellant that he was required by law to permit an inspection under § 503 of the Housing Code:

> Sec. 503 RIGHT TO ENTER BUILDING. Authorized employees of the City departments or City agencies, so far as may be necessary for the performance of their duties, shall, upon presentation of proper credentials, have the right to enter, at reasonable times, any building, structure, or premises in the City to perform any duty imposed upon them by the Municipal code.

Appellant nevertheless refused the inspectors access to his apartment without a search warrant. Thereafter, a complaint was filed charging him with refusing to permit a lawful inspection in violation of § 507 of the Code. * * Appellant was arrested on December 2 and released on bail. When his demurrer to the criminal complaint was denied, appellant filed this petition for a writ of prohibition.

---

* The inspection was conducted pursuant to § 86(3) of the San Francisco Municipal Code, which provides that apartment house operators shall pay an annual license fee in part to defray the cost of periodic inspections of their buildings. The inspections are to be made by the Bureau of Housing Inspection "at least once a year and as often thereafter as may be deemed necessary." The permit of occupancy, which prescribes the apartment units which a building may contain, is not issued until the license is obtained.

* * "Sec. 507 PENALTY FOR VIOLATION. Any person, the owner or his authorized agent who violates, disobeys, omits, neglects, or refuses to comply with, or who resists or opposes the execution of any of the provisions of this Code, or any order of the Superintendent, the Director of Public Works, or the Director of Public Health made pursuant to this Code, shall be guilty of a misdemeanor and upon conviction thereof shall be punished by a fine not exceeding five hundred dollars ($500.00), or by imprisonment, not exceeding six (6) months or by both such fine and imprisonment, unless otherwise provided in this Code, and shall be deemed guilty of a separate offense for every day such violation, disobedience, omission, neglect or refusal shall continue."

Appellant has argued throughout this litigation that § 503 is contrary to the Fourth and Fourteenth Amendments in that it authorizes municipal officials to enter a private dwelling without a search warrant and without probable cause to believe that a violation of the Housing Code exists therein. Consequently, appellant contends, he may not be prosecuted under § 507 for refusing to permit an inspection unconstitutionally authorized by § 503. Relying on *Frank* v. *Maryland, Eaton* v. *Price,* and decisions in other States,† the District Court of Appeal held that § 503 does not violate Fourth Amendment rights because it "is part of a regulatory scheme which is essentially civil rather than criminal in nature, inasmuch as that section creates a right of inspection which is limited in scope and may not be exercised under unreasonable conditions." Having concluded that *Frank* v. *Maryland,* to the extent that it sanctioned such warrantless inspections, must be overruled, we reverse.

I.

The Fourth Amendment provides that, "The right of the people to be secure in their persons, houses, papers, and effects, against unreasonable searches and seizures, shall not be violated, and no Warrants shall issue, but upon probable cause, supported by Oath or affirmation, and particularly describing the place to be searched, and the persons or things to be seized." The basic purpose of this Amendment, as recognized in countless decisions of this Court, is to safeguard the privacy and security of individuals against arbitrary invasions by governmental officials. The Fourth Amendment thus gives concrete expression to a right of the people which "is basic to a free society," *Wolf* v. *Colorado,* 338 U. S. 25, 27. As such, the Fourth Amendment is enforceable against the States through the Fourteenth Amendment. *Ker* v. *California,* 374 U. S. 23, 30.

Though there has been general agreement as to the fundamental purpose of the Fourth Amendment, translation of the abstract prohibition against "unreasonable searches and seizures" into workable guidelines for the decision of particular cases is a difficult task which has for many years divided the members of this Court. Nevertheless, one governing principle, justified by history and by current experience, has consistently been followed: except in certain carefully defined classes of cases, a search of private property without proper consent is "unreasonable" unless it has been authorized by a valid search warrant. See, *e.g., Stoner* v. *California,* 376 U. S. 483; *United States* v. *Jeffers,* 342 U. S. 48; *McDonald* v. *United States,* 335 U. S. 451; *Agnello* v. *United States,* 269 U. S. 20. As the Court explained in *Johnson* v. *United States,* 333 U. S. 10, 14:

> The right of officers to thrust themselves into a home is also a grave concern, not only to the individual but to a society which chooses to dwell

---

† *Givner* v. *State,* 210 Md. 484, 124 A. 2d 764 (1956); *City of St. Louis* v. *Evans,* 337 S. W. 2d 948 (Mo. 1960); *State ex rel. Eaton* v. *Price,* 168 Ohio St. 123, 151 N. E. 2d 523 (1958), aff'd by an equally divided Court, 364 U. S. 263 (1960). See also *State* v. *Rees,* 258 Iowa 813, 139 N. W. 2d 406 (1966); *Commonwealth* v. *Hadley,* 351 Mass. 439, 222 N. E. 2d 681 (1966), appeal docketed Jan. 5, 1967, No. 1179, Misc., O. T. 1966; *People* v. *Laverne,* 14 N. Y. 2d 304, 200 N. E. 2d 441 (1964).

in reasonable security and freedom from surveillance. When the right of privacy must reasonably yield to the right of search is, as a rule, to be decided by a judicial officer, not by a policeman or government enforcement agent.

In *Frank* v. *Maryland,* this Court upheld the conviction of one who refused to permit a warrantless inspection of private premises for the purposes of locating and abating a suspected public nuisance. Although *Frank* can arguably be distinguished from this case on its facts,†† the Frank opinion has generally been interpreted as carving out an additional exception to the rule that warrantless searches are unreasonable under the Fourth Amendment. See *Eaton* v. *Price, supra.* The District Court of Appeal so interpreted *Frank* in this case, and that ruling is the core of appellant's challenge here. We proceed to a reexamination of the factors which persuaded the *Frank* majority to adopt this construction of the Fourth Amendment's prohibition against unreasonable searches.

To the *Frank* majority, municipal fire, health, and housing inspection programs "touch at most upon the periphery of the important interests safeguarded by the Fourteenth Amendment's protection against official intrusion," 359 U. S., at 367, because the inspections are merely to determine whether physical conditions exist which do not comply with minimum standards prescribed in local regulatory ordinances. Since the inspector does not ask that the property owner open his doors to a search for "evidence of criminal action" which may be used to secure the owner's criminal conviction, historic interests of "self-protection" jointly protected by the Fourth and Fifth Amendments‡ are said not to be involved, but only the less intense "right to be secure from intrusion into personal privacy." *Id.,* at 365.

We may agree that a routine inspection of the physical condition of private property is a less hostile intrusion than the typical policeman's search for the fruits and instrumentalities of crime. For this reason alone, *Frank* differed from the great bulk of Fourth Amendment cases which have been considered by this Court. But we cannot agree that the Fourth Amendment interests at stake in these inspection cases are merely "peripheral." It is surely anomalous to say that the individual and his private property are fully protected by the Fourth Amendment only when the individual is suspected of criminal behavior.‡‡ For instance, even

---

†† In *Frank,* the Baltimore ordinance required that the health inspector "have cause to suspect that a nuisance exists in any house, cellar or enclosure" before he could demand entry without a warrant, a requirement obviously met in *Frank* because the inspector observed extreme structural decay and a pile of rodent feces on the appellant's premises. Section 503 of the San Francisco Housing Code has no such "cause" requirement, but neither did the Ohio ordinance at issue in *Eaton* v. *Price,* a case which four Justices thought was controlled by *Frank.* 364 U. S., at 264, 265, n. 2 (opinion of MR. JUSTICE BRENNAN).

‡ See *Boyd* v. *United States,* 116 U. S. 616. Compare *Schmerber* v. *California,* 384 U. S. 757, 766-772.

‡‡ See *Abel* v. *United States,* 362 U. S. 217, 254-256 (MR. JUSTICE BRENNAN, dissenting); *District of Columbia* v. *Little,* 85 U. S. App. D. C. 242, 178 F.2d 13, aff'd, 339 U. S. 1.

the most law-abiding citizen has a very tangible interest in limiting the circumstances under which the sanctity of his home may be broken by official authority, for the possibility of criminal entry under the guise of official sanction is a serious threat to personal and family security. And even accepting *Frank*'s rather remarkable premise, inspections of the kind we are here considering do in fact jeopardize "self-protection" interests of the property owner. Like most regulatory laws, fire, health, and housing codes are enforced by criminal processes. In some cities, discovery of a violation by the inspector leads to a criminal complaint.|| Even in cities where discovery of a violation produces only an administrative compliance order,|| || refusal to comply is a criminal offense, and the fact of compliance is verified by a second inspection, again without a warrant.# Finally, as this case demonstrates, refusal to permit an inspection is itself a crime, punishable by fine or even by jail sentence.

The *Frank* majority suggested, and appellee reasserts, two other justifications for permitting administrative health and safety inspections without a warrant. First, it is argued that these inspections are "designed to make the least possible demand on the individual occupant." 359 U. S. at 367. The ordinances authorizing inspections are hedged with safeguards, and at any rate the inspector's particular decision to enter must comply with the constitutional standard of reasonableness even if he may enter without a warrant.## In addition, the argument proceeds, the warrant process could not function effectively in this field. The decision to inspect an entire municipal area is based upon legislative or administrative assessment of broad factors such as the area's age and condition. Unless the magistrate is to review such policy matters, he must issue a "rubber stamp" warrant which provides no protection at all to the property owner.

In our opinion, these arguments unduly discount the purposes behind the warrant machinery contemplated by the Fourth Amendment. Under the present system, when the inspector demands entry, the occupant has no way of knowing whether enforcement of the municipal code involved requires inspection of his premises, no way of knowing the lawful limits of the inspector's power to search, and no way of knowing whether the inspector himself is acting under proper authorization. These are questions which may be reviewed by a neutral magistrate without any reassessment of the basic agency decision to canvass an area. Yet, only by refusing entry and risking a crimial conviction can the occupant at present

---

|| See New York, NY, Administrative Code § D26-8.0 (1964).

|| || See Washington, DC, Housing Regulations § 2104.

#This is the more prevalent enforcement procedure. See Note, Enforcement of Municipal Housing Codes, 78 Harv. L. Rev. 801, 813-816.

##The San Francisco Code requires that the inspector display proper credentials, that he inspect "at reasonable times," and that he not obtain entry by force, at least when there is no emergency. The Baltimore ordinance in *Frank* required that the inspector "have cause to suspect that a nuisance exists." Some cities notify residents in advance, by mail or posted notice, of impending area inspections. State courts upholding these inspections without warrants have imposed a general reasonableness requirement. See cases cited, footnote †, page 346, *supra*.

challenge the inspector's decision to search. And even if the occupant possesses sufficient fortitude to take this risk, as appellant did here, he may never learn any more about the reason for the inspection than that the law generally allows housing inspectors to gain entry. The practical effect of this system is to leave the occupant subject to the discretion of the official in the field. This is precisely the discretion to invade private property which we have consistently circumscribed by a requirement that a disinterested party warrant the need to search. See cases cited, p. 529, *supra*. We simply cannot say that the protections provided by the warrant procedure are not needed in this context; broad statutory safeguards are no substitute for individualized review, particularly when those safeguards may only be invoked at the risk of a criminal penalty.

The final justification suggested for warrantless administrative searches is that the public interest demands such a rule: it is vigorously argued that the health and safety of entire urban populations is dependent upon enforcement of minimum fire, housing, and sanitation standards, and that the only effective means of enforcing such codes is by routine systematized inspection of all physical structures. Of course, in applying any reasonableness standard, including one of constitutional dimension, an argument that the public interest demands a particular rule must receive careful consideration. But we think this argument misses the mark. The question is not, at this stage at least, whether these inspections may be made, but whether they may be made without a warrant. For example, to say that gambling raids may never be made. In assessing whether the public interest demands creation of a general exception to the Fourth Amendment's warrant requirement, the question is not whether the public interest justifies the type of search in question, but whether the authority to search should be evidenced by a warrant, which in turn depends in part upon whether the burden of obtaining a warrant is likely to frustrate the governmental purpose behind the search. See *Schmerber* v. *California*, 384 U. S. 757, 770-771. It has nowhere been urged that fire, health, and housing code inspection programs could not achieve their goals within the confines of a reasonable search warrant requirement. Thus, we do not find the public need argument dispositive.

In summary, we hold that administrative searches of the kind at issue here are significant intrusions upon the interests protected by Fourth Amendment, that such searches when authorized and conducted without a warrant procedure lack the traditional safeguards which the Fourth Amendment guarantees to the individual, and that the reasons put forth in *Frank* v. *Maryland* and other cases for upholding these warrantless searches are insufficient to justify so substantial a weakening of the Fourth Amendment's protections. Because of the nature of the municipal programs under consideration, however, these conclusions must be the beginning, not the end, of our inquiry. The *Frank* majority gave recognition to the unique character of these inspection programs by refusing to require search warrants; to reject that disposition does not justify ignoring the question that other accommodation between public need and individual rights is essential.

II.

The Fourth Amendment provides that, "no Warrants shall issue, but upon probable cause." Borrowing from more typical Fourth Amendment cases, appellant argues not only that code enforcement inspection programs must be circumscribed by a warrant procedure, but also that warrants should issue only when the inspector possesses probable cause to believe that a particular dwelling contains violations of the minimum standards prescribed by the code being enforced. We disagree.

In cases in which the Fourth Amendment requires that a warrant to search be obtained, "probable cause" is the standard by which a particular decision to search is tested against the constitutional mandate of reasonableness. To apply this standard, it is obviously necessary first to focus upon the governmental interest which allegedly justifies official intrusion upon the constitutionally protected interests of the private citizen. For example, in a criminal invesigation, the police may undertake to recover specific stolen or contraband goods. But that public interest would hardly justify a sweeping search of an entire city conducted in the hope that these goods might be found. Consequently, a search for these goods, even with a warrant, is "reasonable" only when there is "probable cause" to believe that they will be uncovered in a particular dwelling.

Unlike the search pursuant to a criminal investigation, the inspection programs at issue here are aimed at securing city-wide compliance with minimum physical standards for private property. The primary governmental interest at stake is to prevent even the unintentional development of conditions which are hazardous to public health and safety. Because fires and epidemics may ravage large urban areas, because unsightly conditions adversely affect the economic values of neighboring structures, numerous courts have upheld the police power of municipalities to impose and enforce such minimum standards even upon existing structures.* In determining whether a particular inspection is reasonable — and thus in determining whether there is probable cause to issue a warrant for that inspection — the need for the inspection must be weighed in terms of these reasonable goals of code enforcement.

There is unanimous agreement among those most familiar with this field that the only effective way to seek universal compliance with the minimum standards required by municipal codes is through routine periodic inspections of all structures.** It is here that the probable cause debate is focused, for the agency's decision to conduct an area inspection is unavoidably based on its appraisal of conditions in the area as a whole, not on its knowledge of conditions in each particular building. Appellee contends that, if the probable cause standard urged by ap-

---

*See *Abbate Bros.* v. *City of Chicago,* 11 Ill. 2d 337, 142 N. E. 2d 691; *City of Louisville* v. *Thompson,* 339 S. W. 2d 869 (Ky.); *Adamec* v. *Post,* 273 N. Y. 250, 7 N. E. 2d 120; *Paquette* v. *City of Fall River,* 338 Mass. 368, 155 N. E. 2d 775; *Richards* v. *City of Columbia,* 227 S. C. 538, 88 S. E. 2d 683; *Boden* v. *City of Milwaukee,* 8 Wis. 2d 318, 99 N. W. 2d 156.

pellant is adopted, the area inspection will be eliminated as a means of seeking compliance with code standards and the reasonable goals of code enforcement will be dealt a crushing blow.

In meeting this contention, appellant argues first, that his probable cause standard would not jeopardize area inspection programs because only a minute portion of the population will refuse to consent to such inspections, and second, that individual privacy in any event should be given preference to the public interest in conducting such inspections. The first argument, even if true, is irrelevant to the question whether the area inspection is reasonable within the meaning of the Fourth Amendment. The second argument is in effect an assertion that the area inspection is an unreasonable search. Unfortunately, there can be no ready test for determining reasonableness other than by balancing the need to search against the invasion which the search entails. But we think that a number of persuasive factors combine to support the reasonableness of area code-enforcement inspections. First, such programs have a long history of judicial and public acceptance. See *Frank* v. *Maryland*, 359 U. S., at 367-371. Second, the public interest demands that all dangerous conditions be prevented or abated, yet it is doubtful that any other canvassing technique would achieve acceptable results. Many such conditions — faulty wiring is an obvious example — are not observable from outside the building and indeed may not be apparent to the inexpert occupant himself. Finally, because the inspections are neither personal in nature nor aimed at the discovery of evidence of crime, they involve a relatively limited invasion of the urban citizen's privacy. Both the majority and the dissent in *Frank* emphatically supported this conclusion:

> Time and experience have forcefully taught that the power to inspect dwelling places, either as a matter of systematic area-by-area search or, as here, to treat a specific problem, is of indispensable importance to the maintenance of community health; a power that would be greatly hobbled by the blanket requirement of the safeguards necessary for a search of evidence of criminal acts. The need for preventive action is great, and city after city has seen this need and granted the power of inspection to its health officials; and these inspections are apparently welcomed by all but an insignificant few. Certainly, the nature of our society has not vitiated the need for inspections first thought necessary 158 years ago, nor has experience revealed any abuse or inroad on freedom in meeting this need by

---

**See Osgood & Zwerner, Rehabilitation and Conservation, 25 Law & Contemp. Prob. 705, 718 and n. 43; Schwartz, Crucial Areas in Administrative Law, 34 Geo. Wash. L. Rev. 401, 423 and n. 93; Comment, Rent Withholding and the Improvement of Substandard Housing, 53 Calif. L. Rev. 304, 316-317; Note, Enforcement of Municipal Housing Codes, 78 Harv. L. Rev. 801. 807, 851; Note, Municipal Housing Codes, 69 Harv. L. Rev. 1115, 1124-1125. Section 311(a) of the Housing and Urban Development Act of 1965, 79 Stat. 478, 42 U. S. C. § 1468 (1964 Ed., Supp. I), authorizes grants of federal funds "to cities, other municipalities, and counties for the purpose of assisting such localities in carrying out programs of concentrated code enforcement in deteriorated or deteriorating areas in which such enforcement, together with those public improvements to be provided by the locality, may be expected to arrest the decline of the area."

means that history and dominant public opinion have sanctioned. 359 U. S., at 372.

. . . This is not to suggest that a health official need show the same kind of proof to a magistrate to obtain a warrant as one must who would search for the fruits or instrumentalities of crime. Where considerations of health and safety are involved, the facts that would justify an inference of "probable cause" to make an inspection are clearly different from those that would justify such an inference where a criminal investigation has been undertaken. Experience may show the need for periodic inspections of certain facilities without a further showing of cause to believe that substandard conditions dangerous to the public are being maintained. The passage of a certain period without inspection might of itself be sufficient in a given situation to justify the issuance of a warrant. The test of "probable cause" required by the Fourth Amendment can take into account the nature of the search that is being sought. 359 U. S., at 383 (MR. JUSTICE DOUGLAS, dissenting).

Having concluded that the area inspection is a "reasonable" search of private property within the meaning of the Fourth Amendment, it is obvious that "probable cause" to issue a warrant to inspect must exist if reasonable legislative or administrative standards for conducting an area inspection are satisfied with respect to a particular dwelling. Such standards, which will vary with the municipal program being enforced, may be based upon the passage of time, the nature of the building (*e.g.*, a multifamily apartment house), or the condition of the entire area, but they will not necessarily depend upon specific knowledge of the condition of the particular dwelling. It has been suggested that so to vary the probable cause test from the standard applied in criminal cases would be to authorize a "synthetic search warrant" and thereby to lessen the overall protections of the Fourth Amendment. *Frank* v. *Maryland*, 359 U. S. at 373. But we do not agree. The warrant procedure is designed to guarantee that a decision to search private property is justified by a reasonable governmental interest. But reasonableness is still the ultimate standard. If a valid public interest justifies the intrusion contemplated, then there is probable cause to issue a suitably restricted search warrant. Cf. *Oklahoma Press Pub. Co.* v. *Walling*, 327 U. S. 186. Such an approach neither endangers time-honored doctrines applicable to criminal investigations nor makes a nullity of the probable cause requirement in this area. It merely gives full recognition to the competing public and private interests here at stake and, in so doing, best fulfills the historic purpose behind the constitutional right to be free from unreasonable government invasions of privacy. See *Eaton* v. *Price*, 364 U. S., at 273-274 (opinion of MR. JUSTICE BRENNAN).

III.

Since our holding emphasizes the controlling standard of reasonableness, nothing we say today is intended to foreclose prompt inspections, even without a warrant, that the law has traditionally upheld in emergency situations. See *North American Cold Storage Co.* v. *City of Chicago,* 211 U. S. 306 (seizure of unwholesome food); *Jacobson* v. *Massachusetts,* 197 U. S. 11 (compulsory smallpox vaccination); *Compagnie Francaise* v. *Board of Health,* 186 U. S. 380 (health quarantine); *Kroplin* v. *Truax,* 119 Ohio St. 610, 165 N. E. 498 (summary destruction of tubercular cattle). On the other hand, in the case of most routine area inspections, there is no compelling urgency to inspect at a particular time or on a particular day. Moreover, most citizens allow inspections of their property without a warrant. Thus, as a practical matter and in light of the Fourth Amendment's requirement that a warrant specify the property to be searched, it seems likely that warrants should normally be sought only after entry is refused unless there has been a citizen complaint or there is other satisfactory reason for securing immediate entry. Similarly, the requirement of a warrant procedure does not suggest any change in what seems to be the prevailing local policy, in most situations, of authorizing entry, but not entry by force, to inspect.

IV.

In this case, appellant has been charged with a crime for his refusal to permit housing inspector to enter his leasehold without a warrant. There was no emergency demanding immediate access; in fact, the inspectors made three trips to the building in an attempt to obtain appellant's consent to search. Yet no warrant was obtained and thus appellant was unable to verify either the need for or the appropriate limits of the inspection. No doubt, the inspectors entered the public portion of the building with the consent of the landlord, through the building's manager but appellee does not contend that such consent was sufficient to authorize inspection of appellant's premises. Cf. *Stoner* v. *California,* 376 U. S. 483; *Chapman* v. *United States,* 365 U. S. 610; *McDonald* v. *United States,* 335 U. S. 451. Assuming the facts to be as the parties have alleged, we therefore conclude that appellant had a constitutional right to insist that the inspectors obtain a warrant to search and that appellant may not constitutionally be convicted for refusing to consent to the inspection. It appears from the opinion of the District Court of Appeal that under these circumstances a writ of prohibition will issue to the criminal court under California law.

The judgment is vacated and the case is remanded for further proceedings not inconsistent with this opinion. The dissenting opinion for this case and the next case appears on page 357.

It is so ordered.

In the following case of See v. *City of Seattle,* consider again an individual's constitutional rights *vs.* the protection of society as a whole from fire. Consider also the effect of the court's decision on fire department code enforcement efforts.

SEE v. CITY OF SEATTLE

Appeal from the Supreme Court of Washington.
No. 180. Argued February 15, 1967. — Decided June 5, 1967.
387 U.S. 541

MR. JUSTICE WHITE delivered the opinion of the Court.

Appellant seeks reversal of his conviction for refusing to permit a representative of the City of Seattle Fire Department to enter and inspect appellant's locked commercial warehouse without a warrant and without probable cause to believe that a violation of any municipal ordinance existed therein. The inspection was conducted as part of a routine, periodic, city-wide canvass to obtain compliance with Seattle's Fire Code. City of Seattle Ordinance No. 87870. c. 8.01. After he refused the inspector access, appellant was arrested and charged with violating § 8.01.050 of the Code:

> INSPECTION OF BUILDING AND PREMISES. It shall be the duty of the Fire Chief to inspect and he may enter all buildings and premises, except the interiors of dwellings, as often as may be necessary for the purpose of ascertaining and causing to be corrected any conditions liable to cause fire, or any violations of the provisions of this Title, and of any other ordinance concerning fire hazards.

Appellant was convicted and given a suspended fine of $100* despite his claim that § 8.01.050, if interpreted to authorize this warrantless inspection of his warehouse, would violate his rights under the Fourth and Fourteenth Amendments. We noted probable jurisdiction and set this case for argument with *Camara* v. *Municipal Court, ante,* p. 523. 385 U. S. 808. We find the principles enunciated in the *Camara* opinion applicable here and therefore we reverse.

In *Camara,* we held that the Fourth Amendment bars prosecution of a person who has refused to permit a warrantless code-enforcement inspection of his personal residence. The only question which this case presents is whether *Camara* applies to similar inspections of commercial structures which are not used as private residences. The Supreme Court of Washington, in affirming appellant's conviction, suggested that this Court "has applied different standards of reasonableness to searches of dwellings than to places of business," citing *Davis* v. *United States,* 328 U. S. 582. The Washington court held, and appellee here argues, that

---

*Conviction and sentence were pursuant to § 8.01.140 of the Fire Code:

"PENALTY. Anyone violating or failing to comply with any provision of this Title or lawful order of the Fire Chief pursuant hereto shall upon conviction thereof be punishable by a fine not to exceed Three Hundred Dollars ($300.00), or imprisonment in the City Jail for a period not to exceed ninety (90) days, or by both such fine and imprisonment, and each day of violation shall constitute a separate offense."

§ 8.01.050, which excludes "the interiors of dwellings,"** establishes a reasonable scheme for the warrantless inspection of commercial premises pursuant to the Seattle Fire Code.

In *Go-Bart Importing Co.* v. *United States,* 282 U. S. 344; *Amos* v. *United States,* 255 U. S. 313; and *Silverthorne Lumber Co.* v. *United States,* 251 U. S. 385, this Court refused to uphold otherwise unreasonable criminal investigative searches merely because commercial rather than residential premises were the object of the police intrusions. Likewise, we see no justification for so relaxing Fourth Amendment safeguards where the official inspection is intended to aid enforcement of laws prescribing minimum physical standards for commercial premises. As we explained in *Camara,* a search of private houses is presumptively unreasonable if conducted without a warrant. The businessman, like the occupant of a residence, has a constitutional right to go about his business free from unreasonable official entries upon his private commercial property. The businessman, too, has that right placed in jeopardy if the decision to enter and inspect for violation of regulatory laws can be made and enforced by the inspector in the field without official authority evidenced by a warrant.

As governmental regulation of business enterprise has mushroomed in recent years, the need for effective investigative techniques to achieve the aims of such regulation has been the subject of substantial comment and legislation.† Official entry upon commercial property is a technique commonly adopted by administrative agencies at all levels of government to enforce a variety of regulatory laws; thus, entry may permit inspection of the structure in which a business is housed, as in this case, or inspection of business products, or a perusal of financial books and records. This Court has not had occasion to consider the Fourth Amendment's relation to this broad range of investigations.†† However, we have dealt with the Fourth Amendment issues raised by another common investigative technique, the administrative subpoena of corporate books and records. We find strong support in these subpoena cases for our conclusion that warrants are a necessary and a tolerable limitation on the right to enter upon and inspect commercial premises.

**"Dwelling" is defined in the Code as "a building occupied exclusively for residential purposes and having not more than two (2) dwelling units." Such dwellings are subject to the substantive provisions of the Code, but the Fire Chief's right to enter such premises is limited to times "when he has reasonable cause to believe a violation of the provisions of this Title exists therein." § 8.01.040. This provision also lacks a warrant procedure.

†See Antitrust Civil Process Act of 1962, 76 Stat. 548, 15 U. S. C. §§ 1311–1314; H. R. Rep. No. 708, 83d Cong., 1st Sess. (1953) (reporting the "factory inspection" amendments to the Federal Food, Drug, and Cosmetic Act, 67 Stat. 476, 21 U. S. C. § 374); Davis, The Administrative Power of Investigation, 56 Yale L. J. 1111; Handler, The Constitutionality of Investigations by the Federal Trade Commission, I & II, 28 Col. L. Rev. 708, 905; Schwartz, Crucial Areas in Administrative Law, 34 Geo. Wash. L. Rev. 401, 425–430; Note, Constitutional Aspects of Federal Tax Investigations, 57 Col. L. Rev. 676.

††In *United States* v. *Cardiff,* 344 U. S. 174, this Court held that the Federal Food, Drug, and Cosmetic Act did not compel that consent be given to warrantless inspections of establishments covered by the Act. (As a result, the statute was subsequently amended, see preceding footnote †, *supra.*) See also *Federal Trade Comm'n* v. *American Tobacco Co.,* 264 U. S. 298.

It is now settled that, when an administrative agency subpoenas corporate books or records, the Fourth Amendment requires that the subpoena be sufficiently limited in scope, relevant in purpose, and specific in directive so that compliance will not be unreasonably burdensome.‡ The agency has the right to conduct all reasonable inspections of such documents which are contemplated by statute, but it must delimit the confines of a search by designating the needed documents in a formal subpoena. In addition, while the demand to inspect may be issued by the agency, in the form of an administrative subpoena, it may not be made and enforced by the inspector in the field, and the subpoenaed party may obtain judicial review of the reasonableness of the demand prior to suffering penalities for refusing to comply.

It is these rather minimal limitations on administrative action which we think are constitutionally required in the case of investigative entry upon commercial establishments. The agency's particular demand for access will of course be measured, in terms of probable cause to issue a warrant, against a flexible standard of reasonableness that takes into account the public need for effective enforcement of the particular regulation involved. But the decision to enter and inspect will not be the product of the unreviewed discretion of the enforcement officer in the field.‡‡ Given the analogous investigative functions performed by the administrative subpoena and the demand for entry, we find untenable the proposition that the subpoena, which has been termed a "constructive" search, *Oklahoma Press Pub. Co.* v. *Walling,* 327 U. S. 186, 202, is subject to Fourth Amendment limitations which do not apply to actual searches and inspections of commercial premises.

We therefore conclude that administrative entry, without consent, upon the portions of commercial premises which are not open to the public may only be compelled through prosecution or physical force within the framework of a warrant procedure.|| We do not in any way imply that business premises may not reasonably be inspected in many more situations than private homes, nor do we question such accepted regulatory techniques as licensing programs which require inspections prior to operating a business or marketing a product. Any constitutional challenge to such programs can only be resolved, as many have been in the past, on a case-by-case basis under the general Fourth Amendment standard of

---

‡See *United States* v. *Morton Salt Co.,* 338 U. S. 632; *Oklahoma Press Pub. Co.* v. *Walling,* 327 U. S. 186; *United States* v. *Bausch & Lomb Optical Co.,* 321 U. S. 707; *Hale* v. *Henkel,* 201 U. S. 43. See generally 1 Davis, Administrative Law, §§ 3.05–3.06 (1958).

‡‡We do not decide whether warrants to inspect business premises may be issued only after access is refused; since surprise may often be a crucial aspect of routine inspections of business establishments, the reasonableness of warrants issued in advance of inspection will not necessarily vary with the nature of the regulation involved and may differ from standards applicable to private homes.

||*Davis* v. *United States,* 328 U. S. 582, relied upon by the Supreme Court of Washington, held only that government officials could demand access to business premises and, upon obtaining consent to search, could seize gasoline ration coupons issued by the Government and illegally possessed by the petitioner. *Davis* thus involved the reasonableness of a particular search of business premises but did not involve a search warrant issue.

reasonableness. We hold only that the basic component of a reasonable search under the Fourth Amendment — that it not be enforced without a suitable warrant procedure — is applicable in this context, as in others, to business as well as to residential premises. Therefore, appellant may not be prosecuted for exercising his constitutional right to insist that the fire inspector obtain a warrant authorizing entry upon appellant's locked warehouse.

Reversed.

MR. JUSTICE CLARK, with whom MR. JUSTICE HARLAN and MR. JUSTICE STEWART join, dissenting.|| ||

Eight years ago my Brother Frankfurter wisely wrote in *Frank* v. *Maryland*, 359 U. S. 360 (1959):

> "Time and experience have forcefully taught that the power to inspect dwelling places, either as a matter of systematic area-by-area search or, as here, to treat a specific problem, is of indispensable importance to the maintenance of community health; a power that would be greatly hobbled by the blanket requirement of the safeguards necessary for a search of evidence of criminal acts. The need for preventive action is great, and city after city has seen this need and granted the power of inspection to its health officials; and these inspections are apparently welcomed by all but an insignificant few." At 372.

Today the Court renders this municipal experience, which dates back to Colonial days, for naught by overruling *Frank* v. *Maryland* and by striking down hundreds of city ordinances throughout the country and jeopardizing thereby the health, welfare, and safety of literally millions of people.

But this is not all. It prostitutes the command of the Fourth Amendment that "no Warrants shall issue, but upon probable cause" and sets up in the health and safety codes area inspection a newfangled "warrant" system that is entirely foreign to Fourth Amendment standards. It is regrettable that the Court wipes out such a long and widely accepted practice and creates in its place such enormous confusion in all of our towns and metropolitan cities in one fell swoop. I dissent.

I.

I shall not treat in any detail the constitutional issue involved. For me it was settled in *Frank* v. *Maryland, supra.* I would adhere to that decision and the reasoning therein of my late Brother Frankfurter. Time has not shown any need for change. Indeed the opposite is true, as I shall show later. As I read it, the Fourth

---

|| ||This opinion applies also to No. 92, *Camara* v. *Municipal Court of the City and County of San Francisco, ante,* p. 523.

Amendment guarantee of individual privacy is, by its language, specifically qualified. It prohibits only those searches that are "unreasonable." The majority seem to recognize this for they set up a new test for the long-recognized and enforced Fourth Amendment's "probable-cause" requirement for the issuance of paper warrants, in area inspection programs, with probable cause based on area inspection standards as set out in municipal codes, and with warrants issued by the rubber stamp of a willing magistrate.# In my view, this degrades the Fourth Amendment.

II.

Moreover, history supports the *Frank* disposition. Over 150 years of city *in rem* inspections for health and safety purposes have continuously been enforced. In only one case during all that period have the courts denied municipalities this right. See *District of Columbia* v. *Little*, 85 U. S. App. D. C. 242, 178 F.2d 13 (1949), aff'd on other grounds, 339 U. S. 1 (1950). In addition to the two cases in this Court [*Frank, supra,* and *Eaton* v. *Price*, 364 U. S. 263 (1960)], which have upheld the municipal action, not a single state high court has held against the validity of such ordinances. Indeed, since our *Frank* decision five of the States' highest courts have found that reasonable inspections are constitutionally permissible and in fact imperative, for the protection of health, safety, and welfare of the millions who inhabit our cities and towns.##

I submit that under the carefully circumscribed requirements of health and safety codes, as well as the facts and circumstances of these particular inspections, there is nothing unreasonable about the ones undertaken here. These inspections meet the Fourth Amendment's test of reasonableness and are entirely consistent with the Amendment's commands and our cases.

There is nothing here that suggests that the inspection was unauthorized, unreasonable, for any improper purpose, or designed as a basis for a criminal prosecution; nor is there any indication of any discriminatory, arbitrary, or capricious action affecting the appellant in either case. Indeed, Camara was admittedly violating the Code by living in quarters prohibited thereby; and See was operating a locked warehouse — a business establishment subject to inspection.

The majority say, however, that under the present system the occupant has no way of knowing the necessity for the inspection, the limits of the inspector's power, or whether the inspector is himself authorized to perform the search. Each

---

#Under the probable-cause standard laid down by the Court, it appears to me that the issuance of warrants could more appropriately be the function of the agency involved than that of the magistrate. This would also relieve magistrates of an intolerable burden. It is therefore unfortunate that the Court fails to pass on the validity of the use of administrative warrants.

##*DePass* v. *City of Spartanburg*, 234 S. C. 198, 107 S. E. 2d 350 (1959); *City of St. Louis* v. *Evans*, 337 S. W. 2d 948 (Mo. 1960); *Camara* v. *Municipal Court*, 237 Cal. App. 2d 128, 46 Cal. Rptr. 585 (1965), pet. for hearing in Cal. Sup. Ct. den. (Civ. No. 22128) Nov. 19, 1965; *Commonwealth* v. *Hadley*, 351 Mass. 439, 222 N. E. 2d 681, appeal docketed, Jan. 5, 1967, No. 1179, Misc., O. T. 1966; *City of Seattle* v. *See*, 67 Wash. 2d 475, 408 P. 2d 262 (1965).

of the ordinances here is supported by findings as to the necessity for inspections of this type and San Francisco specifically bans the conduct in which appellant Camara is admittedly engaged. Furthermore, all of these doubts raised by the Court could be resolved very quickly. Indeed, the inspectors all have identification cards which they show the occupant and the latter could easily resolve the remaining questions by a call to the inspector's superior or, upon demand, receive a written answer thereto. The record here shows these challenges could have been easily interposed. The inspectors called on several occasions, but still no such questions were raised.* These cases, from the outset, were based on the Fourth Amendment, not on any of the circumstances surrounding the attempted inspection. To say, therefore, that the inspection is left to the discretion of the officer in the field is to reach a conclusion not authorized by this record or the ordinances involved here. The Court says the question is not whether the "inspections may be made, but whether they may be made without a warrant." With due respect, inspections of this type have been made for over a century and a half without warrants and it is a little late to impose a death sentence on such procedures now. In most instances the officer could not secure a warrant — such as in See's case — thereby insulating large and important segments of our cities from inspection for health and safety conditions. It is this situation — which is even recognized by the Court — that should give us pause.

III.

The great need for health and safety inspection is emphasized by the experience of San Francisco, a metropolitan area known for its cleanliness and safety ever since it suffered earthquake and fire back in 1906. For the fiscal year ending June 30, 1965, over 16,000 dwelling structures were inspected, of which over 5,600 required some type of compliance action in order to meet code requirements. And in 1965-1966 over 62,000 apartments, hotels, and dwellings were inspected with similar results. During the same period the Public Works Department conducted over 52,000 building inspections, over 43,000 electrical ones and over 33,000 plumbing inspections. During the entire year 1965-1966 inspectors were refused entry on less than 10 occasions where the ordinance required the householder to so permit.

In Seattle, the site of No. 180, *See* v. *City of Seattle*, fire inspections of commercial and industrial buildings totaled over 85,000 in 1965. In Jacksonville, Florida, over 21,000 fire inspections were carried on in the same year, while in excess of 135,000 health inspections were conducted. In Portland, Oregon, out of 27,000 health and safety inspections over 4,500 violations of regulations were uncovered and the fire marshal in Portland found over 17,000 violations of the fire code in 1965 alone. In Boston over 56,000 code violations were uncovered in 1966 while in Baltimore a somewhat similar situation was reported.

---

*Indeed, appellant Camara was summoned to the Office of the District Attorney — but failed to appear — where he certainly could have raised these questions.

In the larger metropolitan areas such as Los Angeles, over 300,000 inspections (health and fire) revealed over 28,000 hazardous violations. In Chicago during the period November 1965 to December 1966, over 18,000 buildings were found to be rodent-infested out of some 46,000 inspections. And in Cleveland the division of housing found over 42,000 violations of its code in 1965; its health inspectors found over 33,000 violations in commercial establishments alone and over 27,000 dwelling code infractions were reported in the same period. And in New York City the problem is even more acute. A grand jury in Brooklyn conducted a housing survey of 15 square blocks in three different areas and found over 12,000 hazardous violations of code restrictions in those areas alone. Prior to this test there were only 567 violations reported in the three areas. The pressing need for inspection is shown by the fact that some 12,000 additional violations were actually present at that very time.

An even more disastrous effect will be suffered in plumbing violations. These are not only more frequent but also the more dangerous to the community. Defective plumbing causes back siphonage of sewage and other household wastes. Chicago's disastrous amoebic dysentery epidemic is an example. Over 100 deaths resulted. Fire code violations also often cause many conflagrations. Indeed, if the fire inspection attempted in *District of Columbia* v. *Little,* 339 U. S. 1 (1950), had been permitted a two-year-old child's death resulting from a fire that gutted the home involved there on August 6, 1949, might well have been prevented.

Inspections also play a vital role in urban redevelopment and slum clearance. Statistics indicate that slums constitute 20% of the residential area of the average American city, still they produce 35% of the fires, 45% of the major crimes, and 50% of the disease. Today's decision will play havoc with the many programs now designed to aid in the improvement of these areas. We should remember the admonition of MR. JUSTICE DOUGLAS in *Berman* v. *Parker,* 348 U. S. 26, 32 (1954):

> Miserable and disreputable housing conditions may do more than spread disease and crime and immorality. They may also suffocate the spirit by reducing the people who live there to the status of cattle. They may indeed make living an almost insufferable burden.

IV.

The majority propose two answers to this admittedly pressing problem of need for constant inspection of premises for fire, health, and safety infractions of municipal codes. First, they say that there will be few refusals of entry to inspect. Unlike the attitude of householders as to codes requiring entry for inspection, we have few empirical statistics on attitudes where consent must be obtained. It is true that in the required entry-to-inspect situations most occupants welcome the periodic visits of municipal inspectors. In my view this will not be true when consent is necessary. The City of Portland, Oregon, has a voluntary home inspection program. The 1966 record shows that out of 16,171 calls where the occupant was at home, entry was refused in 2,540 cases — approximately one out of six. This is a large percentage and would place an intolerable burden on the inspection service when required to secure warrants. What is more important is that out of the houses inspected 4,515 hazardous conditions were found! Hence, on the same percentage, there would be approximately 840 hazardous situations in the 2,540 in which inspection was refused in Portland.

Human nature being what it is, we must face up to the fact that thousands of inspections are going to be denied. The economics of the situation alone will force this result. Homeowners generally try to minimize maintenance costs and some landlords make needed repairs only when required to do so. Immediate prospects for costly repairs to correct possible defects are going to keep many a door closed to the inspector. It was said by way of dissent in Frank v. Maryland, supra, at 384, that "[o]ne rebel a year" is not too great a price to pay for the right to privacy. But when voluntary inspection is relied upon this "one rebel" is going to become a general rebellion. That there will be a significant increase in refusals is certain and, as time goes on, that trend may well become a frightening reality. It is submitted that voluntary compliance cannot be depended upon.

The Court then addresses itself to the propriety of warrantless area inspections.** The basis of "probable cause" for area inspection warrants, the Court says, begins with the Fourth Amendment's reasonableness requirement; in determining whether an inspection is reasonable "the need for the inspection must be weighed in terms of these reasonable goals of code enforcement." It adds that there are "a number of persuasive factors" supporting "the reasonableness of area code-enforcement inspections." It is interesting to note that the factors the Court relies upon are the identical ones my Brother Frankfurter gave for excusing warrants in Frank v. Maryland, supra. They are: long acceptance historically; the great public interest in health and safety; and the impersonal nature of the inspections — not for evidence of crime — but for the public welfare. Upon this reasoning, the Court concludes that probable cause exists "if reasonable legislative or administrative standards for conducting an area inspection are satisfied with respect to a particular dwelling." These standards will vary, it says, according to the code program and the condition of the area with reference thereto rather than the condition of a particular dwelling. The majority seem to hold that warrants may be obtained after a refusal of initial entry; I can find no such constitutional distinction or command. These boxcar warrants will be identical as to every dwelling in the area, save the street number itself. I daresay they will be printed up in pads of a thousand or more — with space for the street number to be inserted — and issued by magistrates in broadcast fashion as a matter of course.

**It is interesting to note that in each of the cases here the authorities were making periodic area inspections when the refusals to allow entry occurred. Under the holding of the Court today, "probable cause" would therefore be present in each case and a "paper warrant" would issue as a matter of course. This but emphasizes the absurdity of the holding.

I ask: Why go through such an exercise, such a pretense? As the same essentials are being followed under the present procedures, I ask: Why the ceremony, the delay, the expense, the abuse of the search warrant? In my view this will not only destroy its integrity but will degrade the magistrate issuing them and soon bring disrepute not only upon the practice but upon the judicial process. It will be very costly to the city in paperwork incident to the issuance of the paper warrants, in loss of time of inspectors and waste of the time of magistrates and will result in more annoyance to the public. It will also be more burdensome to the occupant of the premises to be inspected. Under a search warrant the inspector can enter any time he chooses. Under the existing procedures he can enter only at reasonable times and invariably the convenience of the occupant is considered. I submit that the identical grounds for action elaborated today give more support — both legal and practical — to the present practice as approved in *Frank* v. *Maryland, supra,* than they do to this legalistic facade that the Court creates. In the Court's anxiety to limit its own holding as to mass searches it hopes to divert attention from the fact that it destroys the health and safety codes as they apply to individual inspections of specific problems as contrasted to area ones. While the latter are important, the individual inspection is often more so; that was true in *District of Columbia* v. *Little* and it may well be in both *Camara* and *See.* Frankly, I cannot understand how the Court can authorize warrants in wholesale fashion in the case of an area inspection, but hold the hand of the inspector when a specific dwelling is hazardous to the health and safety of its neighbors.

As a result of the *Camara* and *See* decisions, fire service inspection procedures were carefully scrutinized. Chief Charles W. Bahme, in the 1967 Edition of his book titled *The Fireman's Law Book,* comments on fire department inspection procedures as follows:[14]

> As a result of the above decision [*Camara* v. *See*] it would be advisable to program all routine building inspections on the basis of an orderly geographical approach; if the inspections are planned so that the firemen will proceed through their districts on a building-to-building — block-to-block — basis, and from one adjoining battalion into the next, then when that rare objection is voiced on the part of some building owner to having his premises inspected, it can be easily shown that his building is not being "singled out" for a shakedown, but was reached in the normal course of the planned sequence of inspections. Even in Seattle, Chief Vickery told the author that he only has two or three persons per hundred thousand inspections who resist the firemen's attempt to enter their premises. In these few instances, there should be no difficulty in obtaining a search warrant where the above inspection procedure has been followed, since there could be no valid accusation that the fire department is trying to persecute the person objecting, and it could be easily established that such entry was required to determine that the premises involved did not create an undue hazard to the adjoining neighbors.

---

[14]Charles W. Bahme, *The Fireman's Law Book,* NFPA, Boston, 1967, pp. 141–142.

The National Fire Protection Association, as a result of the decisions in the *Camara* and *See* cases, published suggested criteria to meet Supreme Court guidelines. The criteria are as follows:[15]

Suggested Criteria to Meet the United States Supreme Court Guidelines

1. Discuss these guidelines and the recent U. S. Supreme Court decisions with your legal counsel. Enforcement procedures contemplated should be discussed to determine how they will apply to your fire prevention program and procedures.

2. Establish adequate identification for the inspection staff. These should include:
   (a) Fire department badge with a number.
   (b) Identification card with a photograph of the individual and of the badge number.

3. Brief the staff on the procedure to follow on initiating an inspection. When requesting entry:
   (a) The inspector should make verbal identification.
   (b) Show an I.D. card.
   (c) Explain the purpose of the inspection.
   (d) Request permission of the person in charge to carry out the inspection.
   (e) Invite the person in charge to accompany the inspector on the inspection tour.

4. The fire prevention inspection authority should develop formal, written inspection procedures spelling out how, when, where, and what inspections are to be carried out. The book should include information on the importance of fire department preplanning techniques, since information in regard to building construction and occupancy is an aid to successful fire fighting. Thus, if entry is denied, the information in writing is available to a magistrate indicating that the inspection request was made as one of an orderly and programmed series of inspections. It would be good policy to brief magistrates in advance in regard to inspectional procedures customarily followed.

> *Example:* The inspector is denied entry into a mercantile occupancy in the 600 block of Main Street, for no apparent reason, during a routine series of block-by-block inspections. The inspector proceeds to a magistrate and requests a search warrant for entry to inspect for firesafety hazards. In order to establish that the particular inspection is a part of an area-wide inspection procedure, the inspectional procedures book giving detailed inspection plans and methods may be presented as evidence.

5. If the inspector is denied entry and has probable cause to believe that an extreme or unusually hazardous operation or condition exists which could mean a loss of life while obtaining a warrant, the inspector should proceed to have the operation halted or the condition corrected even if it means having the person denying entry placed under arrest.

---

[15]Robert W. Grant, *Public Fire Safety Inspections*, NFPA, 1967, pp. 24–25.

Example: A fire inspector obtains information that there is a large gathering of people in a particular building with the number of persons far exceeding the exit capacity, several of the exits blocked by chairs, extremely combustible decorations, and there are live candles being used in the audience. When the inspector makes verbal identification to the building manager, the manager denies entry. The inspector would be justified in halting the performance, and, if the manager interferes, having the manager placed under arrest if necessary.

6. If the inspection authority anticipates problems of entry at a particular address on a routine area inspection, the inspector may go to a magistrate and request a warrant. The warrant may be based on the fact that the area inspection is being carried out and the Inspectional Procedures Book may be presented as evidence. Therefore, if the inspector is refused entry, the inspector may then present the warrant for entry at the address specified. This also provides for the element of surprise where it is anticipated conditions may be changed if the warrant is not sought until entry has been refused.

Example: A manufacturing concern at 203 Main Street has previously refused entry of an authorized fire inspector on a routine area inspection. It was necessary for the inspector to obtain a warrant for entry into the building. During an ensuing period, a program of public relations was carried out, but to no avail with the manufacturing concern. The calendar of routine inspections shows that the time has arrived for the next inspection. The inspection authority may obtain a warrant before approaching the specified occupancy. Therefore, a warrant may be available for immediate entry if necessary.

7. Develop a good inspection records system by address, owner, and occupant. Records should be kept of all inspection activities showing dates of inspection, name of inspector, hazards detected, and corrections secured, as well as other information pertinent to firesafety inspections of a particular occupancy or address. This information can help prove probable cause to issue a warrant. The records should be in a form available to present to a magistrate at the time a warrant is requested.

Example: Inspection of a hotel reveals that the panic hardware on the rear door is chained to prevent the patrons from letting other persons up the rear stairs. The hotel manager removes the chain. Inspections are made the next two weeks and the same condition exists each time. On the fourth inspection, entry is denied. The inspector requests a warrant and presents as probable cause records of previous inspections indicating the need for frequent inspections of the occupancy. The prior record shows that hazardous conditions or ordinance violations have existed.

8. Establish guidelines on what constitutes an extreme emergency situation whereupon the inspector could proceed to enter the building without a warrant and halt the operation or condition, obtaining assistance if necessary.

This could be an emergency situation which is very similar to an actual fire or explosion, except that the ignition may not have occurred. Situations may also exist which immediately affect the safety of occupants and/or the general public.

> *Example:* The fire department receives a call that there is an odor of gas in the vicinity of a fruit ripening plant. When the fire company and fire inspector arrive on the scene, they are denied entry by the manager who says he is able to handle the situation. The inspector would be justified, in the interest of public safety, on insisting on making a personal inspection of the hazard and seeing that the proper precautions have been taken. In all such cases it would be proper procedure to promptly report the action taken to the inspector's superior.

9. Review licensing and permit procedures. Indications should be made on the permits and licenses that inspections will be made to assure that the requirements of the permit or license are being complied with. The refusal of proper entry into a building being inspected for the issuance of a license would justify an automatic denial of license.

10. Fire inspectors should be well trained in the recognition of fire hazards, fire prevention procedures and techniques, and the laws and ordinances that the inspector is expected to enforce.

11. Develop a healthy public relations program for your fire prevention authority. Provide information in a friendly manner on what is expected in regard to cooperation between the public and the fire prevention authority. Copies of all codes and ordinances should be readily available to the public. Fire prevention authorities should explain how their regulations and inspection procedures are preventing the loss of life and property.

12. Train your inspectors to be courteous, friendly, and helpful in their relationship with the public. During inspections, they should be businesslike and should display an objective attitude to the public.

13. Fire inspectors should be enthusiastic salespeople for firesafety and show a genuine concern for the public they serve.

The constitutionality of administrative searches was again addressed by the Supreme Court in 1971, in the following case of *Wyman, Commissioner of New York Department of Social Services, et al.* v. *James.* Compare this ruling with the rulings in *Camara* and *See.*

WYMAN, COMMISSIONER OF NEW YORK DEPARTMENT OF SOCIAL SERVICES, Et Al. v. JAMES

Appeal from the United States District Court for the
Southern District of New York.
400 U.S. 309
No. 69. Argued October 20, 1970. — Decided January 12, 1971.

MR. JUSTICE BLACKMUN delivered the opinion of the Court.

This appeal presents the issue whether a beneficiary of the program for Aid to Families with Dependent Children (AFDC)* may refuse a home visit by the caseworker without risking the termination of benefits.

The New York State and City social services commissioners appeal from a judgment and decree of a divided three-judge District Court holding invalid and unconstitutional in application § 134 of the New York Social Service Law,** § 175 of the New York Policies Governing the Administration of Public Assistance,† and §§ 351.10 and 351.21 of Title 18 of the New York Code of Rules and Regulations,†† and granting injunctive relief. *James* v. *Goldberg,* 303 F. Supp. 935 (SDNY 1969). This Court noted probable jurisdiction but, by a divided vote, denied a requested stay. 397 U. S. 904.

---

*In *Goldberg* v. *Kelly,* 397 U. S. 254, 256 n. 1 (1970), the Court observed that AFDC is a categorical assistance program supported by federal grants-in-aid but administered by the States according to regulations of the Secretary of Health, Education, and Welfare. See New York Social Services Law, §§ 343–362 (1966 and Supp. 1969–1970). Aspects of AFDC have been considered in *King* v. *Smith,* 392 U. S. 309 (1968); *Shapiro* v. *Thompson,* 384 U. S. 618 (1969); *Goldberg* v. *Kelly, supra; Rosado* v. *Wyman,* 397 U. S. 397 (1970); and *Dandridge* v. *Williams,* 397 U. S. 471 (1970).

**"§ 134. Supervision
"The public welfare officials responsible . . . for investigating any application for public assistance and care, shall maintain close contact with persons granted public assistance and care. Such persons shall be visited as frequently as is provided by the rules of the board and/or regulations of the department or required by the circumstances of the case, in order that any treatment or service tending to restore such persons to a condition of self-support and to relieve their distress may be rendered and in order that assistance or care may be given only in such amount and as long as necessary. . . . The circumstances of a person receiving continued care shall be reinvestigated as frequently as the rules of the board or regulations of the department may require."
Section 134-a, as added by Laws 1967, c. 183, effective April 1, 1967, provides:
"In accordance with regulations of the department, any investigation or reinvestigation of eligibility . . . shall be limited to those factors reasonably necessary to insure that expenditures shall be in accord with applicable provisions of this chapter and the rules of the board and regulations of the department and shall be conducted in such manner so as not to violate any civil right of the applicant or recipient. In making such investigation or reinvestigation, sources of information, other than public records, shall be consulted only with the permission of the applicant or recipient. However, if such permission is not granted by the applicant or recipient, the appropriate public welfare official may deny, suspend or discontinue public assistance or care until such time as he may be satisfied that such applicant or recipient is eligible therefor."

†"Mandatory visits must be made in accordance with law that requires that persons be visited at least once every three months if they are receiving . . . Aid to Dependent Children. . . ."

††"Section 351.10. *Required home visits and contacts.* Social investigation as defined and described . . . shall be made of each application or reapplication for public assistance or care as the basis for determination of initial eligibility.
"a. Determination of initial eligibility shall include contact with the applicant and at least one home visit which shall be made promptly in accordance with agency policy. . . ."
"Section 351.21. *Required contacts.* Contacts with recipients and collateral sources shall be adequate as to content and frequency and shall include home visits, office interviews, correspondence, reports on resources and other necessary documentation."
"Section 369.2 of Title 18 provides in part: "(c) *Welfare of child or minor.* A child or minor shall be considered to be eligible for ADC if his home situation is one in which his physical, mental and moral well-being will be safeguarded and his religious faith preserved and protected. (1) In determining the ability of a parent or relative to care for the child so that this purpose is achieved, the home shall be judged by the same standards as are applied to self-maintaining families in the community. When, at the time of application, a home does not meet the usual standards of health and decency but the welfare of the child is not endangered, ADC shall be granted and defined services provided in an effort to improve the situation. When appropriate, consultation or direct service shall be requested from child welfare."

The District Court majority held that a mother receiving AFDC relief may refuse, without forfeiting her right to that relief, the periodic home visit which the cited New York statutes and regulations prescribe as a condition for the continuance of assistance under the program. The beneficiary's thesis, and that of the District Court majority, is that home visitation is a search and, when not consented to or when not supported by a warrant based on probable cause, violates the beneficiary's Fourth and Fourteenth Amendment rights.

Judge McLean, in dissent, thought it unrealistic to regard the home visit as a search; felt that the requirement of a search warrant to issue only upon a showing of probable cause would make the AFDC program "in effect another criminal statute" and would "introduce a hostile arm's length element into the relationship" between worker and mother, "a relationship which can be effective only when it is based upon mutual confidence and trust"; and concluded that the majority's holding struck "a damaging blow" to an important social welfare program. 303 F. Supp., at 946.

I.

The case comes to us on the pleadings and supporting affidavits and without the benefit of testimony which an extended hearing would have provided. The pertinent facts, however, are not in dispute.

Plaintiff Barbara James is the mother of a son, Maurice, who was born in May 1967. They reside in New York City. Mrs. James first applied for AFDC assistance shortly before Maurice's birth. A caseworker made a visit to her apartment at that time without objection. The assistance was authorized.

Two years later, on May 8, 1969, a caseworker wrote Mrs. James that she would visit her home on May 14. Upon receipt of this advice, Mrs. James telephoned the worker that, although she was willing to supply information "reasonable and relevant" to her need for public assistance, any discussion was not to take place at her home. The worker told Mrs. James that she was required by law to visit in her home and that refusal to permit the visit would result in the termination of assistance. Permission was still denied.

On May 13 the City Department of Social Services sent Mrs. James a notice of intent to discontinue assistance because of the visitation refusal. The notice advised the beneficiary of her right to a hearing before a review officer. The hearing was requested and was held on May 27. Mrs. James appeared with an attorney at that hearing.‡ They continued to refuse permission for a worker to visit the James home, but again expressed willingness to cooperate and to permit visits elsewhere. The review officer ruled that the refusal was a proper ground for the termination of assistance. His written decision stated:

---

‡No isssue of procedural due process is raised in this case. Cf. *Goldberg* v. *Kelly,* 397 U. S. 254 (1970), and *Wheeler* v. *Montgomery,* 397 U. S. 280 (1970).

The home visit which Mrs. James refuses to permit is for the purpose of determining if there are any changes in her situation that might affect her eligibility to continue to receive Public Assistance, or that might affect the amount of such assistance, and to see if there are any social services which the Department of Social Services can provide to the family.

A notice of termination issued on June 2.

Thereupon, without seeking a hearing at the state level, Mrs. James, individually and on behalf of Maurice, and purporting to act on behalf of all other persons similarly situated, instituted the present civil rights suit under 42 U. S. C. § 1983. She alleged the denial of rights guaranteed to her under the First, Third, Fourth, Fifth, Sixth, Ninth, Tenth, and Fourteenth Amendments, and under Subchapters IV and XVI of the Social Security Act and regulations issued thereunder. She further alleged that she and her son have no income, resources, or support other than the benefits received under the AFDC program. She asked for declaratory and injunctive relief. A temporary restraining order was issued on June 13, *James* v. *Goldberg,* 302 F. Supp. 478 (SDNY 1969), and the three-judge District Court was convened.

II.

The federal aspects of the AFDC program deserve mention. They are provided for in Subchapter IV, Part A, of the Social Security Act of 1935, 49 Stat. 627, as amended, 42 U. S. C. §§ 601-610 (1964 Ed., Supp. V), Section 401 of the Act, 42 U. S. C. § 601 (1964 Ed., Supp. V), specifies its purpose, namely, "encouraging the care of dependent children in their own homes or in the homes of relatives by enabling each State to furnish financial assistance and rehabilitation and other services . . . to needy dependent children and the parents or relatives with whom they are living to help maintain and strengthen family life. . . ." The same section authorizes the federal appropriation for payments to States that qualify. Section 402, 42 U. S. C. § 602 (1964 Ed., Supp. V), provides that a state plan, among other things, must "provide for granting an opportunity for a fair hearing before the State agency to any individual whose claim for aid to families with dependent children is denied or is not acted upon with reasonable promptness"; must "provide that the State agency will make such reports . . . as the Secretary [of Health, Education, and Welfare] may from time to time require"; must "provide that the State agency shall, in determining need, take into consideration any other income and resources of any child or relative claiming aid"; and must "provide that where the State agency has reason to believe that the home in which a relative and child receiving aid reside in, unsuitable for the child because of the neglect, abuse, or exploitation of such child, it shall bring such condition to the attention of the appropriate court or law enforcement agencies in the State. . . ." Section 405, 42 U. S. C. § 605, provides that:

Whenever the State agency has reason to believe that any payments of aid . . . made with respect to a child are not being or may not be used in the best interests of the child, the State agency may provide for such counseling and guidance services with respect to the use of such payments and the management of other funds by the relative . . . in order to assure use of such payments in the best interests of such child, and may provide for advising such relative that continued failure to so use such payments will result in substitution therefor of protective payments . . . or in seeking the appointment of a guardian . . . or in the imposition of criminal or civil penalties. . . .

III.

When a case involves a home and some type of official intrusion into that home, as this case appears to do, an immediate and natural reaction is one of concern about Fourth Amendment rights and the protection which that Amendment is intended to afford. Its emphasis indeed is upon one of the most precious aspects of personal security in the home: the right of the people to be secure in their persons, houses, papers, and effects. . . .'' This Court has characterized that right as ''basic to a free society.'' *Wolf* v. *Colorado,* 338 U. S. 25, 27 (1949); *Camara* v. *Municipal Court,* 387 U. S. 523, 528 (1967). And over the years the Court consistently has been most protective of the privacy of the dwelling. See, for example, *Boyd,* v. *United States,* 116 U. S. 616, 626-630 (1886); *Mapp* v. *Ohio,* 367 U. S. 643 (1961); *Chimel* v. *California,* 395 U. S. 752 (1969); *Vale* v. *Louisiana,* 399 U. S. 30 (1970). In *Camara,* MR. JUSTICE WHITE, after noting that the ''translation of the abstract prohibition against 'unreasonable searches and seizures' into workable guidelines for the decision of particular cases is a difficult task,'' went on to observe,

Nevertheless, one governing principle, justified by history and by current experience, has consistently been followed: except in certain carefully defined classes of cases, a search of private property without proper consent is ''unreasonable'' unless it has been authorized by a valid search warrant. 387 U. S., at 528-529.

He pointed out, too, that one's Fourth Amendment protection subsists apart from his being suspected of criminal behavior. 387 U. S., at 530.

IV.

This natural and quite proper protective attitude, however, is not a factor in this case, for the seemingly obvious and simple reason that we are not concerned here with any search by the New York social service agency in the Fourth Amendment meaning of that term. It is true that the governing statute and regulations appear to make mandatory the initial home visit and the subsequent periodic ''contacts'' (which may include home visits) for the inception and continuance of aid. It is also true that the caseworker's posture in the home visit is perhaps, in a sense, both rehabilitative and investigative. But this latter aspect, we think, is given too broad a character and far more emphasis than it deserves if it is equated with a search in the traditional criminal law context. We note, too, that the visitation in itself is not forced or compelled, and that the beneficiary's denial of permission is not a criminal act. If consent to the visitation is withheld, no visitation takes place. The aid then never begins or merely ceases, as the case may be. There is no entry of the home and there is no search.

V.

If however, we were to assume that a caseworker's home visit, before or subsequent to the beneficiary's initial qualification for benefits, somehow (perhaps because the average beneficiary might feel she is in no position to refuse consent to the visit), and despite its interview nature, does possess some of the characteristics of a search in the traditional sense, we nevertheless conclude that the visit does not fall within the Fourth Amendment's proscription. This is because it does not descend to the level of unreasonableness. It is unreasonableness which is the Fourth Amendment's standard. *Terry* v. *Ohio,* 392 U. S. 1, 9 (1968); *Elkins* v. *United States,* 364 U. S. 206, 222 (1960). And Mr. Chief Justice Warren observed in *Terry* that "the specific content and incidents of this right must be shaped by the context in which it is asserted." 392 U. S., at 9.

There are a number of factors that compel us to conclude that the home visit proposed for Mrs. James is not unreasonable:

1. The public's interest in this particular segment of the area of assistance to the unfortunate is protection and aid for the dependent child whose family requires such aid for that child. The focus is on the *child* and, further, it is on child who is *dependent*. There is no more worthy object of the public's concern. The dependent child's needs are paramount, and only with hesitancy would we relegate those needs, in the scale of comparative values, to a position secondary to what the mother claims as her rights.

2. The agency, with tax funds provided from federal as well as from state sources, is fulfilling a public trust. The State, working through its qualified welfare agency, has appropriate and paramount interest and concern in seeing and assuring that the intended and proper objects of that tax-produced assistance are the ones who benefit from the aid it dispenses. Surely it is not unreasonable, in the Fourth Amendment sense or in any other sense of that term, that the State have at its command a gentle means, of limited extent and of practical and considerate application, of achieving that assurance.

3. One who dispenses purely private charity naturally has an interest in and expects to know how his charitable funds are utilized and put to work. The public, when it is the provider, rightly expects the same. It might well expect more,

because of the trust aspect of public funds, and the recipient, as well as the caseworker, has not only an interest but obligation.

4. The emphasis of the New York statutes and regulations is upon the home, upon "close contact" with the beneficiary, upon restoring the aid recipient "to a condition of self-support," and upon the relief of his distress. The federal emphasis is no different. It is upon "assistance and rehabilitation," upon maintaining and strengthening family life, and upon "maximum self-support and personal independence consistent with the maintenance of continuing parental care and protection. . . ." 42 U. S. C. § 601 (1964 Ed., Supp. V); *Dandridge* v. *Williams,* 397 U. S. 471, 479 (1970), and *id.,* at 510 (MARSHALL, J., dissenting). It requires cooperation from the state agency upon specified standards and in specified ways. And it is concerned about any possible exploitation of the child.

5. The home visit, it is true, is not required by federal statute or regulation.‡‡ But it has been noted that the visit is "the heart of welfare administration"; that it affords "a personal, rehabilitative orientation, unlike that of most federal programs"; and that the "more pronounced service orientation" effected by Congress with the 1956 amendments to the Social Security Act "gave redoubled importance to the practice of home visiting." Note, Rehabilitation, Investigation and the Welfare Home Visit, 79 Yale L. J. 746, 748 (1970). The home visit is an established routine in States besides New York.||

6. The means employed by the New York agency are significant. Mrs. James received written notice several days in advance of the intended home visit.|| || The date was specified. Section 134-a of the New York Social Services Law, effective April 1, 1967, and set forth in footnote** [page 366], *supra,* sets the tone. Privacy

---

‡‡The federal regulations require only periodic determinations of eligibility. HEW Handbook of Public Assistance Administration, pt. IV, § 2200(d). But they also require verification of eligibility by making field investigations "including home visits" in a selected sample of cases. Pt. II, § 6200(a)(3).

||See, *e.g.,* Ala., Manual for Administration of Public Assistance, pt. I-8(B) (1968 rev.); Ariz., Regulations promulgated pursuant to Rev. Stat. Ann. § 46–203 (1956), Reg. 3-203.6 (1968); Ark. Stat. Ann. § 83–131 (1960); Cal. State Dept. of Social Welfare Handbook, C-012.50 (1964); Colo. Rev. Stat. Ann. § 119–9–1 *et seq.* (Supp. 1967), as amended, Laws 1969, c. 279; Fla. Public Assistance c. 100; Ga. Division of Social Administration — Public Assistance Manual, pt. III, § V(D)(2), pt. VIII(A)(1)(b) (1969); Ill. Rev. Stat. c. 23, § 4–7 (1967); Ind. Ann. Stat. § 52–1247 (1964), Dept. Pub. Welfare, Rules & Regs., Reg. 2–403 (1965); Mich. Public Assistance Manual, Item 243(3)(F) (Rev.) (1967); Miss. Code Ann. § 7177 (1942) (Laws of 1940, c. 294); Mo. Public Assistance Manual, Dept. of Welfare, § III (1969); Nebraska, State Plan and Manual Regulations, pt. IX, §§ 5760, 5771; N.J., Manual of Administration, Division of Public Welfare, Pt. II, §§ 2120, 2122 (1969); N. M. Stat. Ann. § 13–1–13 (1953), Health and Social Services Dept. Manual, §§ 211.5, 272.11; S. C. Dept. of Public Welfare Manual, Vol. IV (D)(2); S. D. Comp. Laws Ann. § 28–7–7 (1967) (formerly S. D. Code § 55.3805); Tenn. Code Ann. § 14–309 (1955), Public Assistance Manual, Vol. II, p. 212 (1968 rev.); Wis. Stat. § 49.19(2) (1967).

|| ||It is true that the record contains 12 affidavits, all essentially identical, of aid recipients (other than Mrs. James) which recite that a caseworker "most often" comes without notice; that when he does, the plans the recipient had for that time cannot be carried out; that the visit is "very embarrassing to me if the caseworker comes when I have company"; and that the caseworker "sometimes asks very personal questions" in front of children.

is emphasized. The applicant-recipient is made the primary source of information as to eligibility. Outside informational sources, other than public records, are to be consulted only with the beneficiary's consent. Forcible entry or entry under false pretenses or visitation outside working hours or snooping in the home are forbidden. The Health, Education, and Welfare *Handbook of Public Assistance Administration,* pt. IV, §§ 2200(a) and 2300; 18 NYCRR §§ 351.1, 351.6, and 351.7. All this minimizes any "burden" upon the homeowner's right against unreasonable intrusion.

7. Mrs. James, in fact, on this record presents no specific complaint of any unreasonable intrusion of her home and nothing that supports an inference that the desired home visit had as its purpose the obtaining of information as to criminal activity. She complains of no proposed vistation at an awkward or retirement hour. She suggests no forcible entry. She refers to no snooping. She describes no impolite or reprehensible conduct of any kind. She alleges only, in general and nonspecific terms, that on previous visits and, on information and belief, on visitation at the home of other aid recipients, "questions concerning personal relationships, beliefs and behavior are raised and pressed which are unnecessary for a determination of continuing eligibility." Paradoxically, this same complaint could be made of a conference held elsewhere than in the home, and yet this is what is sought by Mrs. James. The same complaint could be made of the census taker's questions. See MR. JUSTICE MARSHALL'S opinion, as United States Circuit Judge, in *United States* v. *Rickenbacker,* 309 F.2d 462 (CA2 1962), *cert.* denied, 371 U.S. 962. What Mrs. James appears to want from the agency that provides her and her infant son with the necessities for life is the right to receive those necessities upon her own informational terms, to utilize the Fourth Amendment as a wedge for imposing those terms, and to avoid questions of any kind.#

8. We are not persuaded, as Mrs. James would have us be, that all information pertinent to the issue of eligibility can be obtained by the agency through an interview at a place other than the home, or, as the District Court majority suggested, by examining a lease or a birth certificate, or by periodic medical examinations, or by interviews with school personnel. 303 F. Supp. at 943. Although these secondary sources might be helpful, they would not always assure verification of actual residence or of actual physical presence in the home, which are requisites for

---

#We have examined Mrs. James' case record with the New York City Department of Social Services, which, as an exhibit, accompanied defendant Wyman's answer. It discloses numerous interviews from the time of the initial one on April 27, 1967, until the attempted termination in June 1969. The record is revealing as to Mrs. James' failure ever really to satisfy the requirements for eligibility; as to constant and repeated demands; as to attitude toward the caseworker; as to reluctance to cooperate; as to evasiveness; and as to occasional belligerency. There are indications that all was not always well with the infant Maurice (skull fracture, a dent in the head, a possible rat bite). The picture is a sad and unhappy one.

##§ 406(a) of the Social Security Act, as amended, 42 U. S. C. § 606(a) (1964 Ed., Supp. V); § 349B1 of the New York Social Services Law.

AFDC benefits, ## or of impending medical needs. And, of course, little children, such as Maurice James, are not yet registered in school.

9. The visit is not one by police or uniformed authority. It is made by a caseworker of some training* whose primary objective is, or should be, the welfare, not the prosecution, of the aid recipient for whom the worker has profound responsibility. As has already been stressed, the program concerns dependent children and the needy families of those children. It does not deal with crime or with the actual or suspected perpetrators of crime. The caseworker is not a sleuth but rather, we trust, is a friend to one in need.

10. The home visit is not a criminal investigation, does not equate with a criminal investigation, and despite the announced fears of Mrs. James and those who would join her, is not in aid of any criminal proceeding. If the visitation serves to discourage misrepresentation or fraud, such a by-product of that visit does not impress upon the visit itself a dominant criminal investigative aspect. And if the visit should, by chance, lead to the discovery of fraud and a criminal prosecution should follow,** then, even assuming that the evidence discovered upon the home visitation is admissible, an issue upon which we express no opinion, that is a routine and expected fact of life and a consequence no greater than that which necessarily ensues upon any other discovery by a citizen of criminal conduct.

11. The warrant procedure, which the plaintiff appears to claim to be so precious to her, even if civil in nature, is not without its seriously objectionable features in the welfare context. If a warrant could be obtained (the plaintiff affords us little help as to how it would be obtained), it presumably could be applied for *ex parte,* its execution would require no notice, it would justify entry by force, and its hours for execution† would not be so limited as those prescribed for home visitation. The warrant necessarily would imply conduct either criminal or out of compliance with an asserted governing standard. Of course, the force behind the warrant argument, welcome to the one asserting it, is the fact that it would have to rest upon probable cause, and probable cause in the welfare context, as Mrs. James concedes, requires more than the mere need of the caseworker to see the child in the home and to have assurance that the child is there and is receiving the benefit of the aid that has been authorized for it. In this setting the warrant argument is out of place.

---

*The *amicus* brief submitted on behalf of the Social Services Employees Union Local 371, AFSCME, AFL-CIO, the bargaining representative for the social service staff employed in the New York City Department of Social Services, recites that "caseworkers are either badly trained or untrained" and that "[g]enerally, a caseworker is not only poorly trained, but also young and inexperienced . . . ." Despite this astonishing description by the union of the lack of qualification of its own members for the work they are employed to do, we must assume that the caseworker possesses at least some qualifications and some dedication to duty.

**See, *e.g.,* New York Social Services Law § 145.

†New York Code Crim. Proc. § 801.

It seems to us that the situation is akin to that where an Internal Revenue Service agent, in making a routine civil audit of a taxpayer's income tax return, asks that the taxpayer produce for the agent's review some proof of a deduction the taxpayer has asserted to his benefit in the computation of his tax. If the taxpayer refuses, there is, absent fraud, only a disallowance of the claimed deduction and a consequent additional tax. The taxpayer is fully within his "rights" in refusing to produce the proof, but in maintaining and asserting those rights a tax detriment results and it is a detriment of the taxpayer's own making. So here Mrs. James has the "right" to refuse the home visit, but a consequence in the form of cessation of aid, similar to the taxpayer's resultant additional tax, flows from that refusal. The choice is entirely hers, and nothing of constitutional magnitude is involved.

VI.

Camara v. *Municipal Court,* 387 U. S. 523 (1967), and its companion case, *See* v. *City of Seattle,* 387 U. S. 541 (1967), both by a divided Court, are not inconsistent with our result here. Those cases concerned, respectively, a refusal of entry to city housing inspectors checking for a violation of a building's occupancy permit, and a refusal of entry to a fire department representative interested in compliance with a city's fire code. In each case a majority of this Court held that the Fourth Amendment barred prosecution for refusal to permit the desired warrantless inspection. *Frank* v. *Maryland,* 359 U. S. 360 (1959), a case that reached an opposing result and that concerned a request by a health officer for entry in order to check the source of a rat infestation, was *pro tanto* overruled. Both *Frank* and *Camara* involved dwelling quarters. *See* had to do with a commercial warehouse.

But the facts of the three cases are significantly different from those before us. Each concerned a true search for violations. *Frank* was a criminal prosecution for the owner's refusal to permit entry. So, too, was *See*. *Camara* had to do with a writ of prohibition sought to prevent an already pending criminal prosecution. The community welfare aspects, of course, were highly important, but each case arose in a criminal context where a genuine search was denied and prosecution followed.

In contrast, Mrs. James is not being prosecuted for her refusal to permit the home visit and is not about to be so prosecuted. Her wishes in that respect are fully honored. We have not been told, and have not found, that her refusal is made a criminal act by any applicable New York or federal statute. The only consequence of her refusal is that the payment of benefits ceases. Important and serious as this is, the situation is no different than if she had exercised a similar negative choice initially and refrained from applying for AFDC benefits. If a statute made her refusal a criminal offense, and if this case were one concerning her prosecution under that statute, *Camara* and *See* would have conceivable pertinency.

MR. JUSTICE DOUGLAS, dissenting.

VII.

Our holding today does not mean, of course, that a termination of benefits upon refusal of a home visit is to be upheld against constitutional challenge under all conceivable circumstances. The early morning mass raid upon homes of welfare recipients is not unknown. See *Parrish* v. *Civil Service Comm'n,* 66 Cal. 2d 260, 425 P.2d 223 (1967); Reich, Midnight Welfare Searches and the Social Security Act, 72 Yale L. J. 1347 (1963). But that is not this case. Facts of that kind present another case for another day.

We therefore conclude that the home visitation as structured by the New York statutes and regulations is a reasonable administrative tool; that it serves a valid and proper administrative purpose for the dispensation of the AFDC program; that it is not an unwarranted invasion of personal privacy; and that it violates no right guaranteed by the Fourth Amendment.

Reversed and remanded with directions to enter a judgment of dismissal.

It is so ordered.

MR. JUSTICE WHITE concurs in the judgment and joins the opinion of the Court with the exception of Part IV thereof.

MR. JUSTICE DOUGLAS, dissenting.

We are living in a society where one of the most important forms of property is government largesse which some call the "new property."†† The payrolls of government are but one aspect of that "new property." Defense contracts, highway contracts, and the other multifarious forms of contracts are another part. So are subsidies to air, rail, and other carriers. So are disbursements by government for scientific research.‡ So are TV and radio licenses to use the air space which of course is part of the public domain. Our concern here is not with those subsidies but with grants that directly or indirectly implicate the *home life* of the recipients.

In 1969 roughly 127 billion dollars were spent by the federal, state, and local governments on "social welfare."‡‡ To farmers alone almost four billion dollars were paid, in part for not growing certain crops. Almost 129,000 farmers received $5,000 or more, their total benefits exceeding $1,450,000,000.‖ Those payments were in some instances very large, a few running a million or more a year. But the majority were payments under $5,000 each.

---

††See Reich, The New Property, 73 Yale L. J. 733, 737–739.

‡See Ginzburg, What Science Policy?, Columbia Forum, Fall 1970, p. 12.

‡‡See Appendix I to this opinion.

‖See Appendix II to this opinion.

Yet almost every beneficiary, whether rich or poor, rural or urban, has a "house" — one of the places protected by the Fourth Amendment against "unreasonable searches and seizures." || || The question in this case is whether receipt of largesse from the government makes the *home* of the beneficiary subject to access by an inspector of the agency of oversight, even though the beneficiary objects to the intrusion and even though the Fourth Amendment's procedure for access to one's *house* or *home* is not followed. The penalty here is not, of course, invasion of the privacy of Barbara James, only her loss of federal or state largesse. That, however, is merely rephrasing the problem. Whatever the semantics, the central question is whether the government by force of its largesse has the power to "buy up" rights guaranteed by the Constitution.# But for the assertion of her constitutional right, Barbara James in this case would have received the welfare benefit.

We spoke in *Speiser* v. *Randall,* 357 U. S. 513, of the denial of tax exemptions by a State because of exercise of First Amendment rights.

It cannot be gainsaid that a discriminatory denial of a tax exemption for engaging in speech is a limitation on free speech. . . . To deny an exemption to claimants who engage in certain forms of speech is in effect to penalize them for such speech. Its deterrent effect is the same as if the State were to fine them for this speech. *Id.,* at 518.

Likewise, while second-class mail rates may be granted or withheld by the Government, we would not allow them to be granted "on condition that certain economic or political ideas not be disseminated." *Hannegan* v. *Esquire, Inc.,* 327 U. S. 146, 156.

In *Sherbert* v. *Verner,* 374 U. S. 398, a State providing unemployment insurance required recipients to accept suitable employment when it became available or lose the benefits. An unemployed lady was offered a job requiring her to work Saturdays but she refused because she was a Seventh Day Adventist to whom Saturday was the Sabbath. The State cancelled her unemployment benefits and we reversed, saying:

The ruling forces her to choose between following the precepts of her religion and forfeiting benefits, on the one hand, and abandoning one of the precepts of her religion in order to accept work, on the other hand. Governmental imposition of such a choice puts the same kind of burden upon the free exercise of religion as would a fine imposed against appellant for her Saturday worship.

|| ||"The right of the people to be secure in their persons, houses, papers, and effects, against unreasonable searches and seizures, shall not be violated, and no Warrants shall issue, but upon probable cause, supported by Oath or affirmation, and particularly describing the place to be searched, and the persons or things to be seized."

#See Note, Unconstitutional Conditions, 73 Harv. L. Rev. 1595, 1599.

Nor may the South Carolina court's construction of the statute be saved from constitutional infirmity on the ground that unemployment compensation benefits are not appellant's "right" but merely a "privilege." It is too late in the day to doubt that the liberties of religion and expression may be infringed by the denial of or placing of conditions upon a benefit or privilege. . . . [T]o condition the availability of benefits upon this appellant's willingness to violate a cardinal principle of her religious faith effectively penalizes the free exercise of her constitutional liberties. *Id.*, at 404, 406.

These cases are in the tradition of *United States* v. *Chicago, M., St. P. & P. R. Co.*, 282 U.S. 311, 328-329,## where Mr. Justice Sutherland, writing for the Court, said:*

[T]he rule is that the right to continue the exercise of a privilege granted by the state cannot be made to depend upon the grantee's submission to a condition prescribed by the state which is hostile to the provisions of the federal Constitution.

What we said in those cases is as applicable to Fourth Amendment rights as to those of the First. The Fourth, of course, speaks of "unreasonable" searches and seizures, while the First is written in absolute terms. But the right of privacy which the Fourth protects is perhaps as vivid in our lives as the right of expression sponsored by the First. *Griswold* v. *Connecticut*, 318 U.S. 479, 484. If the regime under which Barbara James lives were enterprise capitalism as, for example, if she ran a small factory geared into the Pentagon's procurement program, she certainly would have a right to deny inspectors access to her *home* unless they came with a warrant.

##And see Hale, Unconstitutional Conditions and Constitutional Rights, 35 Col. L. Rev. 321 (1935); *Frost & Frost Co.* v. *Railroad Comm'n*, 271 U.S. 583, 594.

Flemming v. *Nestor*, 33 U.S. 603, is not in accord with that tradition. There we upheld the right of Congress to strip away accrued social security benefits. Nestor, an alien, came to this country in 1913. From the enactment of the Social Security Act until 1955 Nestor and his employers contributed payments to the fund. In 1955 Nestor became eligible for old-age benefits. One year later he was deported for having been a member of the Communist Party between 1933 and 1939 — a time when it was perfectly legal to be a member. In 1954 Congress passed a law which provided for the loss of social security benefits for anyone deported for having been a member of the Communist Party. Like the law providing for deportation for membership this law, too, was fully retroactive. Thus Nestor was deported after he had retired based on a law condemning membership in the Communist Party at the time when it was legal to be a member, and stripped of his retirement income based on a law which was triggered by that deportation. We upheld the constitutionality of the 1954 law by a 5–4 majority.
The majority stated Nestor's property had not been taken without due process because Nestor had no property rights; his interest was "noncontractual" and could "not be soundly analogized to that of the holder of an annuity." 363 U.S., at 610. The majority then went on to hold social security benefits were only protected from congressional action which is "utterly lacking in rational justification." *Id.*, at 611.
If it was unconstitutional in *Speiser* to condition a tax exemption on a limitation on freedom of speech, it was equally unconstitutional to withhold a social security benefit conditioned on a limitation of freedom of association. A right-privilege distinction was implicitly rejected in *Speiser* and explicitly rejected in *Sherbert*. Today's decision when dealing with a state statute joins *Flemming* as an anomaly in the cases dealing with unconstitutional conditions.

That is the teaching of *Camara* v. *Municipal Court*, 387 U. S. 523, and *See* v. *City of Seattle*, 387 U. S. 541. In those cases we overruled *Frank* v. *Maryland*, 359 U. S. 360, and held the Fourth Amendment applicable to administrative searches of both the *home* and a business. The applicable principle, as stated in *Camara* as "justified by history and by current experience" is that "except in certain carefully defined classes of cases, a search of private property without proper consent is 'unreasonable' unless it has been authorized by a valid search warrant." 387 U. S., at 528-529. In *See* we added that the "businessman, like the occupant of a residence, has a constitutional right to go about his business free from unreasonable official entries upon his private commercial property." *Id.*, at 543. There is not the slightest hint in *See* that the Government could condition a business license on the "consent" of the licensee to the administrative searches we held violated the Fourth Amendment. It is a strange jurisprudence indeed which safeguards the businessman at his place of work from warrantless searches but will not do the same for a mother in her *home*.

Is a search of her home without a warrant made "reasonable" merely because she is dependent on government largesse? Judge Skelly Wright has stated the problem succinctly:

> Welfare has long been considered the equivalent of charity and its recipients have been subjected to all kinds of dehumanizing experiences in the government's effort to police its welfare payments. In fact, over half a billion dollars are expended annually for administration and policing in connection with the Aid to Families with Dependent Children program. Why such large sums are necessary for administration and policing has never been adequately explained. No such sums are spent policing the government subsidies granted to farmers, airlines, steamship companies, and junk mail dealers, to name but a few. The truth is that in this subsidy area society has simply adopted a double standard, one for aid to business and the farmer and a different one for welfare. Poverty, Minorities, and Respect For Law, 1970 Duke L. J. 425, 437-438.

If the welfare recipient was not Barbara James but a prominent, affluent cotton or wheat farmer receiving benefit payments for not growing crops, would not the approach be different? Welfare in aid of dependent children, like social security and unemployment benefits, has an aura of suspicion.** There doubtless are

---

** Juvenal wrote:

"Poverty's greatest curse, much worse than the fact of it, is that it makes men objects of mirth, ridiculed, humbled, embarrassed." Satires 39 (Indiana Univ. Press, 1958).

In the 1837 Term the Court held in *City of New York* v. *Miln*, 11 Pet. 102, that New York could require ships coming in from abroad to report the names, ages, *etc.*, of every person brought to these shores. The Court said: "We think it as competent and as necessary for a state to provide precautionary measures against the moral pestilence of paupers, vagabonds, and possibly convicts; as it is to guard against the physical pestilence, which may arise from unsound and infectious articles imported, or from a ship, the crew of which may be labouring under an infectious disease." *Id.*, at 142.

I regretfully conclude that today's decision is ideologically of the same vintage.

frauds in every sector of public welfare whether the recipient be a Barbara James or someone who is prominent or influential. But constitutional rights — here the privacy of the home — are obviously not dependent on the poverty or on the affluence of the beneficiary. It is the precincts of the home that the Fourth Amendment protects; and their privacy is as important to the lowly as to the mighty.†

> [S]tudies tell us that the typical middle income American reaches retirement age with a whole bundle of interests and expectations: as homeowner, as small investor, and as social security "beneficiary." Of these, his social security retirement benefits are probably his most important resource. Should this, the most significant of his rights, be entitled to a quality of protection inferior to that afforded his other interests? It becomes the task of the rule of law to surround this new "right" to retirement benefits with protections against arbitrary government action, with substantive and procedural safeguards that are as effective in context as the safeguards enjoyed by traditional rights of property in the best tradition of the older law.††

It may be that in some tenements one baby will do service to several women and call each one "mom." It may be that other frauds, less obvious, will be perpetrated. But if inspectors want to enter the precincts of the home against the wishes of the lady of the house, they must get a warrant. The need for exigent action as in cases of "hot pursuit" is not present, for the lady will not disappear; nor will the baby.

I would place the same restrictions on inspectors entering the *homes* of welfare

---

†An individual who refuses to allow the home visit could either be a welfare recipient at the time or an applicant for assistance. In neither case would the outcome of the refusal be different.

If the mother is already a recipient, Social Services Regulations § 351.21, 18 NYCCR § 351.21, requires continuing contacts at home between the recipient and the social worker. Should a recipient refuse a visit then § 175 of the Policies Governing the Administration of Public Assistance ("Mandatory visits must be made in accordance with law that requires that persons be visited. . . .") would require termination. When the decision to "discontinue, suspend or reduce" benefits is made, the recipient would receive a hearing under § 351.26 at which the recipient could present "written and oral relevant evidence and argument to demonstrate why his grant should not be discontinued, suspended or reduced." Since § 134 of the Social Services Law requires visits, the refusal to allow the visit would apparently be dispositive of the matter.

That seems to be conceded here by the commissioner. In light of that fact, the failure of appellee, who went to a hearing and was denied relief, to pursue any further state remedy seems irrelevant as the only question posed was the constitutionality under the Fourth Amendment of the termination of assistance for failure to agree to the warrantless entry into her home.

Except in very limited circumstances (Social Services Regulations §§ 351.10 and 372 [Emergency Assistance]) an initial home visit and investigation is necessary before receiving benefits. Should a potential recipient refuse the initial visit, he would be notified under § 351.14(b) of the reason for the denial. Then he could request a "fair hearing" under Board Rule 85 and Social Services Regulations § 358. Again it appears that refusing the visit would be dispositive of the claim.

The extent to which a person could receive emergency assistance after refusal of a visit is unclear. Social Services Regulations § 372.3 recognizes that emergency assistance could be available to a person while the "fair hearing" is pending. It would seem, however, that implicit in § 372.3 is the notion that if the claim is disposed of, then the emergency assistance would terminate. Also emergency assistance is limited to periods not in excess of 30 consecutive days in any 12-month period. Social Services Regulations § 372.1.

††Jones, The Rule of Law and the Welfare State, 58 Col. L. Rev. 143, 154–155 (1958).

beneficiaries as are on inspectors entering the *homes* of those on the payroll of government, or the *homes* of those who contract with the government, or the *homes* of those who work for those having government contracts. The values of the *home* protected by the Fourth Amendment are not peculiar to capitalism as we have known it; they are equally relevant to the new form of socialism which we are entering. Moreover, as the numbers of functionaries and inspectors multiply, the need for protection of the individual becomes indeed more essential if the values of a free society are to remain.

What Lord Acton wrote Bishop Creighton‡ about the corruption of power is increasingly pertinent today:

> I cannot accept your canon that we are to judge Pope and King unlike other men, with a favourable presumption that they did no wrong. If there is any presumption it is the other way against holders of power, increasing as the power increases. Historic responsibility has to make up for the want of legal responsibility. Power tends to corrupt and absolute power corrupts absolutely. Great men are almost always bad men, even when they exercise influence and not authority: still more when you superadd the tendency or the certainty of corruption by authority.

The bureaucracy of modern government is not only slow, lumbering, and oppressive; it is omnipresent. It touches everyone's life at numerous points. It pries more and more into private affairs, breaking down the barriers that individuals erect to give them some insulation from the intrigues and harassments of modern life.‡‡ Isolation is not a constitutional guarantee; but the sanctity of the sanctuary of the *home* is such — as marked and defined by the Fourth Amendment, *McDonald* v. *United States*, 335 U. S. 451, 453. What we do today is to depreciate it.

I would sustain the judgment of the three-judge court in the present case.

MR. JUSTICE MARSHALL, whom MR. JUSTICE BRENNAN joins, dissenting.

Although I substantially agree with its initial statement of the issue in this case, the Court's opinion goes on to imply that the appellee has refused to provide information germane to a determination of her eligibility for AFDC benefits. The record plainly shows, however, that Mrs. James offered to furnish any information that the appellants desired and to be interviewed at any place other than her home. Appellants rejected her offers and terminated her benefits solely on the

---

‡J. Acton, Essays on Freedom and Power 364 (H. Finer Ed. 1948).

‡‡Mass raids upon the homes of welfare recipients are matters of record. See *Parrish* v. *Civil Service Comm'n*, 66 Cal. 2d 260, 425 P.2d 223, where an inspector was discharged because he refused to engage in such "illegal activity" and was granted relief by way of back pay.

ground that she refused to permit a home visit. In addition, appellants make no contention that any sort of probable cause exists to suspect appellee of welfare fraud or child abuse.

Simply stated, the issue in this case is whether a state welfare agency can require all recipients of AFDC benefits to submit to warrantless "visitations" of their homes. In answering that question, the majority dodges between constitutional issues to reach a result clearly inconsistent with the decisions of this Court. We are told that there is no search involved in this case; that even if there were a search, it would not be unreasonable; and that even if this were an unreasonable search, a welfare recipient waives her right to object by accepting benefits. I emphatically disagree with all three conclusions. Furthermore, I believe that binding regulations of the Department of Health, Education, and Welfare prohibit appellants from requiring the home visit.

I.

The Court's assertion that this case concerns no search "in the Fourth Amendment meaning of that term" is neither "obvious" nor "simple." I should have thought that the Fourth Amendment governs all intrusions by agents of the public upon personal security, *Terry* v. *Ohio,* 392 U. S. 1, 18 n. 15 (1968). As MR. JUSTICE HARLAN has said:

> [T]he Constitution protects the privacy of the home against all unreasonable intrusion of whatever character. . . . "[It applies] to all invasions on the part of the government and its employees of the sanctity of a man's home." *Poe* v. *Ullman,* 367 U. S. 497, 550-551 (1961) (dissenting opinion).

This Court has rejected as "anomalous" the contention that only suspected criminals are protected by the Fourth Amendment, *Camara* v. *Municipal Court,* 387 U. S. 523, 530 (1967). In an era of rapidly burgeoning governmental activities and their concomitant inspectors, caseworkers, and researchers, a restriction of the Fourth Amendment to "the traditional criminal law context" tramples the ancient concept that a man's home is his castle. Only last Term, we reaffirmed that this concept has lost none of its vitality, *Rowan* v. *United States Post Office,* 387 U. S. 728, 738 (1970).

Even if the Fourth Amendment does not apply to each and every governmental entry into the home, the welfare visit is not some sort of purely benevolent inspection. No one questions the motives of the dedicated welfare caseworker. Of course, caseworkers seek to be friends, but the point is that they are also required to be sleuths. The majority concedes that the "visitation" is partially investigative, but claims that this investigative aspect has been given too much emphasis. Emphasis has indeed been given. Time and again, in briefs and at oral argument, appellants emphasized the need to enter AFDC homes to guard against welfare fraud and child abuse, both of which are felonies.|| The New York statutes provide emphasis by requiring all caseworkers to report any evidence of fraud that a home visit uncovers, N. Y. Social Services Law § 145. And appellants have strenuously emphasized the importance of the visit to provide evidence leading to civil forfeitures including elimination of benefits and loss of child custody.

Actually, the home visit is precisely the type of inspection proscribed by *Camara* and its companion case *See* v. *City of Seattle,* 387 U. S. 541 (1967), except that the welfare visit is a more severe intrusion upon privacy and family dignity. Both the home visit and the searches in those cases may convey benefits to the householder. Fire inspectors give frequent advice concerning fire prevention, wiring capacity, and other matters, and obvious self-interest causes many to welcome the fire or safety inspection. Similarly, the welfare caseworker may provide welcome advice on home management and child care. Nonetheless, both searches may result in the imposition of civil penalties — loss or reduction of welfare benefits or an order to upgrade a housing defect. The fact that one purpose of the visit is to provide evidence that may lead to an elimination of benefits is sufficient to grant appellee protection since *Camara* stated that the Fourth Amendment applies to inspections which can result in only civil violations, 387 U. S., at 531. But here the case is stronger since the home visit, like many housing inspections, may lead to criminal convictions.

The Court attempts to distinguish *See* and *Camara* by telling us that those cases involved "true" and "genuine" searches. The only concrete distinction offered is that *See* and *Camara* concerned criminal prosecutions for refusal to permit the search. The *Camara* opinion did observe that one could be prosecuted for a refusal to allow that search; but, apart from the issue of consent, there is neither logic in, nor precedent for, the view that the ambit of the Fourth Amendment depends not on the character of the governmental intrusion but on the size of the club that the State wields against a resisting citizen. Even if the magnitude of the penalty were relevant, which sanction for resisting the search is more severe? For protecting the privacy of her home, Mrs. James lost the sole means of support for herself and her infant son. For protecting the privacy of his commercial warehouse, Mr. See received a $100 suspended fine.

Conceding for the sake of argument that someone might view the "visitation" as a search, the majority nonetheless concludes that such a search is not unreasonable. However, its mode of reaching that conclusion departs from the entire history of Fourth Amendment case law. Of course, the Fourth Amendment test is reasonableness, but in determining whether a search is reasonable, this Court is not free merely to balance, in a totally *ad hoc* fashion, any number of sub-

---

||For example, appellants' Reply Brief offers two specific illustrations of the home visit's efficacy. In the first, a man was discovered in the home and benefits were terminated. In the second, child abuse was discovered.

jective factors. An unbroken line of cases holds that, subject to a few narrowly drawn exceptions, any search without a warrant is constitutionally unreasonable, see, *e.g.*, *Agnello* v. *United States*, 269 U. S. 20, 32 (1925); *Johnson* v. *United States*, 333 U. S. 10, 13-14 (1948); *Chapman* v. *United States*, 365 U. S. 610, 613-615 (1961); *Camara* v. *Municipal Court*, 387 U. S. 523, 528-529 (1967); *Chimel* v. *California*, 395 U. S. 752, 762 (1969); *Vale* v. *Louisiana*, 399 U. S. 30, 34-35 (1970). In this case, no suggestion that evidence will disappear, that a criminal will escape, or that an officer will be injured, justifies the failure to obtain a warrant. Instead, the majority asserts what amounts to three state interests that allegedly render this search reasonable. None of these interests is sufficient to carve out a new exception to the warrant requirement.

First, it is argued that the home visit is justified to protect dependent children from "abuse" and "exploitation." These are heinous crimes, but they are not confined to indigent households. Would the majority sanction, in the absence of probable cause, compulsory visits to all American homes for the purpose of discovering child abuse? Or is this Court prepared to hold as a matter of constitutional law that a mother, merely because she is poor, is substantially more likely to injure or exploit her children? Such a categorical approach to an entire class of citizens would be dangerously at odds with the tenets of our democracy.

Second, the Court contends that caseworkers must enter the homes of AFDC beneficiaries to determine eligibility. Interestingly, federal regulations do not require the home visit. In fact, the regulations specify the recipient himself as the primary source of eligibility information thereby rendering an inspection of the home only one of several alternative secondary sources.|| || The majority's implication that a biannual home visit somehow assures the verification of actual residence or actual physical presence in the home strains credulity in the context of urban poverty. Despite the caseworker's responsibility for dependent children, he is not even required to see the children as a part of the home visit.# Appellants offer scant explanation for their refusal even to attempt to utilize public records, expenditure receipts, documents such as leases, nonhome interviews, personal financial records, sworn declarations, *etc.* — all sources that governmental agencies regularly accept as adequate to establish eligibility for other public benefits. In this setting, it ill behooves appellants to refuse to utilize informational sources less drastic than an invasion of the privacy of the home.

We are told that the plight of Mrs. James is no different from that of a taxpayer who is required to document his right to a tax deduction, but this analogy is

---

|| ||HEW *Handbook of Public Assistance Administration*, pt. IV, § 2200(e)(1).

#Appellants respond by asserting that if the caseworker becomes suspicious concerning the child's absence, further investigation may take place. One certainly would hope that the caseworker would continue his investigation, but the fact remains that the failure to require that the child be seen undercuts the argument that the home visit is designed to protect the child's welfare and necessary to verify his presence in the home.

seriously flawed. The record shows that Mrs. James has offered to be interviewed anywhere other than her home, to answer any questions, and to provide any documentation that the welfare agency desires. The agency curtly refused all these offers and insisted on its "right" to pry into appellee's home. Tax exemptions are also governmental "bounty." A true analogy would be an Internal Revenue Service requirement that in order to claim dependency exemption, a taxpayer *must* allow a specially trained IRS agent to invade the home for the purpose of questioning the occupants and looking for evidence that the exemption is being properly utilized for the benefit of the dependent. If such a system were even proposed, the cries of constitutional outrage would be unanimous.

Appellants offer a third state interest that the Court seems to accept as partial justification for this search. We are told that the visit is designed to rehabilitate, to provide aid. This is strange doctrine indeed. A paternalistic notion that a complaining citizen's constitutional rights can be violated so long as the State is somehow helping him is alien to our Nation's philosophy. More than 40 years ago, Mr. Justice Brandeis warned:

> Experience should teach us to be most on our guard to protect liberty when the Government's purposes are beneficent. *Olmstead* v. *United States,* 277 U. S. 438, 479 (1928) (dissenting opinion).

Throughout its opinion, the majority alternates between two views of the State's interest in requiring the home visit. First we are told that the State's purpose is benevolent so that no search is involved. Next we are told that the State's need to prevent child abuse and to avoid the misappropriation of welfare funds justifies dispensing with the warrant requirement. But when all the State's purposes are considered at one time, I can only conclude that the home visit is a search and that, absent a warrant, that search is unreasonable.##

Although the Court does not agree with my conclusion that the home visit is an unreasonable search, its opinion suggests that even if the visit were unreasonable, appellee has somehow waived her right to object. Surely the majority cannot believe that valid Fourth Amendment consent can be given under the threat of the loss of one's sole means of support. Nor has Mrs. James waived her rights. Had the Court squarely faced the question of whether the State can condition welfare payments on the waiver of clear constitutional rights, the answer would be plain. The decisions of this Court do not support the notion that a State can use welfare benefits as a wedge to coerce "waiver" of Fourth Amendment rights, see Reich, *Midnight Welfare Searches and the Social Security Act,* 72 Yale L. J. 1347,

---

##Since the majority refuses to sanction the warrant procedure in any form, I have not discussed what standard should be required for a warrant to issue. Certainly, if one of the purposes of the welfare search is to obtain evidence of criminal conduct, that is no reason to permit less than a probable cause. And because the home visit is a more severe intrusion than in the housing inspection and there are less drastic means to obtain eligibility information, I would apply the analysis of *Camara* and would be inclined to utilize a traditional probable cause standard.

1349-1350 (1963); Note, *Rehabilitation, Investigation and the Welfare Home Visit,* 79 Yale L. J. 746, 758 (1970). In *Sherbert* v. *Verner,* * this Court did not say, "Aid merely ceases. There is no abridgement of religious freedom." Nor did the Court say in *Speiser* v. *Randall,* ** "The tax is simply increased. No one is compelled to relinquish First Amendment rights." As my Brother DOUGLAS points out, the majority's statement that Mrs. James' "choice [to be searched or to lose her benefits] is entirely hers, and nothing of constitutional magnitude is involved" merely restates the issue. To MR. JUSTICE DOUGLAS' eloquent discussion of the law of unconstitutional conditions, I would add only that this Court last Term reaffirmed *Sherbert* and *Speiser* as applicable to the law of public welfare:

> Relevant constitutional restraints apply as much to the withdrawal of public assistance benefits as to disqualification for unemployment compensation . . . denial of a tax exemption . . . or . . . discharge from public employment. *Goldberg* v. *Kelly,* 397 U. S. 254, 262 (1970).

The Court's examination of the constitutional issues presented by this case has constrained me to respond. It would not have been necessary to reach these questions for I believe that HEW regulations, binding on the States, prohibit the unconsented home visit.†

The federal *Handbook of Public Assistance Administration* provides:

> The [state welfare] agency especially guards against violations of legal rights and common decencies in such areas as entering a home by force, *or without permission,* or under false pretenses; making home visits outside of working hours, and particularly making such visits during sleeping hours. . . .   Part IV, § 2300(a). *Emphasis supplied.)*

Although the tone of this language is descriptive, HEW requirements are stated in terms of principles and objectives, *Handbook,* pt. I, § 4210 (3); and appellants do not contend that this regulation is merely advisory. Instead, appellants respond with the tired assertion that consent obtained by threatening termination of benefits constitutes valid permission under this regulation. There is no reason to suspect that HEW shares this crabbed view of consent. The *Handbook,* itself, insists on careful scrutiny of purported consent, pt. IV, § 2400. Section 2200(a) is

---

*374 U. S. 398 (1963).

**357 U. S. 513 (1958).

†It is a time-honored doctrine that statutes and regulations are first examined by a reviewing court to see if constitutional questions can be avoided, *Ashwander* v. *TVA,* 297 U. S. 288, 346–348 (1936) (Brandeis, J., concurring); see, *e.g., Dandridge* v. *Williams,* 397 U. S. 471 (1970); *King* v. *Smith,* 92 U. S. 309 (1968). The court below chose not to invoke this doctrine, and litigation in this Court has emphasized the constitutional issues. However, the nonconstitutional questions were briefed by an *amicus curiae* and appellants responded fully in their Reply Brief. The parties may prefer a decision on constitutional grounds; but we, of course, are not bound by their litigation strategies.

designed to protect the privacy of welfare recipients, and it would be somewhat ironic to adopt a construction of the regulation that provided that any person who invokes his privacy rights ceases to be a recipient.

Appellants next object that the home visit has long been a part of welfare administration and has never been disapproved by HEW. The short answer to this is that we deal with only the *unconsented* home visit. The general utility and acceptance of the home visit casts little light on whether HEW might prefer not to impose the visit on unwilling recipients. Appellants also remind us that the Federal Government itself requires a limited number of home visits for sampling purposes. However, while there may well be a special need to employ mandatory visits as a part of quality control samples, Mrs. James' home was not a part of such a sample. Furthermore appellants admit that § 2200(a) governs the quality control program; so it is not clear that unconsented home visits are allowed even for sampling purposes. Although there appears to be no regulatory history, appellants tell us § 2200(a) merely permits a recipient to refuse a particular home visit and does not allow him to forbid home visits altogether. I suppose that one could read such a limitation into the section, but given the regulation's explicit language, given that HEW does not require home visits and views the visits as only one of several alternative sources of eligibility information, given HEW's concern for the privacy of its clients, and given the durable principle of this Court that doubtful questions of interpretation should be resolved in a manner which avoids constitutional questions, *United States* v. *Delaware & Hudson Co.,* 213 U. S. 366, 407 (1909), I would conclude that Mrs. James is protected by § 2200(a).

III.

In deciding that the homes of AFDC recipients are not entitled to protection from warrantless search by welfare caseworkers, the Court declines to follow prior case law and employs a rationale that, if applied to the claims of all citizens, would threaten the vitality of the Fourth Amendment. This Court has occasionally pushed beyond established constitutional contours to protect the vulnerable and to further basic human values. I find no little irony in the fact that the burden of today's departure from principled adjudication is placed upon the lowly poor. Perhaps the majority has explained why a commercial warehouse deserves more protection than does this poor woman's home. I am not convinced; and, therefore, I must respectfully dissent.

EMPLOYMENT AND BUSINESS FACILITIES

The question of the right of public officials to search or inspect occupancies versus a citizen's right to refuse to allow such searchers and/or inspections without a proper warrant has been extended to employment and business facilities. The following decision by the Supreme Court addresses the constitutionality of warrantless, "surprise" inspections by public officials that are authorized by some statutes. It should be noted that not all statutes that authorize warrantless searches are automatically unconstitutional. Each statute must be viewed separately with regard to its constitutionality. In the following decision, what are the reasons given for finding the OSHA statute unconstitutional? (Note that only applicable footnotes are included in the decision.)

RAY MARSHALL, SECRETARY OF LABOR, ET. AL., APPELLANTS
v.
BARLOW'S, INC.

On Appeal from the United States District Court
for the District of Idaho.
Supreme Court of the United States
No. 76-1143. May 23, 1978.

MR. JUSTICE WHITE delivered the opinion of the Court.

Section 8 (a) of the Occupational Safety and Health Act of 1970 (OSHA)†† empowers agents of the Secretary of Labor (the Secretary) to search the work area of any employment facility within the Act's jurisdiction. The purpose of the search is to inspect for safety hazards and violations of OSHA regulations. No search warrant or other process is expressly required under the Act.

On the morning of September 11, 1975, an OSHA inspector entered the customer service area of Barlow's, Inc., an electrical and plumbing installation business located in Pocatello, Idaho. The president and general manager, Ferrol G. "Bill" Barlow, was on hand; and the OSHA inspector, after showing his credentials,‡ informed Mr. Barlow that he wished to conduct a search of the working areas of the business. Mr. Barlow inquired whether any complaint had been received about his company. The inspector answered no, but that Barlow's, Inc. had simply turned up in the agency's selection process. The inspector again asked to enter the nonpublic area of the business; Mr. Barlow's response was to inquire whether the inspector had a search warrant. The inspector had none. Thereupon, Mr. Barlow refused the inspector admission to the employee area of his business. He said he was relying on his rights as guaranteed by the Fourth Amendment of the United States Constitution.

---

††"In order to carry out the purposes of this chapter, the Secretary, upon presenting appropriate credentials to the owner, operator, or agent in charge, is authorized —
    "(1) To enter without delay and at reasonable times any factory, plant, establishment, construction site, or other area, workplace or environment where work is performed by an employee of an employer; and
    "(2) To inspect and investigate during regular working hours and at other reasonable times, and within reasonable limits and in a reasonable manner, any such place of employment and all pertinent conditions, structures, machines, apparatus, devices, equipment, and materials therein, and to question privately any such employer, owner, operator, agent, or employee."
84 Stat. 1590, 29 U. S. C. § 657 (a) (1970).

‡This is required by the Act. See previous n. ††, *supra.*

Three months later, the Secretary petitioned the United States District Court for the District of Idaho to issue an order compelling Mr. Barlow to admit the inspector.‡‡ The requested order was issued on December 30, 1975, and was presented to Mr. Barlow on January 5, 1976. Mr. Barlow again refused admission, and he sought his own injunctive relief against the warrantless searches assertedly permitted by OSHA. A three-judge court was convened. On December 30, 1976, it ruled in Mr. Barlow's favor. 424 F. Supp. 437. Concluding that *Camara* v. *Municipal Court,* 387 U. S. 523, 528–529 (1967), and *See* v. *City of Seattle,* 387 U. S. 541, 543 (1967), controlled this case, the court held that the Fourth Amendment required a warrant for the type of search involved here‖ and that the statutory authorization for warrantless inspections was unconstitutional. An injunction against searches or inspections pursuant to § 8 (a) was entered. The Secretary appealed, challenging the judgment, and we noted probable jurisdiction. . . .

I.

The Secretary urges that warrantless inspections to enforce OSHA are reasonable within the meaning of the Fourth Amendment. Among other things, he relies on § 8 (a) of the Act, 29 U. S. C. § 657 (a), which authorizes inspection of business premises without a warrant and which the Secretary urges represents a congressional construction of the Fourth Amendment that the courts should not reject. Regretfully, we are unable to agree.

The Warrant Clause of the Fourth Amendment protects commercial buildings as well as private homes. To hold otherwise would belie the origin of that Amendment, and the American colonial experience. An important forerunner of the first 10 Amendments to the United States Constitution, the Virginia Bill of Rights, specifically opposed "general warrants, whereby an officer or messenger may be commanded to search suspected places without evidence of a fact committed."‖ ‖ The general warrant was a recurring point of contention in the colonies immediately preceding the Revolution.# The particular offensiveness it engendered was acutely felt by the merchants and businessmen whose premises and products were inspected for compliance with the several Parliamentary revenue measures that most irritated the colonists.## "[T]he Fourth Amendment's commands grew

---

‡‡A regulation of the Secretary, 29 CFR § 1903.4, requires an inspector to seek compulsory process if an employer refuses a requested search. See p. 9, *infra,* and n. ‡, p. 387.

‖No *res judicata* bar arose against Mr. Barlow from the December 30, 1975 order authorizing a search, because the earlier decision reserved the constitutional issue. See opinion of District Court, reprinted in jurisdictional statement at 5a.

‖ ‖H. S. Commager, Documents of American History 104, 8th Ed., 1968.

#See, *e. g.,* O. M. Dickerson, "Writs of Assistance as a Cause of the Revolution," The Era of the American Revolution 40, R. Morris, ed., 1939.

##The Stamp Act of 1765, the Townsend Revenue Act of 1767, and the tea tax of 1773 are notable examples. See Commager, *supra,* n. ‖ ‖, at 53, 63. For commentary, see S. E. Morison, H. S. Commager & W. E. Leuchtenburg, I The Growth of the American Republic 143, 149, 159, 1969.

in large measure out of the colonists' experience with the writs of assistance . . . [that] granted sweeping power to customs officials and other agents of the King to search at large for smuggled goods." *United States* v. *Chadwick,* 433 U. S. 7–8 (1977). See also *G. M. Leasing Corporation* v. *United States,* 429 U. S. 338, 355 (1977). Against this background, it is untenable that the ban on warrantless searches was not intended to shield places of business as well as of residence.

This Court has already held that warrantless searches are generally unreasonable, and that this rule applies to commercial premises as well as homes. In *Camara* v. *Municipal Court,* 387 U. S. 523, 528–529 (1967), we held:

> [E]xcept in certain carefully defined classes of cases, a search of private property without proper consent is "unreasonable" unless it has been authorized by a valid search warrant.

On the same day, we also ruled:

> As we explained in *Camara,* a search of private houses is presumptively unreasonable if conducted without a warrant. The businessman, like the occupant of a residence, has a constitutional right to go about his business free from unreasonable official entries upon his private commercial property. The businessman, too, has that right placed in jeopardy if the decision to enter and inspect for violation of regulatory laws can be made and enforced by the inspector in the field without official authority evidenced by a warrant. *See* v. *City of Seattle,* 387 U. S. 541, 543 (1967).

These same cases also held that the Fourth Amendment prohibition against unreasonable searches protects against warrantless intrusions during civil as well as criminal investigations. *See* v. *City of Seattle, supra,* at 543. The reason is found in the "basic purpose of this Amendment . . . [which] is to safeguard the privacy and security of individuals against arbitrary invasions by governmental officials." *Camara, supra,* at 528. If the government intrudes on a person's property, the privacy interest suffers whether the government's motivation is to investigate violations of criminal laws or breaches of other statutory or regulatory standards. It therefore appears that unless some recognized exception to the warrant requirement applies, *See* v. *City of Seattle, supra,* would require a warrant to conduct the inspection sought in this case.

The Secretary urges that an exception from the search warrant requirement has been recognized for "pervasively regulated business[es]," *United States* v. *Biswell,* 406 U. S. 311, 316 (1972), and for "closely regulated" industries "long subject to close supervision and inspection" *Colonnade Catering Corp.* v. *United States,* 397 U. S. 72, 74, 77 (1970). These cases are indeed exceptions, but they represent responses to relatively unique circumstances. Certain industries have such a history of government oversight that no reasonable expectation of privacy, see *Katz* v. *United States,* 389 U. S. 347, 351-352 (1967), could exist for a proprietor over the stock of such an enterprise. Liquor *(Colonnade)* and firearms *(Biswell)* are industries of this type; when an entrepreneur embarks upon such a business, he has voluntarily chosen to subject himself to a full arsenal of governmental regulations.

Industries such as these fall within the "certain carefully defined classes of cases," referenced in *Camara, supra,* at 528. The element that distinguishes these enterprises from ordinary businesses is a long tradition of close government supervision, of which any person who chooses to enter such a business must already be aware. "A central difference between those cases *[Colonnade* and *Biswell]* and this one is that businessmen engaged in such federally licensed and regulated enterprises accept the burdens as well as the benefits of their trade, whereas the petitioner here was not engaged in any regulated or licensed business. The businessman in a regulated industry in effect consents to the restrictions placed upon him." *Almeida-Sanchez* v. *United States,* 413 U. S. 266, 271 (1973).

The clear import of our cases is that the closely regulated industry of the type involved in *Colonnade* and *Biswell* is the exception. The Secretary would make it the rule. Invoking the Walsh-Healy Act of 1936, 41 U. S. C. § 35, *et seq.,* the Secretary attempts to support a conclusion that all businesses involved in interstate commerce have long been subjected to close supervision of employee safety and health conditions. But the degree of federal involvement in employee working circumstances has never been of the order of specificity and pervasiveness that OSHA mandates. It is quite unconvincing to argue that the imposition of minimum wages and maximum hours on employers who contracted with the government under the Walsh-Healy Act prepared the entirety of American interstate commerce for regulation of working conditions to the minutest detail. Nor can any but the most fictional sense of voluntary consent to later searches be found in the single fact that one conducts a business affecting interstate commerce; under current practice and law, few businesses can be conducted without having some effect on interstate commerce.

The Secretary also attempts to derive support for a *Colonnade-Biswell*-type exception by drawing analogies from the field of labor law. In *Republic Aviation Corp.* v. *NLRB,* 324 U. S. 793 (1945), this Court upheld the rights of employees to solicit for a union during nonworking time where efficiency was not compromised. By opening up his property to employees, the employer had yielded so much of his private property rights as to allow those employees to exercise § 7 rights under the National Labor Relations Act. But this Court also held that the private property rights of an owner prevailed over the intrusion of nonemployee organizers, even in nonworking areas of the plant and during nonworking hours. *NLRB* v. *Babcock & Wilcox Co.,* 351 U. S. 105 (1956).

The critical fact in this case is that entry over Mr. Barlow's objection is being sought by a Government agent.* Employees are not being prohibited from reporting OSHA violations. What they observe in their daily functions is undoubtedly beyond the employer's reasonable expectation of privacy. The Government inspector, however, is not an employee. Without a warrant he stands in no better position than a member of the public. What is observable by the public is observable, without a warrant, by the Government inspector as well.** The owner of a business has not, by the necessary utilization of employees in his operation, thrown open the areas where employees alone are permitted to the warrantless scrutiny of Government agents. That an employee is free to report, and the Government is free to use, any evidence of noncompliance with OSHA that the employee observes furnishes no justification for federal agents to enter a place of business from which the public is restricted and to conduct their own warrantless search.†

II.

The Secretary nevertheless stoutly argues that the enforcement scheme of the Act requires warrantless searches, and that the restrictions on search discretion contained in the Act and its regulations already protect as much privacy as a warrant would. The Secretary thereby asserts the actual reasonableness of OSHA-searches, whatever the general rule against warrantless searches might be. Because "reasonableness is still the ultimate standard," *Camara* v. *Municipal Court, supra,* at 539, the Secretary suggests that the Court decide whether a warrant is needed by arriving at a sensible balance between the administrative necessities of OSHA inspections and the incremental protection of privacy of business owners a warrant would afford. He suggests that only a decision exempting OSHA inspections from the Warrant Clause would give "full recognition to the competing public and private interests here at stake." *Camara* v. *Municipal Court, supra,* at 539.

The Secretary submits that warrantless inspections are essential to the proper enforcement of OSHA because they afford the opportunity to inspect without prior notice and hence to preserve the advantages of surprise. While the dangerous conditions outlawed by the Act include structural defects that cannot be quickly hidden or remedied, the Act also regulates a myriad of safety details that may be amenable to speedy alteration or disguise. The risk is that during the interval between an inspector's initial request to search a plant and his procuring a warrant

---

*The Government has asked that Mr. Barlow be ordered to show cause why he should not be held in contempt for refusing to honor the inspection order, and its position is that the OSHA inspector is now entitled to enter at once, over Mr. Barlow's objection.

**Cf. *Air Pollution Variance Bd.* v. *Western Alfalfa Corp.*, 416 U. S. 861 (1974).

†The automobile search cases cited by the Secretary are even less helpful to his position than the labor cases. The fact that automobiles occupy a special category in Fourth Amendment case law is by now beyond doubt, due, among other factors, to the quick mobility of a car, the registration requirements of both the car and the driver, and the more available opportunity for plain-view observations of a car's contents. *Cady* v. *Dombrowski,* 413 U. S. 433, 441–442 (1973); see also *Chambers* v. *Maroney,* 399 U. S. 42, 48–51 (1970). Even so, probable cause has not been abandoned as a requirement for stopping and searching an automobile.

following the owner's refusal of permission, violations of this latter type could be corrected and thus escape the inspector's notice. To the suggestion that warrants may be issued *ex parte* and executed without delay and without prior notice, thereby preserving the element of surprise, the Secretary expresses concern for the administrative strain that would be experienced by the inspection system, and by the courts, should *ex parte* warrants issued in advance become standard practice.

We are unconvinced, however, that requiring warrants to inspect will impose serious burdens on the inspection system or the courts, will prevent inspections necessary to enforce the statute, or will make them less effective. In the first place, the great majority of businessmen can be expected in normal course to consent to inspection without warrant; the Secretary has not brought to this Court's attention any widespread pattern of refusal.†† In those cases where an owner does insist on a warrant, the Secretary argues that inspection efficiency will be impeded by the advance notice and delay. The Act's penalty provisions for giving advance notice of a search, 29 U. S. C. § 666 (f), and the Secretary's own regulations, 29 CFR § 1903.6, indicate that surprise searches are indeed contemplated. However, the Secretary has also promulgated a regulation providing that upon refusal to permit an inspector to enter the property or to complete his inspection, the inspector shall attempt to ascertain the reasons for the refusal and report to his superior, who shall "promptly take appropriate action, including compulsory process, if necessary." 29 CFR § 1903.4.‡ The regulation represents a choice to proceed by process where entry is refused; and on the basis of evidence available from present practice, the Act's effectiveness has not been crippled by providing those owners who wish to refuse an initial requested entry with a time lapse while the inspector obtains the necessary process.‡‡ Indeed, the kind of process sought in this case and

---

††We recognize that today's holding might itself have an impact on whether owners choose and resist requested searches; we can only await the development of evidence not present on this record to determine how serious an impediment to effective enforcement this might be.

‡It is true, as the Secretary asserts, that § 8 (a) of the Act, 29 U. S. C. § 657 (a), purports to authorize inspections without warrant; but it is also true that it does not forbid the Secretary from proceeding to inspect only by warrant or other process. The Secretary has broad authority to prescribe such rules and regulations as he may deem necessary to carry out his responsibilities under this chapter "including rules and regulations dealing with the inspection of an employer's establishment." § 8 (g) (2), 29 U. S. C. § 657 (g)(2). The regulations with respect to inspections are contained in 29 CFR Part 1903. Section 1903.4, referred to in the text, provides as follows:
"Upon a refusal to permit a Compliance Safety and Health Officer, in the exercise of his official duties, to enter without delay and at reasonable times any place of employment or any place therein, to inspect, to review records, or to question any employer, owner, operator, agent, or employee, in accordance with § 1903.3 or to permit a representative of employees to accompany the Compliance Safety and Health Officer during the physical inspection of any workplace in accordance with § 1903.8, the Compliance Safety and Health Officer shall terminate the inspection or confine the inspection to other areas, conditions, structures, machines, apparatus, devices, equipment, materials, records, or interviews concerning which no objection is raised. The Compliance Safety and Health Officer shall endeavor to ascertain the reason for such refusal, and he shall immediately report the refusal and the reason therefor to the Area Director. The Area Director shall immediately consult with the Assistant Regional Director and the Regional Solicitor, who shall promptly take appropriate action, including compulsory process, if necessary."
When his representative was refused admission by Mr. Barlow, the Secretary proceeded in federal court to enforce his right to enter and inspect, as conferred by 29 U. S. C. § 657.

‡‡A change in the language of the Compliance Operations Manual for OSHA (*Continued on next page*)

apparently anticipated by the regulation provides notice to the business operator.|| If this safeguard endangers the efficient administration of OSHA, the Secretary should never have adopted it, particularly when the Act does not require it. Nor is it immediately apparent why the advantages of surprise would be lost if, after being refused entry, procedures were available for the Secretary to seek an *ex parte* warrant and to reappear at the premises without further notice to the establishment being inspected.|| ||

---

‡(*Continued from previous page*) inspectors supports the inference that, whatever the Act's administrators might have thought at the start, it was eventually concluded that enforcement efficiency would not be jeopardized by permitting employers to refuse entry, at least until the inspector obtained compulsory process. The 1972 Manual included a section specifically directed to obtaining "warrants," and one provision of that section dealt with *ex parte* warrants:

"In cases where a refusal of entry is to be expected from the past performance of the employer, or where the employer has given some indication prior to the commencement of the investigation of his intention to bar entry or limit or interfere with the investigation, a warrant should be obtained before the inspection is attempted. Cases of this nature should also be referred through the Area Director to the appropriate Regional Solicitor and the Regional Administrator alerted." OSHA Compliance Operations Manual (Jan. 1972), at V-7.

The latest available manual, incorporating changes as of November 1977, deletes this provision, leaving only the details for obtaining "compulsory process" *after* an employer has refused entry. *OSHA Field Operations Manual,* Vol. V, at V-4–V-5. In its present form, the Secretary's regulation appears to permit establishment owners to insist on "process"; and hence their refusal to permit entry would fall short of criminal conduct within the meaning of 18 U. S. C. §§ 111 and 1114, which make it a crime forcibly to impede, intimidate or interfere with federal officials, including OSHA inspectors, while engaged in or on account of the performance of their officials' duties.

||The proceeding was instituted by filing an "Application for Affirmative Order to Grant Entry and for an Order to show cause why such affirmative order should not issue." The District Court issued the order to show cause, the matter was argued, and an order then issued authorizing the inspection and enjoining interference by Barlow. The following is the order issued by the District Court:

"IT IS HEREBY ORDERED, ADJUDGED AND DECREED that the United States of America, United States Department of Labor, Occupational Safety and Health Administration, through its duly designated representative or representatives, are entitled to, and shall have hereby, entry upon the premises known as Barlow's Inc., 225 West Pine, Pocatello, Idaho, and upon said business premises to conduct an inspection and investigation as provided for in Section 8 of the Occupational Safety and Health Act of 1970 (29 U. S. C. 651, *et seq.*), as part of an inspection program designed to assure compliance with that Act; that the inspection and investigation shall be conducted during regular working hours or at other reasonable times, within reasonable limits and in a reasonable manner, all as set forth in the regulations pertaining to such inspections promulgated by the Secretary of Labor, at 29 C. F. R., Part 1903; that appropriate credentials as representatives of the Occupational Safety and Health Administration, United States Department of Labor, shall be presented to the Barlow's Inc. representative upon said premises and the inspection and investigation shall be commenced as soon as practicable after the issuance of this Order and shall be completed with reasonable promptness; that the inspection and investigation shall extend to the establishment or other area, workplace, or environment where work is performed by employees of the employer, Barlow's Inc., and to all pertinent conditions, structures, machines, apparatus, devices, equipment, materials, and all other things therein (including but not limited to records, files, papers, processes, controls, and facilities) bearing upon whether Barlow's Inc. is furnishing to its employees employment and a place of employment that are free from recognized hazards that are causing or are likely to cause death or serious physical harm to its employees, and whether Barlow's Inc. is complying with the Occupational Safety and Health Standards promulgated under the Occupational Safety and Health Act and the rules, regulations, and orders issued pursuant to that Act; that representatives of the Occupational Safety and Health Administration may, at the option of Barlow's Inc., be accompanied by one or more employee of Barlow's Inc., pursuant to Section 8 (e) of that Act; that Barlow's Inc., its agents, representatives, officers, and employees are hereby enjoined and restrained from in any way whatsoever interfering with the inspection and investigation authorized by this Order and, further, Barlow's Inc. is hereby ordered and directed to, within five working days from the date of this Order, furnish a copy of this Order to its officers and managers, and, in addition, to post a copy of this Order at its employee's bulletin board, located upon the business premises; and Barlow's Inc. is hereby ordered and directed to comply in all respects with this order and allow the inspection and investigation to take place, without delay and forthwith."

|| ||Insofar as the Secretary's statutory authority is concerned, a regulation expressly providing that the Secretary could proceed *ex parte* to seek a warrant or its equivalent would (*Continued on next page*)

Whether the Secretary proceeds to secure a warrant or other process, with or without prior notice, his entitlement to inspect will not depend on his demonstrating probable cause to believe that conditions in violation of OSHA exist on the premises. Probable cause in the criminal law sense is not required. For purposes of an administrative search such as this, probable cause justifying the issuance of a warrant may be based not only on specific evidence of an existing violation# but also on a showing that "reasonable legislative or administrative standards for conducting an . . . inspection are satisfied with respect to a particular [establishment]." *Camara* v. *Municipal Court, supra,* at 538. A warrant showing that a specific business has been chosen for an OSHA search on the basis of a general administrative plan for the enforcement of the Act derived from neutral sources such as, for example, dispersion of employees in various types of industries across a given area, and the desired frequency of searches in any of the lesser divisions of the area, would protect an employer's Fourth Amendment rights.## We doubt that the consumption of enforcement energies in the obtaining of such warrants will exceed manageable proportions.

Finally, the Secretary urges that requiring a warrant for OSHA inspectors will mean that, as a practical matter, warrantless search provisions in other regulatory statutes are also constitutionally infirm. The reasonableness of a warrantless search, however, will depend upon the specific enforcement needs and privacy guarantees of each statute. Some of the statutes cited apply only to a single industry, where regulations might already be so pervasive that a *Colonnade-Biswell* exception to the warrant requirement could apply. Some statutes already envision resort to federal court enforcement when entry is refused, employing specific language in some cases* and general language in others.** In short, we base to-

---

|| ||(*Continued from previous page*) appear to be as much within the Secretary's power as the regulation currently in force and calling for "compulsory process."

#Section 8 (f)(1), 29 U. S. C. § 657 (f)(1), provides that employees or their representatives may give written notice to the Secretary of what they believe to be violations of safety or health standards and may request an inspection. If the Secretary then determines that "there are reasonable grounds to believe that such violation or danger exists, he shall make a special inspection in accordance with the provisions of this Section as soon as practicable." The statute thus purports to authorize a warrantless inspection in these circumstances.

##The Secretary's Brief, p. 9, n. 7, states that the Barlow inspection was not based on an employee complaint but was a "general schedule" investigation. "Such general investigations," he explains, "now called Regional Programmed Inspections, are carried out in accordance with criteria based upon accident experience and the number of employees exposed in particular industries. U. S. Department of Labor, Occupational Safety and Health Administration, Field Operations Manual, *supra,* 1 CCH Employment Safety and Health Guide ¶ 4327.2 (1976)."

*The Mine Safety Act provides "Whenever an operator . . . refuses to permit the inspection or investigation of any mine which is subject to this chapter . . . a civil action for preventive relief, including and application for a permanent or temporary injuction, restraining order, or other order, may be instituted by the Secretary in the district court of the United States for the district. . . ." 30 U. S. C. § 733 (a). "The Secretary may institute a civil action for relief, including a permanent or temporary injunction, restraining order, or any other appropriate order in the district court . . . whenever such operator or his agent . . . refuses to permit the inspection of the mine. . . . Each court shall have jurisdiction to provide such relief as may be appropriate." 30 U. S. C. § 818. Another example is the Air Pollution Control Act, which grants federal district courts jurisdiction "to require compliance" with the Administrator's attempt to inspect under 42 U. S. C. § 1857c-9, when the Administrator has commenced "a civil action for appropriate relief." 42 U. S. C. § 1857c-8 (b)(4).

**Exemplary language is contained in the Animal Welfare Act which provides (*Continued on next page*)

day's opinion on the facts and law concerned with OSHA and do not retreat from a holding appropriate to that statute because of its real or imagined effect on other, different administrative schemes.

Nor do we agree that the incremental protections afforded the employer's privacy by a warrant are so marginal that they fail to justify the administrative burdens that may be entailed. The authority to make warrantless searches devolves almost unbridled discretion upon executive and administrative officers, particularly those in the field, as to when to search and whom to search. A warrant, by contrast, would provide assurances from a neutral officer that the inspection is reasonable under the Constitution, is authorized by statute, and is pursuant to an administrative plan containing specific neutral criteria.† Also, a warrant would then and there advise the owner of the scope and objects of the search, beyond which limits the inspector is not expected to proceed.†† These are important functions for a warrant to perform, functions which underlie the Court's prior decisions that the Warrant Clause applies to inspections for compliance with regulatory statutes.‡ *Camara* v. *Municipal Court, supra; See* v. *City of Seattle, supra.* We conclude that the concerns expressed by the Secretary do not suffice to

---

**(*Continued from previous page*) for inspections by the Secretary of Agriculture; federal district courts are vested with jurisidiction "specifically to enforce, and to prevent and restrain violations of this chapter, and shall have jurisdiction in all other kinds of cases arising under this chapter." 7 U. S. C. § 2146 (c) (Supp. V 1975). Similar provisions are included in other agricultural inspection acts; see, *e. g.*, 21 U. S. C. § 674 (meat product inspection); 21 U. S. C. § 1050 (egg product inspection). The Internal Revenue Code, whose excise tax provisions requiring inspections of businesses are cited by the Secretary, provides "The District Courts . . . shall have such jurisdiction to make and issue in civil actions writs and orders of injunction . . . and such other orders and processes, and to render such . . . decrees as may be necessary or appropriate for the enforcement of the internal revenue laws." 26 U. S. C. § 7402 (a). For gasoline inspections, federal district courts are granted jurisdiction to restrain violations and enforce standards (one of which, 49 U. S. C. § 1677, requires gas transporters to permit entry or inspection). The owner is to be afforded the opportunity for notice and response in most cases, but "failure to give such notice and afford opportunity shall not preclude the granting of appropriate relief [by the district court]." 49 U. S. C. § 1679 (a).

†The application for the inspection order filed by the Secretary in this case represented that "the desired inspection and investigation are contemplated as part of an inspection program designed to assure compliance with the Act and are authorized by Section 8 (a) of the Act." The program was not described, however, or any facts presented that would indicate why an inspection of Barlow's establishment was within the program. The order that issued concluded generally that the inspection authorized was "part of an inspection program designed to assure compliance with the Act."

††Section 8 (a) of the Act, 29 U. S. C. § 657 (a), provides that "in order to carry out the purposes of this chapter" the Secretary may enter any establishment, area, work place or environment "where work is performed by employees of an employer" and "inspect and investigate" any such place of employment and all "pertinent conditions, structures, machines, apparatus, devices, equipment, and materials therein, and . . . question privately any such employer, owner, operator, agent, or employee." Inspections are to be carried out "during regular working hours and at other reasonable times, and within reasonable limits and in a reasonable manner." The Secretary's regulations echo the statutory language in these respects. 29 CFR § 1903.3. They also provide that inspectors are to explain the nature and purpose of the inspection and to "indicate generally the scope of the inspection." 29 CFR § 1903.7 (a). Environmental samples and photographs are authorized, 29 CFR § 1903.7 (b), and inspections are to be performed so as "to preclude unreasonable disruption of the employer's establishment." 29 CFR § 1903.7 (d). The order that issued in this case reflected much of the foregoing statutory and regulatory language.

‡Delineating the scope of a search with some care is particularly important where documents are involved. Section 8 (c) of the Act, 29 U. S. C. § 657 (c), provides that an employer must "make, keep and preserve, and make available to the Secretary [of Labor] or Secretary (*Continued on next page*)

justify warrantless inspections under OSHA or vitiate the general constitutional requirement that for a search to be reasonable a warrant must be obtained.

III.

We hold that Barlow was entitled to a declaratory judgment that the Act is unconstitutional insofar as it purports to authorize inspections without warrant or its equivalent and to an injunction enjoining the Act's enforcement to that extent.‡‡ The judgment of the District Court is therefore affirmed.

So ordered.

MR. JUSTICE BRENNAN took no part in the consideration or decision of this case.

———————————

In this decision, WHITE, J., delivered the opinion of the Court, in which BURGER, C. J., and STEWART, MARSHALL, and POWELL, JJ., joined. STEVENS, J., filed a dissenting opinion, in which BLACKMUN and REHNQUIST, JJ., joined. BRENNAN, J., took no part in the consideration or decision of the case. Following is the dissent in *Marshall* v. *Barlow's Inc.* Read the dissent carefully and be prepared to describe the opposing view.

———————————

‡(*Continued from previous page*) of Health, Education and Welfare'' such records regarding his activities relating to OSHA as the Secretary of Labor may prescribe by regulation as necessary or appropriate for enforcement of the statute or for developing information regarding the causes and prevention of occupational accidents and illnesses. Regulations requiring employers to maintain records of and to make periodic reports on ''work-related deaths, injuries and illnesses'' are also contemplated, as are rules requiring accurate records of employee exposures to potential toxic materials and harmful physical agents.

In describing the scope of the warrantless inspection authorized by the statute, § 8 (a) does not expressly include any *records* among those items or things that may be examined, and § 8 (c) merely provides that the employer is ''to make available'' his pertinent records and to make periodic reports. The Secretary's regulation, 29 CFR § 1903.3, however, expressly includes among the inspector's powers the authority ''to review records required by the Act and regulations published in this chapter, and other records which are directly related to the purpose of the inspection.'' Further, 29 CFR § 1903.7 requires inspectors to indicate generally ''the records specified in § 1903.3 which they wish to review'' but ''such designations of records shall not preclude access to additional records specified in § 1903.3.'' It is the Secretary's position, which we reject, that an inspection of documents of this scope may be effected without a warrant.

The order that issued in this case included among the objects and things to be inspected ''all other things therein (including but not limited to records, files, papers, processes, controls and facilities) bearing upon whether Barlow's, Inc., is furnishing its employees employment and a place of employment that are free from recognizable hazards that are causing or are likely to cause death or serious physical harm to its employees, and whether Barlow's, Inc., is complying with . . .'' the OSHA regulations.

‡‡The injunction entered by the District Court, however, should not be understood to forbid the Secretary from exercising the inspection authority conferred by § 657 pursuant to regulations and judicial process that satisfy the Fourth Amendment. The District Court did not address the issue whether the order for inspection that was issued in this case was the functional equivalent of a warrant, and the Secretary has limited his submission in this case to the constitutionality of a warrantless search of the Barlow establishment authorized by § 8 (a). He has expressly declined to rely on 29 CFR § 1903.4 and upon the order obtained in this case. Tr. of Oral Arg. 19. Of course, if the process obtained here, or obtained in other cases under revised regulations, would satisfy the Fourth Amendment, there would be no occasion for enjoining the inspections authorized by § 8 (a).

MR. JUSTICE STEVENS, with whom MR. JUSTICE BLACKMUN and MR. JUSTICE REHNQUIST join, dissenting.

Congress enacted the Occupational Safety and Health Act to safeguard employees against hazards in the work areas of businesses subject to the Act. To ensure compliance, Congress authorized the Secretary of Labor to conduct routine, nonconsensual inspections. Today the Court holds that the Fourth Amendment prohibits such inspections without a warrant. The Court also holds that the constitutionally required warrant may be issued without any showing of probable cause. I disagree with both of these holdings.

The Fourth Amendment contains two separate clauses, each flatly prohibiting a category of governmental conduct. The first clause states that the right to be free from unreasonable searches "shall not be violated"; || the second unequivocally prohibits the issuance of warrants except "upon probable cause." || || In this case the ultimate question is whether the category of warrantless searches authorized by the statute is "unreasonable" within the meaning of the first clause.

In cases involving the investigation of criminal activity, the Court has held that the reasonableness of a search generally depends upon whether it was conducted pursuant to a valid warrant. See, *e.g., Coolidge* v. *New Hampshire*, 403 U. S. 443. There is, however, also a category of searches which are reasonable within the meaning of the first clause even though the probable cause requirement of the Warrant Clause cannot be satisfied. See *United States* v. *Martinez-Fuerte*, 428 U. S. 543; *Terry* v. *Ohio*, 392 U. S. 1; *South Dakota* v. *Opperman*, 428 U. S. 364; *United States* v. *Biswell*, 406 U. S. 311. The regulatory inspection program challenged in this case, in my judgment, falls within this category.

I.

The warrant requirement is linked "textually . . . to the probable-cause concept" in the Warrant Clause. *South Dakota* v. *Opperman, supra*, at 370, n. 5. The routine OSHA inspections are, by definition, not based on cause to believe there is a violation on the premises to be inspected. Hence, if the inspections were measured against the requirements of the Warrant Clause, they would be automatically and unequivocally unreasonable.

Because of the acknowledged importance and reasonableness of routine inspections in the enforcement of federal regulatory statutes such as OSHA, the Court recognizes that requiring full compliance with the Warrant Clause would invalidate all such inspection programs. Yet, rather than simply analyzing such programs under the "reasonableness" clause of the Fourth Amendment, the Court

---

|| "The right of the people to be secure in their persons, houses, papers, and effects, against unreasonable searches and seizures, shall not be violated . . . ."

|| || ". . . and no Warrants shall issue, but upon probable cause, supported by Oath or affirmation, and particularly describing the place to be searched, and the persons or things to be seized."

holds the OSHA program invalid under the Warrant Clause and then avoids a blanket prohibition on all routine, regulatory inspections by relying on the notion that the "probable cause" requirement in the Warrant Clause may be relaxed whenever the Court believes that the governmental need to conduct a category of "searches" outweighs the intrusion on interests protected by the Fourth Amendment.

The Court's approach disregards the plain language of the Warrant Clause and is unfaithful to the balance struck by the Framers of the Fourth Amendment — "the one procedural safeguard in the Constitution that grew directly out of the events which immediately preceded the revolutionary struggle with England."# This preconstitutional history includes the controversy in England over the issuance of general warrants to aid enforcement of the seditious libel laws and the colonial experience with writs of assistance issued to facilitate collection of the various import duties imposed by Parliament. The Framers' familiarity with the abuses attending the issuance of such general warrants provided the principal stimulus for the restraints on arbitrary governmental intrusions embodied in the Fourth Amendment.

> "[O]ur constitutional fathers were not concerned about warrantless searches, but about overreaching warrants. It is perhaps too much to say that they feared the warrant more than the search, but it is plain enough that the warrant was the prime object of their concern. Far from looking at the warrant as a protection against unreasonable searches, they saw it as an authority for unreasonable and oppressive searches. . . .##

Since the general warrant, not the warrantless search, was the immediate evil at which the Fourth Amendment was directed, it is not surprising that the Framers placed precise limits on its issuance. The requirement that a warrant only issue on a showing of particularized probable cause was the means adopted to circumscribe the warrant power. While the subsequent course of Fourth Amendment jurisprudence in this Court emphasizes the dangers posed by warrantless searches conducted without probable cause, it is the general reasonableness standard in the first clause, not the Warrant Clause, that the Framers adopted to limit this category of searches. It is of course true that the existence of a valid warrant normally satisfies the reasonableness requirement under the Fourth Amendment. But we should not dilute the requirements of the Warrant Clause in an effort to force every kind of governmental intrusion which satisfies the Fourth Amendment definition of a "search" into a judicially developed, warrant-preference scheme.

Fidelity to the original understanding of the Fourth Amendment, therefore, leads to the conclusion that the Warrant Clause has no application to routine,

---

#Landynski, Search and Seizure and the Supreme Court, 19 (1966).
##Taylor, Two Studies in Constitutional Interpretation, 41 (1969).

regulatory inspections of commercial premises. If such inspections are valid, it is because they comport with the ultimate reasonableness standard of the Fourth Amendment. If the Court were correct in its view that such inspections, if undertaken without a warrant, are unreasonable in the constitutional sense, the issuance of a ''new-fangled warrant'' — to use Mr. Justice Clark's characteristically expressive term — without any true showing of particularized probable cause would not be sufficient to validate them.*

II.

Even if a warrant issued without probable cause were faithful to the Warrant Clause, I could not accept the Court's holding that the Government's inspection program is constitutionally unreasonable because it fails to require such a warrant procedure. In determining whether a warrant is a necessary safeguard in a given class of cases, ''the Court has weighed the public interest against the Fourth Amendment interest of the individual. . . .'' *United States* v. *Martinez-Fuerte*, 428 U. S., at 555. Several considerations persuade me that this balance should be struck in favor of the routine inspections authorized by Congress.

Congress has determined that regulation and supervision of safety in the work place furthers an important public interest and that the power to conduct warrantless searches is necessary to accomplish the safety goals of the legislation. In assessing the public interest side of the Fourth Amendment balance, however, the Court today substitutes its judgment for that of Congress on the question of what inspection authority is needed to effectuate the purposes of the Act. The Court states that if surprise is truly an important ingredient of an effective, representative inspection program, it can be retained by obtaining *ex parte* warrants in advance. The Court assures the Secretary that this will not unduly burden enforcement resources because most employers will consent to inspection.

The Court's analysis does not persuade me that Congress' determination that the warrantless inspection power is a necessary adjunct of the exercise of the regulatory power is unreasonable. It was surely not unreasonable to conclude that the rate at which employers deny entry to inspectors would increase if covered businesses, which may have safety violations on their premises, have a right to deny warrantless entry to a compliance inspector. The Court is correct that this problem could be avoided by requiring inspectors to obtain a warrant prior to every inspection visit. But the adoption of such a practice undercuts the Court's explanation of why a warrant requirement would not create undue enforcement problems. For, even if it were true that many employers would not exercise their right to demand a warrant, it would provide little solace to those charged with administration of OSHA; faced with an increase in the rate of refusals and the added costs generated by futile trips to inspection sites where entry is denied, officials may be compelled to adopt a general practice of obtaining warrants in advance.

---

*See v. *City of Seattle*, 387 U. S. 541, 547 (CLARK, J., dissenting).

While the Court's prediction of the effect a warrant requirement would have on the behavior of covered employers may turn out to be accurate, its judgment is essentially empirical. On such an issue, I would defer to Congress' judgment regarding the importance of a warrantless search power to the OSHA enforcement scheme.

The Court also appears uncomfortable with the notion of second-guessing Congress and the Secretary on the question of how the substantive goals of OSHA can best be achieved. Thus, the Court offers an alternative explanation for its refusal to accept the legislative judgment. We are told that, in any event, the Secretary, who is charged with enforcement of the Act, has indicated that inspections without delay are not essential to the enforcement scheme. The Court bases this conclusion on a regulation prescribing the administrative response when a compliance inspector is denied entry. It provides that: "[t]he Area Director shall immediately consult with the Assistant Regional Director and the Regional Solicitor who shall promptly take appropriate action including compulsory process if necessary." 29 CFR § 1903.4. The Court views this regulation as an admission by the Secretary that no enforcement problem is generated by permitting employers to deny entry and delaying the inspection until a warrant has been obtained. I disagree. The regulation was promulgated against the background of a statutory right to immediate entry, of which covered employers are presumably aware and which Congress and the Secretary obviously thought would keep denials of entry to a minimum. In these circumstances, it was surely not unreasonable for the Secretary to adopt an orderly procedure for dealing with what he believed would be the occasional denial of entry. The regulation does not imply a judgment by the Secretary that delay caused by numerous denials of entry would be administratively acceptable.

Even if a warrant requirement does not "frustrate" the legislative purpose, the Court has no authority to impose an additional burden on the Secretary unless that burden is required to protect the employer's Fourth Amendment interests.**
The essential function of the traditional warrant requirement is the interposition of a neutral magistrate between the citizen and the presumably zealous law enforcement officer so that there might be an objective determination of probable cause. But this purpose is not served by the new-fangled inspection warrant. As the Court acknowledges, the inspector's "entitlement to inspect will not depend on his demonstrating probable cause to believe that conditions in violation of OSHA exist on the premises. . . . For purposes of an administrative search such as this, probable cause justifying the issuance of a warrant may be based . . . on a showing that 'reasonable legislative or administrative standards for conducting an

---

** When it passed OSHA, Congress was cognizant of the fact that in light of the enormity of the enforcement task "the number of inspections which it would be desirable to have made will undoubtedly for an unforeseeable period, exceed the capacity of the inspection force. . . ." Committee Print, Legislative History of the Occupational Safety and Health Act of 1970, Senate Committee on Labor and Public Welfare, 92d Cong., 1st Sess., 152 (1971).

. . . inspection are satisfied with respect to a particular [establishment].' " *Ante,* at 12–13. To obtain a warrant, the inspector need only show that "a specific business has been chosen for an OSHA search on the basis of a general administrative plan for the enforcement of the Act derived from neutral sources. . . ." *Ante,* at 13. Thus, the only question for the magistrate's consideration is whether the contemplated inspection deviates from an inspection schedule drawn up by high-level agency officials.

Unlike the traditional warrant, the inspection warrant provides no protection against the search itself for employers whom the government has no reason to suspect are violating OSHA regulations. The Court plainly accepts the proposition that random health and safety inspections are reasonable. It does not question Congress' determination that the public interest in work places free from health and safety hazards outweighs the employer's desire to conduct his business only in the presence of permittees, except in those rare instances when the government has probable cause to suspect that the premises harbor a violation of the law. What purposes, then, are served by the administrative warrant procedure? The inspection warrant purports to serve three functions: to inform the employer that the inspection is authorized by the statute, to advise him of the lawful limits of the inspection, and to assure him that the person demanding entry is an authorized inspector. *Camara* v. *Municipal Court,* 387 U. S., at 532. An examination of these functions in the OSHA context reveals that the inspection warrant adds little to the protection already afforded by the statute and pertinent regulations, and the slight additional benefit it might provide is insufficient to identify a constitutional violation or to justify overriding Congress' judgment that the power to conduct warrantless inspections is essential.

The inspection warrant is supposed to assure the employer that the inspection is in fact routine, and that the inspector has not improperly departed from the program of representative inspections established by responsible officials. But to the extent that harassment inspections would be reduced by the necessity of obtaining a warrant, the Secretary's present enforcement scheme would have precisely the same effect. The representative inspections are conducted " 'in accordance with criteria based upon accident experience and the number of employees exposed in particular industries.' " *Ante,* 13, n. 17. If, under the present scheme, entry to covered premises is denied, the inspector can gain entry only by informing his administrative superiors of the refusal and seeking a court order requiring the employer to submit to the inspection. The inspector who would like to conduct a nonroutine search is just as likely to be deterred by the prospect of informing his superiors of his intention and of making false representations to the court when he seeks compulsory process as by the prospect of having to make bad-faith representations in an *ex parte* warrant proceeding.

The other two asserted purposes of the administrative warrant are also adequately achieved under the existing scheme. If the employer has doubts about the official status of the inspector, he is given adequate opportunity to reassure himself in this regard before permitting entry. The OSHA inspector's statutory right to enter the premises is conditioned upon the presentation of appropriate credentials. 29 U. S. C. § 657 (a) (1). These credentials state the inspector's name, identify him as an OSHA compliance officer, and contain his photograph and signature. If the employer still has doubts, he may make a toll free call to verify the inspector's authority, *Usery* v. *Godfrey Brake & Supply Service, Inc.*, 545 F. 2d 52, 54 (CA8 1976), or simply deny entry and await the presentation of a court order.

The warrant is not needed to inform the employer of the lawful limits of an OSHA inspection. The statute expressly provides that the inspector may enter all areas in a covered business "where work is performed by an employee of an employer," 29 U. S. C. § 657 (a)(1), "to inspect and investigate during regular working hours and at other reasonable times and within reasonable limits and in a reasonable manner . . . all pertinent conditions, structures, machines, apparatus, devices, equipment, and materials therein . . . ." 29 U.S.C. § 657 (a)(2). See also 29 CFR § 1903. While it is true that the inspection power granted by Congress is broad, the warrant procedure required by the Court does not purport to restrict this power but simply to ensure that the employer is apprised of its scope. Since both the statute and the pertinent regulations perform this informational function, a warrant is superfluous.

Requiring the inspection warrant, therefore, adds little in the way of protection to that already provided under the existing enforcement scheme. In these circumstances, the warrant is essentially a formality. In view of the obviously enormous cost of enforcing a health and safety scheme of the dimensions of OSHA, this Court should not, in the guise of construing the Fourth Amendment, require formalities which merely place an additional strain on already overtaxed federal resources.

Congress, like this Court, has an obligation to obey the mandate of the Fourth Amendment. In the past the "Court has been particularly sensitive to the amendment's broad standard of 'reasonableness' where . . . authorizing statutes permitted the challenged searches." *Almeida-Sanchez* v. *United States*, 413 U. S. 266, 290 (WHITE, J., dissenting). In *United States* v. *Martinez-Fuerte, supra,* for example, respondents challenged the routine stopping of vehicles to check for aliens at permanent checkpoints located away from the border. The checkpoints were established pursuant to statutory authority and their location and operation were governed by administrative criteria. The Court rejected respondents' argument that the constitutional reasonableness of the location and operation of the fixed checkpoints should be reviewed in a *Camara* warrant proceeding. The Court observed that the reassuring purposes of the inspection warrant were adequately served by the visible manifestations of authority exhibited at the fixed checkpoints.

Moreover, although the location and method of operation of the fixed checkpoints were deemed critical to the constitutional reasonableness of the challenged stops, the Court did not require Border Patrol officials to obtain a warrant based on a showing that the checkpoints were located and operated in accordance with administrative standards. Indeed, the Court observed that "[t] he choice of checkpoint locations must be left largely to the discretion of Border Patrol officials, to be exercised in accordance with statutes and regulations that may be applicable. . . [and] [m]any incidents of checkpoint operation also must be committed to the discretion of such officials." *Id.*, at 559–560, n. 13. The Court had no difficulty assuming that those officials responsible for allocating limited enforcement resources would be "unlikely to locate a checkpoint where it bears arbitrarily or oppressively on motorists as a class." *Id.*, at 559.

The Court's recognition of Congress' role in balancing the public interest advanced by various regulatory statutes and the private interest in being free from arbitrary governmental intrusion has not been limited to situations in which, for example, Congress is exercising its special power to exclude aliens. Until today we have not rejected a congressional judgment concerning the reasonableness of a category of regulatory inspections of commercial premises.† While businesses are unquestionably entitled to Fourth Amendment protection, we have "recognized that a business, by its special nature and voluntary existence, may open itself to intrusions that would not be permissible in a purely private context." *G. M. Leasing Corp.* v. *United States*, 429 U. S. 388, 353. Thus, in *Colonnade Catering Corp.* v. *United States*, 397 U. S. 72, the Court recognized the reasonableness of a statutory authorization to inspect the premises of a caterer dealing in alcoholic beverages, noting that "Congress has broad authority to design such powers of inspection under the liquor laws it deems necessary to meet the evils at hand." *Id.*, at 76. And in *United States* v. *Biswell*, 406 U. S. 311, the Court sustained the authority to conduct warrantless searches of firearm dealers under the Gun Control Act of 1968 primarily on the basis of the reasonableness of the congressional evaluation of the interests at stake.††

The Court, however, concludes that the deference accorded Congress in *Biswell* and *Colonnade* should be limited to situations where the evils addressed by the regulatory statute are peculiar to a specific industry and that industry is one which

---

†The Court's rejection of a legislative judgment regarding the reasonableness of the OSHA inspection program is especially puzzling in light of recent decisions finding law enforcement practices constitutionally reasonable, even though those practices involved significantly more individual discretion than the OSHA program. See, *e.g.*, *Terry* v. *Ohio*, 392 U. S. 1; *Adams* v. *Williams*, 407 U. S. 143; *Cady* v. *Dombrowski*, 413 U. S. 433; *South Dakota* v. *Opperman*, 428, U. S. 364.

††The Court held:
"In the context of a regulatory inspection system of business premises that is carefully in time, place, and scope, the legality of the search depends . . . on the authority of a valid statute.
"We have little difficulty in concluding that where, as here, regulatory inspections further urgent federal interests, and the possibilities of abuse and a threat to privacy are not of impressive dimensions, the inspection may proceed without a warrant where specifically authorized by statute." *Id.*, at 315, 317.

has long been subject to government regulation. The Court reasons that only in those situations can it be said that a person who engages in business will be aware of and consent to routine, regulatory inspections. I cannot agree that the respect due the congressional judgment should be so narrowly confined.

In the first place, the longevity of a regulatory program does not, in my judgment, have any bearing on the reasonableness of routine inspections necessary to achieve adequate enforcement of that program. Congress' conception of what constitutes urgent federal interests need not remain static. The recent vintage of public and congressional awareness of the dangers posed by health and safety hazards in the work place is not a basis for according less respect to the considered judgment of Congress. Indeed, in *Biswell,* the Court upheld an inspection program authorized by a regulatory statute enacted in 1968. The Court there noted that "[f]ederal regulation of the interstate traffic in firearms is not as deeply rooted in history as is governmental control of the liquor industry, but close scrutiny of this traffic is undeniably" an urgent federal interest, 406 U. S., at 315. Thus, the critical fact is the congressional determination that federal regulation would further significant public interests, not the date that determination was made.

In the second place, I see no basis for the Court's conclusion that a congressional determination that a category of regulatory inspections is reasonable need only be respected when Congress is legislating on an industry-by-industry basis. The pertinent inquiry is not whether the inspection program is authorized by a regulatory statute directed at a single industry but whether Congress has limited the exercise of the inspection power to those commercial premises where the evils at which the statute is directed are to be found. Thus, in *Biswell,* if Congress had authorized inspections of all commercial premises as a means of restricting the illegal traffic in firearms, the Court would have found the inspection program unreasonable; the power to inspect was upheld because it was tailored to the subject matter of Congress' proper exercise of regulatory power. Similarly, OSHA is directed at health and safety hazards in the work place, and the inspection power granted the Secretary extends only to those areas where such hazards are likely to be found.

Finally, the Court would distinguish the respect accorded Congress' judgment in *Colonnade* and *Biswell* on the ground that businesses engaged in the liquor and firearms industry "accept the burdens as well as the benefits of their trade. . . ." *Ante,* at 5. In the Court's view, such businesses consent to the restrictions placed upon them, while it would be fiction to conclude that a businessman subject to OSHA consented to routine safety inspections. In fact, however, consent is fictional in both contexts. Here, as well as in *Biswell,* businesses are required to be aware of and comply with regulations governing their business activities. In both situations, the validity of the regulations depends not upon the consent of those regulated but on the existence of a federal statute embodying a congressional determination that the public interest in the health of the Nation's work force or the limitation of illegal firearms traffic outweighs the businessman's interest in preventing a government inspector from viewing those areas of his premises which relate to the subject matter of the regulation.

The case before us involves an attempt to conduct a warrantless search of the working area of an electrical and plumbing contractor. The statute authorizes such an inspection during reasonable hours. The inspection is limited to those areas over which Congress has exercised its proper legislative authority.‡ The area is also one to which employees have regular access without any suggestion that work performed or the equipment used has any special claim to confidentiality.‡‡ Congress has determined that industrial safety is an urgent federal interest requiring regulation and supervision, and further, that warrantless inspections are necessary to accomplish the safety goals of the legislation. While one may question the wisdom of pervasive governmental oversight of industrial life, I decline to question Congress' judgment that the inspection power is a necessary enforcement device in achieving the goals of a valid exercise of regulatory power.||

I respectfully dissent.

A week after the Supreme Court handed down its decision in the preceding case of *Marshall* v. *Barlow's, Inc.*, it ruled again on the subject of warrantless searches. This time the constitutional question involved warrantless administrative searches to investigate the cause of a fire. Under what circumstances do members of the fire service require a warrant to investigate the cause of a fire?

STATE OF MICHIGAN, PETITIONER
v.
LOREN TYLER AND ROBERT TOMPKINS

---

‡What the Court actually decided in *Camara* v. *Municipal Court*, 387 U. S. 523, and *See* v. *City of Seattle*, 387 U. S. 541, does not require the result it reaches today. *Camara* involved a residence, rather than a business establishment; although the Fourth Amendment extends its protection to commercial buildings, the central importance of protecting residential privacy is manifest. The building involved in *See* was, of course, a commercial establishment, but a holding that a locked warehouse may not be entered pursuant to a general authorization to "enter all buildings and premises, except the interior of dwellings, as often as may be necessary," *id.*, at 541, need not be extended to cover more carefully delineated grants of authority. My view that the *See* holding should be narrowly confined is influenced by my favorable opinion of the dissent written by MR. JUSTICE CLARK and joined by JUSTICES HARLAN and STEWART. As *Colonnade* and *Biswell* demonstrate, however, the doctrine of *stare decisis* does not compel the Court to extend those cases to govern today's holding.

‡‡The Act and pertinent regulation provides protection for any trade secrets of the employer. 29 U. S. C. §§ 664–665; 29 CFR 1903.9.

||The decision today renders presumptively invalid numerous inspection provisions in federal regulatory statutes. *E. g.*, 30 U. S. C. § 813 (Coal Mine Health and Safety Act); 30 U. S. C. § 723, 724 (Metal and Nonmetallic Mine Safety Act); 21 U. S. C. § 603 (Inspection of meat and food products). The fact that some of these provisions apply only to a single industry, as noted above, does not alter this fact. And the fact that some "envision resort to federal court enforcement when entry is refused" is also irrelevant since the OSHA inspection program invalidated here requires compulsory process when a compliance inspector has been denied entry.

On Writ of Certiorari to the
Supreme Court of Michigan.
Supreme Court of the United States
May 31, 1978.

MR. JUSTICE STEWART delivered the opinion of the Court.

The respondents, Loren Tyler and Robert Tompkins, were convicted in a Michigan trial court of conspiracy to burn real property in violation of Mich. Comp. Laws § 750.157a.|| || Various pieces of physical evidence and testimony based on personal observation, all obtained through unconsented and warrantless entries by police and fire officials onto the burned premises, were admitted into evidence at the respondents' trial. On appeal, the Michigan Supreme Court reversed the convictions, holding that "the warrantless searches were unconstitutional and that the evidence obtained was therefore inadmissible." [Citations omitted.] We granted certiorari to consider the applicability of the Fourth and Fourteenth Amendments to official entries onto fire-damaged premises. — U. S. —.

I.

Shortly before midnight on January 21, 1970, a fire broke out at Tyler's Auction, a furniture store in Oakland County, Mich. The building was leased to respondent Loren Tyler, who conducted the business in association with respondent Robert Tompkins. According to trial testimony of various witnesses, the fire department responded to the fire and was "just watering down smoldering embers" when Fire Chief See arrived on the scene around 2 a.m. It was Chief See's responsibility "to determine the cause and make out all reports." Chief See was met by Lt. Lawson, who informed him that two plastic containers of flammable liquid had been found in the building. Using portable lights, they entered the gutted store, which was filled with smoke and steam, to examine the containers. Concluding that the fire "could possibly have been an arson," Chief See called Police Detective Webb, who arrived around 3:30 a.m. Detective Webb took several pictures of the containers and of the interior of the store, but finally abandoned his efforts because of the smoke and steam. Chief See briefly "[l]ooked throughout the rest of the building to see if there was any further evidence, to determine what the cause of the fire was." By 4 a.m. the fire had been extinguished and the fire fighters departed. See and Webb took the two containers to the fire station, where they were turned over to Webb for safekeeping. There was neither consent nor a warrant for any of these entries into the building, nor for the removal of the containers. The respondents challenged the introduction of these containers at trial, but abandoned their objection in the State Supreme Court. [Citations omitted.] Four hours after he had left Tyler's Auction, Chief See

---

|| ||In addition, Tyler was convicted of the substantive offenses of burning real property, Mich. Comp. Laws § 750.73, and burning insured property with intent to defraud, Mich. Comp. Laws § 750.75.

returned with Assistant Chief Somerville, whose job was to determine the "origin of all fires that occur within the Township." The fire had been extinguished and the building was empty. After a cursory examination they left, and Somerville returned with Detective Webb around 9 a.m. In Webb's words, they discovered suspicious "burn marks in the carpet, which [Webb] could not see earlier that morning, because of the heat, steam, and the darkness." They also found "pieces of tape, with burn marks, on the stairway." After leaving the building to obtain tools, they returned and removed pieces of the carpet and sections of the stairs to preserve these bits of evidence suggestive of a fuse trail. Somerville also searched through the rubble "looking for any other signs or evidence that showed how this fire was caused." Again, there was neither consent nor a warrant for these entries and seizures. Both at trial and on appeal, the respondents objected to the introduction of evidence thereby obtained.

On February 16 Sergeant Hoffman of the Michigan State Police Arson Section returned to Tyler's Auction to take photographs.# During this visit or during another at about the same time, he checked the circuit breakers, had someone inspect the furnace, and had a television repairman examine the remains of several television sets found in the ashes. He also found a piece of fuse. Over the course of his several visits, Hoffman secured physical evidence and formed opinions that played a substantial role at trial in establishing arson as the cause of the fire and in refuting the respondents' testimony about what furniture had been lost. His entries into the building were without warrants or Tyler's consent, and were for the sole purpose "of making an investigation and seizing evidence." At the trial, respondents' attorney objected to the admission of physical evidence obtained during these visits, and also moved to strike all of Hoffman's testimony "because it was got in an illegal manner."##

The Michigan Supreme Court held that with only a few exceptions, any entry onto fire-damaged private property by fire or police officials is subject to the warrant requirements of the Fourth and Fourteenth Amendments. "[Once] the blaze [has been] extinguished and the fire fighters have left the premises, a warrant is required to reenter and search the premises, unless there is consent or the premises have been abandoned." [Citations omitted.] Applying this principle, the court ruled that the series of warrantless entries that began after the blaze had been extinguished at 4 a.m. on January 22 violated the Fourth and Fourteenth Amendments.* It found that the "record does not factually support a conclusion that

---

#Sergeant Hoffman had entered the premises with other officials at least twice before, on January 26 and 29. No physical evidence was obtained as a result of these warrantless entries.

##The State's case was substantially buttressed by the testimony of Oscar Frisch, a former employee of the respondents. He described helping Tyler and Tompkins move valuable items from the store and old furniture into the store a few days before the fire. He also related that the respondents had told him there would be a fire on January 21, and had instructed him to place mattresses on top of other objects so that they would burn better.

*Having concluded that warrants should have been secured for the postfire (Continued on next page)

Tyler had abandoned the fire-damaged premises'' and accepted the lower court's finding that '' '[c]onsent for the numerous searches was never obtained from defendant Tyler.' '' *[Citations omitted.]* Accordingly, the court reversed the respondents' convictions and ordered a new trial.

II.

The decisions of this Court firmly establish that the Fourth Amendment extends beyond the paradigmatic entry into a private dwelling by a law enforcement officer in search of the fruits or instrumentalities of crime. As this Court stated in *Camara* v. *Municipal Court,* 387 U. S., at 528, the "basic purpose of this Amendment . . . is to safeguard the privacy and security of individuals against arbitrary invasions by government officials." The officials may be health, fire, or building inspectors. Their purpose may be to locate and abate a suspected public nuisance, or simply to perform a routine periodic inspection. The privacy that is invaded may be sheltered by the walls of a warehouse or other commercial establishment not open to the public. *See* v. *City of Seattle,* 387 U. S. 541; *Marshall* v. *Barlow's, Inc.,* — U. S. __, __ (slip op., at 3-5). These deviations from the typical police search are thus clearly within the protection of the Fourth Amendment.

The petitioner argues, however, that an entry to investigate the cause of a recent fire is outside that protection because no individual privacy interests are threatened. If the occupant of the premises set the blaze, then, in the words of the petitioner's brief, his "actions show that he has no expectation of privacy" because "he has abandoned those premises within the meaning of the Fourth Amendment." And if the fire had other causes, "the occupants of the premises are treated as victims by police and fire officials." In the petitioner's view, "[t]he likelihood that they will be aggrieved by a possible intrusion into what remains of their privacy in badly burned premises is negligible."

This argument is not persuasive. For even if the petitioner's contention that arson establishes abandonment be accepted, its second proposition — that innocent fire victims inevitably have no protection expectations of privacy in whatever remains of their property — is contrary to common experience. People may go on living in their homes or working in their offices after a fire. Even when that is impossible, private effects often remain on the fire-damaged premises. The petitioner may be correct in the view that most innocent fire victims are treated

---

(Continued from previous page) searches, the court explained that different standards of probable cause governed searches to determine the cause of a fire and searches to gather evidence of crime. It then described what standard of probable cause should govern all the searches in this case:

"While it may be no easy task under some circumstances to distinguish as a factual matter between an administrative inspection and a criminal investigation, in the instant case the Court is not faced with that task. Having lawfully discovered the plastic containers of flammable liquid and other evidence of arson before the fire was extinguished, Fire Chief See focused his attention on assembling proof of arson and began a criminal investigation. At that point there was probable cause for issuance of a criminal investigative search warrant." 399 Mich., at 577, 250 N. W. 2nd, at 474 *[citations omitted].*

courteously and welcome inspections of their property to ascertain the origin of the blaze, but "even if true, [this contention] is irrelevant to the question whether the . . . inspection is reasonable within the meaning of the Fourth Amendment." *Camara, supra,* at 536. Once it is recognized that innocent fire victims retain the protection of the Fourth Amendment, the rest of the petitioner's argument unravels. For it is of course impossible to justify a warrantless search on the ground of abandonment by arson when that arson has not yet been proved, and a conviction cannot be used *ex post facto* to validate the introduction of evidence used to secure that same conviction.

Thus, there is no diminution in a person's reasonable expectation of privacy nor in the protection of the Fourth Amendment simply because the official conducting the search wears the uniform of a fire fighter rather than a policeman, or because his purpose is to ascertain the cause of a fire rather than to look for evidence of a crime, or because the fire might have been started deliberately. Searches for administrative purposes, like searches for evidence of crime, are encompassed by the Fourth Amendment. And under that Amendment, "one governing principle, justified by history and by current experience, has consistently been followed: except in certain carefully defined classes of cases, a search of private property without proper consent is 'unreasonable' unless it has been authorized by a valid search warrant." *Camara, supra,* at 528–529. The showing of probable cause necessary to secure a warrant may vary with the object and intrusiveness of the search,** but the necessity for the warrant persists.

The petitioner argues that no purpose would be served by requiring warrants to investigate the cause of a fire. This argument is grounded on the premise that the only fact that need be shown to justify an investigatory search is that a fire of undetermined origin has occurred on those premises. The petitioner contends that this consideration distinguishes this case from *Camara,* which concerned the necessity for warrants to conduct routine building inspections. Whereas the occupant of premises subjected to an unexpected building inspection may have no way of knowing the purpose or lawfulness of the entry, it is argued that the occupant of burned premises can hardly question the factual basis for fire officials wanting access to his property. And whereas a magistrate performs the significant function of assuring that an agency's decision to conduct a routine inspection of a particular dwelling conforms with reasonable legislative or administrative standards, he can do little more than rubber stamp an application to search fire-damaged premises for the cause of the blaze. In short, where the justification for the search is as sim-

---

**For administrative searches conducted to enforce local building, health, or fire codes, " 'probable cause' to issue a warrant to inspect . . . exist[s] if reasonable legislative or administrative standards for conducting an area inspection are satisfied with respect to a particular dwelling. Such standards, which will vary with the municipal program being enforced, may be based upon the passage of time, the nature of the building (*e.g.,* a multifamily apartment house), or the condition of the entire area, but they will not necessarily depend upon specific knowledge of the condition of the particular dwelling." *Camara, supra,* at 538: *Marshall* v. *Barlow's, Inc.,* — U. S. —, — (slip op., at 12–13). See LaFave, Administrative Searches and the Fourth Amendment: The *Camara* and *See* Cases, 1967 Sup. Ct. Rev. 1, 18–20.

ple and as obvious to everyone as the fact of a recent fire, a magistrate's review would be a time-consuming formality of negligible protection to the occupant.

The petitioner's argument fails primarily because it is built on a faulty premise. To secure a warrant to investigate the cause of a fire, an official must show more than the bare fact that a fire has occurred. The magistrate's duty is to assure that the proposed search will be reasonable, a determination that requires inquiry into the need for the intrusion on the one hand, and the threat of disruption to the occupant on the other. For routine building inspections, a reasonable balance between these competing concerns is usually achieved by broad legislative or administrative guidelines specifying the purpose, frequency, scope, and manner of conducting the inspections. In the context of investigatory fire searches, which are not programmatic but are responsive to individual events, a more particularized inquiry may be necessary. The number of prior entries, the scope of the search, the time of day when it is proposed to be made, the lapse of time since the fire, the continued use of the building, and the owner's efforts to secure it against intruders might all be relevant factors. Even though a fire victim's privacy must normally yield to the vital social objective of ascertaining the cause of the fire, the magistrate can perform the important function of preventing harassment by keeping that invasion to a minimum. *[Citations omitted.]*

In addition, even if fire victims can be deemed aware of the factual justification for investigatory searches, it does not follow that they will also recognize the legal authority for such searches. As the Court stated in *Camara*, ''when the inspector demands entry [without a warrant], the occupant has no way of knowing whether enforcement of the municipal code involved requires inspection of his premises, no way of knowing the lawful limits of the inspector's power to search, and no way of knowing whether the inspector himself is acting under proper authorization.'' 387 U. S., at 532. Thus, a major function of the warrant is to provide the property owner with sufficient information to reassure him of the entry's legality. *[Citation omitted.]*

In short, the warrant requirement provides significant protection for fire victims in this context, just as it does for property owners faced with routine building inspections. As a general matter, then, official entries to investigate the cause of a fire must adhere to the warrant procedures of the Fourth Amendment. In the words of the Michigan Supreme Court: ''Where the cause [of the fire] is undetermined, and the purpose of the investigation is to determine the cause and to prevent such fires from occurring or recurring, a . . . search may be conducted pursuant to a warrant issued in accordance with reasonable legislative or administrative standards or, absent their promulgation, judicially prescribed standards; if evidence of wrongdoing is discovered, it may, of course, be used to establish probable cause for the issuance of a criminal investigative search warrant or in prosecution.'' But ''[i]f the authorities are seeking evidence to be used in a criminal prosecution, the usual standard [of probable cause] will apply.'' *[Cita-*

tions omitted.] Since all the entries in this case were "without proper consent" and were not "authorized by a valid search warrant," each one is illegal unless it falls within one of the "certain carefully defined classes of cases" for which warrants are not mandatory. *[Citation omitted.]*

III.

Our decisions have recognized that a warrantless entry by criminal law enforcement officials may be legal when there is compelling need for official action and no time to secure a warrant. *Warden* v. *Hayden*, 387 U. S. 294 (warrantless entry of house by police in hot pursuit of armed robber); *Ker* v. *California*, 374 U. S. 23 (warrantless and unannounced entry of dwelling by police to prevent imminent destruction of evidence). Similarly, in the regulatory field, our cases have recognized the importance of "prompt inspections, even without a warrant. . . . in emergency situations." *Camara, supra,* at 539, citing *North American Cold Storage Co.* v. *City of Chicago*, 211 U. S. 306 (seizure of unwholesome food); *Jacobson* v. *Massachusetts*, 197 U. S. 11 (compulsory smallpox vaccination); *Compagnie Francaise* v. *Board of Health*, 186 U. S. 380 (health quarantine).

A burning building clearly presents an exigency of sufficient proportions to render a warrantless entry "reasonable." Indeed, it would defy reason to suppose that firemen must secure a warrant or consent before entering a burning structure to put out the blaze. And once in a building for this purpose, fire fighters may seize evidence of arson that is in plain view. *[Citation omitted.]* Thus, the Fourth and Fourteenth Amendments were not violated by the entry of the firemen to extinguish the fire at Tyler's Auction, nor by Chief See's removal of the two plastic containers of flammable liquid found on the floor of one of the showrooms.

Although the Michigan Supreme Court appears to have accepted this principle, its opinion may be read as holding that the exigency justifying a warrantless entry to fight a fire ends, and the need to get a warrant begins, with the dousing of the last flame. *[Citation omitted.]* We think this view of the fire fighting function is unrealistically narrow, however. Fire officials are charged not only with extinguishing fires, but with finding their causes. Prompt determination of the fire's origin may be necessary to prevent its recurrence, as through the detection of continuing dangers such as faulty wiring or a defective furnace. Immediate investigation may also be necessary to preserve evidence from intentional or accidental destruction. And, of course, the sooner the officials complete their duties, the less will be their subsequent interference with the privacy and the recovery efforts of the victims. For these reasons, officials need no warrant to remain in a building for a reasonable time to investigate the cause of a blaze after it has been extinguished.† And if the warrantless entry to put out the fire and determine its

---

†The circumstances of particular fires and the role of fire fighters and investigating officials will vary widely. A fire in a single-family dwelling that clearly is extinguished at some identifiable time presents fewer complexities than those likely to attend a fire that spreads through a *(Continued on next page)*

cause is constitutional, the warrantless seizure of evidence while inspecting the premises for these purposes also is constitutional.

IV.

A.

The respondents argue, however, that the Michigan Supreme Court was correct in holding that the departure by the fire officials from Tyler's Auction at 4 a.m. ended any license they might have had to conduct a warrantless search. Hence, they say that even if the firemen might have been entitled to remain in the building without a warrant to investigate the cause of the fire, their departure and re-entry four hours later that morning required a warrant.

On the facts of this case, we do not believe that a warrant was necessary for the early morning re-entries on January 22. As the fire was being extinguished, Chief See and his assistants began their investigation, but visibility was severely hindered by darkness, steam, and smoke. Thus they departed at 4 a.m. and returned shortly after daylight to continue their investigation. Little purpose would have been served by their remaining in the building, except to remove any doubt about the legality of the warrantless search and seizure later that same morning. Under these circumstances, we find that the morning entries were no more than an actual continuation of the first, and the lack of a warrant thus did not invalidate the resulting seizure of evidence.

B.

The entries occurring after January 22, however, were clearly detached from the initial exigency and warrantless entry. Since all of these searches were conducted without valid warrants and without consent, they were invalid under the Fourth and Fourteenth Amendments, and any evidence obtained as a result of those entries must, therefore, be excluded at the respondents' retrial.

V.

In summation, we hold that an entry to fight a fire requires no warrant, and that once in the building, the officials may remain there for a reasonable time to investigate the cause of the blaze. Thereafter, additional entries to investigate the cause of the fire must be made pursuant to the warrant procedures governing administrative searches. See *Camara, supra,* at 534–539, *See* v. *City of Seattle, supra,* at 544–545; *Marshall* v. *Barlow's Inc., supra,* at — (slip op., at 12–13). Evidence of arson discovered in the course of such investigations is admissible at trial, but if

---

†(*Continued from previous page*) large apartment complex or that engulfs numerous buildings. In the latter situations, it may be necessary for officials — pursuing their duty both to extinguish the fire and to ascertain its origin — to remain on the scene for an extended period of time repeatedly entering or re-entering the building or buildings, or portions thereof. In determining what constitutes a "reasonable time to investigate," appropriate recognition must be given to the exigencies that confront officials serving under these conditions, as well as to individuals' reasonable expectations of privacy.

the investigating officials find probable cause to believe that arson has occurred and require further access to gather evidence for a possible prosecution, they may obtain a warrant only upon a traditional showing of probable cause applicable to searches for evidence of crime. *[Citation omitted.]*

These principles require that we affirm the judgment of the Michigan Supreme Court ordering a new trial.††

Affirmed.

MR. JUSTICE BLACKMUN joins the judgment of the Court and Parts I, III, and IV-A of its opinion.

MR. JUSTICE BRENNAN took no part in the consideration or decision of this case.

MR. JUSTICE STEVENS, concurring in part and concurring in the judgment.

Because Part II of the Court's opinion in this case, like the opinion in *Camara* v. *Municipal Court,* 387 U. S. 523, seems to assume that an official search must either be conducted pursuant to a warrant or not take place at all, I cannot join its reasoning.

In particular, I cannot agree with the Court's suggestion that, if no showing of probable cause could be made, "the warrant procedures governing administrative searches," *Ante,* at 12, would have complied with the Fourth Amendment. In my opinion, an "administrative search warrant" does not satisfy the requirements of the Warrant Clause.‡ See *Marshall* v. *Barlow's, Inc.,* — U. S. — (STEVENS, J., dissenting). Nor does such a warrant make an otherwise unreasonable search reasonable.

A warrant provides authority for an unannounced, immediate entry and search.

---

††The petitioner alleges that respondent Tompkins lacks standing to object to the unconstitutional searches and seizures. The Michigan Supreme Court refused to consider the State's argument, however, because the prosecutor failed to raise the issue in the trial court or in the Michigan Court of Appeals. 399 Mich., at 571, 250 N. W. 2d, at 470-471. We read the state court's opinion to mean that in the absence of a timely objection by the State, a defendant will be presumed to have standing. Failure to present a federal question in conformance with state procedure constitutes an adequate and independent ground of decision barring review in this Court, so long as the State has a legitimate interest in enforcing its procedural rule. *Henry* v. *Mississippi,* 379 U. S. 443, 447. See *Safeway Stores* v. *Oklahoma Grocers,* 360 U. S. 334, 342, n. 7; *Cardinale* v. *Louisiana,* 394 U. S. 437, 438. The petitioner does not claim that Michigan's procedural rule serves no legitimate purpose. Accordingly, we do not entertain the petitioner's standing claim which the state court refused to consider because of procedural default.

‡The Warrant Clause of the Fourth Amendment provides that ". . . no Warrants shall issue, but upon probable cause, supported by Oath or affirmation, and particularly describing the place to be searched, and the persons or things to be seized."

No notice is given when an application for a warrant is made and no notice precedes its execution; when issued, it authorizes entry by force.‡‡ In my view, when there is no probable cause to believe a crime has been committed and when there is no special enforcement need to justify an unannounced entry,||the Fourth Amendment neither requires nor sanctions an abrupt and peremptory confrontation between sovereign and citizen.|| || In such a case, to comply with the constitutional requirement of reasonableness, I believe the sovereign must provide fair notice of an inspection.#

The Fourth Amendment interests involved in this case could have been protected in either of two ways — by a warrant, if probable cause existed; or by fair notice, if neither probable cause not a special law enforcement need existed. Since the entry on February 16 was not authorized by a warrant and not preceded by advance notice, I concur in the Court's judgment and in Parts I, III, and IV of its opinion.

MR. JUSTICE WHITE, with whom MR. JUSTICE MARSHALL joins, concurring in part and dissenting in part.

I join in all but Part IV-A of the opinion, from which I dissent. I agree with the Court that:

> [A]n entry to fight a fire requires no warrant, and that once in the building, officials may remain there for a reasonable time to investigate the cause of the blaze. Thereafter, additional entries to investigate the cause of the fire must be made pursuant to the warrant procedures governing administrative searches. *Ante,* at 11–12.

---

‡‡See *Wyman* v. *James,* 400 U. S. 309, 323–324. As the Court observed in *Wyman,* a warrant is not simply a device providing procedural protections for the citizen; it also grants the government increased authority to invade the citizen's privacy. See *Miller* v. *United States,* 357 U. S. 301, 307–308.

||In this case, there obviously was a special enforcement need justifying the initial entry to extinguish the fire, and I agree that the search on the morning after the fire was a continuation of that entirely legal entry. A special enforcement need can, of course, be established on more than a case-by-case basis, especially if there is relevant legislative determination of need. See *Marshall* v. *Barlow's Inc.,* — U. S. —, — (STEVENS, J., dissenting).

|| ||The Fourth Amendment ensures "[t]he right of the people to be secure in their persons, houses, papers, and effects, against unreasonable searches and seizures." *(Emphasis added.)* Surely this broad protection encompasses the expectation that the government cannot demand immediate entry when it neither has probable cause to suspect illegality nor any other pressing enforcement concern. Yet under the rationale in Part II of the Court's opinion, the less reason an officer has to suspect illegality, the less justification he need give the magistrate in order to conduct an unannounced search. Under this rationale, the police will have no incentive — indeed a disincentive — to establish probable cause before obtaining authority to conduct an unannounced search.

#See LaFave, Administrative Searches and the Fourth Amendment: The *Camara* and *See* Cases, 1967 Sup. Ct. L. Rev. 1. The requirement of giving notice before conducting a routine administrative search is hardly unprecedented. It closely parallels existing procedures for administrative subpoenas, see, *e.g.,* 15 U. S. C. § 1312, and is, as Professor LaFave points out, embodied in English law and practice. See LaFave, *supra,* at 31–32.

The Michigan Supreme Court found that the warrantless searches, at 8 and 9 a.m. were not, in fact, continuations of the earlier entry under exigent circumstances## and therefore ruled inadmissible all evidence derived from those searches. The Court offers no sound basis for overturning this conclusion of the state court that the subsequent re-entries were distinct from the original entry. Even if, under the Court's "reasonable time" criterion, the firemen might have stayed in the building for an additional four hours — a proposition which is by no means clear — the fact remains that the firemen did not choose to remain and continue their search, but instead locked the door and departed from the premises entirely. The fact that the firemen were willing to leave demonstrates that the exigent circumstances justifying their original warrantless entry were no longer present. The situation is thus analogous to that in *GM Leasing Corp.* v. *United States,* 429 U. S. 338, 358–359 (1977):

> The agents' own action . . . in their delay for two days following their first entry, and for more than one day following the observation of materials being moved from the office, before they made the entry during which they seized the records, is sufficient to support the District Court's implicit finding that there were no exigent circumstances.

To hold that some subsequent re-entries are "continuations" of earlier ones will not aid firemen, but confuse them, for it will be difficult to predict in advance how a court might view a re-entry. In the end, valuable evidence may be excluded for failure to seek a warrant that might have easily been obtained.

Those investigating fires and their causes deserve a clear demarcation of the constitutional limits of their authority. Today's opinion recognizes the need for speed and focuses attention on fighting an ongoing blaze. The fire truck need not stop at the courthouse in rushing to the flames. But once the fire has been extinguished and the firemen have left the premises, the emergency is over. Further intrusion on private property can and should be accompanied by a warrant indicating the authority under which the firemen presume to enter and search.

There is another reason for holding that re-entry after the initial departure required a proper warrant. The state courts found that at the time of the first re-entry a criminal investigation was underway and that the purpose of the officers in re-entering was to gather evidence of crime. Unless we are to ignore these findings, a warrant was necessary. *Camara* v. *Municipal Court,* 387 U. S. 523 (1967), and *See* v. *City of Seattle,* 387 U. S. 541 (1967), did not differ with *Frank* v. *Maryland,* 359 U. S. 360 (1959), that searches for criminal evidence are of special

---

##The Michigan Supreme Court recognized that "if there are exigent circumstances, such as reason to believe that the destruction of evidence is imminent or that a further entry of the premises is necessary to prevent the recurrence of the fire, no warrant is required and evidence discovered is admissible." 399 Mich. 564, 578, 250 N. W. 2d 467, 474 (1977). It found, however, that "In the circumstances of this case there were no exigent circumstances justifying the searches made hours, days or weeks after the fire was extinguished." *Id.,* at 579, 250 N. W. 2d, at 475.

significance under the Fourth Amendment.

MR. JUSTICE REHNQUIST, dissenting.

I agree with my Brother STEVENS, for the reasons expressed in his dissenting opinion in *Marshall* v. *Barlow's, Inc.,* — U. S. —,—,— (1978) (STEVENS, J., dissenting), that the "Warrant Clause has no application to routine, regulatory inspections of commercial premises." Since in my opinion the searches involved in this case fall within that category, I think the only appropriate inquiry is whether they were reasonable. The Court does not dispute that the entries which occurred at the time of the fire and the next morning were entirely justified, and I see nothing to indicate that the subsequent searches were not also eminently reasonable in light of all the circumstances.

In evaluating the reasonableness of the later searches, their most obvious feature is that they occurred after a fire which had done substantial damage to the premises, including the destruction of most of the interior. Thereafter the premises were not being used and very likely could not have been used for business purposes, at least until substantial repairs had taken place. Indeed, there is no indication in the record that after the fire Tyler ever made any attempt to secure the premises. As a result, the fire department was forced to lock up the building to prevent curious bystanders from entering and suffering injury. And as far as the record reveals, Tyler never objected to this procedure or attempted to reclaim the premises for himself.

Thus, regardless of whether the premises were technically "abandoned" within the meaning of the Fourth Amendment, cf. *Abel* v. *United States,* 362 U. S. 217, 241 (1960); *Hester* v. *United States,* 265 U. S. 57 (1924), it is clear to me that no purpose would have been served by giving Tyler notice of the intended search or by requiring that the search take place during the hours which in other situations might be considered the only "reasonable" hours to conduct a regulatory search. In fact, as I read the record, it appears that Tyler not only had notice that the investigators were occasionally entering the premises for the purpose of determining the cause of the fire, but he never voiced the slightest objection to these searches and actually accompanied the investigators on at least one occasion. App. 54–57. In fact, while accompanying the investigators during one of these searches, Tyler himself suggested that the fire very well may have been caused by arson. App. 56. This observation, coupled with all the other circumstances, including Tyler's knowledge of, and apparent acquiescence in, the searches, would have been taken by any sensible person as an indication that Tyler thought the searches ought to continue until the culprit was discovered; at the very least they indicated that he had no objection to these searches. Thus, regardless of what sources may serve to inform one's sense of what is reasonable, in the circumstances of this case I see nothing to indicate that these searches were in any way unreasonable for purposes of the Fourth Amendment.

Since the later searches were just as reasonable as the search the morning immediately after the fire in light of all these circumstances, the admission of evidence derived therefrom did not, in my opinion, violate respondents' Fourth and Fourteenth Amendment rights. I would accordingly reverse the judgment of the Supreme Court of Michigan which held to the contrary.

SUMMARY

The field of administrative law is a vast and complex one. Many administrative agencies — created by the legislature or by executive order — function as regulatory branches of government in virtually every area of American life. Because administrative agencies make rules that can have the effect of law and also have the power to adjudicate, the field of administrative law reaches into the three major branches of American government: legislative, executive, and judicial. Because regulatory or administrative agencies are empowered to regulate the economic, employment, safety, and myriad other activities of individuals and businesses as well as public service organizations such as the fire service, administrative law has become a vital branch of American law.

ACTIVITIES

1. Give three examples of how the growth of administrative agencies affected the growth of administrative law.
2. Explain why administrative agencies have been referred to as "the fourth branch of American government."
3. Identify two of the major constitutional questions that have arisen with regard to administrative agencies and their powers.
4. Explain the effect the decisions in *See* v. *City of Seattle* and *Camara* v. *Municipal Court City and County of San Francisco* have on fire service procedures. Include in your explanation the constitutional issues that were raised.
5. Explain how the Fourth and Fourteenth Amendments to the Constitution of the United States affect the procedure of warrantless searches and inspections. Refer to Appendix A for assistance with your answers.
6. Describe the inspection procedures followed in your community before and after the *Camara* and *See* decisions. What guidelines for fire service inspection would you propose be followed as a result of these decisions?
7. Using the court cases discussed in this chapter, discuss the differences and similarities in constitutional rights and search decisions with regard to residential occupancies and business establishments.
8. What are the constitutional rights of fire-damage victims?
9. You are chief of your city's fire station. Your station has burned down, and the cause of the fire is suspected arson.
   (a) Explain why you would or would not need to obtain a search warrant to

conduct your investigation. (Apply what you have learned from Chapter II, "Municipal Corporations," to help form your decision.)

(b) If a department member were suspected of setting the fire, could you search the member's area in the station dormitory? Why or why not?

10. What role does the time of a fire investigation play in search rulings?

BIBLIOGRAPHY

Carrow, Milton M., *The Background of Administrative Law*, Associated Lawyers Publishing Company, Newark, NJ, 1948.

Davis, Kenneth Culp, *Administrative Law and Government*, West Publishing Co., St. Paul, MN, 1960.

Dickinson, John, *Administrative Justice and the Supremacy of Law In the United States*, © Harvard University Press, Russell and Russell, Inc., New York, 1955.

Freund, Ernst, *et al., The Growth of American Administrative Law*, © Bar Association of St. Louis, Thomas Law Book Company, St. Louis, 1923.

Mashaw, Jerry L. and Merrill, Richard A., *Introduction to the American Public Law System*, West Publishing Co., St. Paul, MN, 1975.

Parker, Reginald, *Administrative Law*, The Bobbs-Merrill Company, Indianapolis, 1952.

Schwartz, Bernard, *An Introduction to American Administrative Law*, Oceana Publications, Inc., Dobbs Ferry, NY, 1962.

Woll, Peter, *Administrative Law*, University of California Press, Berkeley, 1963.

For an in-depth study of the topics presented in this chapter, the following is recommended:

Gellhorn, Walter and Byse, Clark, *Administrative Law*, The Foundation Press, Mineola, NY, 1974.

VIII

ADMINISTRATIVE PROCEDURE

by Vincent M. Brannigan, J.D.

This chapter was written by Professor Vincent Brannigan, who holds a J.D. degree from Georgetown University and is a member of the Maryland Bar. He has served with the National Bureau of Standards Center for Fire Research and the National Fire Prevention and Control Administration, and was Legal Assistant to the Administrative Law Judge of the Consumer Product Safety Commission. He is now a faculty member of the Department of Fire Protection Engineering and the Department of Textiles and Consumer Economics at the University of Maryland.

OVERVIEW

The nature of an administrative agency affects the procedures it will follow. An agency's procedures are evolved from an agency's duties and functions. Administrative agencies have two major powers: rule-making — similar to legislative powers, and decision-making — similar to judicial powers. Administrative procedures are based on the particular aspect of administrative process with which the agency is concerned. However, all agencies have certain procedures in common, and definite safeguards that are afforded to the individual or individuals with whom the agency deals.

THE PURPOSE OF ADMINISTRATIVE AGENCIES

In preceding chapters of this book, the current American governmental system has been explained. This system is usually considered to be threefold: (1) the legislative branch, which makes the law, (2) the executive branch, which enforces the law, and (3) the judiciary branch, which interprets the law. There are, however, a number of problems — particularly those present in the fire protection field — that do not lend themselves to effective solution without adding further governmental organizations and agencies.

FREEING THE LEGISLATURE FROM TRIVIA

A primary purpose of establishing an administrative agency to handle a policy-making area is to free the legislature from trivia, *i.e.*, from having to deal on a day-to-day basis with highly technical and/or economic determinations.

Legislatures are in session, after all, only a limited amount of time: in many states they meet only a part of the year. If they were required to analyze and decide every detail of every program, and revise and change those details to fit changing economic or technical circumstances, it would be virtually impossible for them to get anything else done.

BRINGING EXPERTISE TO THE FIELD

A second major reason for establishing an administrative agency in a specific field is that the subject matter of that field might be so involved or scientific that it would be impossible to expect the legislature to become expert in the subject matter of the field. This is particularly true in technical areas such as fire protection. Legislators are selected on their ability to win the confidence of the electorate and not on any innate or inherent technical, scientific, or other specialized or expert ability in a single field.

The term "specialist" refers to an organization or agency that spends virtually all of its time working in a specialized area. The term "expert" indicates an agency or an individual that has developed an understanding or appreciation or ability to deal with a special field. An expert has significantly greater knowledge of a field or area than is normally found in the general population. Some specialists are experts, some are not. Some experts are specialists, and some are not. The problem of bringing expertise to a field is particularly acute in areas where it is obvious that expertise can only be learned by long application to study and work in the field. There are some areas which, by their very nature, cannot be taught in colleges or universities or learned very quickly. There are areas, such as the fire protection field, in which there may not be any private nongovernmental demand for services; therefore, the government must hire and maintain employees in this field in order to develop the necessary expertise.

There are other forms of problems that require arbitrary determination by experts so that all persons dealing with the field can come to similar conclusions when using similar decision-making methods. Administrative agencies can be used to define these arbitrary decisions in such a way that both public and private entities can use the decision-making system. This is usually found in the standards-making operations of government — for example, by the National Bureau of Standards (NBS). The National Bureau of Standards defines the standards of weights and measures.

PROVIDING FOR EMERGENCY ACTION

A third purpose for establishing administrative agencies is to ensure that there is a sufficiently authoritative presence available to handle emergency action. There is no doubt that the executive branch of government can simply be given emergency powers to deal with a problem; however, the allocation of power to the executive branch does not necessarily mean that there will be a functioning system of expert, specialized individuals available to act in policy-making determinations, especially when the legislature is not in session. Many types of emergencies require far more than merely enforcing existing laws and statutes. They require immediate determination of new policies and the application of long-developed expertise to new and different technological problems. For example, the severe blizzard of February 6, 1978 resulted in high tides that broke through sea walls and heavily damaged the South Shore area of Massachusetts. Two days later, agents from the National Flood Insurance Program were on the scene to assist in the emergency and to set up offices to help the flooded homeowners. The expertise of the agents of this agency made them able to deal quickly and efficiently with the emergency.

PROVIDING CONTINUITY IN DECISION-MAKING

The final purpose of administrative agencies is to provide for continuity of decision-making. Such continuity is important because our concept of legislature has no provisions for arriving at the same conclusion at different times when faced with the same type problem. However, the turnover of the legislature at annual or biannual elections is usually looked upon as a way of guaranteeing that each problem can be reviewed over and over again by different individuals with different inclinations, different responsibilities to constituents, and different conclusions derived from the same set of facts. The executive branch, while expected to maintain a consistent attitude towards the need to enforce the laws, does change in our system on a regular basis. This can lead to varying interpretations of the enforcement obligations of the executive, and different priorities for the applications of the executive enforcement function.

The judiciary is normally thought of as the place where continuity of decision-making should reside in the government. Judges commonly refer to prior decisions in order to arrive at current decisions. However, such practice does not provide for a proper concept of continuity in the decision-making process. Continuity is possible only when the facts are similar to prior cases, when there is no new knowledge, and when the risk level that was acceptable in the prior case is still acceptable. If each of these factors were present, similar decisions could be reached.

On the other hand, the process of making the decision should sufficiently take into account changed circumstances to allow for a different decision even where the facts are the same.

In administrative agencies, at least in theory, proper decisions should be based on a logical analysis of the problem supplemented by the agency's expertise. This is not the same as changing a policy based on the popular will. It is a different function. An administrative agency, therefore, can be defined more or less by its function. To be truly an administrative agency, an agency should be directly involved in making policy and making decisions using its specialized personnel and its expert personnel, and should be relied on by the government for policy conclusions within its scope of operation.

ORGANIZATION OF AN ADMINISTRATIVE AGENCY

Administrative agencies evolved over a period of time, and are still evolving today. There are many differences of opinion among experts in the field as to precisely what powers, responsibilities, and decisions can be given to administrative agencies. In particular, some states are much more restrictive than the federal government when it comes to the concept of the "independent" regulatory agency. However, all agencies have certain common features that are evident. All agencies must be established by statute from an authority having jurisdiction over the subject matter of the agency. This concept of jurisdiction involves both territorial and subject-matter determination.

"DELEGATION DOCTRINE"

The first problem that hinders the establishment of an administrative agency is that some legislatures, by their own constitutions or statutes, do not have the power to delegate their decision-making powers to an agency. This concept, long known as the "delegation doctrine," can be traced to our democratic form of government. Legislators are selected by the people and, having been selected by the people, are expected to make the decisions — not to pass decision-making powers on to other groups. This is particularly true when the group that has been delegated the decision-making power (by the legislature) can make regulations that have the force and effect of law, the violation of which can lead to penalties to the citizen. The delegation doctrine has always been much more important at the state level than at the federal level. Since the 1930s, no federal agency has been significantly challenged as possessing an unlawful delegation of federal authority. However, states take varying views of the delegation doctrine concept.

ADMINISTRATIVE DISCRETION

The heart of the controversy over delegation is the extent to which discretionary authority has been given to an administrative agency to carry out its function. The ability to make decisions — which is the concept of a discretionary function — or to refuse to make decisions, is what gives an administrative agency its reason for existence.

There are two kinds of duties normally described as belonging to governmental agencies: (1) ministerial duties, in which a governmental official can be required by law to perform a function, and (2) discretionary duties, in which the governmental official is given (by law) the discretion to determine how this function will be carried out. An agency without discretion to determine how it will perform its function cannot properly be characterized as an administrative agency. An example of a typical agency without discretion is a county or state recording office where land records, security agreements, and other documents are recorded. Such an agency is given no discretion in determining whether or not it will accept a document for its records.

A common agency with discretion might be a county liquor licensing board, which is required to determine whether or not an individual applying for a license is or is not of good character. Another example would be the Department of Social and Health Services in the state of Washington. This agency has, as its responsibility, the issuing of licenses to certain occupancies, such as hotels.

The most important point about discretion, however, is that the scope of the agency's discretion, as well as the method of exercising that discretion, is normally enacted as part of the agency's enabling statute. The agency is thus given a restricted scope in which to exercise its discretion. For example, the Consumer Product Safety Commission (CPSC) is given the discretion to declare certain products to be substantial hazards; however, the Commission's jurisdiction is limited to products generally used in and around the home, thus excluding a number of common products.

The scope of discretion for some agencies is quite broad. For example, the Federal Trade Commission (FTC) has jurisdiction over most interstate and foreign commerce, and the scope of its discretion is defined by a concept of unfair and deceptive acts or practices in commerce. The agency is given the discretion to determine what is unfair and deceptive.

ADMINISTRATORS

The enabling act or statute will always indicate at what level in the agency final decisions within the agency's discretion will be made. Normally, the persons charged with this discretionary function will be the administrators of the agency. There are a number of different types of administrative agencies that are characterized by the exercise of agency discretion, and there are agencies in which a single individual is given all of the discretionary authority. This individual is usually also given the power to organize and operate the administrative agency. It is, however, not required that these functions be performed by the same person. Occasionally an official is given the discretionary authority and the power to make final rulings, although another individual within the same agency (but superior in rank) has the power to overrule the subordinate official.

The other major form of administrative agency is based on the concept of a group decision. Such an agency is usually referred to as a commission or a board. The Consumer Product Safety Commission (CPSC), for example, has five voting members. All final agency exercises of discretionary function are subject to a vote of the five commissioners. The chairperson of the commission is the executive officer and operating chief of the agency, but when it comes to exercising the agency's discretionary decision-making powers, the chairperson has only one vote out of five.

Each system, individual administrators and groups, has advantages and disadvantages. Individual administrators can perhaps change the policies of their agencies in a way that would be faster than the group system. The group system, on the other hand, can provide continuity (if individual terms are staggered) in the decision-making process and allow for a wider range of expertise or points of view. Group decision-makers, however, can be paralyzed by the inability of any single point of view to result in a majority point of view of the group.

TENURE OF ADMINISTRATORS

The type of term administrators serve varies. One type of term is a civil service term. Many fire officials fall into this category. They serve for a certain term on good behavior until they either resign or are fired "for cause," or until they reach the age of mandatory retirement. Normally, an individual removed "for cause" from an agency has the right to appeal to either an appropriate civil service organization or to the court, and will usually be entitled to a hearing on the cause.

A second type of term commonly found among administrators is a fixed term of years. For example, each member of the Federal Reserve Board — an administrative agency charged with certain functions regarding the nation's banking system — serves a 15-year term. Each commissioner of the Consumer Product Safety Commission serves a 7-year term. Long terms are thought by many to encourage independence in the decision-maker from the political process. They are criticized by others for encouraging that same independence and lack of accountability to the political process. No single length of term or even the concept of term has been agreed upon as any sort of ideal. If the administrative agency is a group administration, all members can serve terms expiring at the same time, or the terms can be staggered to ensure that there will be at least some continuity in the agency at all times.

A third sort of tenure is basically at the will of an appointing authority. It is not unusual, however, that confirmation of an appointment to an agency be required by the legislature. For example, if the President nominates the secretary of a department, the appointment wanted still requires confirmation by the Senate. (This is thought to be part of the system of checks and balances — a way in which the Senate can control, at least in part, the ideological inclinations of the persons appointed to head the regulatory agencies.) Many federal administrators who are secretaries of departments or subordinate officials serve at the will of the President. These administrators can be removed by the President without any explanation or reason. This can be done also in local or state administrative agencies. Normally, the disadvantage of this type of system is considered to be a reduction in the independence of the decision-making process, balanced perhaps by a corresponding increase in the responsibility of the administrator both to the people and to the political process.

POWERS OF THE ADMINISTRATIVE AGENCIES

An administrative agency is defined not only by the scope of its discretion and the area subject to its jurisdiction, but also by the powers granted by the legislature in the enabling legislation. Certain of these powers might be described as generic; that is, all administrative agencies have at least one power or some form of the powers described in this section. An agency that does not have any of these powers is probably not an administrative agency. Such an agency is more likely a simple executive enforcement department. An administrative agency is concerned with the making of policy and the making of rules, rather than merely the enforcement of rules or standards or statutes.

RULE-MAKING

The first major power of an administrative agency is often referred to as rule-making. Rule-making is the authority of an administrative agency to promulgate regulations which, if in accordance with the enabling statute and the Constitution, will carry the force and effect of law.

Rule-making is a legislative type of function. However, the difference between an administrative agency and a legislature is that the legislature, when enacting the enabling legislation for the agency, normally sets certain boundaries on the ability of the agency to make rules. Over the years, the courts have set down certain standards that the rules must meet in order to have the force and effect of law. This area will be covered in a later section titled "Judicial Review of Agency Proceedings."

Interpretive Rules: There are two types of rules that agencies normally enact. The first is referred to as an interpretive rule. An interpretive rule is usually a determination by the agency of the meaning of a term in its own enabling act or statute. For example, if the enabling statute included the term "multiple dwellings," townhouses may be included in the definition of multiple dwellings if townhouses (even though single-property dwelling units) represented a common fire problem. Usually interpretive rules can be promulgated or changed by the agency with a minimum of procedural requirements. For example, if an agency is required to give a hearing on a certain matter by its enabling statute, it promulgates the rules and regulations for the conduct of the hearing — serving the parties, who can represent whom, and other similar issues.

Legislative Rules: The second type of rule-making, legislative rules, results in the agency actually carrying out the legislative function assigned to it by its own statute. For example, the promulgation or adoption of a building or fire code by an administrative agency is a legislative rule-making function. The agency makes rules that direct nonagency persons to do or refrain from doing certain things, and that set the penalties or other actions that will follow a violation of the rules.

Rule-making is one of the processes by which an agency sets the policy that it feels will best carry out the mandate given to it in its enabling legislation. A person cannot be held to violate a rule until the rule is properly promulgated, and conduct prior to the promulgation of the rule cannot be made into any sort of violation of the rule.

ADJUDICATION

The second major power of an administrative agency, which is given to some but not all administrative agencies, is the power to make policy and carry out its enabling legislation by a process known as adjudication. Adjudication is a hybrid procedure developed during the late 1930s and early 1940s to give an agency the necessary flexibility in carrying out its statutory authority. Certain types of conduct are set down by the legislature in the statute as being violative of the basic statute, but they are only set down in the most vague and general terms. The administrative agency fleshes out the basic statutory policy by bringing individual adjudications against individual respondents. The respondents are charged at an appropriate hearing with violation of the statute and, if the charges are proved, they are required to make remedial action. The basic important concept of adjudication is that it is a means by which an agency develops its policy. It allows the agency to change its policy over a period of time when responding to changed circumstances. It also allows the agency to develop new concepts to deal with new hazards on a case-by-case basis rather than attempting to legislate for an entire problem by the rule-making procedure. Most federal agencies and many state agencies have this power to make policy by adjudication.

EMERGENCY POWERS

A third type of power normally accorded to certain administrative agencies is the power to take emergency action when an agency, in its discretion, perceives a hazard or other circumstance that violates either its rule or its statutes. Normally, an agency will have a wide discretion to take emergency action, but must be prepared to defend the correctness of its action. For example, the Occupational Health and Safety Administration (OSHA) has, at its discretion, the authority to close a business should OSHA agents determine a life or firesafety hazard. However, OSHA must be prepared to defend this decision and give proof for such a closing. The power to take emergency action is given to agencies in order to carry out certain defined statutory intentions to prevent situations from deteriorating prior to determinations in hearings on the issue.

AGENCY PROCEDURES

In carrying out its statutory function, an agency will usually set down rules of procedure for carrying out the functions assigned to that agency. Most agencies have both formal and informal procedures, or "on-the-record" and "off-the-record" procedures.

FORMAL PROCEDURES

When an agency is making a formal record proceeding, it uses many of the same legal techniques involved in any court-type proceeding.

Oaths: The first power is the power to take testimony on oath. This is thought of as a guarantee of truth since the deliberate false statement under oath is a criminal offense.

Subpoenas: A second crucial power is the power to subpoena information or individuals. Often an agency will not have in its own files or obtainable by its own researchers or analysts sufficient information to make a decision. A subpoena allows the agency to use compulsory process to obtain this information. Subpoenas are of two types: (1) one to compel the production of a person for questioning, and (2) another to compel the production of documents. The use of compulsory authority to compel the production of documents or other information is often combined with the concept of discovery. Discovery is a pretrial procedure by which two sides in a lawsuit inform themselves of the other's position in order to prevent "surprises" or a decision on the case based on the inability of one side to see the types of documents and information in the possession of the other side. Some agencies have the power to enforce their own subpoenas or discovery sanctions. Other agencies can issue subpoenas, but if the defendant refuses to comply they must go to a court for enforcement of the subpoena.

Transcripts: A third element of a record-type proceeding is transcripts of the proceedings which, in connection with the exhibits presented at the proceeding, constitute the record. Transcripts are made either stenographically by a trained court reporter or by the transcription of recording tapes. In either case they are usually presented to the parties at the proceeding for any corrections or clarifications prior to becoming the official record in the proceeding. Exhibits permitted in administrative proceedings can include almost any combination of physical evidence or documentary evidence in support of the position of one of the parties.

Affidavits: A particular type of documentary evidence often permitted to become part of the record of the proceeding is an affidavit. An affidavit is a written statement made under oath by an individual not present at the proceedings. Affidavits made under certain circumstances where both sides can cross-examine the witness making the affidavit are referred to as depositions.

Cross-examinations: One area of particular controversy in record-type proceedings is the power to cross-examine the witnesses in the proceeding. Cross-examination is quite common in adjudicative-type proceedings, but is controversial in rule-making proceedings. The purpose of cross-examination is to test the truthfulness and accuracy of a witness' statement by confronting the witness with instances of the witness' bias or prejudice, contrary statements made by the witness at different times, logical errors in the witness' own testimony, or to elicit statements that reflect unfavorably on the truthfulness of the witness in the proceeding. Cross-examination is considered a test of the truth of the witness' statements and, in criminal cases, is considered a part of fundamental due process protected by the U.S. Constitution.

INFORMAL PROCEDURES

In addition to formal, record-type proceedings, there are less formal, or nonrecord proceedings in which the agency makes determinations in accordance with a procedure involving notice to the public of the agency's intent to act and an invitation to submit documents or testimony to the agency in an attempt to influence the agency's decision-making. This can be done in both rule-making and adjudication, although it is more common in interpretive rule-making. Submissions are either written or oral, and usually become public information.

AGENCY PROCEEDINGS

Since the agency has a number of different types of powers, it is common for certain proceedings to be conducted by the agency as a means of exercising the powers given to it by the enabling statute.

INFORMAL RULE-MAKING

The first type of proceeding commonly used is an informal rule-making proceeding, often used for interpretive rules and sometimes for legislative rules. Usually the first step in this type of proceeding is the posting of a notice of proposed rule-making in a place designated by law for providing public notice. In the federal government this notice is placed in the *Federal Register,* a daily publication containing all the information designed to give notice to the American people of the activities of the government. Usually, the notice of proposed rule-making will contain a draft rule, an agency rationale in support of it, and an invitation for oral or written comment. Also included would be a deadline for the receipt of comments, specification of the form of the comment, and notation of where to deliver the comment. Comments are either in written or oral form, and are solicited for the guidance and assistance of the agency in its decision making.

The comments submitted to the agency can involve proposed policy changes, proposed word changes in the rule, or suggestions that the rule be eliminated. The agency then reviews the comments received and promulgates a final rule based on its own opinion and the comments it has received. However, there is no requirement in this type of rule-making for the agency to explain why it did what it did or to show why the agency was correct in its promulgation and a commentator was incorrect. It is an informal proceeding and, as such, can therefore only be challenged in the court.

FORMAL RULE-MAKING

There are two types of formal rule-making generally used in administrative proceedings: (1) rule-making on the record, and (2) formal rule-making. The difference between these two types of rule-making is significant. If an agency is required to make a finding on the record, then the reviewing court will examine the evidence presented at the hearing and determine whether there is sufficient evidence in the record to support the agency's determination in the case. If there is insufficient support in the record, the reviewing court will reverse the agency. If the finding is not required to be made on the record the agency can explain its decision in terms of its expertise, which will usually satisfy a reviewing court if the agency's explanation is sufficient.

In any case, the procedure is approximately the same. The first step in formal rule-making, as in informal rule-making, is public notice of the rule-making process. The notice will normally contain the proposed rule in a much more detailed form than in informal rule-making. The notice is required to be published in a manner designated by either the enabling statutes or another state statute providing for the notice to the public.

The next step is a hearing. At this point, there is a significant difference between rule-making on the record and ordinary rule-making. In rule-making on the record, the agency will present evidence consisting of witnesses, affidavits, and exhibits in support of the rule. It may then permit cross-examination of its witnesses or other examination of its evidence by any other person at the hearing. This is normally followed by a record proceeding, in which any interested party can present witnesses and evidence in support of its position on the rule. These interested parties are also permitted to file legal arguments, called briefs, on the legal issues of the case or rule. The responses by the outside parties are generally referred to as "comments." The agency will usually evaluate the comments and respond to them in another published document. In response to the comments and to the hearing, the agency will publish proposed final rules and allow a second round of comments (though perhaps not hearings) on the proposed final rule.

Following another comment period, the agency promulgates the final rule. The purpose of this rather long and cumbersome rule-making procedure is to ensure that the agency not only makes the best rules possible, but also that no person is deprived of a chance to have input into the rule-making process.

Not all rule-making proceedings — even those on the record — follow this extensive guideline. Often, however, procedural guidelines are mandated by either a federal or state administrative procedure act. The purpose is to create a record in support of the rule that will allow a reviewing court to determine whether or not the rule was properly promulgated.

ADJUDICATION

As previously stated, agencies have the power to develop their policies both by rule-making and adjudication. Since adjudication normally involves specific consequences to an individual defendant or respondent, its procedures are different from those of rule-making.

The first requirement normally present in an adjudication is notice to the respondent. This usually involves individual notice of a type acceptable to courts in their own proceedings, and will involve service of the complaint or other notice of the proceeding on the individual or corporate respondent in a way that complies with the state court procedure. This individual legal notice is considered a fundamental part of due process in our society. Very rarely will substitute notice of any sort be permitted. The notice will contain the entire complaint against the respondent. Usually, it will also include a description of the enabling act — or at least a reference to it — as the source of the authority for bringing the complaint, and a concise description of the type of conduct of the respondent that is thought to be in violation of the statute. Generally, the notice need not include a detailed claim of the policy of the agency that supports the adjudication: that is a matter for the hearing itself. The notice will often contain a copy of the rules of the agency (if necessary) or a reference to where the rules can be obtained. It will also include a notice of the time, date, and place of the hearing.

In an adjudication, the hearing itself contains the elements that allow it to be referred to as a trial-type hearing. However, the precise nature of the hearing can vary with the type of adjudicatory action to be taken by the agency. Not all elements of a hearing are required in every case. There are a few, however, that in a trial-type hearing are normally present. Usually, the general elements of a trial-type hearing are as follows: the burden of proof of the accusation contained in the notice is the responsibility of the administrative agency. The agency must support its position with a preponderance of the credible evidence introduced at the proceeding. However, the side that proposes a particular fact bears the burden of proof as to the existence of that fact. This is done so that the decision-maker can arrive at appropriate decisions. If the decision-maker is required to make a finding based on the preponderance of the evidence, either the proposing side has carried its burden of proof, in which case it wins, or it has not, in which case it loses on that particular fact issue. The administrative agency bears the burden of proving all facts necessary to support its complaint. The "preponderance of the evidence" is the usual standard of proof in a civil proceeding, which contrasts with the standard of proof in criminal proceedings: "beyond a reasonable doubt." The fact that the standard of proof is less in a civil proceeding reflects a social determination that it is more important to come to a decision that is closest to being correct in a civil proceeding than it is in a criminal proceeding.

A key element of a trial-type hearing is an impartial decision-maker who derives no benefit from determining the case in favor of one side over the other. Obviously a decision-maker whose income or promotion opportunities depend on a finding for one side or the other will not be an impartial decision-maker. This has resulted in the creation of a new and different type of decision-maker at the federal level — the administrative law judge. Administrative law judges are qualified for their post by the Civil Service Commission, and are hired by an agency. They are required to take testimony and find facts in adjudicative proceedings, but have a significant degree of independence guaranteed by federal government statutes setting up these proceedings. Although they are employees of an agency, they are in no way under the control of the agency in the decisions they reach in any particular case. In addition, an administrator cannot fire an administrative law judge because of a decision in a case, nor can the judge's promotion or personnel file be affected in any significant way by the handling of a case. If an administrative law judge were to be dismissed "for cause," a trial-type hearing before the Civil Service Commission would result. Their degree of independence and the protection they receive provides them with a significant degree of insulation from the agency administrator in the trial of the case.

In addition to an impartial decision-maker, trial-type hearings include the right to hear all testimony in favor of the government's position and, often (though not always), the right to cross-examine the evidence presented to the decision-maker. These rights prevent the agency from submitting confidential documents or other secret information that will affect the decision to the decision-maker. If cross-examination is allowed, these rights also allow the respondent to test the credibility of the government witnesses. In addition to cross-examination of the government witnesses, there will normally be a right to present the respondent's own witnesses at the proceeding. The testimony of these witnesses becomes part of the record and can be used by the judge in deciding the case. If the respondent has the right to cross-examine the government's witnesses, the government will have the right to cross-examine the respondent's witnesses.

Normally, the right to be represented by counsel at the hearing is accorded to all respondents in adjudication. Another right that is not as frequently present in adjudication is the right to discover other documents in the agency that might help the respondent in presenting a case (called "discovery"). In the federal government and many states, certain documents are available to all citizens under the Freedom of Information Act whether or not the citizens are involved in an adjudicative proceeding.

Related to discovery is the power to use subpoenas or another compulsory process for obtaining witnesses for the respondent. If the government has the power to subpoena witnesses for its side, it would normally be considered due process to require that the respondent have a similar power to subpoena witnesses for its case. Usually, a respondent will have the right to determine whether to consent to the order and avoid a hearing. Consenting to the order prevents the hearing itself.

Different agencies come to different types of decisions, based on their procedures. In some federal agencies, for example, the administrative law judge writes a recommended decision that is then reviewed by the administrator or commission that has been given the decision-making authority by Congress. The administrator or the commission determines whether or not to endorse the decision of the administrative law judge, and proceeds accordingly. In other agencies, the administrative law judge's decision becomes final unless an appeal is made by either the agency staff or the respondent or other party to the proceeding. If neither side appeals, administrators in that case can put the case on their own calendar for decision. In a review by the agency or an appeal to the agency, there is usually no further right to present evidence by either side, unless for some reason the administrative law judge excluded the evidence at the lower level. Usually the agency decides only legal issues or issues of agency policy where it may differ from the interpretation of the agency's policy that the administrative law judge presented in the decision. On occasion, however, the reviewing administrators may decide for their own reasons or own purposes to take new evidence on the issues in controversy in the case. In that case, they may either hear the evidence themselves or remand the case to the administrative law judge with directions to reopen the proceedings.

Having satisfied itself that there is sufficient evidence on which to base its decision, the agency then issues a decision that incorporates its own findings as to the facts, its conclusions as to the law, and the policies of the agency, which the adjudication is based on. It can, of course, find in favor of the respondent either because the case was not proved or the policy would not be served by a finding against the respondent in a particular case.

There are obvious differences in the consequences of rule-making proceedings and adjudications in terms of agency policy. A person can be prosecuted either administratively or in the courts for violation of an agency rule. When an agency has set policy by adjudication, all this allows the agency to do is either incorporate the policy determined in the adjudication into a rule, or enforce the same policy in subsequent adjudication against other respondents. The enunciation of a policy does not create a precedent in the legal sense; it binds no other persons than the respondent or other party to the proceeding.

Intervention: Intervention is a concept developed in administrative law — particularly in adjudication — to allow other parties interested in the subject matter of the adjudication to present evidence and their position in the case. Intervention is always a matter of statute or agency rule, and there is no constitutional right to intervene in a proceeding. Some agencies encourage intervention in order to give them a wide variety of opinions in the adjudication. This is particularly true where the agency is bringing an action against a commercial or industrial respondent and the intervenors are consumers. Less frequently, intervenors will be competitors or other commercial institutions in the same line of business who wish to be part of the agency policy-making proceeding.

Intervention varies. Sometimes it is in the nature of an *amicus curiae*, a traditional legal term dealing with outside parties who were allowed as friends of the court to file legal briefs and opinions in a pending case. Some intervenors are only granted the right to express their opinion in the case. In this instance, the opinion does not bind the intervening party. In other types of intervention, the intervening party becomes a full, actual participant in the hearing for all purposes, including calling witnesses and presenting evidence. In this situation, the intervenor is subject to the agency's jurisdiction and agrees to abide by the agency decision in the adjudication. One of the more obvious advantages of being in this position is that it gives the intervenor the right to appeal the decision of the agency to the court. However, if the intervenor is not granted party status, there is no right to appeal the decision.

ENFORCEMENT PROCEEDINGS

There is another type of hearing process used by an agency that does not involve the determination of agency policy. Although referred to by many different names, this new class of proceeding is usually called an enforcement proceeding. Some agencies do not enforce their regulations in in-house agency proceedings. Often they are required to go to court in order to enforce the proceeding at all. However, particularly at the federal level, some agencies do conduct in-house enforcement hearings. Those agencies that do have their own enforcement proceedings also have a typical type of procedure.

Enforcement proceedings are brought pursuant to an existing rule, order, or statute. The conduct that is alleged to be violative is usually clearly spelled out in the existing statute or rule. The primary difference between an enforcement proceeding and a criminal-type court proceeding for violation of a law lies in the fact that enforcement proceedings generally result in remedial action or, occasionally, in a civil penalty.

An enforcement proceeding commences with a notice to a respondent of the violation of the order or rule. This notice is served personally on the respondent, the same way a notice of adjudication is served. The respondent is entitled to a trial-type hearing similar to an adjudication. In an enforcement proceeding, the issue is whether or not the respondent has complied with a rule or statute, not whether the rule or statute is or is not appropriate or suitable. For example, certain rules were promulgated in the early 1970s with regard to the flammability of carpets. The Consumer Product Safety Commission was later given the authority to enforce these rules. If the Consumer Product Safety Commission found evidence of a violation of a rule, enforcement proceedings would be initiated before the agency's administrative law judge.

The advantages of conducting enforcement proceedings within the agency are quite clear: the individual making the decision and giving the remedial order is more familiar with the subject matter of the case and is, at the very least, a specialist and perhaps an expert. This also means that interpretations of (perhaps) doubtful or confusing portions of the statute are made first in the agency. In addition, the agency sees quite clearly when an existing rule or order has resulted in confusion and receives direct feedback regarding how it should restate or change its own regulation.

There are other methods of proceeding in the area of agency enforcement of rules. One example is seen following the issuance of a citation by a fire department official. The individual cited by the fire department official could be required to appeal the citation to a superior of the official — often the fire marshal in the department. Proceeding this way should be examined in some depth and against the framework of administrative law already set out. In typical fire department practice, an order is made out by the fire marshal or fire inspector on the scene in order to correct a situation. Sometimes, but not often, the order corresponds to the notice required for a proper enforcement proceeding. However, a defect in this order could be giving the order to an individual who is not the proper receiver of legal notices. For example, if the law puts a burden on the owner of a structure, serving the occupant may be inadequate unless local law clearly makes the occupant the owner's agent for such service. Another flaw could be an insufficient description of the rule or statute violated by the respondent. If the citation is defective in this respect, it is unlikely that the citation can be used as grounds for proceeding against the respondent.

Typically, the first appeal from the inspector's technical decision is directly to the fire marshal. In such a case the fire marshal must determine whether to afford a trial-type hearing to the respondent. Of particular interest at this point is whether the fire marshal places the burden of proof of the defect or violation on the inspector. If the inspector is not required to testify or otherwise give evidence at the hearing, then it is probable that at no time in the administrative enforcement process does the respondent get a hearing in the legal sense of the term. The fact that a hearing is not adequate does not mean that it is a waste of time. For example, an administrative agency can be structured so that the respondent is required to make any technical defense — i.e., a defense based on the factual situation as opposed to the law — to the fire marshal first, or in any subsequent court proceedings the respondent will not be permitted to raise a technical defense. A fire department with this type of enforcement system would do well to determine whether or not a proper fair hearing could be provided before an impartial magistrate in order to expedite the administrative proceedings and avoid further court proceedings.

In some cases no further administrative action is taken after the issuance of a citation and its confirmation by the fire marshal. There are two possibilities after the case has reached this point. One is that an enforcement action is started in the courts. It is important to remember that there is a difference between an enforcement action started in the courts and a court action to enforce a previously properly determined administrative order. As a rule of thumb, if the respondent was not accorded a fair hearing from the administrative agency that fair hearing will be accorded before the courts in the enforcement proceeding. It is also possible that the case will not be brought to the court. First, the fire department may not have the authority to institute a court proceeding. In many jurisdictions that authority resides solely with the district attorney, the state's attorney, the county attorney, or some other legal functionary who is entitled to sue on behalf of a governmental agency. If there is no cooperation with the fire department at this point, it is unlikely that the proceeding can go forward. In the absence of a properly developed record using an administrative tribunal it is possible that, for example, the district attorney will not be impressed by the quality of the evidence available in support of the case.

JUDICIAL REVIEW OF AGENCY PROCEEDINGS

The questions can be asked, "Why should the courts be involved at all in areas committed to administrative agencies? Doesn't the fact that administrative agencies have both legislative and judicial authority make a role for the court absolutely unnecessary?" One response is that in administrative proceedings agencies are trying to get someone outside of the government to do something. Often, when trying to force people to do something that they don't want to do, the first question that has to be answered is, "If I don't do it, what can you do to me?" Traditionally the answer has been fines and imprisonments for failure to do what the government required. The imposition of these penalties has historically been a judicial function: the legislature is specifically prohibited by the Constitution from imposing penalties and punishments on individuals. The judiciary has always been the protector of certain basic constitutional rights which, if infringed, give the individual a right to redress from the government. With very rare exception, all actions by an administrative agency are subject to one form or another of judicial review by the court. Usually this review is provided for by statute. Occasionally it is merely implied from the way the governmental entity was set up.

STANDARD OF JUDICIAL REVIEW

Since a large number of different administrative agency practices have been described, it is not surprising that there is a large number of different types of judicial review of the agency action. The concept of the standard of proof in a fact-finding proceeding has already been described, *i.e.,* how much does the agency have to prove? In the civil cases common in agencies, the answer is determined by the preponderance of the evidence. There is also a standard of judicial review for each of the varying kinds of agency action. It is important to remember the difference inherent in most determinations between findings of fact where the agency is asserting either the existence or the nonexistence of a factual set of circumstances underlying its decision and conclusions of law that are the agency's interpretation of its own and other statutes. Courts may be willing to defer greatly towards administrative agencies in the findings of fact, but, in the conclusions of law, interpreting the law has always been the province of the courts. The three standards of review or items used in the standard of review are significant because they determine how much proof the administrative agency needs to provide in favor of what it does in order to be affirmed by the court.

The first concept is in review of an agency's fact-finding process in the federal government. The standard of review used is whether there is substantial evidence in the record, considered as a whole, in favor of the agency's position. Substantial evidence does not necessarily mean the same as preponderance of the evidence. Substantial evidence refers to the fact that there is a significant amount of evidence in the record that favors the position that the administrator or commission has reached. This does not mean that, if given the same evidence, the court would come to the same conclusion. The court defers in part to the agency's expertise and ability to interpret the evidence in front of it and formulate its own policy. This standard is used for most reviews of agency action. It is probably the standard that exists in most states.

The second standard of review, used less often but very much in favor of the agency, is that in certain cases the agency findings will be upheld unless there is clear evidence of an abuse of discretion by the agency. This standard is often used when reviewing interpretive rules of the agency. In other words, when an agency is issuing rules that interpret the facts related to its enabling statute — unless there is a clear abuse of the agency's discretion — the courts will normally defer to the agency's own interpretation of its statute.

The third standard of review is not properly a standard at all, but a rule of evidence that often has the same effect as a standard of review. This rule is a legal presumption that the courts have that administrators do their job properly and in accordance with the law and that the agency is presumed to have acted in a regular fashion and is not required to prove that, for example, it followed its procedures correctly. It is up to the person challenging the agency's determination to prove to the contrary. This is the presumption of administrative regularity and is of great help to agencies because it allows them to avoid proving many of the day-to-day mundane characteristics of the agency time and time again in the courts. Under this presumption, if an official is required to review certain documents, for example, the official will be presumed by the courts to have reviewed the documents unless evidence is shown to the contrary.

The standards of review just described refer to factual issues. An agency's legal interpretation of its own statute, for example, is not viewed under the abuse of discretion standard but by a different standard dealing with legal consideration. When reviewing an agency's interpretation of the law, reviewing courts give varying amounts of credence to the agency's own interpretation of its statute. Some agencies are deferred to greatly by the courts; with others, the courts tend to ignore the agency's legal opinion of its own statute because the agency, in interpreting its own statute, is reading the law the same way the courts are.

REVIEW OF AGENCY JURISDICTION

One is always entitled to a judicial review of an agency action with regard to whether the agency is acting within the jurisdiction of its enabling statute. If the agency is not within the jurisdiction of its enabling statute its actions are *ultra vires* and will not be enforced by the court.

The first aspect that the court will normally review is whether the enabling statute supports the administrative proceeding. There are four types of defects that might be present: (1) the agency has no jurisdiction over the respondent, (2) the place where the alleged violation took place or is taking place is not within the territorial jurisdiction of the agency, (3) the type of proceeding brought by the agency is not provided for in the enabling statute, and (4) the type of relief requested is not within the jurisdiction of the agency. These are all statutory jurisdictional defects. One of the most important considerations concerning the jurisdictional defects in agency proceedings is the fact that these defects are not waived by failure to raise them before the agency itself. If a defect is jurisdictional in nature, it can be raised at any point in the review.

In addition to jurisdictional defects, a defense in a court proceeding or judicial review of an administrative agency action might be that the agency's enabling statute itself violates the appropriate constitution. This may be either a federal or a state constitution. Occasionally, when dealing with municipal corporations, it may be alleged that the statute violates the general powers of the municipal corporation to set up an agency. This constitutional-type argument is also one that can be raised at any point in the proceeding. If an agency action is found to be outside the agency's jurisdiction or the act is found unconstitutional, then the entire proceeding is null and void. These objections are usually raised in the court in the context of an attack on an agency determination adverse to a particular claimant in an adjudicative proceeding, although occasionally they are raised as defenses in rule-making actions.

REVIEW OF AGENCY PROCEDURES

In addition to the reviewing court looking at the facts and the law of the agency, it will also review — under certain circumstances — the procedures used by the agency in reaching its determination. However, if procedural defects are not raised at the agency level, it is possible that the court will consider the procedural rights to have been waived.

There are two major types of objections to the procedures used by an agency in an administrative proceeding. Similar to the statute, the complaint is first that the procedures are unconstitutional, perhaps because they do not provide for a sufficient hearing or that the procedures are somehow stacked against the respondent and do not provide due process. The second is that the procedures used in the specific case or adopted by the agency by rule do not comply with the statute, *i.e.*, the statute simply mandates a certain type of hearing or certain type of proceeding and the agency has not complied.

Another defense is that the procedure, though constitutional and though complied with by the agency, is inadequate to provide the reviewing court with sufficient information on which to base its decision, which therefore entitles the respondent to an additional hearing or other proceeding in the court to raise an objection. For example, in the proceedings before the agency, if neither side had compulsory process because the municipality had no such authority, then before the reviewing court — which has compulsory process — a respondent might claim that the facts in the case would show differently if the court would subpoena a witness on its behalf who was not willing to appear in the agency proceeding. This is an attack in the review on the procedures of the agency and is asking the reviewing court to provide an additional forum.

REVIEW OF RULE-MAKING

The judicial review of agency proceedings must be broken down in terms of review of adjudications, rule-making, and enforcement proceedings. In a review of rule-making, the first question is, "Who can appeal an agency rule?" This is not as simple to answer as it appears, particularly in the federal arena. When an agency makes a rule it is generally balancing certain costs against certain benefits. It may result in particular segments of society bearing certain costs and others receiving certain benefits.

Standing: The concept of the right to appear in court in a case is referred to as standing. Standing is a concept not easily defined. It has usually meant a sufficiently concrete stake in a proceeding to render the court review something other than a mere theoretical exercise with regard to the person bringing the review. Sometimes the complaints of individuals who wish to review actions of the government are considered simply so tenuous or weak that they will not be heard in the courts. Other claims of individuals who might be considered to have very small interests in the proceeding have been given standing by the court. These have included both people who thought that the government went too far in its rule and, particularly in the environmental area, individuals who feel that the government has not gone far enough. Standing is the type of area that a state can regulate by statute, particularly when it means expanding standing but even when it contracts it to only those with an actual recognizable dollar interest in the proceeding. Whether this is advisable or not is not always clear. Standing is very simply the right to bring the lawsuit to review the rule.

Timeliness: The second element of the reviewing lawsuit is timeliness. How long does one have after the rule is published in a final form to bring the lawsuit challenging the rule? This is often a very short period of time — sixty days is common in the federal courts. If the reviewing action is not brought within sixty days of the final date of the rule, then virtually all judicial review of the rule is foreclosed except for the basic jurisdictional and constitutional defects previously mentioned. Procedural defects or challenges to the rule itself are waived. Occasionally an administrative agency fails to finalize the rule in the proper form. If it fails to properly finalize the rule, then the time period of sixty days may never run. If the rule is not published in final form in the *Federal Register,* the time period simply may never have run out on a person challenging the statute. It is particularly important for local governmental entities to be sure that they have complied with the appropriate rule.

Venue: The third element of a review of rule-making is the venue of the reviewing court. Venue is a concept of place: "Where is the rule being reviewed?" Venue is territorial in nature, and requires that reviewing courts also have appropriate jurisdiction. If a review is brought to the wrong court, the court will not hear the case. This is important, particularly when the time period for review is short. Time lost in going to the wrong court may preclude filing in the right court on time. Sometimes a rule can be reviewed in several different places, *e.g.,* any of the eleven federal appeal courts can review a CPSC standard. The first petition filed determines the venue for review; sometimes this is a matter of seconds.

Standard of Review: The next major issue in review of rule-making is determining the standard of review of an agency rule in the court. This is not an easy question to answer since it varies significantly between the federal government and the states. The federal government is quite clear: according to the Administrative Procedure Act as interpreted by the Supreme Court, the standard is substantial evidence on the record considered as a whole. The states may vary from this standard. There is not really anything constitutional or often even statutory about the standard of review. It simply depends on what the courts consider to be required to do justice under the circumstances.

New Evidence: One question that is often raised is, "What happens to an individual who wished to present evidence in the rule-making proceeding and was denied that opportunity by the administrator of the agency?" The usual activity of the court will be to send the case back to the agency to hear the new evidence, if it was offered at a time that made it "timely" in the first place. If it is newly discovered evidence, the court will decide whether or not the evidence might change the administrator's mind and may send the case back to the administrator for further review. Normally, if an individual was denied the opportunity to present evidence to the agency, the court will either hear the evidence on appeal or will remand it to the agency for an appropriate hearing if it determines that the individual had a right to have the evidence considered. Burden of proof is normally on the agency to justify its refusal to hear the evidence.

Intervention on Appeal: Occasionally a party or an individual who did not appeal the original rule-making wishes to be heard at the appellate level in the court. This area is up to the discretion of the court as to whether or not it will permit an additional party who did not appeal the rule originally to intervene in the case on review in the courts. If the proposed intervening party proves that the party was unable to intervene before the agency, the court usually permits intervention.

REVIEW OF ADJUDICATION

The review of an adjudication by an administrative agency follows a different set of rules from a review of rule-making, but many of the factors are the same if a proper hearing was extended at the agency level. The only people who can appeal an adjudication to the court, *i.e.*, the only people with standing, are parties to the adjudication. This is usually limited to respondents and intervenors. Outside parties, no matter how interested in the proceeding and if they have not become intervenors, cannot appeal an agency determination. The review of the adjudication must be timely, *i.e.*, it must come within the time period required by the statute. The standard of review of an agency adjudication is very similar to the standard of review of a rule: it is the setting of agency policy and it is the standard of substantial evidence on the record as a whole as a standard of proof and reasonable relationship to the statute as a question of law. If the respondent claims not to have had a fair hearing in the adjudication, the court may extend the respondent the right to a hearing in the court and the right to present any evidence that was not made part of the record, or the court may send the proceeding back for further hearings in the administrative agency. On occasion, in an appeal of an adjudication by respondents, the court may allow intervention at the review level. It will do this usually only if it will facilitate the original review. In an adjudication, the court will determine whether or not the agency policy is in accordance with the statute. It will not base its decision on whether it agrees with the agency policy. If it is in accordance with the statute that is the function — the determination of agency policy is the function — that has been committed to administrative discretion. It will also review whether or not the remedy proposed is in accordance with the statute and is designed to achieve the policy objective in the agency.

REVIEW OF ENFORCEMENT

Since enforcement actions are very similar to civil trials, the court's function is much more like an appellate reviewing court in a regular civil trial. Some reviewing courts, in enforcement proceedings where the basic question at issue is whether or not the individual violated the law, will afford the defendant a trial *de novo*, *i.e.*, an entirely new trial with the burden of proof on the agency. This is done in those proceedings where the individual who is the respondent in the enforcement proceeding was not granted a full fair trial within the agency, and is particularly true within those areas where either the Constitution of the federal government or the state requires a jury trial, since jury trials by definition are never available in administrative proceedings. It is important to remember that in an enforcement proceeding pursuant to a rule or a statute, the wisdom of the rule or the statute may not be attacked. The time for challenging the unreasonableness of the rule had expired after the review period of the rule itself.

Discriminatory Prosecution: One of the most interesting areas of enforcement proceeding review is the concept of discriminatory prosecution. In such a case the respondent claims to have been singled out for prosecution while others similarly engaged were ignored. This claim can also be raised in an adjudicative proceeding. There are two components to such a defense. One is that it was the intent of the agency to single out and punish the respondent for doing something that was permitted to others. This type of conduct would appear to be a violation of the Constitutional guarantee of equal protection of the laws. On the other hand, it may not have been possible for the administrative agency to pursue all wrongdoers — even all those guilty of precisely the same behavior at the same time. The courts have often put the respondent in a position of having to prove that the agency was biased. However, an agency that makes a practice of enforcing the law in a way that deprives certain classes of individuals of their constitutional rights can be stopped by appropriate legal proceedings. This is particularly useful to remember in the fire protection field since it was a fire code enforced in a discriminatory manner that led to the enunciation of this principle.

REMEDIES IN ADMINISTRATIVE PROCEEDINGS

What remedies can be requested in administrative proceedings, whether they are enforcement proceedings or adjudication? As stated earlier in this text, injunctive proceedings, *i.e.,* proceedings to order someone to do something, have classically been considered part of the equity power of the government. Administrative proceedings have involved both equity and law-type orders and penalties, and can be thought of as a mix of the two.

CEASE AND DESIST ORDERS

The classic administrative penalties or remedies available in administrative proceedings are cease and desist orders and abatement orders. A cease and desist order normally requires the respondent to stop doing something that has been found by the agency to be a violation of the law.

ABATEMENT

An abatement, particularly an abatement of nuisance order, is an order to eliminate a hazardous condition, such as one that presents a firesafety hazard.

REPAIR

Usually an adjudication would end in a cease and desist order or an abatement order. However, cease and desist orders and abatement orders are relatively limited remedies. They may not cover certain types of problems. Among the most significant remedy currently available is the concept of requiring the individual to actually repair the hazardous situation created by the individual's action. This is not necessarily the same as a requirement to abate a nuisance or to cease and desist a violation of the law. For example, a landlord of an apartment building that is in violation of the area's fire code requirements for apartments can abate the nuisance by evicting the tenants. However, although the problem of the fire hazard to the tenants has been solved by removing them from unsafe conditions, it has generated the problem of shelter for the evicted tenants. An order to repair the premises also solves the fire problem, without promoting another problem.

RECALL

It has become necessary to develop more flexible enforcement techniques in administrative proceedings. The Consumer Product Safety Commission was given the authority, in its Act, to require a manufacturer of a defective consumer product that either fails to meet a commission rule or is found to be a substantial product hazard in an adjudication to repair all of the products that have been sold or to replace the products with a similar product or to refund the purchase price of the product. These are typically referred to as recall actions.

DAMAGES

Some agencies have been given the power to assess damages against a corporation or individual and order the damages paid to the people defrauded or otherwise injured by the respondent. This differs significantly from common law damage actions since the agency is bringing the action on behalf of the individuals who have lost money. The damages that can be recovered by the agency might be two kinds: contract damages — often the purchase price and consequential damages, or those further damages that stem from a defective product. For example, if a landlord maintained a building that failed to meet code requirements and tenants had leases in that building, and the landlord subsequently closed the building to abate the nuisance, the landlord might be required to pay the difference between the cost in the defective building and the cost in a building that met the codes or standards.

CIVIL PENALTIES

An additional administrative area that has been developing quite recently is the concept of civil penalties against a respondent in an administrative proceeding. These penalties are placed for the purpose of discouraging wrongdoers from violating the law without requiring those burden of proofs and other defenses available in criminal action.

LICENSE REVOCATION

A final type of administrative action that exists and could be used particularly in the fire protection field is the use of license revocation proceedings. Licensing by the state is a function by which the state extends to someone the legal right to carry on a trade or perform some other type of operation. Drivers' licenses, medical licenses, apartment licenses, and plumbers' licenses are all various types of licenses used by the state. It is popularly thought that the state can simplify certain matters by requiring a license to perform a function and simply lifting the license if the function is not properly performed, thus achieving the goals of the statute. This procedure is based on an interpretation of the law that assumes that licenses are privileges extended by the state, the removal of which is simply the removal of a state privilege and not one subject to either due process or judicial review. The courts have — in this area and in others — discarded the concept of privilege. In a related area, administrative agencies are often called upon to determine whether a particular contractor will be allowed to do business with the state or whether a particular individual receiving welfare will continue to receive it. At one time these types of decisions were clearly considered by the state to be nonreviewable by the courts and not requiring any due process for the individual. In the late sixties and early seventies this concept changed radically. The right — privilege dichotomy — was replaced by the concept of an entitlement, particularly in the welfare area. Broadly speaking, what the courts determined was that even if a particular activity could not be characterized as a right, the state could not take it away from an individual without affording due process to that individual. Entitlement almost certainly applies to licenses, welfare, and other state operations. This does not mean, however, that these cannot be used as control techniques; it simply means that due process must be afforded. Due process would normally be considered as either the right specified under adjudications or, more properly, enforcement proceedings.

EXTRA-AGENCY PROCEEDINGS

It is important to look at some of the ways of administrative agencies' interaction with other portions of the legal system; in particular, with the decision that to use administrative processes rather than other legal processes leads to certain significant legal consequences.

CRIMINAL PROCEEDINGS

Most administrative processes are remedial in nature. Penalties are normally limited to violations of clear agency rules or statutes, and even then the fines are limited to civil penalties. No further punishment is possible. An alternative to these administrative penalties is criminal proceedings. Some administrative agencies are charged by law with determining whether or not to refer a case to the prosecutor for criminal penalties. The prosecutor is not given the discretion to start such a case. A local law might provide, for example, that violations of the fire code shall be investigated by the fire department and referred to the prosecutor, and without such referral no prosecution shall commence. This is not common, but can be the practical arrangement of the local prosecutorial force. In other settings — particularly the federal government — it is the defined agency responsibility to refer criminal proceedings to the Department of Justice.

Felonies: Three types of criminal proceedings can be defined. The first are felonies. At the common law, felonies were a certain small number of significant serious crimes that were punishable by death. Our statutory definitions of "felony" are not consistent among the states and the federal government, but most commonly felonies are considered to be those crimes that are punishable by imprisonment for a term greater than one year. The typical procedure in a felony is for the case to be reviewed by a grand jury and an indictment handed down, or for the prosecutor in certain states to file an information charging the individual with the felony. Felony defendants are always entitled to a jury trial. A felony is normally prosecuted by the appropriate county, state, or federal prosecutor.

Misdemeanors: Below the felony in order of significance is the misdemeanor. Many fire codes and other violations are misdemeanors. Misdemeanor prosecutions are usually initiated by filing an information. Rarely are individuals extradited to face misdemeanor charges. Persons charged with misdemeanors are entitled to a jury trial unless the time of imprisonment is so short that it qualifies as a so-called minor offense.

Ordinance Violations: The most common criminal-type prosecution is prosecution for violation of an ordinance in which a penalty is sought for rather than merely a remedial order. Ordinance violations are initiated by filing an appropriate information with the court. The person being tried for the ordinance violation is often not entitled to a jury trial, since it is considered a minor offense. In some states individuals are entitled to jury trials. Ordinance violations are often restricted to traffic and other minor circumstances.

CRIMINAL PROSECUTIONS

In all criminal prosecutions, including ordinance violations, there is a presumption that the defendant is innocent. In addition, the defendant has an absolute right to remain silent, and does not need to provide any information that will assist in prosecution. The defendant is entitled to have the facts that prove guilt proved beyond a reasonable doubt, and is entitled to have the criminal statutes strictly construed in the defendant's favor. The defendant is also, if indigent, entitled to a court-appointed attorney in misdemeanor and felony proceedings. These factors make the criminal law an often unwieldy tool for enforcement in the fire protection field.

CRIMINAL ENFORCEMENT

The fact that criminal proceedings can lead to imprisonment clearly merits their consideration for particularly recalcitrant defendants. The fact that an agency has initiated administrative proceedings is no bar whatever to a parallel or simultaneous criminal proceeding. A defendant can be ordered to repair the damage and be sent to jail. The only limitation is: if the adjudicative proceedings are used to develop the evidence for the criminal proceeding, the evidence may be disallowed if obtained in violation of the person's rights against self-incrimination. Other general problems with criminal enforcement include the fact that the defendants specified by a law may turn out to be corporations who have all of the protection of the criminal law and the advantage that they cannot be sent to prison. Occasionally, major corporate officials can be convicted for the acts of the corporation, but a great deal depends on how the actual statute is phrased.

A second problem with criminal prosecution is that personal jurisdiction over the defendants must be obtained in order to bring the action. A third problem is that local prosecutorial policy may be in the hands of someone outside the administrative agency who does not consider that the cases as the agency recommends are attractive enough for prosecution. And the final problem is that a criminal penalty may be an inadequate remedy to solve the problem in the community. It may be satisfying, and it may, in fact, even deter other individuals from committing the same crime; however, it may not solve the immediate problem facing the community.

INJUNCTIVE PROCEEDINGS

Another type of agency proceeding is an injunctive proceeding. Injunctions were the equity courts' remedies to right wrongs when no other legal recourse was possible. An advantage of an injunctive proceeding is that, in an emergency, a court will often take the agency's word for the existence of a hazardous condition and in an *ex parte,* or one-sided, proceeding, issue a temporary restraining order to prevent the respondent from continuing to allow a hazardous condition to exist. Temporary restraining orders are backed by the power of the court, not of the agency. In a hearing before the court, usually granted in a very short time, the respondent is allowed to make any defense and the government is called upon to make its case, with the burden of proof falling to the government. If the decision is made in favor of the government, the government will issue a permanent injunction or extend it with a continual series of temporary injunctions. The advantage of this form of proceeding is that it is extremely flexible and relates specifically to the precise problem at hand in the case. The disadvantage is that the decision-maker, though advised by the agency, is not an expert and must rely on the agency to provide the technical and supportive information in order to reach a decision. In the event that the agency does not have sufficient credibility with the court for its determination to be respected, it is possible that the court will not grant sufficient deference to the agency's expertise to grant the injunction. The other problem is that injunctions are future-oriented; they do not necessarily solve problems arising from the past action of the respondent, and if it is the agency's job to remedy those past problems, this may be an inadequate action.

DAMAGE PROCEEDINGS

A third type of action sometimes permitted (as previously mentioned) is for the agency to bring a lawsuit on behalf of individuals who have been harmed in a damage proceeding. The advantage of this proceeding is that there is actual restitution to the person harmed. The action can be brought, for example, for antitrust laws, on behalf of either named individuals or individuals to be ascertained, or by the state as *parens patriae* for all of the people. In other words, the state recovers the money and uses it to benefit the public good. This approach is past-oriented, but it does deprive the wrongdoer of some of the fruits of the wrongful activity.

SUMMARY

Administrative agencies are similar to legislatures and the courts in that they also possess governmental authority. This authority can be legislative, judicial, or executive, and can, in many ways, be compared to the powers and authority of the three branches of government created by the Constitution. Although citizens are generally more familiar with civil procedures and criminal procedures, it is likely that most citizens will not come in direct, personal contact with these procedures. However, administrative procedure and process — procedures of which most citizens are unaware — is an area of law that touches everyday life and affects citizens almost daily. This "mysterious" area of governmental authority is not so mysterious once it is understood in terms of the context in which it functions, and also in terms of the framework and guidelines in which it is set. For members of the fire service, a knowledge of administrative process and procedure is vital. There are many administrative agencies within the fire service and, in fact, a fire department can function as an administrative agency itself. In addition, rules, regulations, and decisions of administrative agencies related to the interests, needs, and functions of the fire service affect the professional lives of all of the members of the fire service daily — in fire prevention, fire protection, employment considerations, and so forth.

ACTIVITIES

1. List and describe four reasons why administrative agencies were established.
2. One of the characteristics of an administrative agency is that it possesses discretionary powers and duties. Describe this function of administrative agencies and discuss with your classmates how problems could arise in this particular area.
3. Identify and describe the three basic powers of administrative agencies.
4. Compare a formal "on the record" administrative agency procedure with an informal administrative procedure.
5. Describe a violation of a fire-service-related administrative law. Describe the preparation and steps necessary to bring this matter before the proper administrative agency and successfully win your case.
6. Describe the advantages of holding enforcement proceedings within an administrative agency.

7. Why is it important for members of the fire service to know the correct administrative procedures to follow?
8. List and describe the various standards of review used by the courts in decisions involving administrative agencies.
9. Identify and describe the seven kinds of remedies usually requested in administrative proceedings. Give an example of when each remedy would be most effective.
10. With a classmate, debate some of the advantages and disadvantages of an administrative agency's resorting to criminal enforcement procedures.

BIBLIOGRAPHY

Carrow, Milton M., *The Background of Administrative Law*, Associated Lawyers Publishing Company, Newark, NJ, 1948.

Davis, Kenneth Culp, *Administrative Law and Government*, West Publishing Co., St. Paul, MN, 1960.

Dickinson, John, *Administrative Justice and the Supremacy of Law in the United States*, © Harvard University Press, 1927, Russell and Russell, Inc., New York, NY, 1955.

Freund, Ernst, *et al.*, *The Growth of American Administrative Law*, © Bar Association of St. Louis, Thomas Law Book Company, St. Louis, MO, 1923.

Mashaw, Jerry L. and Merrill, Richard A., *Introduction to the American Public Law System*, West Publishing Co., St. Paul, MN, 1975.

Parker, Reginald, *Administrative Law*, The Bobbs-Merrill Company, Indianapolis, IN, 1952.

Schwartz, Bernard, *An Introduction to American Administrative Law*, Oceana Publications, Inc., Dobbs Ferry, NY, 1962.

Woll, Peter, *Administrative Law*, University of California Press, Berkeley, CA, 1963.

For an in-depth study of the topics presented in this chapter, the following is recommended:

Robinson, Glen O. and Gellhorn, Ernest, *The Administrative Process*, West Publishing Co., St. Paul, MN, 1974.

APPENDIX A

ORIGIN OF THE CONSTITUTION OF THE UNITED STATES

The War of Independence was conducted by delegates from the original 13 states, and was called the Congress of the United States of America; it was generally known as the Continental Congress. In 1777 the Continental Congress submitted to the legislatures of the states the Articles of Confederation and Perpetual Union, which were ratified by New Hampshire, Massachusetts, Rhode Island, Connecticut, New York, New Jersey, Pennsylvania, Delaware, Virginia, North Carolina, South Carolina, and Georgia, and finally, in 1781, by Maryland.

The first article of the document read: "The stile of this confederacy shall be the United States of America." This did not establish a sovereign nation, because the states delegated only those powers they could not handle individually, such as power to wage war, establish a uniform currency, make treaties with foreign nations and contract debts for general expenses (such as paying the army). Taxes for the payment of such debts were levied by the individual states. Under the Articles, the president signed himself "President of the United States in Congress assembled," but here the United States were considered to be only a cooperating group. Canada was invited to join the union on equal terms but did not act.

When the war was won, however, it became evident that a stronger federal union was needed to protect the mutual interests of the states. The Congress left the initiative to the legislatures. Virginia in Jan. 1786 appointed commissioners to meet with representatives of other states, with the result that delegates from Virginia, Delaware, New York, New Jersey, and Pennsylvania met at Annapolis. Alexander Hamilton prepared for their call by asking delegates from all states to meet in Philadelphia in May 1787 "to render the Constitution of the Federal government adequate to the exigencies of the union." Congress endorsed the plan Feb.

21, 1787. Delegates were sent by all states except Rhode Island.

The convention met May 14, 1787. George Washington was chosen president (presiding officer). The states had appointed 65 delegates, but 10 did not attend. The work was done by 55, not all of whom were present at all sessions. Of the 55 attending delegates, 16 failed to sign, and 39 actually signed Sept. 17, 1787, some with reservations. Some historians have said 74 delegates (9 more than the 65 actually certified) were named and 19 failed to attend. These 9 additional persons refused the appointment, were never delegates, and were never counted as absentees. Washington sent the Constitution to Congress with a covering letter and that body, Sept. 28, 1787, sent it to the legislatures, "in order to be submitted to a convention of delegates chosen in each state by the people thereof."

The Constitution was ratified by the state conventions as follows: Delaware, Dec. 7, 1787, unanimous; Pennsylvania, Dec. 12, 1787, 43 to 23; New Jersey, Dec. 18, 1787, unanimous; Georgia, Jan. 2, 1788, unanimous; Connecticut, Jan. 9, 1788, 128 to 40; Massachusetts, Feb. 6, 1788, 187 to 168; Maryland, Apr. 28, 1788, 63 to 11; South Carolina, May 23, 1788, 149 to 73; New Hampshire, June 21, 1788, 57 to 46; Virginia, June 25, 1788, 89 to 79; New York, July 26, 1788, 30 to 27. Nine states were necessary to establish the operation of the Constitution "between the states so ratifying the same;" New Hampshire was the 9th state. The government did not declare the Constitution in effect until the first Wednesday in Mar. 1789 which was Mar. 4. After that, North Carolina ratified it Nov. 21, 1789, 197 to 77; and Rhode Island May 29, 1790, 34 to 32. Vermont in convention ratified it Jan. 10, 1791, and by act of Congress approved Feb. 19, 1791, was admitted into the Union as the 14th state, Mar. 4, 1791.

CONSTITUTION OF THE UNITED STATES

The Original 7 Articles

PREAMBLE

We, the people of the United States, in order to form a more perfect Union, establish justice, insure domestic tranquility, provide for the common defense, promote the general welfare, and secure the blessings of liberty to ourselves and our posterity, do ordain and establish this Constitution for the United States of America.

ARTICLE I.

Section 1 — Legislative powers; in whom vested:

All legislative powers herein granted shall be vested in a Congress of the United States, which shall consist of a Senate and House of Representatives.

Section 2 — House of Representatives, how and by whom chosen. Qualifications of a Representative. Representatives and direct taxes, how apportioned. Enumeration. Vacancies to be filled. Power of choosing officers, and of impeachment.

1. The House of Representatives shall be composed of members chosen every second year by the people of the several States, and the electors in each State shall have the qualifications requisite for electors of the most numerous branch of the State Legislature.

2. No person shall be a Representative who shall not have attained to the age of twenty-five years, and been seven years a citizen of the United States, and who shall not, when elected, be an inhabitant of that State in which he shall be chosen.

3. *(Representatives and direct taxes shall be apportioned among the several States which may be included within this Union, according to their respective numbers, which shall be determined by adding to the whole number of free persons, including those bound to service for a term of years, and excluding Indians not taxed, three-fifths of all other persons.) (The previous sentence was superseded by Amendment XIV, section 2.)* The actual enumeration shall be made within three years after the first meeting of the Congress of the United States, and within every subsequent term of ten years, in such manner as they shall by law direct. The number of Representatives shall not exceed one for every thirty thousand, but each State shall have at least one

Representative; and until such enumeration shall be made, the State of New Hampshire shall be entitled to choose three, Massachusetts eight, Rhode Island and Providence Plantations one, Connecticut five, New York six, New Jersey four, Pennsylvania eight. Delaware one, Maryland six, Virginia ten, North Carolina five, South Carolina five, and Georgia three.

4. When vacancies happen in the representation from any State, the Executive Authority thereof shall issue writs of election to fill such vacancies.

5. The House of Representatives shall choose their Speaker and other officers; and shall have the sole power of impeachment.

Section 3 — Senators, how and by whom chosen. How classified. Qualifications of a Senator, President of the Senate, his right to vote. President pro tem., and other officers of the Senate, how chosen. Power to try impeachments. When President is tried, Chief Justice to preside. Sentence.

1. The Senate of the United States shall be composed of two Senators from each State. *(chosen by the Legislature thereof,) (The preceding five words were superseded by Amendment XVII, section 1.)* for six years; and each Senator shall have one vote.

2. Immediately after they shall be assembled in consequence of the first election, they shall be divided as equally as may be into three classes. The seats of the Senators of the first class shall be vacated at the expiration of the second year, of the second class at the expiration of the fourth year, and of the third class at the expiration of the sixth year, so that one-third may be chosen every second year; *(and if vacancies happen by resignation, or otherwise, during the recess of the Legislature of any State, the Executive thereof may make temporary appointments until the next meeting of the Legislature, which shall then fill such vacancies.) (The words in parenthesis were superseded by Amendment XVII, section 1.)*

3. No person shall be a Senator who shall not have attained to the age of thirty years, and been nine years a citizen of the United States, and who shall not, when elected, be an inhabitant of that State for which he shall be chosen.

4. The Vice President of the United States shall be President of the Senate, but shall have no vote, unless they be equally divided.

5. The Senate shall choose their other officers, and also a President pro tempore, in the absence of the Vice President, or when he shall exercise the office of President of the United States.

6. The Senate shall have the sole power to try all impeachments. When sitting for that purpose, they shall be on oath or affirmation. When the President of the United States is tried, the Chief Justice shall preside; and no person shall be convicted without the concurrence of two-thirds of the members present.

7. Judgment in cases of impeachment shall not extend further than to removal from office, and disqualification to hold and enjoy any office of honor, trust or profit under the United States: but the party convicted shall nevertheless be liable and subject to indictment, trial, judgment and punishment, according to law.

Section 4 — Times, etc., of holding elections, how prescribed. One session in each year.

1. The times, places and manner of holding elections for Senators and Representatives; shall be prescribed in each State by the Legislature thereof; but the Congress may at any time by law make or alter such regulations, except as to the places of choosing Senators.

2. The Congress shall assemble at least once in every year, and such meeting shall *(be on the first Monday in December,) (The words in parenthesis were superseded by Amendment XX, section 2).* unless they shall by law appoint a different day.

Section 5 — Membership, quorum, adjournments, rules. Power to punish or expel, Journal. Time of adjournments, how limited, etc.

1. Each House shall be the judge of the elections, returns and qualifications of its own members, and a majority of each shall constitute a quorum to do business; but a smaller number may adjourn from day to day, and may be authorized to compel the attendance of absent members, in such manner, and under such penalties as each House may provide.

2. Each House may determine the rules of its proceedings, punish its members for disorderly behavior, and, with the concurrence of two-thirds, expel a member.

3. Each House shall keep a journal of its proceedings, and from time to time publish the same, excepting such parts as may in their judgment require secrecy; and the yeas and nays of the members of either House on any question shall, at the desire of one-fifth of those present, be entered on the journal.

4. Neither House, during the session of Congress, shall, without the consent of the other, adjourn for more than three days, nor to any other place than that in which the two Houses shall be sitting.

Section 6 — Compensation, privileges, disqualifications in certain cases.

1. The Senators and Representatives shall receive a compensation for their services, to be ascertained by law, and paid out of the Treasury of the United States. They shall in all cases, except treason, felony and breach of the peace, be privileged from arrest during their attendance at the session of their respective Houses, and in going to and returning from the same; and for any speech or debate in either House, they shall not be questioned in any other place.

2. No Senator or Representative shall, during the time for which he was elected, be appointed to any civil office under the authority of the United States, which shall have been created, or the emoluments whereof shall have been increased during such time; and no person holding any office under the United States, shall be a member of either House during his continuance in office.

Section 7 — House to originate all revenue bills. Veto. Bill may be passed by two-thirds of each House, notwithstanding, etc. Bill, not returned in ten days, to become a law. Provisions as to orders, concurrent resolutions, etc.

1. All bills for raising revenue shall originate im the House of Representatives; but the Senate may propose or concur with amendments as on other bills.

2. Every bill which shall have passed the House of Representatives and the Senate, shall, before it becomes a law, be presented to the President of the United States; if he approves he shall sign it, but if not he shall return it, with his objections to that House in which it shall have originated, who shall enter the objections at large on their journal, and proceed to reconsider it. If after such reconsideration two-thirds of that House shall agree to pass the bill, it shall be sent, together with the objections, to the other House, by which it shall likewise be reconsidered, and if

approved by two-thirds of that House, it shall become a law. But in all such cases the votes of both Houses shall be determined by yeas and nays, and the names of the persons voting for and against the bill shall be entered on the journal of each House respectively. If any bill shall not be returned by the President within ten days (Sundays excepted) after it shall have been presented to him, the same shall be a law, in like manner as if he had signed it, unless the Congress by their adjournment prevent its return, in which case it shall not be a law.

3. Every order, resolution, or vote to which the concurrence of the Senate and House of Representatives may be necessary (except on a question of adjournment) shall be presented to the President of the United States; and before the same shall take effect, shall be approved by him, or being disapproved by him, shall be repassed by two-thirds of the Senate and House of Representatives, according to the rules and limitations prescribed in the case of a bill.

Section 8 — Powers of Congress.

The Congress shall have power

1. To lay and collect taxes, duties, imposts and excises, to pay the debts and provide for the common defense and general welfare of the United States; but all duties, imposts and excises shall be uniform throughout the United States;

2. To borrow money on the credit of the United States;

3. To regulate commerce with foreign nations, and among the several States, and with the Indian tribes;

4. To establish a uniform rule of naturalization, and uniform laws on the subject of bankruptcies throughout the United States;

5. To coin money, regulate the value thereof, and of foreign coin, and fix the standard of weights and measures;

6. To provide for the punishment of counterfeiting the securities and current coin of the United States;

7. To establish post-offices and post-roads;

8. To promote the progress of science and useful arts, by securing for limited times to authors and inventors the exclusive right to their respective writings and discoveries;

9. To constitute tribunals inferior to the Supreme Court;

10. To define and punish piracies and felonies committed on the high seas, and offenses against the law of nations;

11. To declare war, grant letters of marque and reprisal, and make rules concerning captures on land and water;

12. To raise and support armies, but no appropriation of money to that use shall be for a longer term than two years;

13. To provide and maintain a navy;

14. To make rules for the government and regulation of the land and naval forces;

15. To provide for calling forth the militia to execute the laws of the Union, suppress insurrections and repel invasions;

16. To provide for organizing, arming, and disciplining the militia, and for governing such part of them as may be employed in the service of the United States, reserving to the States respectively, the appointment of the officers, and the authority of training and militia according to the discipline prescribed by Congress;

17. To exercise exclusive legislation in all cases whatsoever, over such district (not exceeding ten miles square) as may, by cession of particular States, and the acceptance of Congress, become the seat of the Government of the United States, and to exercise like authority over all places purchased by the consent of the Legislature of the State in which the same shall be, for the erection of forts, magazines, arsenals, dockyards, and other needful buildings; — And

18. To make all laws which shall be necessary and proper for carrying into execution the foregoing powers, and all other powers vested by this Constitution in the Government of the United States, or in any department or officer thereof,

Section 9 — Provision as to migration or importation of certain persons. Habeas corpus, bills of attainder, etc. Taxes, how apportioned. No export duty. No commercial preference. Money, how drawn from Treasury, etc. No titular nobility. Officers not to receive presents, etc.

1. The migration or importation of such persons as any of the States now existing shall think proper to admit, shall not be prohibited by the Congress prior to the year one thousand eight hundred and eight, but a tax or duty may be imposed on such importation, not exceeding ten dollars for each person.

2. The privilege of the writ of habeas corpus shall not be suspended, unless then in cases of rebellion or invasion the public safety may require it.

3. No bill of attainder or ex post facto law shall be passed.

4. No capitation, or other direct, tax shall be laid, unless in proportion to the census or enumeration herein before directed to be taken. *(Modified by Amendment XVI.)*

5. No tax or duty shall be laid on articles exported from any State.

6. No preference shall be given by any regulation of commerce or revenue to the ports of one State over those of another; nor shall vessels bound to, or from, one State, be obliged to enter, clear, or pay duties in another.

7. No money shall be drawn from the Treasury, but in consequence of appropriations made by law; and a regular statement and account of the receipts and expenditures of all public money shall be published from time to time.

8. No title of nobility shall be granted by the United States; and no person holding any office of profit or trust under them, shall, without the consent of the Congress, accept of any present, emolument, office, or title, of any kind whatever, from any king, prince, or foreign state.

Section 10 — States prohibited from the exercise of certain powers.

1. No State shall enter into any treaty, alliance, or confederation; grant letters of marque and reprisal; coin money; emit bills of credit; make anything but gold and silver coin a tender in payment of debts; pass any bill of attainder, ex post facto law, or law impairing the obligation of contracts, or grant any title of nobility.

2. No State shall, without the consent of the Congress, lay any imposts or duties on imports or exports, except what may be absolutely necessary for executing its inspection laws; and the net produce of all duties and imposts, laid by any State on imports or exports, shall be for the use of the Treasury of the United States; and all such laws shall be subject to the revision and control of the Congress.

3. No State shall, without the consent of Congress, lay any duty of tonnage, keep troops, or ships of war in time of peace, enter into any agreement or compact with another State, or with a foreign power, or engage in war, unless actually invaded, or in such imminent danger as will not admit of delay.

ARTICLE II.

Section 1 — President: his term of office. Electors of President; number and how appointed. Electors to vote on same day. Qualification of President. On whom his duties devolve in case of his removal, death, etc. President's compensation. His oath of office.

1. The Executive power shall be vested in a President of the United States of America. He shall hold his office during the term of four years, and together with the Vice President, chosen for the same term, be elected as follows

2. Each State shall appoint, in such manner as the Legislature thereof may direct, a number of electors, equal to the whole number of Senators and Representatives to which the State may be entitled in the Congress: but no Senator or Representative, or person holding an office of trust or profit under the United States, shall be appointed an elector.

(The electors shall meet in their respective States, and vote by ballot for two persons, of whom one at least shall not be an inhabitant of the same State with themselves. And they shall make a list of all the persons voted for, and of the number of votes for each; which list they shall sign and certify, and transmit sealed to the seat of the Government of the United States, directed to the President of the Senate. The President of the Senate shall, in the presence of the Senate and House of Representatives, open all the certificates, and the votes shall then be counted. The person having the greatest number of votes shall be the President, if such number be a majority of the whole number of electors appointed; and if there be more than one who have such majority, and have an equal number of votes, then the House of Representatives shall immediately choose by ballot one of them for President; and if no person have a majority, then from the five highest on the list the said House shall in like manner choose the President. But in choosing the President, the votes shall be taken by States, the representation from each State having one vote; a quorum for this purpose shall consist of a member or members from two-thirds of the States, and a majority of all the States shall be necessary to a choice. In every case, after the choice of the President, the person having the greatest number of votes of the electors shall be the Vice President. But if there should remain two or more who have equal votes, the Senate shall choose from them by ballot the Vice President.)

(This clause was superseded by Amendment XII.)

3. The Congress may determine the time of choosing the electors, and the day on which they shall give their votes; which day shall be the same throughout the United States.

4. No person except a natural born citizen, or a citizen of the United States, at the time of the adoption of this Constitution, shall be eligible to the office of President; neither shall any person be eligible to that office who shall not have attained to the age of thirty-five years, and been fourteen years a resident within the United States.

(For qualification of the Vice President, see Amendment XII.)

5. In case of the removal of the President from office, or of his death, resignation, or inability to discharge the powers and duties of the said office, the same shall devolve on the Vice President, and the Congress may by law provide for the case of removal, death, resignation or inability both of the President and Vice President, declaring what officer shall then act as President, and such officer shall act accordingly, until the disability be removed, or a President shall be elected.

(This clause has been modified by Amendment XX, sections 3 and 4).

6. *The President shall, at stated times, receive for his services, a compensation, which shall neither be increased nor diminished during the period for which he shall have been elected, and he shall not receive within that period any other emolument from the United States, or any of them.*

7. *Before he enter on the execution of his office, he shall take the following oath or affirmation:*
"I do solemnly swear (or affirm) that I will faithfully execute the office of President of the United States, and will to the best of my ability, preserve, protect and defend the Constitution of the United States."

Section 2 — President to be Commander-in-Chief. He may require opinions of cabinet officers, etc., may pardon. Treaty-making power. Nomination of certain officers. When President may fill vacancies.

1. The President shall be Commander-in-Chief of the Army and Navy of the United States and of the militia of the several States, when called into the actual service of the United States; he may require the opinion, in writing, of the principal officer in each of the executive departments, upon any subject relating to the duties of their respective offices, and he shall have power to grant reprieves and pardons for offenses against the United States, except in cases of impeachment.

2. He shall have power, by and with the advice and consent of the Senate, to make treaties, provided two-thirds of the Senators present concur; and he shall nominate, and by and with the advice and consent of the Senate, shall appoint ambassadors, other public ministers and consuls, judges of the Supreme Court, and all other officers of the United States, whose appointments are not herein otherwise provided for, and which shall be established by law: but the Congress may by law vest the appointment of such inferior officers, as they think proper, in the President alone, in the courts of law, or in the heads of departments.

3. The President shall have power to fill up all vacancies that may happen during the recess of the Senate, by granting commissions, which shall expire at the end of their next session.

Section 3 — President shall communicate to Congress. He may convene and adjourn Congress, in case of disagreement, etc. Shall receive ambassadors, execute laws, and commission officers.

He shall from time to time give to the Congress information of the state of the Union, and recommend to their consideration such measures as he shall judge necessary and expedient; he may, on extraordinary occasions, convene both Houses, or either of them, and in case of disagreement between them, with respect to the time of adjournment, he may adjourn them to such time as he shall think proper; he shall receive ambassadors and other public ministers; he shall take care that the laws be faithfully executed, and shall commission all the officers of the United States.

Section 4 — All civil offices forfeited for certain crimes.

The President, Vice President, and all civil officers of the United States, shall be removed from office on impeachment for, and conviction of, treason, bribery, or other high crimes and misdemeanors.

ARTICLE III.

Section 1 — Judicial powers, Tenure. Compensation.

The judicial power of the United States, shall be vested in one Supreme Court, and in such inferior courts as the Congress may from time to time ordain and establish. The judges, both of the Supreme and inferior courts, shall hold their offices during good behavior, and shall at stated times, receive for their services, a compensation, which shall not be diminished during their continuance in office.

Section 2 — Judicial power; to what cases it extends. Original jurisdiction of Supreme Court; appellate jurisdiction. Trial by jury, etc. Trial, where.

1. The judicial power shall extend to all cases, in law and equity, arising under this Constitution, the laws of the United States, and treaties made, or which shall be made, under their authority; to all cases affecting ambassadors, other public ministers and consuls; to all cases of admiralty and maritime jurisdiction; to controversies to which the United States shall be a party; to controversies between two or more States; between a State and citizens of another State; between citizens of different States, between citizens of the same State claiming lands under grants of different States, and between a State, or the citizens thereof, and foreign states, citizens or subjects.

(This section is modified by Amendment XI.)

2. In all cases affecting ambassadors, other public ministers and consuls, and those in which a State shall be party, the Supreme Court shall have original jurisdiction. In all the other cases before mentioned, the Supreme Court shall have appellate jurisdiction, both as to law and fact, with such exceptions, and under such regulations as the Congress shall make.

3. The trial of all crimes, except in cases of impeachment, shall be by jury; and such trial shall be held in the State where the said crimes shall have been committed; but when not committed within any State, the trial shall be at such place or places as the Congress may by law have directed.

Section 3 — Treason Defined, Proof of Punishment of.

1. Treason against the United States, shall consist only in levying war against them, or in adhering to their enemies, giving them aid and comfort. No persons shall be convicted of treason unless on the testimony of two witnesses to the same overt act, or on confession in open court.

2. The Congress shall have power to declare the punishment of treason, but no attainder of treason shall work corruption of blood, or forfeiture except during the life of the person attainted.

ARTICLE IV.

Section 1 — Each State to give credit to the public acts, etc., of every other State.

Full faith and credit shall be given in each State to the public acts, records, and judicial proceedings of every other State. And the Congress may by general laws prescribe the manner in which such acts, records and proceedings shall be proved, and the effect thereof.

Section 2 — Privileges of citizens of each State. Fugitives from justice to be delivered up. Persons held to service having escaped, to be delivered up.

1. The citizens of each State shall be entitled to all privileges and immunities of citizens in the several States.

2. A person charged in any State with treason, felony, or other crime, who shall flee from justice, and be found in another State, shall on demand of the Executive authority of the State from which he fled, be delivered up, to be removed to the State having jurisdiction of the crime.

(3. No person held to service or labor in one State, under the laws thereof, escaping into another, be discharged from such service or labor, but shall be delivered up on claim of the party to whom such service or labor may be due.) (This clause was superseded by Amendment XIII.)

Section 3 — Admission of new States, Power of Congress over territory and other property.

1. New States may be admitted by the Congress into this Union; but no new State shall be formed or erected within the jurisdiction of any other State; nor any State be formed by the junction of two or more States, or parts of States, without the consent of the Legislatures of the States concerned as well as of the Congress.

2. The Congress shall have power to dispose of and make all needful rules and regulations respecting the territory or other property belonging to the United States; and nothing in this Constitution shall be so construed as to prejudice any claims of the United States, or of any particular State.

Section 4 — Republican form of government guaranteed. Each state to be protected.

The United States shall guarantee to every State in this Union a Republican form of government, and shall protect each of them against invasion; and on application of the Legislature, or of the Executive (when the Legislature cannot be convened) against domestic violence.

ARTICLE V.

Constitution: how amended; proviso.

The Congress, whenever two-thirds of both Houses shall deem it necessary, shall propose amendments to this Constitution, or, on the application of the Legislatures of two-thirds of the several States, shall call a convention for proposing amendments, which, in either case, shall be valid to all intents and purposes, as part of this Constitution, when ratified by the Legislatures of three-fourths of the several states, or by conventions in three-fourths thereof, as the one or the other mode of ratification may be proposed by the Congress; provided that no amendment which may be made prior to the year one thousand eight hundred and eight shall in any manner affect the first and fourth clauses in the Ninth Section of the First Article; and that no State, without its consent, shall be deprived of its equal suffrage in the Senate.

ARTICLE VI.

Certain debts, etc., declared valid, Supremacy of Constitution, treaties, and laws of the United States. Oath to support Constitution, by whom taken. No religious test.

1. All debts contracted and engagements entered into, before the adoption of this Constitution, shall be as valid against the United States under this Constitution, as under the Confederation.

2. This Constitution, and the laws of the United States which shall be made in pursuance thereof; and all treaties made, or which shall be made, under the authority of the United States, shall be the supreme law of the land; and the judges in every State shall be bound thereby, any thing in the Constitution or laws of any State to the contrary notwithstanding.

3. The Senators and Representatives before mentioned, and the members of the several State Legislatures, and all executive and judicial officers, both of the United States and of the several States, shall be bound by oath or affirmation, to support this Constitution; but no religious test shall ever be required as a qualification to any office or public trust under the United States.

ARTICLE VII.

What ratification shall establish Constitution.

The ratification of the Conventions of nine States, shall be sufficient for the establishment of this Constitution between the States so ratifying the same.

Done in convention by the unanimous consent of the States present the Seventeenth day of September in the year of our Lord one thousand seven hundred and eighty seven, and of the independence of the United States of America the Twelfth. In witness whereof we have hereunto subscribed our names.

George Washington, President and deputy from Virginia.
New Hampshire — John Langdon, Nicholas Gilman.
Massachusetts — Nathaniel Gorham, Rufus King.
Connecticut — Wm. Saml. Johnson, Roger Sherman.
New York — Alexander Hamilton.
New Jersey — Will Livingston, David Brearley, Wm. Paterson, Jonah Dayton.
Pennsylvania — B. Franklin, Thomas Mifflin, Robt. Morris, Geo. Clymer, Thos. FitzSimons, Jared Ingersoll, James Wilson, Gouv. Morris.
Delaware — Geo. Read, Gunning Bedford Jun., John Dickinson, Richard Bassett, Jacob Broom.
Maryland — James McHenry, Daniel of Saint Thomas Jenifer, Danl. Carroll.
Virginia — John Blair, James Madison Jr.
North Carolina — Wm. Blount, Rich'd. Dobbs Spaight, Hugh Williamson.
South Carolina — J. Rutledge, Charles Cotesworth Pinckney, Charles Pinckney, Pierce Butler.
Georgia — William Few, Abr. Baldwin.
Attest: William Jackson, Secretary.

TEN ORIGINAL AMENDMENTS: THE BILL OF RIGHTS

In force Dec. 15, 1791

(The First Congress, at its first session in the City of New York, Sept. 25, 1789, submitted to the states 12 amendments to clarify certain individual and state rights not named in the Constitution. They are generally called the Bill of Rights.

(Influential in framing these amendments was the Declaration of Rights of Virginia, written by George Mason (1725-1792) in 1776. Mason, a Virginia delegate to the Constitutional Convention, did not sign the Constitution and opposed its ratification on the ground that it did not sufficiently oppose slavery or safeguard individual rights.

(In the preamble to the resolution offering the proposed amendments, Congress said: "The conventions of a number of the States having at the time of their adopting the Constitution, expressed a desire, in order to prevent misconstruction or abuse of its powers, that further declaratory and restrictive clauses should be added, and as extending the ground of public confidence in the government will best insure the beneficent ends of its institution, be it resolved," etc.

(Ten of these amendments now commonly known as one to 10 inclusive, but originally 3 to 12 inclusive, were ratified by the states as follows: New Jersey, Nov. 20, 1789; Maryland, Dec. 19, 1789; North Carolina, Dec. 22, 1789; South Carolina, Jan. 19, 1790; New Hampshire, Jan. 25, 1790; Delaware, Jan. 28, 1790; New York, Feb. 24, 1790; Pennsylvania, Mar. 10, 1790; Rhode Island, June 7, 1790; Vermont, Nov. 3, 1791; Virginia, Dec. 15, 1791; Massachusetts, Mar. 2, 1939; Georgia, Mar. 8, 1939; Connecticut, Apr. 19, 1939. These original 10 ratified amendments follows as Amendments I to X inclusive.

(Of the two original proposed amendments which were not ratified by the necessary number of states, the first related to apportionment of Representatives; the second, to compensation of members.)

AMENDMENT I.

Religious establishment prohibited. Freedom of speech, of the press, and right to petition.

Congress shall make no law respecting an establishment of religion, or prohibiting the free exercise thereof; or abridging the freedom of speech, or of the press; or the right of the people peaceably to assemble, and to petition the Government for a redress of grievances.

AMENDMENT II.

Right to keep and bear arms.

A well-regulated militia, being necessary to the security of a free State, the right of the people to keep and bear arms, shall not be infringed.

AMENDMENT III.

Conditions for quarters for soldiers.

No soldier shall, in time of peace be quartered in any house, without the consent of the owner, nor in time of war, but in a manner to be prescribed by law.

AMENDMENT IV.

Right of search and seizure regulated.

The right of the people to be secure in their persons, houses, papers, and effects, against unreasonable searches and seizures, shall not be violated, and no warrants shall issue, but upon probable cause, supported by oath or affirmation, and particularly describing the place to be searched, and the persons or things to be seized.

AMENDMENT V.

Provisions concerning prosecution. Trial and punishment — private property not to be taken for public use without compensation.

No person shall be held to answer for a capital, or otherwise infamous crime, unless on a presentment or indictment of a Grand Jury, except in cases arising in the land or naval forces, or in the militia, when in actual service in time of war or public danger; nor shall any person be subject for the same offense to be twice put in jeopardy of life or limb; nor shall be compelled in any criminal case to be a witness against himself, nor be deprived of life, liberty, or property, without due process of law; nor shall private property be taken for public use without just compensation.

AMENDMENT VI.

Right to speedy trial, witnesses, etc.

In all criminal prosecutions, the accused shall enjoy the right to a speedy and public trial, by an impartial jury of the State and district wherein the crime shall have been committed, which district shall have been previously ascertained by law, and to be informed of the nature and cause of the accusation; to be confronted with the witnesses against him; to have compulsory process for obtaining witnesses in his favor, and to have the assistance of counsel for his defense.

AMENDMENT VII.

Right of trial by jury.

In suits at common law, where the value in controversy shall exceed twenty dollars, the right of trial by jury shall be preserved, and no fact tried by a jury shall be otherwise reexamined in any court of the United States, than according to the rules of the common law.

AMENDMENT VIII.

Excessive bail or fines and cruel punishment prohibited.

Excessive bail shall not be required, nor excessive fines imposed, nor cruel and unusual punishments inflicted.

AMENDMENT IX.

Rule of construction of Constitution.

The enumeration in the Constitution, of certain rights, shall not be construed to deny or disparage others retained by the people.

AMENDMENT X.

Rights of States under Constitution.

The powers not delegated to the United States by the Constitution, nor prohibited by it to the States, are reserved to the States respectively, or to the people.

AMENDMENTS SINCE THE BILL OF RIGHTS

AMENDMENT XI.

Judicial powers constructed.

The judicial power of the United States shall not be construed to extend to any suit in law or equity, commenced or prosecuted against one of the United States by citizens of another State, or by citizens or subjects of any foreign state.

(This amendment was proposed to the Legislatures of the several States by the Third Congress on March 4, 1794, and was declared to have been ratified in a message from the President to Congress, dated Jan. 8, 1798.

(It was on Jan. 5, 1798, that Secretary of State Pickering received from 12 of the States authenticated ratifications, and informed President John Adams of that fact.

(As a result of later research in the Department of State, it is now established that Amendment XI became part of the Constitution on Feb. 7, 1795, for on that date it had been ratified by 12 States as follows:

(1. New York, Mar. 27, 1794. 2. Rhode Island, Mar. 31, 1794, 3. Connecticut, May 8, 1794. 4. New Hampshire, June 16, 1794. 5. Massachusetts, June 26, 1794. 6. Vermont, between Oct. 9, 1794, and Nov. 9, 1794. 7. Virginia, Nov. 18, 1794. 8. Georgia, Nov. 29, 1794. 9. Kentucky, Dec. 7, 1794. 10. Maryland, Dec.

26, 1794. 11. Delaware, Jan. 23, 1795. 12. North Carolina, Feb. 7, 1795.

(On June 1, 1796, more than a year after Amendment XI had become a part of the Constitution (but before anyone was officially aware of this), Tennessee had been admitted as a State; but not until Oct. 16, 1797, was a certified copy of the resolution of Congress proposing the amendment sent to the Governor of Tennessee (John Sevier) by Secretary of State Pickering, whose office was then at Trenton, New Jersey, because of the epidemic of yellow fever at Philadelphia; it seems, however, that the Legislature of Tennessee took no action on Amendment XI, owing doubtless to the fact that public announcement of its adoption was made soon thereafter.

(Besides the necessary 12 States, one other, South Carolina, ratified Amendment XI, but this action was not taken until Dec. 4, 1797; the two remaining States, New Jersey and Pennsylvania, failed to ratify.)

AMENDMENT XII.

Manner of choosing President and Vice-President.

(Proposed by Congress Dec. 9, 1803; ratification completed June 15, 1804.)

The Electors shall meet in their respective States and vote by ballot for President and Vice-President, one of whom, at least, shall not be an inhabitant of the same State with themselves; they shall name in their ballots the person voted for as President, and in distinct ballots the person voted for as Vice-President, and they shall make distinct lists of all persons voted for as President, and of all persons voted for as Vice-President, and of the number of votes for each, which lists they shall sign and certify, and transmit sealed to the seat of the Government of the United States, directed to the President of the Senate; the President of the Senate shall, in the presence of the Senate and House of Representatives, open all the certificates and the votes shall then be counted; — The person having the greatest number of votes for President, shall be the President, if such number be a majority of the whole number of Electors appointed; and if no person have such majority, then from the persons having the highest numbers not exceeding three on the list of those voted for as President, the House of Representatives shall choose immediately, by ballot, the President. But in choosing the President, the votes shall be taken by States, the representation from each State having one vote; a quorum for this purpose shall consist of a member or members from two-thirds of the States, and a majority of all the States shall be necessary to a choice. (And if the House of Representatives shall not choose a President whenever the right of choice shall devolve upon them, before the fourth day of March next following, then the Vice-President shall act as President, as in case of the death or other constitutional disability of the President.) (The words in parentheses were superseded by Amendment XX, section 3.) The person having the greatest number of votes as Vice-President, shall be the Vice-President, if such number be a majority of the whole number of Electors appointed, and if no person have a majority, then from the two highest numbers on the list, the Senate shall choose the Vice-President; a quorum for the purpose shall consist of two-thirds of the whole number of Senators, and a majority of the whole number shall be necessary to a choice. But no person constitutionally ineligible to the office of President shall be eligible to that of Vice-President of the United States.

THE RECONSTRUCTION AMENDMENTS

(Amendments XIII, XIV, and XV are commonly known as the Reconstruction Amendments, inasmuch as they followed the Civil War, and were drafted by Republicans who were bent on imposing their own policy of reconstruction on the South. Post-bellum legislatures there — Mississippi, South Carolina, Georgia, for example — had set up laws which, it was charged, were contrived to perpetuate Negro slavery under other names.)

AMENDMENT XIII.

Slavery abolished.

(Proposed by Congress Jan. 31, 1865; ratification completed Dec. 18, 1865. The amendment, when first proposed by a resolution in Congress, was passed by the Senate, 38 to 6, on Apr. 8, 1864, but was defeated in the House, 95 to 66 on June 15, 1864. On reconsideration by the House, on Jan. 31, 1865, the resolution passed, 119 to 56. It was approved by President Lincoln on Feb. 1, 1865, although the Supreme Court had decided in 1798 that the President has nothing to do with the proposing of amendments to the Constitution, or their adoption.)

1. Neither slavery nor involuntary servitude, except as a punishment for crime whereof the party shall have been duly convicted, shall exist within the United States or any place subject to their jurisdiction.

2. Congress shall have power to enforce this article by appropriate legislation.

AMENDMENT XIV.

Citizenship rights not to be abridged.

(The following amendment was proposed to the Legislatures of the several states by the 39th Congress, June 13, 1866, and was declared to have been ratified in a proclamation by the Secretary of State, July 28, 1868.

(The 14th amendment was adopted only by virtue of ratification subsequent to earlier rejections. Newly constituted legislatures in both North Carolina and South Carolina (respectively July 4 and 9, 1868), ratified the proposed amendment, although earlier legislatures had rejected the proposal. The Secretary of State issued a proclamation, which, though doubtful as to the effect of attempted withdrawals by Ohio and New Jersey, entertained no doubt as to the validity of the ratification by North and South Carolina. The following day (July 21, 1868), Congress passed a resolution which declared the 14th Amendment to be a part of the Constitution and directed the Secretary of State so to promulgate it. The Secretary waited, however, until the newly constituted Legislature of Georgia had ratified the amendment, subsequent to an earlier rejection, before the promulgation of the ratification of the new amendment.)

1. All persons born or naturalized in the United States, and subject to the jurisdiction thereof, are citizens of the United States and of the State wherein they reside. No State shall make or enforce any law which shall abridge the privileges or immunities of citizens of the United States; nor shall any State deprive any person of life, liberty, or property, without due process of law; nor deny to any person within its jurisdiction the equal protection of the laws.

2. Representatives shall be apportioned among the several States according to their respective numbers, counting the whole number of persons in each State, excluding Indians not taxed. But when the right to vote at any election for the choice of Electors for President and Vice-President of the United States, Representatives in Congress, the executive and judicial officers of a State, or the members of the Legislature thereof, is denied to any of the male inhabitants of such State, being twenty-one years of age, and citizens of the United States, or in any way abridged, except for participation in rebellion, or other crime, the basis of representation therein shall be reduced in the proportion which the number of such male citizens shall bear to the whole number of male citizens twenty-one years of age in such State.

3. No person shall be a Senator or Representative in Congress, or Elector of President and Vice-President, or hold any office, civil or military, under the United States, or under any State, who, having previously taken an oath, as a member of Congress, or as an officer of the United States, or as a member of any State Legislature, or as an executive or judicial officer of any State, to support the Constitution of the United States, had engaged in insurrection or rebellion against the state, or given aid or comfort to the enemies thereof. But Congress may by a vote of two-thirds of each House, remove such disability.

4. The validity of the public debt of the United States, authorized by law, including debts incurred for payment of pensions and bounties for services in suppressing insurrection or rebellion, shall not be questioned. But neither the United States nor any State shall assume or pay any debt or obligation incurred in aid of insurrection or rebellion against the United States, or any claim for the loss or emancipation of any slave; but all such debts, obligations and claims, shall be held illegal and void.

5. The Congress shall have power to enforce, by appropriate legislation, the provisions of this article.

AMENDMENT XV.

Race no bar to voting rights.

(The following amendment was proposed to the legislatures of the several States by the 40th Congress, Feb. 26, 1869, and was declared to have been ratified in a proclamation by the Secretary of State, Mar. 30, 1870.)

1. The right of citizens of the United States to vote shall not be denied or abridged by the United States or by any State on account of race, color, or previous condition of servitude.

2. The Congress shall have power to enforce this article by appropriate legislation.

AMENDMENT XVI.

Income taxes authorized.

(Proposed by Congress July 12, 1909; ratification completed Feb. 3, 1913.)

The Congress shall have power to lay and collect taxes on incomes, from whatever sources derived, without apportionment among the several States, and without regard to any census or enumeration.

AMENDMENT XVII.

United States Senators to be elected by direct popular vote.

(Proposed by Congress May 13, 1912; ratification completed Apr. 8, 1913.)

1. The Senate of the United States shall be composed of two Senators from each State, elected by the people thereof, for six years; and each Senator shall have one vote. The electors in each State shall have the qualifications requisite for electors of the most numerous branch of the State Legislatures.

2. When vacancies happen in the representation of any State in the Senate, the executive authority of such State shall issue writs of election to fill such vacancies: Provided, That the Legislature of any State may empower the Executive thereof to make temporary appointments until the people fill the vacancies by election as the Legislature may direct.

3. This amendment shall not be so construed as to affect the election or term of any Senator chosen before it becomes valid as part of the Constitution.

AMENDMENT XVIII.

Liquor prohibition amendment.

(Proposed by Congress Dec. 18, 1917; ratification completed Jan. 16, 1919. Repealed by Amendment XXI, effective Dec. 5, 1933.)

(1. After one year from the ratification of this article the manufacture, sale, or transportation of intoxicating liquors, within, the importation thereof into, or the exportation thereof from the United States and all territory subject to the jurisdiction thereof, for beverage purposes is hereby prohibited.

(2. The Congress and the several States shall have concurrent power to enforce this article by appropriate legislation.

(3. This article shall be inoperative unless it shall have been ratified as an amendment to the Constitution by the Legislatures of the several States, as provided in the Constitution, within seven years from the date of the submission hereof to the States by the Congress.)

(The total vote in the Senates of the various States was 1,310 for, 237 against — 84.6% dry. In the lower houses of the States the vote was 3,782 for, 1,035 against — 78.5% dry.

(The amendment ultimately was adopted by all the States except Connecticut and Rhode Island.)

AMENDMENT XIX.

Giving nationwide suffrage to women.

(Proposed by Congress June 4, 1919; ratification certified by Secretary of State, Aug. 26, 1920.)

1. The right of citizens of the United States to vote shall not be denied or abridged by the United States or by any State on account of sex.

2. Congress shall have power to enforce this Article by appropriate legislation.

AMENDMENT XX.

Terms of President and Vice President to begin on Jan. 20; those of Senators, Representatives, Jan. 3.

(Proposed by Congress Mar. 2, 1932; ratification completed Jan. 23, 1933.)

1. The terms of the President and Vice President shall end at noon on the 20th day of January, and the terms of Senators and Representatives at noon on the 3rd day of January, of the years in which such terms would have ended if this article had not been ratified; and the terms of their successors shall then begin.

2. The Congress shall assemble at least once in every year, and such meeting shall begin at noon on the 3rd day of January, unless they shall by law appoint a different day.

3. If, at the time fixed for the beginning of the term of the President, the President elect shall have died, the Vice President elect shall become President. If a President shall not have been chosen before the time fixed for the beginning of his term, or if the President elect shall have failed to qualify, then the Vice President elect shall act as President until a President shall have qualified; and the Congress may by law provide for the case wherein neither a President elect nor a Vice President shall have qualified, declaring who shall then act as President, or the manner in which one who is to act shall be selected, and such person shall act accordingly until a President or Vice President shall have qualified.

4. The Congress may by law provide for the case of the death of any of the persons from whom the House of Representatives may choose a President whenever the right of choice shall have de-

volved upon them, and for the case of the death of any of the persons from whom the Senate may choose a Vice President whenever the right of choice shall have devolved upon them.

5. Sections 1 and 2 shall take effect on the 15th day of October following the ratification of this article (Oct., 1933).

6. This article shall be inoperative unless it shall have been ratified as an amendment to the Constitution by the Legislatures of three-fourths of the several States within seven years from the date of its submission.

AMENDMENT XXI.

Repeal of Amendment XVIII.

(Proposed by Congress Feb. 20, 1933; ratification completed Dec. 5, 1933.)

1. The eighteenth article of amendment to the Constitution of the United States is hereby repealed.

2. The transportation or importation into any State, Territory, or Possession of the United States for delivery or use therein of intoxicating liquors, in violation of the laws thereof, is hereby prohibited.

3. This article shall be inoperative unless it shall have been ratified as an amendment to the Constitution byconventions in the several States, as provided in the Constitution, within seven years from the date of the submission hereof to the States by the Congress.

AMENDMENT XXII.

Limiting Presidential terms of office.

(Proposed by Congress Mar. 21, 1947; ratification completed Feb. 27, 1951.)

1. No person shall be elected to the office of the President more than twice, and no person who has held the office of President, or acted as President, for more than two years of a term to which some other person was elected President shall be elected to the office of the President more than once. But this Article shall not apply to any person holding the office of President when this Article was proposed by the Congress, and shall not prevent any person who may be holding the office of President, or acting as President, during the term within which this Article becomes operative from holding the office of President or acting as President during the remainder of such term.

2. This article shall be inoperative unless it shall have been ratified as an amendment to the Constitution by the Legislatures of three-fourths of the several States within seven years from the date of its submission to the States by the Congress.

AMENDMENT XXIII.

Presidential vote for District of Columbia.

(Proposed by Congress June 17, 1960; ratification completed Mar. 29, 1961.)

1. The District constituting the seat of Government of the United States shall appoint in such manner as the Congress may direct:

A number of electors of President and Vice President equal to the whole number of Senators and Representatives in Congress to which the District would be entitled if it were a State, but in no event more than the least populous State, they shall be in addition to those appointed by the States, but they shall be considered, for the purposes of the election of President and Vice President, to be electors appointed by a State; and they shall meet in the District and perform such duties as provided by the twelfth article of amendment.

2. The Congress shall have power to enforce this article by appropriate legislation.

AMENDMENT XXIV.

Barring poll tax in federal elections.

(Proposed by Congress Aug. 27, 1962; ratification completed Jan. 23, 1964.)

1. The right of citizens of the United States to vote in any primary or other election for President or Vice President, for electors for President or Vice President, or for Senator or Representative in Congress, shall not be denied or abridged by the United States or any State by reason of failure to pay any poll tax or other tax.

2. The Congress shall have power to enforce this article by appropriate legislation.

AMENDMENT XXV.

Presidential disability and succession.

(Proposed by Congress July 6, 1965; ratification completed Feb. 10, 1967.)

1. In case of the removal of the President from office or of his death or resignation, the Vice President shall become President.

2. Whenever there is a vacancy in the office of the Vice President, the President shall nominate a Vice President who shall take office upon confirmation by a majority vote of both houses of Congress.

3. Whenever the President transmits to the President pro tempore of the Senate and the Speaker of the House of Representatives his written declaration that he is unable to discharge the powers and duties of his office, and until he transmits to them a written declaration to the contrary, such powers and duties shall be discharged by the Vice President as Acting President.

4. Whenever the Vice President and a majority of either the principal officers of the executive departments or of such other body as Congress may by law provide, transmit to the President pro tempore of the Senate and the Speaker of the House of Representatives their written declaration that the President is unable to discharge the powers and duties of his office, the Vice President shall immediately assume the powers and duties of the office as Acting President.

Thereafter, when the President transmits to the President pro tempore of the Senate and the Speaker of the House of Representatives his written declaration that no inability exists, he shall resume the powers and duties of his office unless the Vice President and a majority of either the principal officers of the executive department or of such other body as Congress may by law provide, transmit within four days to the President pro tempore of the Senate and the Speaker of the House of Representatives their written declaration that the President is unable to discharge the powers and duties of his office. Thereupon Congress shall decide the issue, assembling within forty-eight hours for that purpose if not in session. If the Congress, within twenty-one days after receipt of the latter written declaration, or, if Congress is not in session, within twenty-one days after Congress is required to assemble, determines by two-thirds vote of both houses that the President is unable to discharge the powers and duties of his office, the Vice President shall continue to discharge the same as Acting President; otherwise, the President shall resume the powers and duties of his office.

AMENDMENT XXVI.

Lowering voting age to 18 years.

(Proposed by Congress Mar. 23, 1971; ratification completed June 30, 1971.)

1. The right of citizens of the United States, who are 18 years of age or older, to vote shall not be denied or abridged by the United States or any state on account of age.

2. The Congress shall have the power to enforce this article by appropriate legislation.

PROPOSED EQUAL RIGHTS AMENDMENT

(Proposed by Congress Mar. 22, 1972; ratification completed, as of mid-1977, by 35 states, not ratified by 6, defeated in 9; needed total of 38 for adoption before deadline, Mar. 22, 1979.)

1. Equality of rights under the law shall not be denied or abridged by the United States or by any State on account of sex.

2. The Congress shall have the power to enforce, by appropriate legislation, the provisions of this article.

3. This amendment shall take effect two years after the date of ratification.

APPENDIX B

FEDERAL LAWS PERTAINING TO
EQUAL EMPLOYMENT

The Civil Rights Act of 1964, as amended by the Equal Employment Act of 1972, contains many of the provisions basic to understanding and applying the federal standards of fair employment practices. Following are printed sections of this act that are of particular interest to members of the fire service.

[¶ 3021]   **GENERAL PROHIBITION — PARTICIPA-
TION IN OR DENIAL OF BENE-
FITS — DISCRIMINATION ON
BASIS OF RACE, COLOR
OR NATIONAL
ORIGIN**

Sec. 601. No person in the United States shall, on the ground of race, color, or national origin, be excluded from participation in, be denied the benefits of, or be subjected to discrimination under any program or activity receiving Federal financial assistance. [July 2, 1964, P. L. 88-352, Title VI, § 601, 78 Stat. 252, 42 U.S.C. § 2000d.]

[¶ 3023]   **JUDICIAL REVIEW — ADMINISTRATIVE
PROCEDURE ACT**

Sec. 603. Any department or agency action taken pursuant to section 2000d-1 of this title shall be subject to such judicial review as may otherwise be provided by law for similar action taken by such department or agency on other grounds. In the case of action, not otherwise subject to judicial review, terminating or refusing to grant or to continue financial assistance upon a finding of failure to comply with a requirement imposed pursuant to section 2000d-1 of this title, any person aggrieved (including any State or political subdivision thereof and any agency of either) may obtain judicial review of such action in accordance with section 1009 of Title 5, and such action shall not be deemed committed to unreviewable agency discretion within the meaning of that section. [July 2, 1964, P. L. 88-352, Title VI, § 603, 78 Stat. 253, 42 U.S.C. § 2000d-2.]

[¶ 3050]   **DISCRIMINATION BECAUSE OF RACE,
COLOR, RELIGION, SEX, OR
NATIONAL ORIGIN**

[¶ 3051]   **[Unlawful Practices of Employers]**

Sec. 703. (a) It shall be an unlawful employment practice for an employer —
(1) to fail or refuse to hire or to discharge any individual, or otherwise to discriminate against any individual with respect to his compensation, terms, conditions, or privileges of

employment, because of such individual's race, color, religion, sex, or national origin; [July 2, 1964, P. L. 88-352, Title VII, 78 Stat. 255, 42 U.S.C. § 2000e-2(a) (1)] or

(2) to limit, segregate, or classify his employees or applicants for employment in any way which would deprive or tend to deprive any individual of employment opportunities or otherwise adversely affect his status as an employee, because of such individual's race, color, religion, sex, or national origin. [As amended March 24, 1972, P. L. 92-261, Title VII, 86 Stat. 109, 42 U.S.C. § 2000e-2(a) (2).]

[¶ 3055]   [Religion, Sex, or National Origin as
Occupational Qualification]

(e) Notwithstanding any other provision of this title, (1) it shall not be an unlawful employment practice for an employer to hire and employ employees, for an employment agency to classify, or refer for employment any individual, for a labor organization to classify its membership or to classify or refer for employment any individual, or for an employer, labor organization, or joint labor-management committee controlling apprenticeship or other training or retraining programs to admit or employ any individual in any such program, on the basis of his religion, sex, or national origin in those certain instances where religion, sex, or national origin is a bona fide occupational qualification reasonably necessary to the normal operation of that particular business or enterprise, and (2) it shall not be an unlawful employment practice for a school, college, university, or other educational institution or institution of learning to hire and employ employees of a particular religion if such school, college, university, or other educational institution or institution of learning is, in whole or in substantial part, owned, supported, controlled, or managed by a particular religion or by a particular religious corporation, association, or society, or if the curriculum of such school, college, university, or other educational institution or institution of learning is directed toward the propagation of a particular religion. [July 2, 1964, P. L. 88-352, Title VII, 78 Stat. 255, 42 U.S.C. § 2000e-2(e).]

[¶ 3058]   [Seniority, Merit or Incentive System — Different
Locations — Equal Pay Law]

(h) Notwithstanding any other provisions of this title, it shall not be an unlawful employment practice for an employer to apply different standards of compensation, or different terms, conditions, or privileges of employment pursuant to a bona fide seniority or merit system, or a system which measures earnings by quantity or quality of production or to employees who work in different locations, provided that such differences are not the result of an intention to discriminate because of race, color, religion, sex, or national origin, nor shall it be an unlawful employment practice for an employer to give and to act upon the results of any professionally developed ability test provided that such test, its administration or action upon the results is not designed, intended or used to discriminate because of race, color, religion, sex or national origin. It shall not be an unlawful employment practice under this title for any employer to differentiate upon the basis of sex in determining the amount of the wages or compensation paid or to be paid to employees of such employer if such differentiation is authorized by the provisions of section 6(d) of the Fair Labor Standards Act of 1938, as amended (29 U.S.C. 206(d)). [July 2, 1964, P. L. 88-352, Title VII, 78 Stat. 255, 42 U.S.C. § 2000e-2(h).]

[¶ 3060]   **[Imbalance in Number or Percentage
of Employees]**

(j) Nothing contained in this title shall be interpreted to require any employer, employment agency, labor organization, or joint labor-management committee subject to this title
to grant preferential treatment to any individual or to any group because of the race, color,
religion, sex, or national origin of such individual or group on account of an imbalance
which may exist with respect to the total number or percentage of persons of any race, color,
religion, sex, or national origin employed by any employer, referred or classified for
employment by any employment agency or labor organization, admitted to membership or
classified by any labor organization, or admitted to, or employed in, any apprenticeship or
other training program, in comparison with the total number or percentage of persons of
such race, color, religion, sex, or national origin in any community, State, section, or other
area, or in the available work force in any community, State, section, or other area. [July 2,
1964, P. L. 88-352, Title VII, 78 Stat. 255, 42 U.S.C. § 2000e-2(j).]

[¶ 3062]   **[Opposition to Unlawful Practices — Filing of
Charges — Participation in Proceedings]**

Sec. 704. (a) It shall be an unlawful employment practice for an employer to discriminate
against any of his employees or applicants for employment, for an employment agency, or
joint labor-management committee controlling apprenticeship or other training or retraining, including on-the-job training programs, to discriminate against any individual, or for a
labor organization to discriminate against any member thereof or applicant for membership, because he has opposed any practice made an unlawful employment practice by this
title, or because he has made a charge, testified, assisted, or participated in any manner in
an investigation, proceeding, or hearing under this title. [As amended March 24, 1972,
P. L. 92-261, Title VII, 86 Stat. 109, 42 U.S.C. § 2000e-3(a).]

[¶ 3063]   **[Notices and Advertisements for Employment]**

(b) It shall be an unlawful employment practice for an employer, labor organization,
employment agency, or joint labor-management committee controlling apprenticeship or
other training or retraining, including on-the-job training programs, to print or publish or
cause to be printed or published any notice or advertisement relating to employment by
such an employer or membership in or any classification or referral for employment by such
a labor organization, or relating to any classification or referral for employment by such an
employment agency, or relating to admission to, or employment in, any program established to provide apprenticeship or other training by such a joint labor-management committee, indicating any preference, limitation, specification, or discrimination, based on
race, color, religion, sex, or national origin, except that such a notice or advertisement may
indicate a preference, limitation, specification, or discrimination based on religion, sex, or
national origin when religion, sex, or national origin is a bona fide occupational qualifica-

tion for employment. [As amended March 24, 1972, P. L. 92-261, Title VII, 86 Stat. 109, 42 U.S.C. § 2000e-3(b).]

[¶ 3064]   EQUAL EMPLOYMENT OPPORTUNITY
                            COMMISSION

[¶3065]   [Creation of Commission — Membership]

Sec. 705. (a) There is hereby created a Commission to be known as the Equal Employment Opportunity Commission, which shall be composed of five members, not more than three of whom shall be members of the same political party. Members of the Commission shall be appointed by the President by and with the advice and consent of the Senate for a term of five years. Any individual chosen to fill a vacancy shall be appointed only for the unexpired term of the member whom he shall succeed, and all members of the Commission shall continue to serve until their successors are appointed and qualified, except that no such member of the Commission shall continue to serve (1) for more than sixty days when the Congress is in session unless a nomination to fill such vacancy shall have been submitted to the Senate, or (2) after the adjournment *sine die* of the session of the Senate in which such nomination was submitted. The President shall designate one member to serve as Chairman of the Commission, and one member to serve as Vice Chairman. The Chairman shall be responsible on behalf of the Commission for the administrative operations of the Commission, and, except as provided in subsection (b), shall appoint, in accordance with the provisions of title 5, United States Code, governing appointments in the competitive service, such officers, agents, attorneys, administrative law judges, and employees as he deems necessary to assist it in the performance of its functions and to fix their compensation in accordance with the provisions of chapter 51 and subchapter III of chapter 53 of title 5, United States Code, relating to classification and General Schedule pay rates: Provided, That assignment, removal, and compensation of administrative law judges shall be in accordance with sections 3105, 3344, 5362, and 7521 of title 5, United States Code. [As amended March 24, 1972, P. L. 92-261, Title VII, 86 Stat. 109, 42 U.S.C. § 2000e-4(a), and further amended by P. L. 95-251, March 27, 1978, 92 Stat. 183.]

[¶ 3071]   [Powers]

(g) The Commission shall have power —
    (1) to cooperate with and, with their consent, utilize regional, State, local, and other agencies, both public and private, and individuals;
    (2) to pay to witness whose depositions are taken or who are summoned before the Commission or any of its agents the same witness and mileage fees as are paid to witnesses in the courts of the United States;
    (3) to furnish to persons subject to this title such technical assistance as they may request to further their compliance with this title or an order issued thereunder;
    (4) upon the request of (i) any employer, whose employees or some of them, or (ii) any labor organization, whose members or some of them, refuse or threaten to refuse to cooperate in effectuating the provisions of this title, to assist in such effectuation by conciliation or such other remedial action as is provided by this title;
    (5) to make such technical studies as are appropriate to effectuate the purposes and policies of this title and to make the results of such studies available to the public;
    (6) to intervene in a civil action brought under section 706 by an aggrieved party

against a respondent other than a government, governmental agency or political subdivision. [As amended March 24, 1972, by P. L. 92-261, Title VII, 86 Stat. 110, 42 U.S.C. § 2000e-4(g).]

[¶ 3076]   **[EEOC Authority]**

Sec. 706. (a) The Commission is empowered, as hereinafter provided, to prevent any person from engaging in any unlawful employment practice as set forth in section 703 or 704 of this title. [As amended March 24, 1972, P. L. 92-261, Title VII, 86 Stat. 104, 42 U.S.C. § 2000e-5(a).]

[¶ 3077]   **[Investigation — Secrecy of Information — .**
Findings — Conciliation]

(b) Whenever a charge is filed by or on behalf of a person claiming to be aggrieved, or by a member of the Commission, alleging that an employer, employment agency, labor organization, or joint labor-management committee controlling apprenticeship or other training or retraining, including on-the-job training programs, has engaged in an unlawful employment practice, the Commission shall serve a notice of the charge (including the date, place and circumstances of the alleged unlawful employment practice) on such employer, employment agency, labor organization, or joint labor-management committee (hereinafter referred to as the "respondent") within ten days, and shall make an investigation thereof. Charges shall be in writing under oath or affirmation and shall contain such information and be in such form as the Commission requires. Charges shall not be made public by the Commission. If the Commission determines after such investigation that there is not reasonable cause to believe that the charge is true, it shall dismiss the charge and promptly notify the person claiming to be aggrieved and the respondent of its action. In determining whether reasonable cause exists, the Commission shall accord substantial weight to final findings and orders made by State or local authorities in proceedings commenced under State or local law pursuant to the requirements of subsections (c) and (d). If the Commission determines after such investigation that there is reasonable cause to believe that the charge is true, the Commission shall endeavor to eliminate any such alleged unlawful employment practice by informal methods of conference, conciliation, and persuasion. Nothing said or done during and as a part of such informal endeavors may be made public by the Commission, its officers or employees, or used as evidence in a subsequent proceeding without the written consent of the persons concerned. Any person who makes public information in violation of this subsection shall be fined not more than $1,000 or imprisoned for not more than one year, or both. The Commission shall make its determination on reasonable cause as promptly as possible and, so far as practicable, not later than one hundred and twenty days from the filing of the charge or, where applicable under subsection (c) or (d), from the date upon which the Commission is authorized to take action with respect to the charge. [As amended March 24, 1972, P. L. 92-261, Title VII, 86 Stat. 104, 42 U.S.C. § 2000e-5(b).]

[¶ 3078]   **[Deferral to State Jurisdiction]**

(c) In the case of an alleged unlawful employment practice occurring in a State, or political subdivision of a State, which has a State or local law prohibiting the unlawful employment practice alleged and establishing or authorizing a State or local authority to grant or seek relief from such practice or to institute criminal proceedings with respect thereto upon receiving notice thereof, no charge may be filed under subsection (a) by the person aggrieved before the expiration of sixty days after proceedings have been commenced under the State or local law, unless such proceedings have been earlier terminated, provided that such sixty-day period shall be extended to one hundred and twenty days during the first year after the effective date of such State or local law. If any requirement for the commencement of such proceedings is imposed by a State or local authority other than a requirement of the filing of a written and signed statement of the facts upon which the proceeding is based, the proceeding shall be deemed to have been commenced for the purposes of this subsection at the time such statement is sent by registered mail to the appropriate State or local authority. [As amended March 24, 1972, P. L. 92-261, Title VII, 86 Stat. 104, 42 U.S.C. § 2000e-5(c).]

For more information concerning procedure for filing charges, time limit for filing charges, court jurisdiction, and temporary injunctions, see ¶ 3079 - 3081D.

[¶ 3082]   [Injunctions — Reinstatement — Back Pay]

(g) If the court finds that the respondent has intentionally engaged in or is intentionally engaging in an unlawful employment practice charged in the complaint, the court may enjoin the respondent from engaging in such unlawful employment practice, and order such affirmative action as may be appropriate, which may include, but is not limited to, reinstatement or hiring of employees, with or without back pay (payable by the employer, employment agency, or labor organization, as the case may be, responsible for the unlawful employment practice), or any other equitable relief as the court deems appropriate. Back pay liability shall not accrue from a date more than two years prior to the filing of a charge with the Commission. Interim earnings or amounts earnable with reasonable diligence by the person or persons discriminated against shall operate to reduce the back pay otherwise allowable. No order of the court shall require the admission or reinstatement of an individual as a member of a union, or the hiring, reinstatement, or promotion of an individual as an employee, or the payment to him of any back pay, if such individual was refused admission, suspended, or expelled, or was refused employment or advancement or was suspended or discharged for any reason other than discrimination on account of race, color, religion, sex, or national origin or in violation of section 704(a). [As amended March 24, 1972, P. L. 92-261, Title VII, 86 Stat. 107, 42 U.S.C. § 2000e-5(g).]

[¶ 3087]   CIVIL ACTIONS BY ATTORNEY GENERAL

[¶ 3087A]   [Attorney General's Complaint]

Sec. 707. (a) Whenever the Attorney General has reasonable cause to believe that any person or group of persons is engaged in a pattern or practice of resistance to the full enjoyment of any of the rights secured by this title, and that the pattern or practice is of such a nature and is intended to deny the full exercise of the rights herein described, the Attorney

General may bring a civil action in the appropriate district court of the United States by filing with it a complaint (1) signed by him (or in his absence the Acting Attorney General), (2) setting forth facts pertaining to such pattern or practice, and (3) requesting such relief, including an application for a permanent or temporary injunction, restraining order or other order against the person or persons responsible for such pattern or practice, as he deems necessary to insure the full enjoyment of the rights herein described. [July 2, 1964, P. L. 88-352, Title VII, 78 Stat. 261, 42 U.S.C. § 2000e-6(a).]

[¶ 3089]   EFFECT ON STATE LAWS

[¶3089A]   [Liability Under State Law]

Sec. 708. Nothing in this title shall be deemed to exempt or relieve any person from any liability, duty, penalty, or punishment provided by any present or future law of any State or political subdivision of a State, other than any such law which purports to require or permit the doing of any act which would be an unlawful employment practice under this title. [July 2, 1964, P. L. 88-352, Title VII, 78 Stat. 262, 42 U.S.C. § 2000e-7.]

For information concerning recordkeeping, relations with state and local agencies, confidential information, and investigatory powers and procedure, see ¶ 3093 - 3097, and Sec. 11 of the National Labor Relations Act.

[¶ 3100A]   [Posting of Commission Notices]

Sec. 711. (a) Every employer, employment agency, and labor organization, as the case may be, shall post and keep posted in conspicuous places upon its premises where notices to employees, applicants for employment, and members are customarily posted a notice to be prepared or approved by the Commission setting forth excerpts from or, summaries of, the pertinent provisions of this title and information pertinent to the filing of a complaint. [July 2, 1964, P. L. 88-352, Title VII, 78 Stat. 265, 42 U.S.C. § 2000e-10(a).]

[¶ 3105]   [Good-Faith Compliance as Defense in
                                    Agency and Court Proceedings]

(b) In any action or proceeding based on any alleged unlawful employment practice, no person shall be subject to any liability or punishment for or on account of (1) the commission by such person of an unlawful employment practice if he pleads and proves that the act or omission complained of was in good faith, in conformity with, and in reliance on any written interpretation or opinion of the Commission, or (2) the failure of such person to publish and file any information required by any provision of this title if he pleads and proves that he failed to publish and file such information in good faith, in conformity with the instructions of the Commission issued under this title regarding the filing of such information. Such a defense, if established, shall be a bar to the action or proceeding, notwithstanding that (A) after such act or omission, such interpretation or opinion is modified or rescinded or is determined by judicial authority to be invalid or of no legal effect, or (B) after publishing or filing the description and annual reports, such publication or filing is determined by judicial authority not to be in conformity with the requirements of this title. [July 2, 1964, P. L. 88-352, Title VII, 78 Stat. 265, 42 U.S.C. § 2000e-12(b).]

[¶ 3106]   **FORCIBLY RESISTING THE COMMIS-
SION OR ITS REPRESENTATIVES**

[¶ 3106A]   **[Penalty for Forcible Interference]**

Sec. 714. The provisions of sections 111 and 1114, title 18, United States Code, shall apply to officers, agents, and employees of the Commission in the performance of their official duties. Notwithstanding the provisions of sections 111 and 1114 of title 18, United States Code, whoever in violation of the provisions of section 1114 of such title kills a person while engaged in or on account of the performance of his official functions under this Act shall be punished by imprisonment for any term of years or for life. [As amended March 24, 1972, P. L. 92-261, Title VII, 86 Stat. 110, 42 U.S.C. § 2000e-13.]

APPENDIX C

ADMINISTRATIVE PROCEDURE ACT

Administrative law is a "child" of the 20th century. Although the first administrative agency was created in 1789, it and others like it were created primarily to correct abuses. In 1920, however, passage of the Transportation Act signalled a new direction and new powers for administrative agencies.

The Transportation Act conferred on the Interstate Commerce Commission — originally created in 1887, to guard against abuses — the responsibility for developing and fostering adequate transportation service. Thus, in addition to correcting abuses, the Interstate Commerce Commission became responsible for ensuring against abuses, something administrative agencies had never done before.

The numerous agencies created in the 1930s to enforce President Franklin Delano Roosevelt's New Deal policies extended administrative control over many previously unregulated aspects of American life. This action, in turn, gave rise to many criticisms, concerns, and legal questions concerning the fairness and relatively uncontrolled nature of agency powers and procedures. Procedural irregularities of administrative agencies caused confusion and alarm, and highlighted the need for guidelines regulating how administrative agencies would and could function.

On February 16, 1939, President Roosevelt asked the Attorney General to appoint a committee to investigate the "need for procedural reform in the field of administrative law." A "Blue Ribbon" committee was duly appointed. This committee interviewed agency executives and members, and also lawyers involved in agency practices.

A detailed report on each agency was produced and used as a basis for subsequent discussions. Public hearings were then held in order to elicit more opinions and recommendations. The final report of the committee is still a primary source of information on the federal administrative process. The ultimate result of the work begun by the committee was the Administrative Procedure Act, passed by a unanimous vote of both Houses of Congress in 1946.

The Administrative Procedure Act (APA) also became a model which the States adopted and/or adapted, thus making the Act the basis on which administrative agencies at all levels of government function. Although some variations in agency functions occur — chiefly because agencies are created with different purposes and different powers — a careful reading of the Administrative Procedure Act will acquaint the reader with the way administrative agencies work.

§ 551. Definitions

For the purpose of this subchapter —

(1) "agency" means each authority of the Government of the United States, whether or not it is within or subject to review by another agency, but does not include —

(A) the Congress;

(B) the courts of the United States;

(C) the governments of the territories or possessions of the United States;

(D) the government of the District of Columbia;

or except as to the requirements of Section 552 of this title —

(E) agencies composed of representatives of the parties or of representatives of organizations of the parties to the disputes determined by them;

(F) courts martial and military commissions;

(G) military authority exercised in the field in time of war or in occupied territory; or

(H) functions conferred by sections 1738, 1739, 1743, and 1744 of title 12; chapter 2 of title 41; or sections 1622,

1884, 1891-1902, and former section 1641(b) (2), of title 50, appendix;

(2) "person" includes an individual, partnership, corporation, association, or public or private organization other than an agency;

(3) "party" includes a person or agency named or admitted as a party, or property seeking and entitled as of right to be admitted as a party, in an agency proceeding, and a person or agency admitted by an agency as a party for limited purposes;

(4) "rule" means the whole or a part of an agency statement of general or particular applicability and future effect designed to implement, interpret, or prescribe law or policy or describing the organization, procedure, or practice requirements of an agency and includes the approval or prescription for the future of rates, wages, corporate or financial structures or reorganization thereof, prices, facilities, appliances, services or allowances therefor or of valuations, costs, or accounting, or practices bearing on any of the foregoing;

(5) "rule making" means agency process for formulating, amending, or repealing a rule;

(6) "order" means the whole or a part of a final disposition, whether affirmative, negative, injunctive, or declaratory in form, of an agency in a matter other than rule making but including licensing;

(7) "adjudication" means agency process for the formulation of an order;

(8) "license" includes the whole or a part of an agency permit, certificate, approval, registration, charter, membership, statutory exemption or other form of permission;

(9) "licensing" includes agency process respecting the grant, renewal, denial, revocation, suspension, annulment, withdrawal, limitation, amendment, modification, or conditioning of a license;

(10) "sanction" includes the whole or a part of an agency —

(A) prohibition, requirement, limitation, or other condition affecting the freedom of a person;

(B) withholding of relief;

(C) imposition of penalty or fine;

(D) destruction, taking, seizure, or withholding of property;

(E) assessment of damages, reimbursement, restitution, compensation, costs, charges, or fees;

(F) requirement, revocation, or suspension of a license; or

(G) taking other compulsory or restrictive action;

(11) "relief" includes the whole or a part of an agency —

(A) grant of money, assistance, license, authority, exemption, exception, privilege, or remedy;

(B) recognition of a claim, right, immunity, privilege, exemption, or exception; or

(C) taking of other action on the application or petition of, and beneficial to, a person;

(12) "agency proceeding" means an agency process as defined by paragraphs (5), (7), and (9) of this section; and

(13) "agency action" includes the whole or a part of an agency rule, order, license, sanction, relief, or the equivalent of denial thereof, or failure to act.

§ 552. Public information; agency rules, opinions, orders, records, and proceedings

(a) Each agency shall make available to the public information as follows:

(1) Each agency shall separately state and currently publish in the Federal Register for the guidance of the public —

(A) descriptions of its central and field organization and the established places at which, the employees (and in the case of a uniformed service, the members) from whom, and the methods whereby, the public may obtain information, make submittals or requests, or obtain decisions;

(B) statements of the general course and method by which its functions are channeled and determined, including the nature and requirements of all formal and informal procedures available;

(C) rules of procedure, descriptions of forms available or the places at which forms may be obtained, and instructions as to the scope and contents of all papers, reports, or examinations;

(D) substantive rules of general applicability adopted as authorized by law, and statements of general policy or interpretations of general applicability formulated and adopted by the agency; and

(E) each amendment, revision, or repeal of the foregoing.

Except to the extent that a person has actual and timely notice of the terms thereof, a person may not in any manner be required to resort to, or be adversely affected by, a matter required to be published in the Federal Register and not so published. For the purpose of this paragraph, matter reasonably available to the class of persons affected thereby is deemed published in the Federal Register when incorporated by reference therein with the approval of the Director of the Federal Register.

(2) Each agency, in accordance with published rules, shall make available for public inspection and copying —

(A) final opinions, including concurring and dissenting opinions, as well as orders, made in the adjudication of cases;

(B) those statements of policy and interpretations which have been adopted by the agency and are not published in the Federal Register; and

(C) administrative staff manuals and instructions to staff that affect a member of the public;

unless the materials are promptly published and copies offered for sale. To the extent required to prevent a clearly unwarranted invasion of personal privacy, an agency may delete identifying details when it makes available or publishes an opinion, statement of policy, interpretation, or staff manual or instruction. However, in each case the justification for the deletion shall be explained fully in writing. Each agency also shall maintain and make available for public inspection and copying a current index providing identifying information for the public as to any matter issued, adopted, or promulgated after July 4, 1967, and required by this paragraph to be made available or published. A final order, opinion, statement of policy, interpretation, or staff manual or instruction that affects a member of the public may be relied on, used, or cited as precedent by an agency against a party other than an agency only if —

(i) it has been indexed and either made available or published as provided by this paragraph; or

(ii) the party has actual and timely notice of the terms thereof.

(3) Except with respect to the records made available under paragraphs (1) and (2) of this subsection, each agency, on request for identifiable records, made in accordance with published rules stating the time, place, fees to the extent authorized by statute, and procedure to be followed, shall make the records promptly available to any person. On complaint, the district court of the United States in the district in which the complainant resides, or has his principal place of business, or in which the agency records are situated, has jurisdiction to enjoin the agency from withholding agency records and to order the production of any agency records improperly withheld from the complainant. In such a case the court *shall determine the matterde novo* and the burden is on the agency to sustain its action. In the event of noncompliance with the order of the court, the district court may punish for contempt the responsible employee, and in the case of a uniformed service, the responsible member. Except as to causes the court considers of greater importance, proceedings before the district court, as authorized by this paragraph, take precedence on the docket over all other causes and shall be assigned for hearing and trial at the earliest practicable date and expedited in every way.

(4) Each agency having more than one member shall maintain and make available for public inspection a record of the final votes of each member in every agency proceeding.

(b) This section does not apply to matters that are—

(1) specifically required by Executive order to be kept secret in the interest of the national defense or foreign policy;

(2) related solely to the internal personnel rules and practices of an agency;

(3) specifically exempted from disclosure by statute;

(4) trade secrets and commercial or financial information obtained from a person and privileged or confidential;

(5) inter-agency or intra-agency memo-

randums or letters which would not be available by law to a party other than an agency in litigation with the agency;

(6) personnel and medical files and similar files the disclosure of which would constitute a clearly unwarranted invasion of personal privacy;

(7) investigatory files compiled for law enforcement purposes except to the extent available by law to a party other than an agency;

(8) contained in or related to examination, operating, or condition reports prepared by, on behalf of, or for the use of an agency responsible for the regulation or supervision of financial institutions; or

(9) geological and geophysical information and data, including maps, concerning wells.

(c) This section does not authorize withholding of information or limit the availability of records to the public, except as specifically stated in this section. This section is not authority to withhold information from Congress.

§ 553. Rule making

(a) This section applies, accordingly to the provisions thereof, except to the extent that there is involved —

(1) a military or foreign affairs function of the United States; or

(2) a matter relating to agency management or personnel or to public property, loans, grants, benefits, or contracts.

(b) General notice of proposed rule making shall be published in the Federal Register, unless persons subject thereto are named and either personally served or otherwise have actual notice thereof in accordance with law. The notice shall include —

(1) a statement of the time, place, and nature of public rule making proceedings;

(2) reference to the legal authority under which the rule is proposed; and

(3) either the terms or substance of the proposed rule or a description of the subjects and issues involved.

Except when notice or hearing is required

by statute, this subsection does not apply —

(A) to interpretative rules, general statements of policy, or rules of agency organization, procedure, or practice; or

(B) when the agency for good cause finds (and incorporates the finding and a brief statement of reasons therefor in the rules issued) that notice and public procedure thereon are impracticable, unnecessary, or contrary to the public interest.

(c) After notice required by this section, the agency shall give interested persons an opportunity to participate in the rule making through submission of written data, views, or arguments with or without opportunity for oral presentation. After consideration of the relevant matter presented, the agency shall incorporate in the rules adopted a concise general statement of their basis and purpose. When rules are required by statute to be made on the record after opportunity for an agency hearing, sections 556 and 557 of this title apply instead of this subsection.

(d) The required publication or service of a substantive rule shall be made not less than 30 days before its effective date, except —

(1) a substantive rule which grants or recognizes an exemption or relieves a restriction;

(2) interpretative rules and statements of policy; or

(3) as otherwise provided by the agency for good cause found and published with the rule.

(e) Each agency shall give an interested person the right to petition for the issuance, amendment, or repeal of a rule.

§ 554. Adjudications

(a) This section applies, according to the provisions thereof, in every case of adjudication required by statute to be determined on the record after opportunity for an agency hearing, except to the extent that there is involved —

(1) a matter subject to a subsequent trial of the law and the facts *de novo* in a court;

(2) the selection or tenure of an employee, except a hearing examiner appointed under section 3105 of this title;

(3) proceedings in which decisions rest solely on inspections, tests, or elections;

(4) the conduct of military or foreign affairs functions;

(5) cases in which an agency is acting as an agent for a court; or

(6) the certification of worker representatives.

(b) Persons entitled to notice of an agency hearing shall be timely informed of —

(1) the time, place, and nature of the hearing;

(2) the legal authority and jurisdiction under which the hearing is to be held; and

(3) the matters of fact and law asserted. When private persons are the moving parties, other parties to the proceeding shall give prompt notice of issues controverted in fact or law; and in other instances agencies may by rule require responsive pleading. In fixing the time and place for hearings, due regard shall be had for the convenience and necessity of the parties or their representatives.

(c) The agency shall give all interested parties opportunity for —

(1) the submission and consideration of facts, arguments, offers of settlement, or proposals of adjustment when time, the nature of the proceeding, and the public interest permit; and

(2) to the extent that the parties are unable so to determine a controversy by consent, hearing the decision on notice and in accordance with sections 556 and 557 of this title.

(d) The employee who presides at the reception of evidence pursuant to section 556 of this title shall make the recommended decision or initial decision required by section 557 of this title, unless he becomes unavailable to the agency. Except to the extent required for the disposition of *ex parte* matters as authorized by law, such an employee may not —

(1) consult a person or party on a fact in issue, unless on notice and opportunity for all parties to participate; or

(2) be responsible to or subject to the supervision or direction of an employee or agent engaged in the performance of investigative or prosecuting functions for an agency.

An employee or agent engaged in the performance of investigative or prosecuting functions for an agency in a case may not, in that or a factually related case, participate or advise in the decision, recommended decision, or agency review pursuant to section 557 of this title, except as witness or counsel in public proceedings. This subsection does not apply —

(A) in determining applications for initial licenses;

(B) to proceedings involving the validity or application of rates, facilities, or practices of public utilities or carriers; or

(C) to the agency or a member or members of the body comprising the agency.

(e) The agency, with like effect as in the case of other orders, and in its sound discretion, may issue a declaratory order to terminate a controversy or remove uncertainty.

§ 555. Ancillary matters

(a) This section applies, according to the provisions thereof, except as otherwise provided by this subchapter.

(b) A person compelled to appear in person before an agency or representative thereof is entitled to be accompanied, represented, and advised by counsel or, if permitted by the agency, by other qualified representative. A party is entitled to appear in person or by or with counsel or other duly qualified representative in an agency proceeding. So far as the orderly conduct of public business permits, an interested person may appear before an agency or its responsible employees for the presentation, adjustment, or determination of an issue, request, or controversy in a proceeding, whether interlocutory, summary, or otherwise, or in connection with an agency function. With due regard for the convenience and necessity of the parties or their representatives and within a reasonable time, each agency shall proceed to conclude a matter presented to it. This subsection does not grant or deny a person who is not a lawyer the right to appear for or represent others before an agency or in an agency proceeding.

(c) Process, requirement of a report, inspection, or other investigative act or demand may not be issued, made, or enforced except as authorized by law. A person compelled to submit data or evidence is entitled to retain or, on payment of lawfully prescribed costs, procure a copy or transcript thereof, except that in a nonpublic investigatory proceeding the witness may for good cause be limited to inspection of the official transcript of his testimony.

(d) Agency subpenas authorized by law shall be issued to a party on request and, when required by rules of procedure, on a statement or showing of general relevance and reasonable scope of the evidence sought. On contest, the court shall sustain the subpena or similar process or demand to the extent that it is found to be in accordance with law. In a proceeding for enforcement, the court shall issue an order requiring the appearance of the witness or the production of the evidence or data within a reasonable time under penalty of punishment for contempt in case of contumacious failure to comply.

(e) Prompt notice shall be given of the denial in whole or in part of a written application, petition, or other request of an interested person made in connection with any agency proceeding. Except in affirming a prior denial or when the denial is self-explanatory, the notice shall be accompanied by a brief statement of the grounds for denial.

§ 556. Hearings; presiding employees; powers and duties; burden of proof; evidence; record as basis of decision

(a) This section applies, according to the provision thereof, to hearings required by section 553 or 554 of this title to be conducted in accordance with this section.

(b) There shall preside at the taking of evidence —

(1) the agency;

(2) one or more members of the body which comprises the agency; or

(3) one or more hearing examiners appointed under section 3105 of this title. This subchapter does not supersede the conduct of specified classes of proceedings, in whole or in part, by or before boards or other employees specially provided for by or designated under statute. The functions of

presiding employees and of employees participating in decisions in accordance with section 557 of this title shall be conducted in an impartial manner. A presiding or participating employee may at any time disqualify himself. On the filing in good faith of a timely and sufficient affidavit of personal bias or other disqualification of a presiding or participating employee, the agency shall determine the matters as a part of the record and decision in the case.

(c) Subject to published rules of the agency and within its powers, employees presiding at hearings may —

(1) administer oaths and affirmations;

(2) issue subpoenas authorized by law;

(3) rule on offers of proof and receive relevant evidence;

(4) take depositions or have depositions taken when the ends of justice would be served;

(5) regulate the course of the hearing;

(6) hold conferences for the settlement or simplification of the issues by consent of the parties;

(7) dispose of procedural requests or similar matters;

(8) make or recommend decisions in accordance with section 557 of this title; and

(9) take other action authorized by agency rule consistent with this subchapter.

(d) Except as otherwise provided by statute, the proponent of a rule or order has the burden of proof. Any oral or documentary evidence may be received, but the agency as a matter of policy shall provide for the exclusion of irrelevant, immaterial, or unduly repetitious evidence. A sanction may not be imposed or rule or order issued except on consideration of the whole record or those parts thereof cited by a party and supported by and in accordance with the reliable, probative, and substantial evidence. A party is entitled to present his case or defense by oral or documentary evidence, to submit rebuttal evidence, and to conduct such cross-examination as may be required for a full and true disclosure of the facts. In rule making or determining claims for money or benefits or applications for initial licenses an agency may, when a party will not to be prejudiced thereby, adopt procedures for the submission of all or part of the evidence in written form.

(e) The transcript of testimony and exhibits, together with all papers and requests filed in the proceeding, constitutes the exclusive record for decision in accordance with section 557 of this title and, on payment of lawfully prescribed costs, shall be made available to the parties. When an agency decision rests on official notice of a material fact not appearing in the evidence in the record, a party is entitled, on timely request, to an opportunity to show the contrary.

§ 557. Initial decisions; conclusiveness; review by agency; submissions by parties; contents of decisions; record

(a) This section applies, according to the provisions thereof, when a hearing is required to be conducted in accordance with section 556 of this title.

(b) When the agency did not preside at the reception of the evidence, the presiding employee or, in cases not subject to section 554(d) of this title, an employee qualified to preside at hearings pursuant to section 556 of this title, shall initially decide the case unless the agency requires, either in specific cases or by general rule, the entire record to be certified to it for decision. When the presiding employee makes an initial decision, that decision then becomes the decision of the agency without further proceedings unless there is an appeal to, or review on motion of, the agency within time provided by rule. On appeal from or review of the initial decision, the agency has all the powers which it would have in making the initial decision except as it may limit the issues on notice or by rule. When the agency makes the decision without having presided at the reception of the evidence, the presiding employee or an employee qualified to preside at hearings pursuant to section 556 of this title shall first recommend a decision, except that in rule making or determining application for initial licenses—

(1) instead thereof the agency may issue a tentative decision or one of its responsible employees may recommend a decision; or

(2) this procedure may be omitted in a case in which the agency finds on the record that due and timely execution of its functions imperatively and unavoidably so requires.

(c) Before a recommended, initial, or tentative decision, or a decision on agency review of the decision of subordinate employees, the parties are entitled to a reasonable opportunity to submit for the consideration of the employees participating in the decision —

(1) proposed findings and conclusions; or

(2) exceptions to the decisions or recommended decisions of subordinate employees or to tentative agency decisions; and

(3) supporting reasons for the exceptions or proposed findings or conclusions. The record shall show the ruling on each finding, conclusion, or exception presented. All decisions, including initial, recommended, and tentative decisions, are a part of the record and shall include a statement of —

(A) findings and conclusions, and the reasons or basis therefor, on all the material issues of fact, law, or discretion presented on the record; and

(B) the appropriate rule, order, sanction, relief, or denial thereof.

§ 558. Imposition of sanctions; determination of applications for licenses; suspension, revocation and expiration of licenses

(a) This section applies, according to the provisions thereof, to the exercise of a power or authority.

(b) A sanction may not be imposed or substantive rule or order issued except within jurisdiction delegated to the agency and as authorized by law.

(c) When application is made for a license required by law, the agency, with due regard for the rights and privileges of all the interested parties or adversely affected persons and within a reasonable time, shall set and complete proceedings required to be conducted in accordance with sections 556 and 557 of this title or other proceedings required by law and shall make its decision. Except in cases of willfulness or those in which public health, interest, or safety requires otherwise, the withdrawal, suspension, revocation, or annulment of a license is lawful only if, before the institution of agency proceedings therefor, the licensee has been given—

(1) notice by the agency in writing of facts or conduct which may warrant the action; and

(2) opportunity to demonstrate or achieve compliance with all lawful requirements.

When the licensee has made timely and sufficient application for a renewal or a new license in accordance with agency rules, a license with reference to an activity of a continuing nature does not expire until the application has been finally determined by the agency.

§ 559. Effect on other laws; effect of subsequent statute

This subchapter, chapter 7, and sections 1305, 3105, 3344, 4301(2) (E), 5362, and 7521, and the provisions of section 5335(a) (B) of this title that relate to hearing examiners, do not limit or repeal additional requirements imposed by statute or otherwise recognized by law. Except as otherwise required by law, requirements or privileges relating to evidence or procedure apply equally to agencies and persons. Each agency is granted the authority necessary to comply with the requirements of this subchapter through the issuance of rules or otherwise. Subsequent statute may not be held to supersede or modify this subchapter, chapter 7, sections 1305, 3105, 3344, 4301(2) (E), 5362, or 7521, or the provisions of section 5335(a) (B) of this title that relate to hearing examiners, except to the extent that it does so expressly.

§ 701. Application; definitions

(a) This chapter applies, according to the provisions thereof, except to the extent that—

(1) statutes preclude judicial review; or

(2) agency action is committed to agency discretion by law.

(b) For the purpose of this chapter—

(1) "agency" means each authority of the Government of the United States, whether or not it is within or subject to review by another agency, but does not include—

(A) the Congress;

(B) the courts of the United States;

(C) the governments of the territories or possessions of the United States;

(D) the government of the District of Columbia;

(E) agencies composed of representatives of the parties or of representatives of organizations of the parties to the disputes determined by them;

(F) courts martial and military commissions;

(G) military authority exercised in the field in time of war or in occupied territory; or

(H) functions conferred by sections 1738, 1739, 1743, and 1744 of title 12; chapter 2 of title 41; or sections 1622, 1884, 1891-1902, and former section 1641(b) (2), of title 50, appendix; and

(2) "person," "rule," "order," "license," "sanction," "relief," and "agency action" have the meanings given them by section 551 of this title.

§ 702. Right of review

A person suffering legal wrong because of agency action, or adversely affected or aggrieved by agency action within the meaning of a relevant statute, is entitled to judicial review thereof.

§ 703. Form and venue of proceeding

The form of proceeding for judicial review is the special statutory review proceeding relevant to the subject matter in a court specified by statute or, in the absence or inadequacy thereof, any applicable form of legal action, including actions for declaratory judgments or writs of prohibitory or mandatory injunction or *habeas corpus,* in a court of competent jurisdiction. Except to the extent that prior, adequate, and exclusive opportunity for judicial review is provided by law, agency action is subject to judicial review in civil or criminal proceedings for judicial enforcement.

§ 704. Actions reviewable

Agency action made reviewable by statute and final agency action for which there is no other adequate remedy in a court are subject to judicial review. A preliminary, procedural, or intermediate agency action or ruling not directly reviewable is subject to review on the review of the final agency action. Except as otherwise expressly required by statute, agency action otherwise final is final for the purposes of this section whether or not there has been presented or determined an application for a declaratory order, for any form of reconsideration, or, unless the agency otherwise requires by rule and provides that the action meanwhile is inoperative, for an appeal to superior agency authority.

§ 705. Relief pending review

When an agency finds that justice so requires, it may postpone the effective date of action taken by it, pending judical review. On such conditions as may be required and to the extent necessary to prevent irreparable injury, the reviewing court, including the court to which a case may be taken on appeal from or on application for *certiorari* or other writ to a reviewing court, may issue all necessary and appropriate process to postpone the effective date of an agency action or to preserve status or rights pending conclusion of the review proceedings.

§ 706. Scope of review

To the extent necessary to decision and when presented, the reviewing court shall decide all relevant questions of law, interpret constitutional and statutory provisions, and determine the meaning or applicability of the terms of an agency action. The reviewing court shall —

(1) compel agency action unlawfully withheld or unreasonably delayed; and

(2) hold unlawful and set aside agency action, findings, and conclusions found to be—

(A) arbitrary, capricious, an abuse of discretion, or otherwise not in accordance with law;

(B) contrary to constitutional right, power, privilege, or immunity;

(C) in excess of statutory jurisdiction, authority, or limitations, or short of statutory right;

(D) without observance of procedure required by law;

(E) unsupported by substantial evidence in a case subject to section 556 and 557 of this title or otherwise reviewed on the record of an agency hearing provided by statute; or

(F) unwarranted by the facts to the extent that the facts are subject to trial *de novo* by the reviewing court.

In making the foregoing determinations, the court shall review the whole record or those parts of it cited by a party, and due account shall be taken of the rule or prejudical error.

§ 3105. Appointment of hearing examiners

Each agency shall appoint as many hearing examiners as are necessary for proceedings required to be conducted in accordance with sections 556 and 557 of this title. Hearing examiners shall be assigned to cases in rotation so far as practicable, and may not perform duties inconsistent with their duties and responsibilities as hearing examiners.

§ 7521. Removal

A hearing examiner appointed under section 3105 of this title may be removed by the agency in which he is employed only for good cause established and determined by the Civil Service Commission on the record after opportunity for hearing.

§ 5362. Hearing examiners

Hearing examiners appointed under section 3105 of this title are entitled to pay prescribed by the Civil Service Commission independently of agency recommendations or ratings and in accordance with subchapter III of this chapter and chapter 51 of this title.

§ 3344. Details; hearing examiners

An agency as defined by section 551 of this title which occasionally or temporarily is insufficiently staffed with hearing examiners appointed under section 3105 of this title may use hearing examiners selected by the Civil Service Commission from and with the consent of other agencies.

§ 1305. Hearing examiners

For the purpose of sections 3105, 3344, 4301(2) (E), 5362, and 7521 and the provisions of section 5335(a) (B) of this title that relate to hearing examiners, the Civil Service Commission may investigate, require reports by agencies, issue reports, including an annual report to Congress, prescribe regulations, appoint advisory committees as necessary, recommend legislation, subpena witnesses and records, and pay witness fees as established for the courts of the United States.

APPENDIX D

Federal Agencies Involved in Fire Protection

Civil Service Commission (CSC)

Consumer Product Safety Commission (CPSC)

Department of Agriculture
- *Farmers Home Administration (FHA)*
- *U. S. Forest Service (USFS)*

Department of Commerce (DOC)
- *Economic Development Administration (EDA)*
- *Maritime Administration (MARAD)*
- *National Bureau of Standards (NBS)*
- *National Fire Prevention and Control Administration (NFPCA)*

Department of Defense (DOD)
- *Defense Civil Preparedness Agency (DCPA)*
- *Military Departments (Air Force, Army, Navy, Marines)*

Department of Health, Education, and Welfare (HEW)
- *National Institute of Occupational Safety and Health (NIOSH)*
- *National Institutes of Health (NIH)*
- *Office of Education*
- *Public Health Service (PHS)*
- *Social Security Administration (SSA)*

Department of Housing and Urban Development (HUD)
- *Federal Disaster Assistance Administration (FDAA)*
- *Federal Housing Administration (FHA)*

Department of the Interior
- *Bureau of Indian Affairs (BIA)*
- *Bureau of Land Management (BLM)*
- *Bureau of Mines (BUMINES)*
- *Mining Enforcement and Safety Administration (MESA)*
- *National Park Service (NPS)*

Department of Justice
- *Law Enforcement Assistance Administration (LEAA)*

Department of Labor (DOL)
- *Bureau of Labor Statistics (BLS)*
- *Occupational Safety and Health Administration (OSHA)*

Department of Transportation (DOT)
- *Federal Aviation Administration (FAA)*
- *Federal Highway Administration (FHWA)*
- *Federal Railroad Administration (FRA)*
- *Materials Transportation Bureau (MTB)*
- *National Highway Traffic Safety Administration (NHTSA)*
- *Urban Mass Transportation Administration (UMTA)*
- *U.S. Coast Guard (USCG)*

Department of the Treasury
- *Bureau of Alcohol, Tobacco, and Firearms*

Environmental Protection Agency (EPA)

Federal Communications Commission (FCC)

Federal Trade Commission (FTC)

General Services Administration (GSA)
- *Federal Supply Service*
- *Public Buildings Service*

National Academy of Engineering (NAE) (Quasi-government)

National Academy of Sciences (NAS) (Quasi-government)

National Aeronautics and Space Administration (NASA)

National Research Council (Quasi-government)

National Science Foundation (NSF)

National Transportation Safety Board (NTSB)

Nuclear Regulatory Commission (NRC) and Energy Research and Development Administration (ERDA)

Small Business Administration (SBA)

Veterans Administration (VA)

APPENDIX E

HEADNOTES FOR CASES INCLUDED IN THIS TEXT

A headnote is a brief summary of a case; it is customarily prepared by a judge or court reporter. Some of the decisions in this text include such headnotes. They should be considered as guides or reminders to cases, for they may be misleading if the decision itself is not carefully read and the legal points and reasoning carefully considered.

Following are brief summaries of each case in this text. These summaries are intended as reminders of the major considerations in each case. The summaries are listed in the order in which they appear in this text. In addition, the section of the chapter in which the cases appear is also indicated to serve as a reminder of the subject under legal consideration.

CHAPTER II — MUNICIPAL CORPORATIONS

Page 25. Municipal Corporations and Quasi-municipal Corporations

State ex rel. Koontz, State Tax Commissioner v. *Board of Park Commissioners of City of Huntingdon et al.*

Supreme Court of Appeals of West Virginia. April 27, 1948. 47 S.E. 2d 689

The State Tax Commissioner argued that the Board of Park Commissioners did not have the authority to levy taxes on all taxable property within its jurisdiction. The case was brought to the Circuit Court, which ruled in favor of the Board. On an appeal to the Supreme Court of Appeals, the decision was essentially affirmed, based on the reasoning that the Board was not a true municipal corporation, and was therefore within its rights to levy taxes for the purpose of the system of parks.

Page 38. Role of Fire Protection in Metropolitan Areas

Cullor et ux. v. *Jackson Township, Putnam County et al.*

Supreme Court of Missouri, Division No. 2. June 9, 1952. 249 S.W. 2d 393

Richard A. Cullor and wife sued Jackson Township, Putnam County, for flood damages to their property caused by water runoff from a public road under repair and reconstruction. The Circuit Court dismissed the case on defendant's motion, and the Cullors appealed to the Supreme Court of Missouri. The Supreme Court ruled that the town was free from liability because it was a quasi-corporation exercising purely governmental functions in road repair and reconstruction.

Page 45. Doctrine of Sovereign Immunity

Steitz et al. v. *City of Beacon.*

Court of Appeals of New York. Dec. 7, 1945. 64 N.E. 2d 704

Charles Steitz and others brought action against the City of Beacon for fire damages to the plaintiffs' property. The Supreme Court of New York granted defendant's motion to dismiss the complaint on the grounds that it failed to state facts sufficient to constitute a cause of action. Plaintiffs appealed to the Court of Appeals of New York. Defendants asked the higher court to concur with lower court judgments, and stated there was no statute justifying imposition of liability upon the city. The lower court judgment was affirmedtby the Court of Appeals.

Page 54. *Ultra Vires* Actions

Strickfaden et al. v. *Green Creek Highway Dist. et al.*

Supreme Court of Idaho. July 10, 1926. 248 P. 456

Charles H. Strickfaden and another brought action for damages for personal injuries against Green Creek Highway District and others. At the conclusion of the plaintiffs' case, the Court granted a nonsuit in favor of three Highway District Commissioners; the jury returned a verdict in favor of the respondent and cross-appellant and against the Highway District, but not one of the District's Commissioners. Appellants appealed from the judgment based on the verdict, and respondents appealed from the judgment of nonsuit in favor of one of the commissioners. The judgment of the lower court was affirmed.

Page 64. Municipal Officers, Agents, and Employees

Dan R. Hudson, as Chairman of the Personnel Board of Jefferson County, Alabama, et al. v. *Billy Gray et al.*
Supreme Court of Alabama. March 26, 1970. Rehearing
Denied May 15, 1970. 234 So. 2d 564

Declaratory judgment action in equity by members of the city fire department against the chairman of the personnel board of Jefferson County and others to determine the right of fire fighters to secure signatures to a petition addressed to the City on the adoption of an ordinance regarding the duty hours of fire fighters. The Circuit Court of Jefferson County judged in favor of the fire fighters, and respondents appealed. The Supreme Court held that securing signatures to a petition by fire department members was not taking part in a "political campaign."

Page 72. Volunteer Fire Companies

Shindledecker v. *Borough of New Bethlehem et al. and three other cases.*
Superior Court of Pennsylvania. June 30, 1941. 20 A. 2d 867

Proceedings under the Workmen's Compensation Act by three members of a volunteer fire company and the widow of a deceased member to recover compensation for personal injuries and death. The Workmen's Compensation Board awarded judgment to plaintiffs, and the Borough of New Bethlehem and its insurance carrier appealed. The judgment of the Board was reversed.

CHAPTER III — THE LAW OF TORTS
Page 88. Legal Duty

Horsham Fire Company No. 1 v. *Fort Washington Fire Company No. 1, Appellant.*
Supreme Court of Pennsylvania. Jan. 3, 1956. 119 A. 2d 71

Action for damages resulting from a collision at an intersection between the fire trucks of different volunteer fire companies that were responding to the same fire alarm. A verdict was reached for the plaintiff company, and the defendant company moved for judgment notwithstanding verdict and a new trial. The Court of Common Pleas refused judgment notwithstanding verdict, granted a new trial, and the defendant company appealed. The Supreme Court held that the question of reckless disregard of the rights of others on the highway was a question for a jury, even though both the plaintiff and the defendant company had certain privileges under the Motor Vehicle Code.

Page 93. Legal Duty

H. C. Johnson, Appellant, v. *Oren Brown, Respondent.*
Supreme Court of Nevada. Oct. 29, 1959. 345 P. 2d 754

Action against the driver of a fire engine for injuries sustained by a truck passenger when the truck was struck by the fire engine. The Second Judicial District Court rendered judgment for the passenger, and the fire fighter appealed. The Supreme Court held that the fire fighter was not driving with due regard for the safety of all persons using the street, and was liable to respondent for injuries sustained.

Page 105. Proximate Cause

Gilbert v. *New Mexico Const. Co. et al.*

Supreme Court of New Mexico. Feb. 26, 1935. Rehearing
Denied May 13, 1935. 44 P. 2d 489

Plaintiff sued the City of Roswell and the New Mexico Construction Company for fire loss to his residence, claiming that low pressure in the city's water main was insufficient for fire fighters to extinguish a fire on his premises. Trial to the Court resulted in judgment against the construction company, and it appealed. The judgment of the lower court was affirmed.

Page 111. Assumption of Risk

Thomas E. Bartels, Kathleen M. Bartels, Judith A. Bartels, and Georgine L. Bartels, minors, by their mother and next friend, Kathleen P. Bartels, and Kathleen P. Bartels, Widow of George E. Bartels, Respondents, v. *Continental Oil Company, a corporation, Appellant.*

Supreme Court of Missouri, Division No. 2. Nov. 9, 1964. 384 S.W. 2d 667

Action for the death of a fire fighter caused when an inadequately vented oil storage tank at a fire "rocketed" into the air and engulfed the fire fighter in flames. The Circuit Court gave judgment to the fire fighter's heirs, and the storage tank owner appealed. The Supreme Court held that the existence of a hidden danger and the failure of the owner to warn the fire fighter of such danger could not be accepted as a usual peril of the fire fighting profession.

Page 116. Doctrine of Rescue

Walker Hauling Company, Inc., et al. v. *Helen P. Johnson, Executrix.*

Court of Appeals of Georgia, Division No. 1. Oct. 26, 1964. Rehearing
Denied Nov. 12, 1964. 139 S.E. 2d 496

Action by an alleged skilled fire fighter for injuries sustained while fighting a bulk petroleum storage plant fire. The Court of Appeals held that if a skilled fire fighter was injured while voluntarily and without rashness or recklessness helping fight a bulk petroleum storage plant fire proximately caused by negligence of the plant owner and the common carrier, the plant owner and the carrier were liable.

Page 120. Contributory Negligence

Campbell et al. v. *Pure Oil Co.*

Supreme Court of New Jersey, Atlantic County. Nov. 12, 1937. 194 A. 873

Fire fighter Frederick Campbell and others brought action against the Pure Oil Company for injuries they sustained in a sudden and unexpected explosion while fighting a fire on the premises. Plaintiffs alleged that the defendant had negligently permitted large quantities of oily, greasy, and gaseous substances to accumulate on the premises, and had permitted storage tanks and containers to become so out of repair that their contents escaped from them. On appeal, the Supreme Court ruled that "the bare facts alleged in the complaint do not justify the conclusion, as a matter of law, the plaintiffs (fire fighters) either knowingly assumed the risk of resulting in their injury or were guilty of contributory negligence.''

Page 127. Public Officers and Employees

Ernest M. Krauth, Plaintiff-Appellant, v. *Israel Geller and Buckingham Homes, Inc., a corporation of the State of New Jersey, also known as Buckingham Builders, Defendants-Respondents.*

Supreme Court of New Jersey. Argued Nov. 9, 1959.
Decided Jan. 11, 1960. 157 A. 2d 129

Action by a fire fighter for injuries sustained during a fire in a fall from a balcony on which a railing had not yet been installed. The plaintiff obtained a judgment on the jury verdict, and the Appellate Division reversed the decision with directions that judgment be entered for the defendant. The fire fighter appealed. The Supreme Court held that the evidence did not establish wanton misconduct of the owner.

Page 132. Public Officers and Employees

Gino Dini et al., Appellants, v. *Irving Naiditch et al., Appellees.*

Supreme Court of Illinois. Sept. 30, 1960. As Modified on
Denial of Rehearing Nov. 30, 1960. 170 N.E. 2d 881

Actions for injury to and deaths of fire fighters in a fire on the defendants property. The Supreme Court entered judgments for the defendants, and appeals were taken. The Supreme Court held that a wife could maintain an action for loss of consortium due to the negligent injury of her husband. The Court also held that a fire fighter performing duty on private property is an invitee, not a licensee, and recovery from a landowner may be based on a failure to exercise reasonable care.

CHAPTER IV — EMPLOYER AND EMPLOYEE RELATIONSHIPS

Page 152. Master and Servant Doctrine.

Miller v. *City of New York.*

Supreme Court, Appellate Division, Second Department. April 1932. 257 N.Y.S. 33

Plaintiff sued for death of his wife by reasons of alleged negligence in the operation of automobile belonging to the fire department. The jury rendered a verdict for $10,000, finding in favor of the plaintiff on the issues of negligence and contributory negligence. On appeal, the lower court judgment was unanimously affirmed.

Page 162. Workmen's Compensation and Related Laws

City of Brunswick v. *Edenfield.*

Court of Appeals of Georgia, Division No. 2. Jan. 17, 1953. 74 S.E. 2d 133

Proceedings on claim for workmen's compensation for death of claimant's husband while serving as a municipal fire fighter. The Superior Court awarded compensation to the claimant, and the City of Brunswick appealed. The Court of Appeals held that where it was customary for the chief of a fire department to hire and discharge fire fighters at will, without confirmation or approval of such action by any other official, and where fire fighters were not subject to civil service, a fire fighter who was so hired by the chief was an "employee" of the city and not an "officer." Therefore, compensation was payable upon his death while in performance of duties.

Page 166. Workmen's Compensation and Related Laws

City of Huntington v. *Fisher.*

Supreme Court of Indiana. April 6, 1942. 40 N.E. 2d 699

The appellee, widow of a fire fighter killed while discharging his duty, was granted compensation by the Industrial Board under the Workmen's Compensation Law. The City brought this action in the Appellate Court, questioning the lawfulness of the award. The Appellate Court held the appellee not entitled to compensation. On appeal, the Supreme Court reversed the lower court decision and remanded the case to the Appellate Court for further proceedings.

Page 168. Fair Labor Standards Act of 1938

Bell et al. v. *Porter et al.*

Circuit Court of Appeals, Seventh Circuit. Dec. 10, 1946. Writ of *Certiorari* Denied April 7, 1947. 159 F. 2d 117

Action by Virgil Bell and others, whose duties were to protect the Elwood Ordnance plant from fire, under the Fair Labor Standards Act against defendants to recover overtime compensation. The District Court rendered judgment for plaintiffs, and defendants appealed. On appeal, the Circuit Court of Appeals reversed the lower court decisions.

Page 172. Fair Labor Standards Act of 1938

Bridgeman et al. v. *Ford, Bacon and Davis, Inc.*

District Court, E. D. Arkansas, W.D. Feb. 25, 1946. 64 F. Supp. 1006

Action by plaintiff and 38 other men, employed as fire fighters by the defendant, under the Fair Labor Standards Act of 1938 to recover overtime compensation. The District Court ruled that no overtime compensation was due under the Fair Labor Standards Act of 1938 based on the plaintiffs' allegations.

Page 186. Appointment and Promotion

State ex rel. O'Connell v. *Roark et al.*

Supreme Court of Florida, Division B. March 15, 1946. 25 So. 2d 275

Appeal from the Court of Record on the Relation of Charles B. O'Connell against George J. Roark and others to secure appointment to the position of second assistant fire chief in the City of Pensacola Fire Department. The lower court denied the request and the Supreme Court affirmed the lower court's decision.

Page 189. Removal, Suspension, Demotion

Robertson v. *City of Rome.*

Court of Appeals of Georgia, Division No. 2. May 15, 1943. 25 S.E. 2d 925

A fire fighter, who was accused of conduct unbeeoming an officer, was refused counsel in a hearing before the Civil Service Board. The Court of Appeals determined that the Board erred in refusing counsel for the plaintiff.

Page 191. Removal, Suspension, Demotion

Kennett et al. v. *Barber*

Supreme Court of Florida, en Banc. June 10, 1947. 31 So. 2d 44
Rehearing Denied July 7, 1947.

Proceeding in *mandamus* by Harry C. Barber against D. C. Kennett, chief of the Miami Beach Fire Department, and others, to compel reinstatement of the plaintiff to the fire department after he was removed for conduct unbecoming a city employee. The Supreme Court reversed the lower court directive to reinstate Barber.

Page 200. Civil Rights Legislation

James H. Blair, Director, Division on Civil Rights, Department of Law and Public Safety, State of New Jersey, Complainant-Respondent, v. *The Mayor and Council, Borough of Freehold et al., Respondents-Appellants.*

Superior Court of New Jersey, Appellate Division. Argued Oct. 4, 1971. Decided Dec. 13, 1971. 285 A. 2d 46

Proceeding on an appeal from an order of the Director of the Division on Civil Rights that determined that the membership and admission procedures for entry into the Borough of Freehold Volunteer Fire Department violated the provisions of the Law Against Discrimination. The order was appealed by both the mayor and council of the Borough, and the fire department. The Superior Court, Appellate Division, affirmed the decision of the Director of Civil Rights.

Page 202. Civil Rights Legislation

The Vulcan Society of the New York City Fire Department, Inc., et al., Plaintiffs-Appellees-Appellants, v. *Civil Service Commission of the City of New York et al., Defendants-Appellants-Appellees, Nicholas M. Cianciotto et al., Intervenors-Defendants-Appellants-Appellees.*

United States Court of Appeals, Second Circuit. Argued Oct. 15, 1973. Decided Nov. 21, 1973. 490 F. 2d 387

Class action by minority individuals and others who had applied for employment with the New York City Fire Department against the Civil Service Commission, the City's department of personnel, and others. Action was based on a claim that procedures used to select New York City fire fighters discriminated against blacks and Hispanics in violation of the equal protection clause of the Fourteenth Amendment. Both plaintiffs and defendants appealed the District Court judgment. The Court of Appeals held that the lower court finding that written and physical examinations for positions of New York City fire fighters had had a racially disproportionate impact was not clearly erroneous and that in combination with defects in preparation and content of written examinations, the use of a merely qualifying physical examination rendered the fire department's selection procedures insufficiently job-related to withstand constitutional attack.

Page 210. Civil Rights Legislation

Boston chapter, N.A.A.C.P., Inc., et al., Plaintiffs-Appellees, v. *Nancy B. Beecher et al., Defendants-Appellees, Director and Commissioners of Civil Service, Defendants-Appellants.*

United States Court of Appeals, First Circuit. Argued May 7, 1974. Decided Sept. 18, 1974. 504 F. 2d 1017

Actions instituted by the United States and by NAACP, based on violations of civil rights with respect to the hiring of fire fighters. The United States District Court entered judgment against defendants, and they appealed. The Court of Appeals held that uncontroverted expert testimony that black and Spanish-surnamed candidates typically performed more poorly on written, multiple-choice-type tests (as those used for fire fighters), could be found in conjunction with other evidence to establish a *prima facie* case that the test had a racially discriminatory impact.

CHAPTER V — CRIMINAL LAW

Page 227. Criminal Intent.

Commonwealth v. *Louis L. Lamothe, Junior.*

Supreme Judicial Court of Massachusetts. Middlesex. Argued
Dec. 4, 1961. Decided Dec. 29, 1961. 179 N.E. 2d 245

Prosecution for attempting to burn a building. On appeal by the defendant,
the Supreme Judicial Court held that the necessary element of malice could be inferred from the willful act of setting a fire by the defendant.

Page 230. Attempt

Maner v. *State.*

Court of Appeals of Georgia, Division No. 1. Sept. 4, 1931. 159 S.E. 902

W. A. Maner was charged with the offense of attempt to commit arson. The
Court of Appeals affirmed the lower court ruling that Maner's acts of strewing excelsior along and upon barrels and boxes stored in a building and pouring alcohol
and gasoline over the excelsior were declared overt acts by the commission of the
crime of arson.

Page 232. Attempt

State v. *Taylor.*

Supreme Court of Oregon. Jan. 2, 1906. 84 P. 82

Moses Taylor was convicted of an attempt to commit arson and he appealed.
On appeal, the Supreme Court affirmed the lower court judgment. The Supreme
Court held that the furnishing of overalls and horse and the payment of money to
others by the defendant for the purpose of arson justified his conviction of an attempt to commit arson.

Page 244. Elements of Arson

State of Hawaii, Plaintiff-Appellee, v. *Duprie K. Dudoit, also known as Dupree K.
Dudoit, Defendant-Appellant.*

Supreme Court of Hawaii. Sept. 17, 1973. 514 P. 2d 373

The defendant was convicted before the First Circuit Court, City and County of
Honolulu, of the crime of arson and he appealed. On appeal, the Supreme Court
reversed the decision on the grounds that the state had the burden of overcoming
the presumption that the fire was of natural or acccidental origin and the evidence
presented by the state did not support the conviction.

Page 249. Willful and Malicious Intent

Peter J. Borza, III v. *State of Maryland.*

Court of Special Appeals of Maryland. March 21, 1975. Certiorari
Denied June 25, 1975. 335 A. 2d 142

Defendant was convicted of statutory arson in the Criminal Court, and he appealed. The Court of Special Appeals held that testimony of the asserted accomplice was sufficiently corroborated by evidence of the defendant's presence
and evidence that the burned store was being operated at a loss by the defendant.
The Court also held that the accomplice's confession of involvement to an FBI
agent was admissible to show abandonment of the conspiracy to commit arson,
even though the accomplice remained silent for over two years after the fire.

CHAPTER VI — CRIMINAL PROCEDURE

Page 272. Determining That a Crime Has Been Committed

State v. *Isensee.*

Supreme Court of North Dakota. Aug. 24, 1933. 249 N.W. 898

The defendant was convicted of the crime of willfully setting fire to and burning personal property with intent to defraud the insurer, and he appealed to the Supreme Court from the judgment of his conviction and from the order denying his motion for a new trial. Defendant contended that the statute for arson embraced more than one subject and was therefore violative of the State Constitution. The Supreme Court ruled that, although the statute was not invalid, certain evidence was properly a matter for the jury to consider and therefore the judgment was reversed and a new trial ordered.

Page 281. Gathering Evidence

Coffin v. *United States.*

Supreme Court of the United States, 1895. 156 U.S. 432, 15 S.Ct. 394

The Supreme Court of the United States reversed the conviction of the defendant and ordered a new trial based on the ruling that, although the jury was instructed regarding the doctrine of "presumption of innocence," the fact that the jury was not instructed regarding the separate doctrine of "proof beyond a reasonable doubt" violated the defendant's constitutional rights.

Page 288. Evidence

Joyce Annette Jenkins v. *Commonwealth of Virginia.*

Supreme Court of Virginia. April 23, 1976. 223 S.E. 2d 880

Defendant was convicted before the Circuit Court of burning her automobile with intent to injure the insurer of the vehicle, and a writ of error was granted. The Supreme Court held the evidence from an analysis made weeks after the fire of floor matting from the burned vehicle, which disclosed evidence of residue from fuel oils, was insufficient to sustain a conviction. The lower court's judgment was reversed.

Page 293. Circumstantial Evidence

State v. *Isensee.*

Supreme Court of North Dakota. Aug 24, 1933. 249 N.W. 898

See summary, page 272.

Page 303. Criminal Search and Seizure

State v. *Steven P. Slezak.*

Supreme Court of Rhode Island. Jan. 30, 1976. 350 A. 2d 605

Defendant was convicted in the Superior Court for the possession of cannabis, and he appealed. Even though the arresting police officers were in possession of a valid search warrant when they searched the defendant's premises, the officers entered without knocking on the door and announcing themselves and the purpose of their presence. The Superior Court held that the entry of the police was unreasonable and illegal, and the subsequent search was unlawful. The conviction was therefore reversed.

Page 306. The "Exclusionary Rule"

State of Indiana, Appellant, v. *Loran O. Buxton, Appellee.*

Supreme Court of Indiana. March 20, 1958. 148 N.E. 2d 547

Prosecution for arson against the owner of a restaurant that was damaged by fire. The Circuit Court sustained an objection to admission of evidence by the State, the owner was acquitted, and the state appealed. The Supreme Court held that where the investigation of the cause of the fire by the deputy state fire marshal and a state trooper was not made at the time of the fire and was made without a search warrant, items seized as evidence were inadmissible in court.

Page 312. Inspection, Investigation, and Search

State of Iowa, Petitioner, v. *Warren J. Rees, Judge, Respondent.*

Supreme Court of Iowa. Jan. 11, 1966. Rehearing Denied
March 8, 1966. 139 N.W. 2d 406

On appeal, the Supreme Court held that the defendant, indicted for arson, did not meet the burden of demonstrating that the evidence against him had been illegally procured by officers acting under explicit statutory authority, that there was no claim of unnecessary or arbitrary force, subterfuge, coercion, or objection, and nothing unreasonable or violative of any constitutional rights was present in the statutory procedure allowing such entry.

CHAPTER VII — ADMINISTRATIVE LAW

Page 344. Occupancies

Camara v. *Municipal Court of the City and County of San Francisco.*

Supreme Court of the United States. Argued Feb. 15, 1967.
Decided June 5, 1967. 387 U.S. 523

Appellant was charged with violating the San Francisco Housing Code for refusing three times a warrantless inspection by city and housing inspectors of the quarters he leased and that allegedly violated the apartment's building occupancy permit. While awaiting trial, defendant sued in a State Superior Court for a writ of prohibition, claiming that the inspection ordinance was unconstitutional for failure to require a warrant for inspections. The Superior Court denied the writ, the District Court of Appeal affirmed holding that the ordinance did not violate the Fourth Amendment, and the Supreme Court of California denied a petition for hearing. Defendant then presented the federal constitutional question to the Supreme Court of the United States, on an appeal from the District Court of Appeal. The Supreme Court ruled that a warrantless code-enforcement inspection once entry has been refused cannot be justified under the Fourth Amendment.

Page 354. Occupancies

See v. *City of Seattle.*

Appeal from the Supreme Court of Washington. Argued Feb. 15, 1967.
Decided June 5, 1967. 387 U.S. 541

Appellant sought reversal of his conviction for refusing to permit a representative of the Seattle Fire Department to enter and inspect his locked commercial warehouse without a warrant and without probable cause to believe that a violation of any municipal ordinance existed therein. The Supreme Court held that a suitable warrant procedure is required by the Fourth Amendment to effect uncontested administrative entry and inspection of private commercial premises.

Page 365. Occupancies

Wyman, Commissioner of New York Department of Social Services, et al. v. *James.*

Appeal from the United States District Court for the Southern District
of New York. Argued Oct. 20, 1970. Decided Jan. 12, 1971. 400 U.S. 309

This appeal from the United States District Court for the Southern District of
New York presented the issue of whether plaintiff Barbara James, a beneficiary of
the program for Aid to Families with Dependent Children (AFDC), may refuse a
home visit by AFDC's caseworker without risking termination of her benefits. The
District Court majority held that a mother receiving AFDC relief may, without
forfeiting her right to benefits, refuse the periodic home visit. The New York
State and City Department of Social Services and Commissioner appealed. The
Supreme Court of the United States held that the home visitation provided for by
New York law in connection with the AFDC program is a reasonable ad-
ministrative tool and does not violate any right guaranteed by the Fourth and
Fourteenth Amendments.

Page 387. Employment and Business Facilities

Marshall, Secretary of Labor, et al. v. *Barlow's, Inc.*

Appeal from the United States District Court for the District of Idaho.
Argued Jan. 9, 1978. Decided May 23, 1978. 98 S.Ct. 1816

An appeal from the United States District Court for the District of Idaho by the
Secretary of Labor and others for judgment in favor of Barlow's, Inc. The District
Court ruled that the president and general manager of Barlow's, Inc. was within
his rights guaranteed by the Fourth Amendment to refuse a warrantless inspection
of his business working area by an OSHA inspector. The Secretary appealed to the
Supreme Court of the United States, challenging the judgment of the District
Court. The Supreme Court held the inspection without a warrant or its equivalent
pursuant to § 8(a) of OSHA violates the Fourth Amendment.

Page 405. Employment and Business Occupancies

State of Michigan, Petitioner, v. *Loren Tyler and Robert Tompkins.*

United States Supreme Court. May 31, 1978. 98 S.Ct. 1942

The respondents, Loren Tyler and Robert Tompkins, were convicted in a
Michigan Trial Court of conspiracy to burn real property. Various pieces of
physical evidence and testimony based on personal observation — all obtained
through unconsented and warrantless entries by police and fire officials on the
burned premises — were admitted as evidence at the trial. On appeal, the
Michigan Supreme Court reversed the convictions, holding that "the warrantless
searches were unconstitutional and that the evidence obtained was therefore inadmissible." The Supreme Court of the United States considered the applicability
of the Fourth and Fourteenth Amendments to official entries on fire-damaged
premises. It affirmed the judgment of the Michigan Supreme Court ordering a
new trial.

GLOSSARY OF TERMS

The study of law is a specialized one involving complex issues and questions. The language that has developed around this study is no less complicated. Today, however, members of the fire service are finding it more and more necessary to become familiar with fundamental legal thought, much of which is embedded in the definitions of legal terms. For this reason, this glossary of the legal terms used in this book has been compiled. The student should remember that these words are defined in their legal senses only: many of them also have everyday meanings that have not been included.

A

Abatement   A lessening or reduction, as in a lawsuit. In common law, it is the complete ending of a suit; in equity, it is merely a suspension, and the suit may be revived.

Abet   To actively assist another individual in the actual or attempted commission of a crime.

Abettor   One who abets.

Ab Initio   Latin, "from the beginning." It is commonly used with reference to the validity of contracts, deeds, marriages, *etc.,* to signify that the thing in question was void from the beginning.

Abrogate   To annul, revoke, set aside, cancel, or repeal.

Accomplice   One who voluntarily helps to commit or attempts to commit a crime.

Accrue   To accumulate, as in the amount of time allowed for the prosecution of a suit before the Statute of Limitations takes effect.

Accusation   A formal charge against one alleged to have broken the law.

Accuse   To initiate formal proceedings against an individual, charging that he or she has broken the law.

Acquit   To exonerate; to formally clear of guilt one who has been accused.

Action   An ordinary proceeding in a court of justice.

Actionable   Any conduct the result of which is sufficient grounds for a legal proceeding.

Actus Reus   Latin, "guilty act." Every crime is made up of an *actus reus* and a *mens rea*. The first is a physical action, and the second is the necessary state of mind. For example, the *actus reus* of murder is homicide — that a body be killed. The *mens rea* of murder is malice — that the killer wished for and planned the act.

Adduce   To offer as example or proof; to cite.

Ad Hoc   Latin, "toward this." An expression used to describe something existing for a particular purpose, and which will not exist once that purpose is accomplished.

Adjudicate   To settle through the use of a judge; to act as a judge.

Adjudication   The determination of the solution to a controversy, based on evidence presented.

Advisement   Careful consideration.

Affidavit   A voluntary, written statement given under oath and sworn to before some person legally authorized to administer it, such as a notary public or an officer of the court.

A Fortiori   Latin, "from the strongest." Reasoning that maintains that because a certain fact is true, then of necessity a certain second conclusion is true because it was implied in the first conclusion. For example, a man not guilty of larceny is, *a fortiori,* not guilty of robbery.

Agency   A relation in which one person acts on behalf of another, with the latter's consent.

Agent   One who, by mutual consent, acts on behalf of another.

Allegation   An accusation or charge.

Allege   To assert without proof. An alleged criminal is one who is, as yet, unconvicted.

Amicus Curiae   Latin, "friend of the

court." One who gives information to the court on some matter of law that is subject to question.

Appeal   A complaint to a superior court of an injustice done by an inferior court, in hopes that the original decision will be reversed.

Appellant   The party who appeals a decision. (*See also Appellee.*)

Appellate Court   *See Court of Appeals.*

Appellee   In an appeal, the party who argues against the reversal of the decision, having originally won the case.

Arguendo   Latin, "in arguing." For the sake of argument. An assumption made *arguendo* will not invalidate the rest of the argument if it is indeed untrue.

Arraign   To accuse of a wrong; to call into custody a person against whom an indictment has been handed down.

Arraignment   The initial step in the criminal process in which the accused is formally charged with an offense, given a copy of the accusation, and informed of his or her Constitutional rights.

Arrest   Seizure and detention of an alleged or suspected offender to answer for a crime. (*See also False Arrest.*)

Arson   The fraudulent and malicious burning of property.

Arsonist   One who commits the crime of arson.

Assault   An intentional attempt to do bodily injury to another by violence. The victim need not actually be touched in order for assault to be proved. (*See also Battery.*)

Assumption of Risk   In tort law, a defense used by the defendant in a negligence suit in which it is claimed that the plaintiff had prior knowledge of a dangerous condition and thus is responsible for voluntarily exposing himself or herself to that danger

Attainder   *See Bill of Attainder.*

Attempt   An endeavor to commit an offense, carried beyond mere preparation, but falling short of actual commission of the offense.

Attorney   One who is licensed to practice law.

Aver   To assert as a fact.

Averment   A positive statement of fact in a pleading, as opposed to a statement merely argued or inferred.

B

Bail   Security, usually money, given to ensure the appearance of the defendant at every stage of a proceeding.

Bar   The place in court that counsellors or advocates occupy while pleading a case, and where the prisoner is brought to be arraigned or sentenced.

Battery   The willful and unlawful beating of another. (*See also Assault.*)

Bigamy   The criminal offense of having two or more wives or husbands at the same time.

Bill of Attainder   A legislative act, in any form, that applies to individuals or members of a group so as to inflict punishment on them without a judicial trial. Such acts are forbidden by the Constitution.

Bind   To subject to legal duties or obligations.

Binding   As used in statute, commonly means obligatory.

Bind Over   To order that a defendant be placed in custody until the decision of the proceeding (usually criminal) is made.

Breach   To break or violate a law, duty, or right, either by commission or omission.

Breach of Contract   A failure to perform any contractual duty.

Breaking   In the crime of burglary, the forcible and substantial attempt to enter, as by breaking or taking out a window, picking a lock or opening it with a key, or unloosening any other fastening that the owner has provided.

Bribery   The voluntary giving of something of value in order to influence performance of an official duty.

Brief   A written argument, concentrating upon legal points and authorities, used by a party's lawyer to convey to the court the essential facts of the case.

Burden of Proof   The duty of a party in a lawsuit to prove disputed facts or allegations, either to avoid dismissal of the suit or to prevail in the suit.

Burglary   The entering of a building at night with the intent to commit a felony or steal property of value.

Bylaws Rules made by a corporation for its own government.

C

Canon Law The rules and regulations governing a church, usually the rules governing the Roman Catholic Church. Canon law has no standing in United States courts.

Cause That which brings about a result. (*See also Proximate Cause.*)

Cause of Action A claim in law and fact sufficient to demand judicial attention.

Certiorari A common law writ used as a means of gaining appellate review.

Charge 1. In criminal law, the substantive offense underlying an accusation or indictment. 2. At the close of the presentation of a case, an address given by the presiding judge that instructs the jury on matters of law and sums up the testimony presented.

Charter A document, issued by the government, that establishes a corporate entity.

Circumscribe To constrict the range or activity of something definitely and clearly.

Circumstantial Evidence Indirect evidence; secondary facts from which a principal fact may be rationally inferred.

Citation 1. A reference to a source of legal authority. 2. An official call or notice to appear in court.

Civil Law That branch of law pertaining to noncriminal matters, such as suits between individuals.

Civil Liberties Immunity from government restrictions on or interference with the rights reserved to individuals by the Constitution.

Civil Rights Rights awarded, defined, and limited by law.

Civil Servant One who works for the government.

Claim The assertion of a right to property or money.

Claimant One who makes a claim.

Common Law The system of jurisprudence, developed in England and adopted in the United States, that is based on judicial precedent rather than laws enacted.

Comparative Negligence The proportional sharing of compensation for injuries between plaintiff and defendant. Damages recovered by the plaintiff are reduced in proportion to the plaintiff's fault.

Compensation 1. Remuneration for work done. 2. Recompense for injury received.

Complaint An accusation or charge against a person alleged to have committed an injury or offense.

Confession An admission of guilt or other incriminating statement made by the accused.

Consent 1. To agree, as to something proposed to be done. 2. An agreement of this kind.

Consign To send or transmit goods to someone for sale.

Consignee The person to whom goods are shipped or otherwise transmitted.

Consortium The legal right of one spouse to the company, affection, and service of the other.

Constitutional Law That branch of law dealing with the interpretation of the Constitution.

Construe To understand or explain the sense or intention of.

Contempt An act or omission tending to obstruct or interfere with the orderly administration of justice or to impair the dignity of the court. Those found in contempt may be punished by fine or imprisonment or both.

Contract An agreement by which a person undertakes to do or not do a certain thing, and for breach of which the law gives remedy. (*See also Breach of Contract.*)

Contractor One who is a party to a contract. (*See also Independent Contractor.*)

Contravene To oppose or act contrary to.

Contributory Negligence Conduct on the part of the plaintiff that ignores the exercise of due care, and hence cooperates with the negligence of the defendant to bring about the plaintiff's harm.

Conversion The wrongful appropriation of property belonging to another.

Convey To transfer property from one individual to another.

Coroner A public official who investigates the causes and circumstances of deaths occurring within a specific jurisdiction and reports the findings of the investigation in a Coroner's Inquest.

Corporate Law That branch of the law dealing with corporations.

Corporation An artificial person or being endowed by law with legal rights and the capacity for perpetual succession, and which is entirely distinct from the individuals who compose it.

Corpus Delicti Latin, "the body of the crime." The objective proof that a crime has been committed.

Corpus Juris Latin, "body of law." A series of texts that contained much of the Roman civil and canon law.

Corroborate To support with evidence or authority.

Counsel A lawyer who assists a client and pleads for this client in court.

Court A tribunal established for the public administration of justice. (*See also Court of Appeals, Court of Claims, Court of Equity, and Court of Record.*)

Court of Appeals A court having jurisdiction to review prior decisions and either uphold or reverse these decisions.

Court of Claims A federal court created in 1855 by Congress to determine claims concerning federal laws, regulations, and contracts.

Court of Equity A court having jurisdiction in cases in which adequate or complete remedy cannot be had according to existing laws. Injunction is a form of specific remedy often used in equity.

Court of Record Any court in which proceedings are permanently recorded and which may fine or imprison those in contempt of its authority.

Covert Not openly shown or engaged in; secret.

Crime Any act that has been deemed contrary to the public good, and hence may be prosecuted in a criminal proceeding.

Criminal One who has been convicted of a violation of criminal law.

Criminal Law That branch of law that concerns crimes, their perpetrators, and their punishments.

Criminal Procedure Legal method concerning criminal law; the mechanics of the legal process of arresting, prosecuting, and convicting alleged criminals.

Cross-examination The process of cross-examining.

Cross-examine To test a witness by asking a series of questions designed to check or discredit the answers to previous questions.

Culpable Deserving of moral blame. Culpability implies fault rather than guilt, or disregard of consequences which may follow an act.

Custody 1. Ownership of property. 2. Restraint and physical control over a person so as to ensure that person's appearance at a hearing. 3. Actual imprisonment, resulting from a criminal conviction.

D

Damages Monetary compensation that the law awards to one who has been injured by another's action.

Decision A judgment given by a competent and qualified tribunal.

Deed A writing under seal used to transfer property, usually real estate.

Defamation The offense of injuring a person's character, fame, or reputation, either by writing or in words. (*See also Libel and Slander.*)

Defendant The party denying, resisting, opposing, or contesting an action.

Defer 1. To put off or delay. 2. To delegate responsibility or powers to another.

Defraud To deprive a person of property or right by fraud, deceit, or artifice.

Demur To object to the pleading of the opposite party as insufficient to support an action or defense and request that it be dismissed.

Demurrer A formal allegation that the facts of a pleading, even if they are true, are not legally sufficient for a case to proceed any further.

De Novo Latin, "anew; from the beginning." Any legal proceeding, such as a trial or hearing, that is held a second time but is treated as though it had never been held before.

Deposition A written declaration of fact, made under oath by a witness in the presence of a judicial officer, and which is taken down in writing before the trial.

Detention A period of temporary custody prior to disposition by a court.

Dicta Plural of dictum.

Dictum A statement or observation in a judicial opinion not necessary to the decision of the case; it is not legally binding on subsequent court decisions. (*See also Holding.*)

Direct Evidence Evidence that applies immediately to the fact being proved, as, for example, if A saw B set A's house on fire.

Discovery A modern pretrial procedure by which one party discloses vital information concerning a case to the opposing party.

Discretion The reasonable exercise of a power or right to act in an official capacity and make choices from among several possible courses of action.

Dismiss To remove a case from a court, or terminate a case without a complete trial.

Doctrine A principle of law established through past decisions.

Doctrine of Rescue *See Rescue Doctrine.*

Double Jeopardy Capability of being tried twice for the same offense, which is prohibited by the Constitution.

Doubt *See Reasonable Doubt.*

Due Care A conception used in tort law to signify the standard of caution or legal duty one person owes to another. It is the degree of care that a person of ordinary prudence and reason would exercise under the same circumstances.

Due Process of Law The phrase introduced in the Constitution when it guaranteed that no person should be "deprived of life, liberty, or property without due process of law." Though open to considerable interpretation, it generally means that the application of the law to an individual case must be fundamentally fair and consistent.

E

Elicit 1. To derive by logical processes. 2. To draw forth or bring out.

Embezzlement The fraudulent appropriation for one's own use or benefit of property or money entrusted to one by another.

Employee One who works for another, usually for wages or salary.

Employer One who uses or engages the services of others, usually for wages or a salary.

Empower To give official or legal authority to.

Enact To authorize as law, as by a legislature.

Enjoin 1. To command or instruct with authority. 2. To forbid or suspend.

Equity The body of law in England that developed in reaction to the common law's inability to handle all situations. Equity qualifies or corrects the law in extreme cases, often with an injunction or other means of relief.

Escrow A writing, such as a deed, temporarily deposited with a neutral third party by the agreement of two parties who have entered into a contract. The third party holds the document until the conditions of the contract are met, at which time the third party will deliver the writing to the grantee.

Estoppel An impediment by which someone is not allowed to allege or deny a fact, based on that person's previous allegation or denial to the contrary.

Et Seq. *Et sequentia,* Latin for "and the following." It is most commonly used to designate page reference numbers.

Evidence All the means by which any alleged matter of fact is proved or disproved. (*See also Direct Evidence, Circumstantial Evidence, and Hearsay Evidence.*)

Exclusionary Rule The principle that provides that otherwise admissible evidence may not be used in a criminal trial if obtained through illegal police conduct.

Execution 1. The performance of what is required in order to give validity to something, as a contract. 2. The authorized killing of a person as punishment for an offense.

Executive That branch of the federal government consisting of the president, cabinet, and the government agencies they control.

Exempt Free from some requirement or liability to which others are subject.

Ex Necessitate Latin, "out of necessity."

Ex Parte Latin, "of or from one side or party." A court order handed down upon the application of one party to an action without notice to the other party of the action. Thus, an *ex parte* injunction is one granted without the adverse party having been previously notified.

Ex Post Facto Latin, "after the fact." A term describing laws passed that punish or increase the punishment of crimes committed before the laws were passed. Such laws are expressly forbidden by the Constitution.

Ex Rel *Ex relatione,* Latin for "upon relation or report."

F

False Arrest Unlawful arrest. It may be a criminal offense or the basis of a civil action.

False Imprisonment The unjustified detention of a person.

Federal Of or pertaining to the United States government, as opposed to state or local governments.

Federal Employer's Liability Act *See Workmen's Compensation Laws.*

Federal Fair Labor Standards Act *See Workmen's Compensation Laws.*

Federal Tort Claims Act An act passed in 1946 that confers exclusive jurisdiction on United States' district courts to hear claims against the United States. In effect, it waives the privilege of sovereign immunity of the federal government, allowing individuals to sue the government in federal court.

Felon One convicted of committing a felony.

Felonious Of, relating to, or having the quality of a felony.

Felony A high crime, such as homicide, arson, rape, burglary, and larceny, for which the punishment is usually death or imprisonment for more than one year.

Feudalism A system of government in effect in medieval times in England having as its basis the relation of lord to vassal with all land held by the lord (the vassal was merely a tenant of the land), and having as its characteristics homage and the service of tenants to lords in military matters and in court, wardship, and forfeiture.

Fiduciary 1. A person having a duty to act primarily for the benefit of another in certain matters. 2. Founded upon or having the quality of a trust or confidence.

File To place among official records as prescribed by law.

Finding A decision of a court on an issue of fact, with the purpose of answering a question raised in a pleading.

Fine A sum of money imposed as punishment for an offense.

Fiscal Of or pertaining to the public finance and financial transactions.

Forensic 1. Belonging to or suitable for use in discussion in court or public discussion and debate. 2. Of the nature of an argument.

Fraud Any act of cunning, deception, or artifice used to circumvent, cheat, or deceive another.

Fraudulent Characterized by or based upon fraud.

G

Gambling Playing a game of chance for money or property. This activity is illegal in many states.

Government The organization, machinery or agency through which a political unit exercises authority and performs functions.

Grand Jury A body (generally 23 persons) whose duty it is to inquire into complaints and accusations in criminal cases and, if it finds sufficient evidence, to issue indictments against the persons complained of.

Grant To give, consent, allow, or transfer something to another with or without compensation, such as a gift of land.

Grantee One to whom a grant is made.

Grantor One who makes a grant.

Gravamen The material part or substance of a complaint or charge. The gravamen of a complaint that one robbed a bank and then left the country would be the fact that one robbed the bank.

Gross Negligence Failure to use even slight care to exercise ordinary prudence.

Guilt The fact of having committed a crime or civil wrong.

Guilty The condition of having been found by a judge or jury to have committed a crime or civil wrong.

H

Habeas Corpus Latin, "you have the body." Often called "the great writ of liberty," it obtains a judicial determination of the legality of the custody of an individual. In criminal cases it requires that an individual be brought before the court and charged in order to avoid false imprisonment.

Hearing A proceeding in which evidence is taken in order to decide issues of fact and reach a decision based on those facts. It takes place before a judge or magistrate sitting without a jury, and is often used to determine whether further legal action is warranted.

Hearsay Evidence Evidence based not on a witness's personal knowledge but on matters told to the witness by someone else; hearsay is inadmissible in court as evidence.

Holding 1. In procedure, any ruling of the court. 2. In property law, property one owns and is in possession of.

Homicide Any killing of one human being by another. (*See also Murder, Manslaughter, and Justifiable Homicde.*

I

Ibid Latin, "in the same place, time, or manner." A term used in source citations to mean "on the same page" or "in the same book," and thus avoid unnecessary repetition of source data.

Illegal Not authorized by or according to the law; against the law.

Immunity The right of exemption from a duty or penalty. Immunity from prosecution might be granted to a witness who answers questions that otherwise might incriminate that witness. (*See also Sovereign Immunity Doctrine.*)

Impeach 1. To charge a public official with wrongdoing while in office. Once accused, the official may be convicted or acquitted. 2. To accuse or challenge, such as a witness or some item of evidence.

Imprison To confine or put in prison.

Impute To assign to a person or other entity the legal responsibility for the act of another.

Imputable Blamable; capable of being assigned the legal responsibility for.

Incorporate 1. To combine or unite so as to form a whole. 2. To legally form a corporation.

Incriminate To charge with or involve in a crime or fault. The Fifth Amendment to the Constitution states that one may refuse to answer questions or give evidence that might incriminate oneself.

Independent Contractor One who agrees to do a piece of work for another, and who retains control of the means and method of doing the work. Neither party to the contract may terminate the contract at will.

Indicia Indications, signs, or circumstances that tend to support something as probable, but that do not prove it.

Indict To formally accuse or charge with some offense.

Indictment A formal written accusation charging one or more persons with a crime that is submitted to a grand jury by the public prosecuting attorney under oath. If accepted by the grand jury, it is the formal accusation of an alleged criminal.

Injunction A writ, issued to remedy a situation, that requires a party to refrain from doing some particular act or activity.

Injury Any wrong or damage done to a person, either to that person's body, rights, representation, or property.

Innocence The quality of being free from guilt.

Innocent 1. Free from guilt or blame. 2. Not convicted of criminal or civil charges.

Inquest A judicial inquiry or examination, most commonly used to refer to the inquiry made by a coroner's jury. (*See also Coroner.*)

In Rem Latin, "of the thing." A term signifying proceedings against a thing rather than a person.

Intent   The state of mind in which a person knows and wishes to bring about the result of an action. Intent must exist at the time a crime is committed for the act to be a crime.

Intention   A determination to act in a particular way so as to bring about certain results.

Interstate Commerce   Dealings and traffic between inhabitants of different states. It is under the jurisdiction of the federal government.

Involuntary Manslaughter   Homicide in which the death of the victim usually results from gross negligence or recklessness but contains no element of malice, as in an automobile accident. (*See also Voluntary Manslaughter.*)

Ipso Facto   Latin, ''by the fact itself; in and of itself.''

J

Jeopardize   To expose to death, loss, injury, hazard, or danger.

Jeopardy   Exposure to death, loss, injury, hazard or danger. (*See also Double Jeopardy.*)

Judex   In Roman law, a private person appointed by the praetor, with the consent of the parties, to try and decide a case.

Judge   A public official authorized to preside over and decide questions brought before a court.

Judgment   The final determination of a court upon matters submitted to it.

Judical   Of or relating to a judgment, the judiciary, or the administration of justice.

Judicature   That department of the government intended to interpret and administer the law.

Judiciary Act of 1789   The congressional act that established the federal judicial system.

Jurisdiction   1. The limits or territory within which authority may be exercised. 2. The power, right, or authority to interpret and apply the law.

Jurisprudence   The study of the structure of legal systems.

Jury   A group of people summoned and sworn to decide on the facts in issue at a trial.

Justice   1. The maintenance or administration of what is right or just, especially by the impartial adjustment of conflicting claims or the assignment of merited rewards or punishments. 2. A judge.

Justifiable Homicide   The killing of one human being by another in order to fulfill the commandment of the law, execute public justice, or maintain lawful self-defense.

Juvenile   Of or relating to children or young people.

L

Larceny   The taking of another's property unlawfully, with the intention of depriving the owner of its use.

Last Clear Chance   A doctrine holding that a defendant may still be liable for the injuries caused to a plaintiff, even though the plaintiff is guilty of contributory negligence. The doctrine applies if the defendant, having observed the plaintiff in a position of danger, failed to avoid injuring the plaintiff when in a position to do so.

Law   The legislative pronouncement of the rules that should guide the actions of individuals in a society. (*See also Common Law, Constitutional Law, Corporate Law, and Criminal Law.*)

Lawsuit   *See Suit.*

Lease   A contract by which one conveys real estate, equipment, or facilities for a specified time and a specified rent.

Legal   1. Of or concerning the law. 2. Allowed or authorized by the law.

Legal Duty   Something that the law specifies must be done by a particular person or group.

Legislation   1. A law or laws enacted. 2. The exercise of the power and function of making laws that have the force of authority.

Legislative   Concerning that branch of the government charged with enacting laws, levying and collecting taxes, and making financial appropriations.

Legislature   An organized body of persons having the power to make laws for a specific political unit.

Lessee One who leases property from a landlord.

Levy To raise, collect, or assess, as in to levy a tax.

Liable To be legally responsible for.

Liability A legal obligation to do something or refrain from doing it. (*See also Vicarious Liability.*)

Libel A malicious defamation of one living or dead, expressed in print, writing, pictures, or effigy, and intended to mar the reputation of that person.

License A right granted to a person that gives that person permission to do something not legally allowed without the permission.

Licensee One who has been granted a license.

Lien A hold or claim upon the property of another as security for a debt.

Litigation A civil action in a court; a judicial contest to determine legal rights and duties held in a civil court.

M

Magistrate A public official entrusted with some part of the administration of the law.

Mala in Se Latin, "evil in and of themselves." Offenses that are bad in themselves, based on principles of natural and moral law. Reckless driving is, for example, a crime *Malum in se.* (*See also* Mala Prohibita.)

Mala Prohibita Latin, "wrong because they are prohibited." Offenses that are unlawful simply because statutes forbid them. Exceeding the speed limit, for example, is an offense *malum prohibitum.* (*See also* Mala in Se.)

Malfeasance The doing of an act that is wrongful and unlawful. (*See also Misfeasance and Nonfeasance.*)

Malice Evil intent; the state of mind accompanying the intentional doing of a wrongful act.

Malpractice Failure to perform a professional duty or exercise professional skill by one rendering a professional service (such as a physician or attorney) that results in loss, injury, or damage.

Malum in Se Singular of *mala in se.*

Malum Prohibitum Singular of *mala prohibita.*

Mandamus Latin, "we order; we compel." An emergency writ issued by a court to compel an official to perform an act recognized by the law as a legal duty.

Mandate 1. A judicial command. 2. To issue a judicial command.

Mandatory Prescribed; containing a command.

Manslaughter The unlawful killing of another without malice, as in an automobile accident or in the sudden heat of passion, upon provocation. (*See also Involuntary Manslaughter and Voluntary Manslaughter.*)

Mayhem Willful and permanent crippling, disfiguration, or mutilation of any part of the body.

Mens Rea Latin, "a guilty mind." The mental state accompanying a forbidden act. (*See also* Actus Rea.)

Mischief Offenses of the law that annoy or irritate, but are usually too minor to be designated crimes.

Misdemeanor A class of criminal offenses not as serious as felonies and punished by less severe penalties.

Misfeasance Doing an act in a wrongful manner, the proper performance of which would have been lawful. (*See also Malfeasance and Nonfeasance.*)

Mistrial A trial terminated and declared void prior to the verdict due to some extraordinary circumstance (such as death or illness), or because of a fundamental error prejudicial to the defendant that cannot be righted (such as improper news coverage), or, most commonly, because the jury cannot reach a verdict.

Mitigate To become less severe or harsh. Mitigating circumstances do not free a person from guilt of a crime, but they may lessen it. Murder, for example, may be only manslaughter if the homicide was committed without malice and in the sudden heat of passion upon extreme provocation.

Modus Operandi Latin, "the method of operation." The way an act is accomplished; for example, "the *modus operandi* of the murderer was strangulation."

Mortgage A conveyance of lands or other

subjects of property as security for a debt on the condition that if the debt is paid, the conveyance shall be void.

Motion   A request to the court asking that the court rule in favor of the requestor.

Municipal   Of or relating to the affairs of a municipality.

Municipality   A political unit, usually urban (as a city), having corporate status and, usually, the right of self-government.

Murder   The malicious killing of one human being by another; it requires premeditated intent to kill plus an element of hatred.

N

Negligence   Failure to exercise that degree of caution which a person of ordinary prudence would exercise in similar circumstances. (*See also Contributory Negligence, Comparative Negligence, Gross Negligence, Ordinary Negligence, and Slight Negligence.*)

Nolo Contendere   Latin, ''I do not wish to contend.'' A plea that a defendant in a criminal prosecution may enter that amounts to an admission of guilt without actually pleading guilty.

Nonfeasance   The complete failure to perform a required duty. (*See also Malfeasance and Misfeasance.*)

N.O.V.   *Non obstante veredicto,* Latin for ''notwithstanding the verdict.'' A judgment *n.o.v.* is one that reverses the decision of the jury, granted when it is obvious that the judgment had no support or was not lawful.

Nuisance   Anything that renders the enjoyment of life and property uncomfortable. (*See also Public Nuisance and Private Nuisance.*)

Null and Void   Invalid; having no legal force.

O

Oath   A declaration of truth which, if made by one who knows it to be false, renders that person guilty of perjury.

Obscenity   Material that appeals to lewd, wanton, and lascivious interest and that lacks literary, artistic, political, or scientific value.

Offense   A violation of the law.

Opinion   The reasoning behind a court's decision of a case.

Ordinance   A local law applying to persons and things in that jurisdiction.

Ordinary Negligence   Failure to use normal care that an ordinary person would use in similar circumstances.

Overrule   1. To overturn or make void the holding of a prior case. 2. To deny, as by the court, any motion or point raised during a case in court.

P

Parens Patriae   Latin, ''the father of his country.'' In England, the monarch. In the United States, the state.

Parity   The quality of being equal or equivalent.

Party   1. Either litigant (plaintiff or defendant) in a lawsuit. 2. A person directly interested in the subject matter of a case. 3. A person who enters into a contract, lease, deed, *etc.*

Penalty   The punishment specified for a particular crime or offense, often involving sacrifice of personal freedom, rights, property, or money.

Pending   1. Not yet decided, as a lawsuit. 2. Awaiting.

Penitentiary   A public institution in which offenders against the law are confined for punishment.

Per Curiam   Latin, ''by the court.'' An opinion written by the court whose author is not named.

Perjury   The criminal offense of making false statements under oath; lying.

Perpetrate   To cause to happen or to commit, as a crime.

Perpetrator   One who causes something to happen, as a crime.

Personnel   A body of persons employed, usually in an office, organization, or factory.

Petition   A formal written request asking that a certain thing be done.

Petty   Of secondary rank or importance, as in a petty offense.

Plaintiff The one who initially brings a lawsuit; the complainant.

Plea An answer or allegation of fact that a defendant, in an action at law, opposes to the plaintiff's declaration.

Pleadings Statements in logical and legal form made by the parties of a civil or criminal suit that constitute the reasons for supporting or defeating the suit.

Praetor An ancient Roman magistrate ranking below a consul and having chiefly judicial functions.

Precedent A previously decided case recognized as an authority for the decision of future cases.

Premeditated Having been intended; having considered a matter beforehand for some length of time, however short. Premeditation is an element of intent.

Prima facie Latin, "at first view." Not requiring further proof to establish existence, validity, or credibility.

Principal 1. The most important. 2. In criminal law, any person concerned in the commission of an offense, regardless of whether that person profits.

Prison Jail; penitentiary.

Private Nuisance Interference with a person's interest in the private use and enjoyment of the individual's land.

Privilege An advantage not enjoyed by all; an exemption from some burden or necessity that a certain group, class, or person has.

Probation A procedure in which the sentence of a convicted offender is suspended and the offender is allowed freedom as long as he maintains good behavior under the supervision of a supervisor.

Procedure Legal method; the body of rules and practice by which legal justice is awarded. (*See also Criminal Procedure.*)

Proceeding 1. The succession of events by which a case is begun, carried through, and terminated. 2. A general term for any of these events.

Profit 1. The excess of returns over expenditure in a transaction. 2. Any valuable return.

Promulgate 1. To put into action or force, as in a law. 2. To make known or public.

Proof The cogency of evidence that compels acceptance of that evidence as fact or truth. (*See also Burden of Proof.*)

Property Something that can be owned or possessed, has exchangeable value, or adds to one's wealth.

Proprietary Characterized by having legal right to or ownership of something.

Prosecute 1. To institute legal proceedings. 2. To pursue a lawsuit or criminal trial until legally decided.

Prosecution 1. The act of pursuing a lawsuit or criminal trial. 2. The party initiating a criminal trial, *i.e.*, the state.

Prosecutor A public official whose duty it is to prepare and conduct the case against persons accused of committing crimes.

Prostitution The act of engaging in promiscuous sexual relations for money.

Proximate 1. Very close to; near. 2. Imminent, about to happen.

Proximate Cause That which in natural and unbroken sequence produces an event.

Public Nuisance A nuisance that offends the public at large or a segment of the public, and interferes with the general health, safety, or peace.

Pursuant to According to; in conformance with.

Purview 1. The body or enacting part of a statute. 2. The limit, purpose, or scope of a statute.

Q

Quash To annul, overthrow, or vacate by judicial decision.

Quasi Latin, "as it were; approximately."

Quo Ad Hoc Latin, "as to this." With respect to this.

R

Rape The act of unlawful sexual intercourse between persons not married to each other accomplished through use of force by the man and implying resistance and lack of consent by the woman.

Ratio Decedendi Latin, "the reason for deciding." The principle that a case establishes; the reasoning behind the decision.

Reasonable Doubt A term for the lack of

certainty required in the mind of a juror for that juror to determine a criminal defendant innocent.

Reasonable Person Standard A phrase used to denote a hypothetical person who has "those qualities of attention, knowledge, intelligence, and judgment which a society requires of its members for the protection of their own interest and the interests of others."

Record 1. To preserve in writing, printing, on tape, or film. 2. A precise history of a suit from beginning to end, including transcriptions of all proceedings.

Recovery 1. The fact of obtaining a final legal judgment in one's favor. 2. The amount of a judgment rendered in one's favor that the court deems one to have "lost."

Redress 1. To remove the cause of a grievance or complaint. 2. Compensation for an injury or loss.

Remand To send back, as for further deliberation.

Remit 1. To submit or refer for consideration, judgment, decision, or action. 2. To give relief from, as suffering. 3. To refrain from exacting, as a tax.

Remuneration A sum of money paid as compensation for loss, injury, or service.

Render 1. To hand down, as in a judgment. 2. To agree upon and report, as a verdict. 3. To cause to be or become.

Repair 1. To compensate for. 2. To return to a healthy or sound state.

Rescind To take back, annul, or cancel, as a contract.

Rescue Doctrine A rule of torts that holds a defendant liable for the victim's rescuer should the rescuer injure himself or herself during a reasonable rescue attempt.

Respondeat Superior Latin, "let the superior reply." The doctrine governing the relationships between masters and servants that holds employers (masters) responsible for the wrongful actions of their employees (servants). It applies to civil cases.

Respondent In equity, the party who answers a bill or other pleading.

Restitution The act of making good, or giving the equivalent for, any loss, injury, or damage.

Rights Things to which one has a just claim, such as power, privilege, or property. (*See also Civil Rights.*)

Riot 1. Public violence, tumult, or disorder. 2. An occurrance of such disorder.

Risk *See Assumption of Risk.*

Robbery Forcible stealing; taking the property of another by violence or terror.

S

Search and Seizure A police practice in which a person or place is searched and evidence useful to an investigation and prosecution of a crime is seized. The Constitution specifies that search and seizure must be reasonable.

Search Warrant An order issued by a judge that directs certain law enforcement officers to conduct a search of certain premises for certain things or persons suspected to be there, and bring them before the court.

Sentence The punishment ordered to be inflicted upon a convicted criminal.

Slander Oral defamation of another; words spoken which tend to damage the reputation of another.

Slight Negligence Failure to use great care to prevent another's injury, harm, or loss.

Solicitation An offense that consists of enticing, inciting, or begging another to commit a felony.

Sovereign Immunity A doctrine that precludes bringing a suit against the government (sovereign) without the government's (sovereign's) consent when it is engaged in a governmental function.

Standing The legal right of a person to challenge in a court the conduct of another, especially the conduct of the government.

Stare Decisis Latin, "to stand by that which was decided." The principle that common law courts seldom abandon principles established in former decisions.

Statute A law enacted by the legislature.

Statute of Limitations An act specifying the time within which parties must take judicial action to prosecute a case. Nearly every action is subject to a Statute of Limitations except murder.

Subpoena A writ issued on court authority that compels a witness to appear in court at a judicial proceeding, the disobedience of which is punishable as contempt of court.

Succession The order in which or the conditions under which one person after another inherits a property, title, or throne.

Sue 1. To seek justice from another by legal process. 2. To bring an action.

Suit A comprehensive term that applies to any proceeding in a civil court in which a person seeks remedy for an injury or loss.

Summary Judgment A decision rendered by a judge in response to a motion claiming that there is no factual dispute in a case, and that hence there is no need to send the case to a jury because the dispute involves only a question of law, not questions of fact.

Summons An order requiring the defendant in a case to appear in court for the trial.

Supra Latin, "above." A term that, in a written work, directs the reader to a preceding portion of the work.

Sustain To support or approve as true, legal, or just.

T

Tenant One who holds land by any kind of title or right, whether permanent or temporary.

Tenet A principle, belief, or doctrine generally held to be true.

Testify To give evidence; to make a statement based on personal knowledge or belief.

Testimony A statement made by a witness under oath, usually related to a legal proceeding.

Title The evidence of one's right to property or the extent of one's interest in property.

Tort A private or civil wrong resulting from a breach of a legal duty, as distinguished from a crime, which is a public wrong.

Tort-feasor One who commits a tort.

Tortious A term describing conduct that subjects the doer to liability for a tort.

Trespass 1. Any transgression or offense, whether it relates to one's person or property. 2. Wrongful entry onto the land of another.

Tribunal An officer or body having authority to adjudicate matters.

Trust 1. A property interest held by one person for the benefit of another. 2. A combination of firms formed by a legal agreement.

U V W

Ultra Vires Latin, "beyond the powers." That which is beyond the power authorized by law for an entity. It is often applied to corporations that exceed the scope of their authorized powers.

V. *Versus,* Latin for "against." A word used in the title of a case between the parties who oppose each other in the case, as in Marbury *v.* Madison.

Vagrancy A general term for a class of minor offenses such as idleness, roaming, wandering, and loitering.

Valid True; having legal efficacy or force.

Venue A term for the county or judicial district in which a case may be tried; it is determined by statute in most jurisdictions.

Verdict The opinion of the jury, or the judge acting as jury, on a question of fact.

Veredicto Latin, "verdict."

Vested Absolute; fixed.

Vicarious Liability Responsibility of one person for the actions of another. It is present in the concept of *respondeat superior,* in which an employer may be held liable for the unlawful actions of an employee.

Violation An infringement or transgression, especially of the law.

Void Empty of significance; having no legal force or effect.

Voluntary Manslaughter A homicide in which there is an intentional killing committed under circumstances that, although they do not justify the homicide, lessen the seriousness of it. An example would be a killing in which the victim provoked the killer.

Warrant 1. A written order, issued by an

authority, or authorized person, directing the arrest of a certain person or persons. 2. A writ directing the doing of a certain act, as a search.

Witness 1. One who testifies in a case under oath. 2. One who is present at a transaction so as to be able to swear that it has taken place.

Workmen's Compensation Laws A group of statutes that in general establishes the liability of an employer for the injury, sickness, and death of employees arising during the course of employment, regardless of fault or negligence on the part of the employers. The Federal Employer's Liability Act and Federal Fair Labor Standards Act are among these.

Writ A mandatory order issued by an authorized court or tribunal in the name of the state that compels a person to do something.

Wrong 1. An injurious, unfair, or illegal act. 2. To commit an injurious, unfair, or illegal act.

SUBJECT INDEX

A

Attempt
  as a crime, 229
  court case regarding, in arson, 230–236
  defined, 229
  of arson, 230
  proof of, 229
  to commit a crime, 229
Attractive nuisance, 103

B

Battery, defined, 103
Bigamy, as crime, 240
Bills of attainder, 223
Board of fire commissioners
  empowered by, 78
  fire commissioner in, 78
  in a fire department, 78
  power of, 78
Boroughs
  as municipal corporations, 31
  described, 31
Breach of contract, as municipal liability, 54
Breaking, defined, 238
Breaking and entering, in burglary, 238
Bribery, as crime, 240
Bureau, fire (see fire bureau)
Bureau of Census
  definition of metropolitan area by, 38
  special district reports by, 35, Table 2.1
Burglary
  as a felony, 224
  as offense against property, 238
  breaking and entering in, 238
  defined, 238
  requirements for, 239
Burning
  needed for arson, 243
  willful and malicious intent in, 248

C

CAB (see Civil Aeronautics Board)
Camara case, 344–353
Canon law, 4
Case abstracts, Appendix E
Cause of action, in tort, 100
Cause of fire, court case regarding investigation of, 405–417
Cause, proximate, 106
Cease and desist order, as remedy in administrative proceedings, 441
Certiorari, described, 14

Charter, for a corporation, 21–22
Chain of command, in fire department, 78, Table 2.3
"Checks and balances," 338
Chief
  appointment of, 78
  duties of, 79
  in a fire department, 78
Chief magistrate, in Roman law, 5
Circumstantial evidence
  court case regarding, 293–300
  for arson, 291
  rule of thumb for, 292
Cities
  as municipal corporations, 29
  described, 29
  origin of, 24
Civil Aeronautics Board, 339
Civil law, Corpus Juris Civilis as basis of, 5
Civil-law state, 222
Civil liberties, defined, 197
Civil penalties, as remedy in administrative proceedings, 442
Civil procedure, steps in, 17
Civil rights
  court case regarding, 198–199
  defined, 197
  described, 197
  legislation, 199
  to own and enjoy property as, 197
  to vote as, 197
Civil Rights Act of 1957
  protection of individual employment rights by, 195
  provisions of, 200
Civil suit, steps in a, 17
Civil service, 183
Civil Service Commission
  functions of, 184
  in fire service recruitment, 184
  personnel administration by, 184
  responsibilities of, 184
Civil service employment, as government employment, 183
Civil service term, of administrators, 424
Code of Hammurabi, 4
Code, Mosaic, 4
Code of Federal Regulations, 340
College of pontiffs, 5
Commission, of a crime, defined, 225
Commission on Civil Rights, 200
Common law
  affecting the fire service, 9
  as unwritten law, 6

D